CUSTOMIZING AUTOCAD 2010

Sham Tickoo

DELMAR
CENGAGE Learning™

Australia • Brazil • Japan • Korea • Mexico • Singapore • Spain • United Kingdom • United States

DELMAR
CENGAGE Learning™

Customizing AutoCAD 2010:
Sham Tickoo

Vice President, Career and Professional Editorial: Dave Garza

Director of Learning Solutions: Sandy Clark

Acquisitions Editor: Stacy Masucci

Managing Editor: Larry Main

Senior Product Manager: John Fisher

Senior Editorial Assistant: Dawn Daugherty

Vice President Marketing, Career and Professional: Jennifer Baker

Executive Marketing Manager: Deborah S. Yarnell

Marketing Manager: Jimmy Stephens

Associate Marketing Manager: Mark Pierro

Production Director: Wendy Troeger

Production Manager: Mark Bernard

Art Director: David Arsenault

Content Project Manager: Angela Iula

Technology Project Manager: Joe Pliss

Production Technology Analyst: Tom Stover

Technical Editors:
Vivek Singh
Swapna D.
(CADCIM Technologies)

Copy Editor:
Anju Jethwani
(CADCIM Technologies)

For product information and technology assistance, contact us at
Career & Professional Group Customer Support, 1-800-648-7450

For permission to use material from this text or product,
submit all requests online at **cengage.com/permissions**
Further permissions questions can be emailed to
permissionrequest@cengage.com

Library of Congress Control Number: 2009930673
ISBN-13: 978-1-4390-5875-6
ISBN-10: 1-4390-5875-X

Delmar Cengage Learning
5 Maxwell Drive
Clifton Park, NY 12065-2919
USA

Notice to the Reader
Publisher does not warrant or guarantee any of the products described herein or perform any independent analysis in connection with any of the product information contained herein. Publisher does not assume, and expressly disclaims, any obligation to obtain and include information other than that provided to it by the manufacturer. The reader is expressly warned to consider and adopt all safety precautions that might be indicated by the activities described herein and to avoid all potential hazards. By following the instructions contained herein, the reader willingly assumes all risks in connection with such instructions. The publisher makes no representations or warranties of any kind, including but not limited to, the warranties of fitness for particular purpose or merchantability, nor are any such representations implied with respect to the material set forth herein, and the publisher takes no responsibility with respect to such material. The publisher shall not be liable for any special, consequential, or exemplary damages resulting, in whole or part, from the readers' use of, or reliance upon, this material.

Printed in the United States of America
1 2 3 4 5 6 7 13 12 11 10 09

DEDICATION

To teachers, who make it possible to disseminate knowledge
to enlighten the young and curious minds
of our future generations

To students, who are dedicated to learning new technologies
and making the world a better place to live in

THANKS

To the faculty and students of the MET Department of
Purdue University Calumet for their cooperation

To engineers of CADCIM Technologies and
Santosh Tickoo for their valuable help

Table of Contents

Chapter 7: Shapes and Text Fonts

Chapter 8: Working with AutoLISP

Chapter 9: Working with Visual LISP

Chapter 10: Visual LISP: Editing the Drawing Database

Chapter 11: Creating Programmable Dialog Boxes Using the Dialog Control Language

Chapter 12: DIESEL: A String Expression Language

Chapter 13: Visual Basic for Application

Chapter 14: Accessing External Database

Chapter 15: Geometry Calculator

Chapter 16: Image Tile Menus

Chapter 17: Button Menus

Chapter 18: Tablet Menus

Chapter 19: Customizing Screen Menus

Preface

AutoCAD, developed by Autodesk Inc., is the most popular PC-CAD system available in the market. Today, over 7 million people use AutoCAD and other AutoCAD based design products. 100% of the Fortune 100 firms and 98% of the Fortune 500 firms are Autodesk customers. AutoCAD's open architecture has allowed third-party developers to write an application software that has significantly added to its popularity. For example, the author of this book has developed a software package "**SMLayout**" for sheet metal products that generates a flat layout of various geometrical shapes such as transitions, intersections, cones, elbows, and tank heads. Several companies in Canada and the United States are using this software package with AutoCAD to design and manufacture various products. AutoCAD also facilitates customization, which enables users to increase their efficiency and improve their productivity.

The purpose of this textbook is to unravel the customizing power of AutoCAD and explain it in a way that is easy to understand. Each customizing technique is thoroughly explained with examples and illustrations that makes it easy for the users to comprehend the customizing concepts of AutoCAD. After reading this textbook, you will be able to generate template drawings, write script files, edit existing menus, create customized menus, create customized **Ribbon**, Ribbon panels, Ribbon tabs and Contextual ribbon tabs, write shape and text files, create new linetypes and hatch patterns, define new commands, write programs in the AutoLISP programming language, edit the existing drawing database, create your own dialog boxes using DCL, customize the status bar using DIESEL, and edit the Program Parameter file (ACAD.PGP). In the process, you will discover some new and unique applications of AutoCAD that may have a significant effect on your drawings. You will also get a better idea of why AutoCAD has become such a popular software package and an international standard in PC-CAD.

To use this textbook, you do not need to be an AutoCAD expert or a programmer. If you are already familiar with the basic AutoCAD commands, you will face no problem in understanding the material presented in this textbook. The book contains a detailed description of various customizing techniques that can be used to customize your system. Every chapter has several examples to illustrate some possible applications of these customizing techniques. The exercises at the end of each chapter provide a challenge to the user to solve the problems on his/her own. In a class situation, these exercises can be assigned to students to test their understanding of the material explained in the chapter. The chapters on AutoLISP programming are written with an assumption that the user has no programming background. Therefore, all commands have been thoroughly explained in a way that has made the programming easy to understand and interesting to learn. All chapters, except Editing the Drawing Database, is independent and can be read in any order and used without reading the rest of the book. The users needs only to read the chapter on AutoLISP before the chapter on Editing the Drawing Database. However, for a better understanding of customizing techniques, it is recommended to start from Chapter 1 and then progress through the chapters. For the convenience of the readers, AutoCAD Release 2010 features have been marked by an asterisk (*) sign at the end of the feature. The following is the summary of each chapter:

Chapter 1: Template Drawings
This chapter explains how to create a Template drawing and how to standardize the information that is common to all drawings. It also describes how to create a Template drawing with paper space and predefined viewports.

Chapter 2: Script Files and Slide Shows
This chapter introduces the users to script files and utilize them to group AutoCAD commands in a predetermined sequence to perform a given operation. This chapter also explains how to use script files to create a slide show that can be used for product presentation.

Chapter 3: Creating Linetypes and Hatch Patterns
This chapter explains how to create a new linetype and edit the linetype file, *acad.lin*. This chapter also describes the techniques of creating a new hatch pattern and the effect of hatch scale and hatch angle on hatch.

Chapter 4: Customizing the ACAD.PGP File
This chapter explains the use of AutoCAD's Program Parameter file (*acad.pgp*) to define aliases for the operating system commands and some of the AutoCAD commands.

Chapter 5: Customizing Menus and Toolbars
This chapter explains how to write pull-down, shortcut, and partial menus. The chapter also explains how to customize the toolbars. Several examples have been given with thorough explanation.

Chapter 6: Customizing Ribbon, Workspaces, and Palettes
This chapter explains how to customize **Ribbon**, workspaces, **Quick Access** toolbar, **Quick Properties** panel, rollover tooltips, and **Tool Palettes**, and double-click actions.

Chapter 7: Shapes and Text Fonts
This chapter defines shapes and explains the method to create shape and text fonts. It also contains a detailed description of special codes and their applications to create shapes and text fonts.

Chapter 8: Working with AutoLISP
This chapter explains different AutoLISP functions and how to use these functions to write a program. It also introduces the users to basic programming techniques and the use of relational and conditional statements in a program.

Chapter 9: Working with Visual LISP
This chapter explains how to start Visual LISP in AutoCAD, use Visual LISP text editor, run Visual LISP programs, use Visual LISP console and formatter, and how to debug Visual LISP programs.

Chapter 10: Visual LISP: Editing the Drawing Database
This chapter describes those AutoLISP functions that allow a user to edit the drawing database.

Chapter 11: Creating Programmable Dialog Boxes Using the Dialog Control Language
This chapter is an introduction to Dialog Control Language and its application in customizing the existing dialog boxes and writing new dialog boxes. It also explains the use of AutoLISP in controlling the dialog boxes.

Chapter 12: DIESEL: A String Expression Language
This chapter describes the DIESEL string expression language and its application in customizing the status bar by altering the value of the **MODEMACRO** system variable.

Chapter 13: Visual Basic for Application
This chapter describes how to install AutoCAD preview VBA, load and run sample VBA projects, utilize the visual basic editor, use AutoCAD objects and object properties, and apply and use AutoCAD methods.

Chapter 14: Accessing External Database
This chapter explains how to create a database from the information associated with AutoCAD objects like blocks. The information extracted can then be arranged in rows and columns.

Chapter 15: Geometry Calculator
The geometry calculator is an ADS application that can be used as an online calculator. This chapter explains how the calculator can be used to evaluate vector, real, and integer expressions. It can also access the existing geometry by using the first three characters of the standard AutoCAD object snap functions like MID, CEN, END, and so on.

Chapter 16: Image Tile Menus
This chapter explains the Image tile menus and how to write them. It also discusses submenus and how to make slides for the Image tile menu.

Chapter 17: Button Menus
This chapter deals with buttons and auxiliary menus and how to assign AutoCAD commands to different buttons of a multi-button pointing device.

Chapter 18: Tablet Menus
This chapter explains how to write a tablet menu, and how to load other menus from the tablet menu. Advantages of the tablet menu, design of the tablet menu, and how AutoCAD assigns commands to different blocks of the tablet menu are also discussed.

Chapter 19: Customizing Screen Menus

This chapter explains how to customize the screen menu, screen shortcut menu, and multiple screen shortcut menus. The chapter also explains some advanced concepts of customization. Several examples have been given with thorough explanation.

Author's Web Site

For Faculty: Please contact the author at **stickoo@calumet.purdue.edu** to access the Web site that contains the following:

1. Program listings and drawings used in this textbook.
2. Syllabus, hints, and answers for the review questions and exercises given in every chapter.

For Students: You can download the files related to this textbook by accessing the author's Web site **www.cadcim.com/students/students.htm**.

Publisher's Web site

You can also get learning materials from the publisher's website:
www.cengage.com/cad/autodeskpress

Chapter 1

Template Drawings

Learning Objectives

After completing this chapter, you will be able to:

- *Create template drawings.*
- *Load template drawings using dialog boxes and the Command line.*
- *Do an initial drawing setup.*
- *Customize drawings with layers and dimensioning specifications.*
- *Customize drawings with layouts, viewports, and paper space.*

CREATING TEMPLATE DRAWINGS

One way to customize AutoCAD is to create template drawings that contain initial drawing setup information and if desired, visible objects and text. When the user starts a new drawing, the settings associated with the template drawing are automatically loaded. If you start a new drawing from the scratch, AutoCAD loads default setup values. For example, the default limits are (0.0,0.0), (12.0,9.0) and the default layer is 0 with white color and a continuous linetype. Generally, these default parameters need to be reset before generating a drawing on the computer using AutoCAD. A considerable amount of time is required to set up the layers, colors, linetypes, lineweights, limits, snaps, units, text height, dimensioning variables, and other parameters. Sometimes, border lines and a title block may also be needed.

In production drawings, most of the drawing setup values remain the same. For example, the company title block, border, layers, linetypes, dimension variables, text height, LTSCALE, and other drawing setup values do not change. You will save considerable time if you save these values and reload them when starting a new drawing. You can do this by creating template drawings that contain the initial drawing setup information configured according to the company specifications. They can also contain a border, title block, tolerance table, block definitions, floating viewports in the paper space, and perhaps some notes and instructions that are common to all drawings.

STANDARD TEMPLATE DRAWINGS

AutoCAD comes with standard template drawings like Acad.dwt, Acadiso.dwt, Ansi a (portrait) -color dependent plot styles.dwt, Din a1 -named plot styles.dwt, and so on. The ansi, din, and iso template drawings are based on the drawing standards developed by ANSI (American National Standards Institute), DIN (German), and ISO (International Organization for Standardization). When you start a new drawing with **STARTUP** system variable set to 1, the **Create New Drawing** dialog box will be displayed. To load the template drawing, choose the **Use a Template** button; the list of standard template drawings is displayed. From this list, you can select any template drawing according to your requirements. If you want to start a drawing with the default settings, select the **Start from Scratch** button in the **Create New Drawing** dialog box. The following are some of the system variables, with the default values that are assigned to the new drawing.

System Variable Name	Default Value
CHAMFERA	0.0000
CHAMFERB	0.0000
COLOR	Bylayer
DIMALT	Off
DIMALTD	2
DIMALTF	25.4
DIMPOST	None
DIMASO	On
DIMASZ	0.18
FILLETRAD	0.0000
GRID	0.5000

GRIDMODE	0	
ISOPLANE	Left	
LIMMIN	0.0000,0.0000	
LIMMAX	12.0000,9.0000	
LTSCALE	1.0	
MIRRTEXT	0	(Text not mirrored like other objects)
TILEMODE	1 (On)	

Example 1

Create a drawing template using the **Advanced Setup** wizard of the **Use a Wizard** button with the following specifications and save it with the name *proto1.dwt*.

Units	Engineering with precision 0'-0.00"
Angle	Decimal degrees with precision 0
Angle Direction	Counterclockwise
Area	144'x96'

Step 1

Set the value of the **STARTUP** system variable to 1 to show the **Startup** dialog box. Choose the **New** button from the **Quick Access** toolbar to display the **Create New Drawing** dialog box. Choose the **Use a Wizard** button and choose the **Advanced Setup** option, as shown in Figure 1-1. Choose **OK**. The **Units** page of the **Advanced Setup** dialog box is displayed.

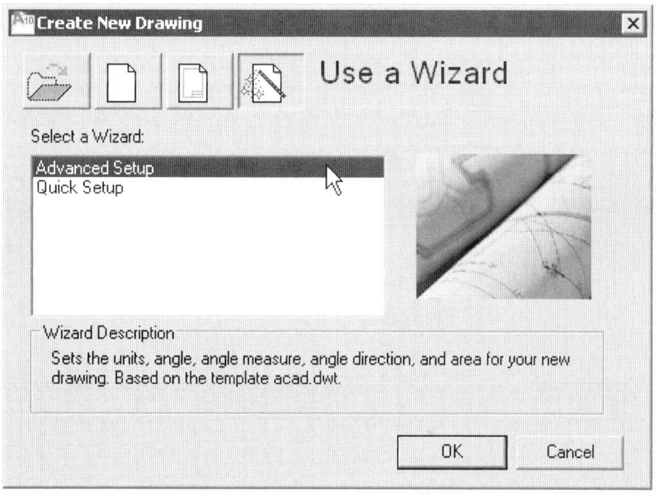

Figure 1-1 *The* ***Advanced Setup*** *wizard option of the* ***Create New Drawing*** *dialog box*

Step 2

Select the **Engineering** radio button. Select **0'-0.00"** precision from the **Precision** drop-down list, refer to Figure 1-2 and then, choose the **Next** button; the **Angle** page of the **Advanced Setup** dialog box is displayed.

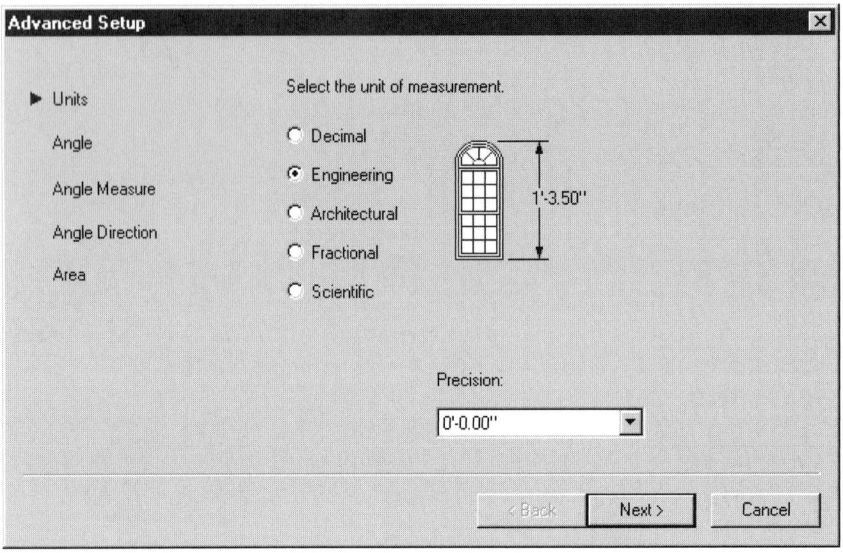

*Figure 1-2 The **Units** page of the **Advanced Setup** dialog box*

Step 3

In the **Angle** page, select the **Decimal Degrees** radio button and select **0** from the **Precision** drop-down list, as shown in Figure 1-3. Choose the **Next** button; the **Angle Measure** page of the **Advanced Setup** dialog box is displayed.

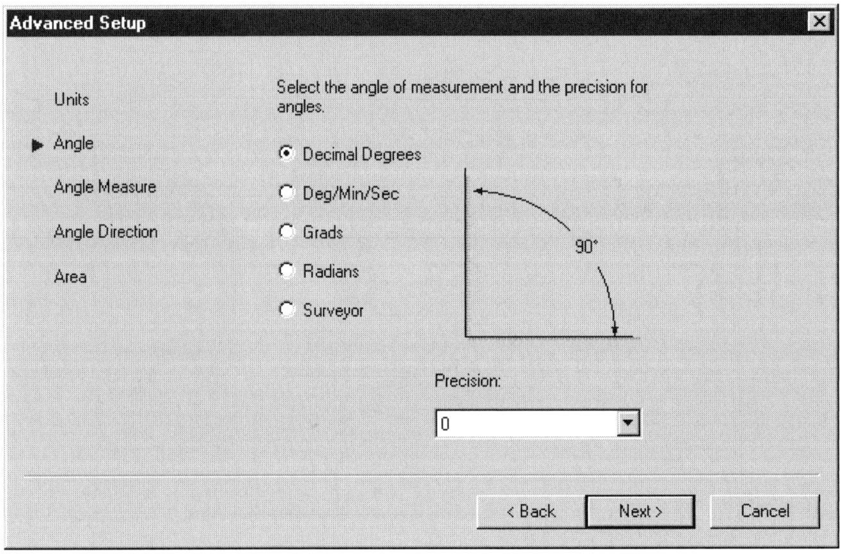

*Figure 1-3 The **Angle** page of the **Advanced Setup** dialog box*

Step 4

In the **Angle Measure** page, select the **East** radio button. Choose the **Next** button to display the **Angle Direction** page.

Step 5

Select the **Counter-Clockwise** radio button and then choose the **Next** button. The **Area** page is displayed. Specify the area as 144' and 96' by entering the value of the width and length as **144'** and **96'** in the **Width** and **Length** edit boxes and then choose the **Finish** button. Use the **All** option of the **ZOOM** command to display new limits on the screen.

Step 6

Now, save the drawing as *proto1.dwt* using the **Save** button from the **Quick Access** toolbar. You need to select **AutoCAD Drawing Template (*dwt)** from the **Files of type** drop-down list in the dialog box. Choose the **Save** button; the **Template Options** dialog box will be displayed on the screen, as shown in Figure 1-4. Enter the description about the template in the **Description** edit box and choose the **OK** button. Now, the drawing will be saved as *proto1.dwt* on the default drive.

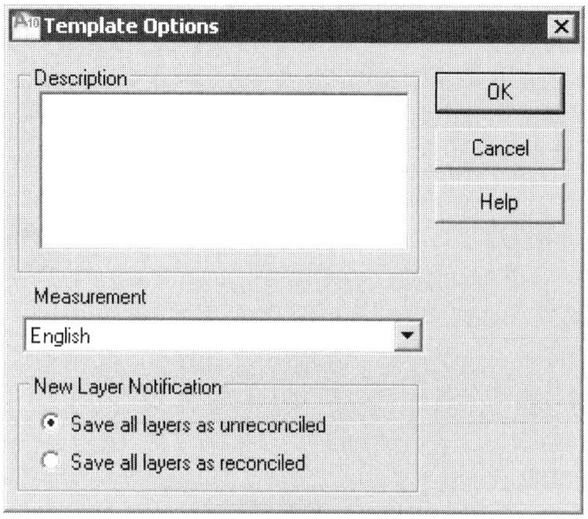

Figure 1-4 *The* ***Template Options*** *dialog box*

Note

To customize only units and area, you can use the ***Quick Setup*** *option in the* ***Create New Drawing*** *dialog box.*

Example 2

Create a drawing template using the following specifications. The template should be saved with the name *proto2.dwt*.

Limits	18.0,12.0
Snap	0.25
Grid	0.50
Text height	0.125
Units	3 digits to the right of decimal point
	Decimal degrees
	2 digits to the right of decimal point
	0 angle along positive *X* axis (east)
	Angle positive if measured counterclockwise

Step 1
Starting a new drawing

Start AutoCAD and choose **Start from Scratch** from the **Create New Drawing** dialog box. From the **Default Settings** area, select the **Imperial (feet and inches)** radio button, as shown in Figure 1-5. Choose **OK** to open a new file.

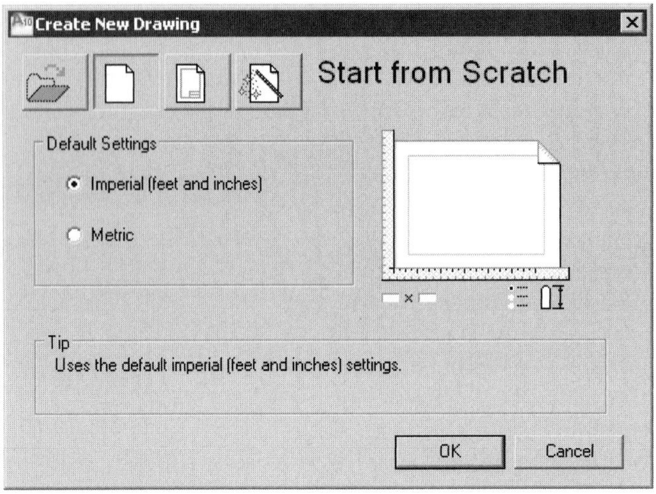

*Figure 1-5 The **Start from Scratch** button of the **Create New Drawing** dialog box*

Step 2
Setting limits, snap, grid, and text size

The **LIMITS** command can be invoked by entering **LIMITS** at the Command prompt.

Command: **LIMITS**
Reset Model space limits:
Specify lower left corner or [ON/OFF] <0.0000,0.0000>: Enter
Specify upper right corner <12.0000,9.0000>: **18,12**

After setting limits, the next step is to expand the drawing display area. Use the **ZOOM** command with the **All** option to display new limits on the screen.

Now, right-click on the **Snap Mode** or **Grid Display** button in the Status Bar to display a shortcut menu. Choose the **Settings** option from the shortcut menu to display the **Drafting Settings** dialog box. You can also choose the **Osnap Settings** button from the **Object Snap** toolbar to display the **Drafting Settings** dialog box. Choose the **Snap and Grid** tab. Enter **0.25** and **0.25** in the **Snap X spacing** and **Snap Y spacing** edit boxes, respectively. Enter **0.5** and **0.5** in the **Grid X spacing** and **Grid Y spacing** edit boxes, respectively. Then choose **OK**.

Note
*You can also use **SNAP** and **GRID** commands to set these values.*

The size of the text can be changed by entering **TEXTSIZE** at the Command prompt.

Command: **TEXTSIZE**
Enter new value for TEXTSIZE <0.2000>: **0.125**

Step 3
Setting units

Choose **Drawing Utilities > Units** from the **Application Menu** or enter **UNITS** at the Command prompt to invoke the **Drawing Units** dialog box, as shown in Figure 1-6. In the **Length** area, select **0.000** from the **Precision** drop-down list. In the **Angle** area, select **Decimal Degrees** from the **Type** drop-down list and **0.00** from the **Precision** drop-down list. Also make sure the **Clockwise** check box in the **Angle** area is not selected.

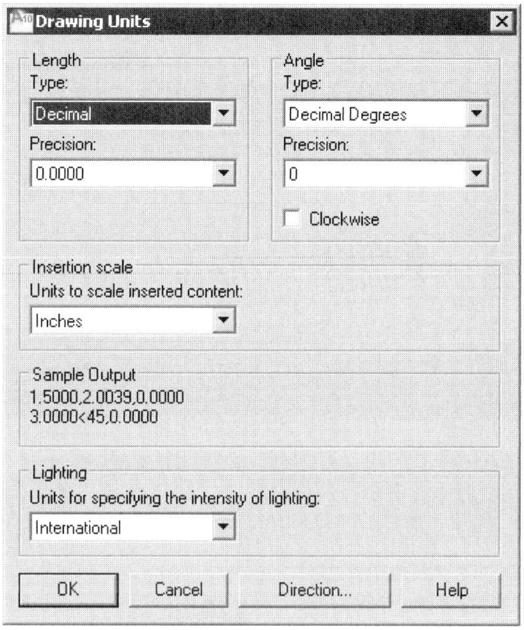

Figure 1-6 *The* ***Drawing Units*** *dialog box*

Choose the **Direction** button from the **Drawing Units** dialog box to display the **Direction Control** dialog box (Figure 1-7) and then select the **East** radio button. Exit both the dialog boxes.

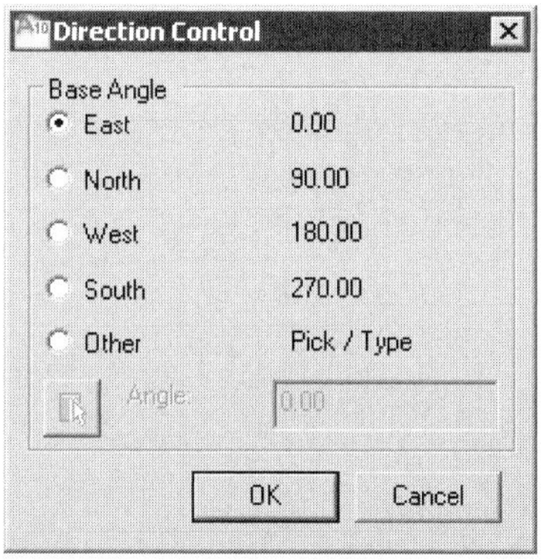

Figure 1-7 *The* ***Direction Control*** *dialog box*

Step 4

Now, save the drawing as *proto2.dwt* using the **Save** button from the **Quick Access** toolbar. You must select **AutoCAD Drawing Template (*dwt)** from the **Files of type** drop-down list in the dialog box. Choose the **Save** button; the **Template Options** dialog box will be displayed on the screen. Enter the description about the template in the **Description** edit box and choose the **OK** button from the **Template Options** dialog box. The drawing will be saved as *proto2.dwt* on the default drive. You can also save this drawing to some other location using the **Save Drawing As** dialog box.

LOADING A TEMPLATE DRAWING

You can use the template drawing to start a new drawing file. To use the preset values of the template drawing, start AutoCAD or select the **New** button from the **Quick Access** toolbar. The dialog box that appears will depend on whether you have set the **STARTUP** system variable to **1** or **0**. If you have set this value as **1**, the **Create New Drawing** dialog box will appear. Choose the **Use a Template** option. All templates that are saved in the default **Template** directory will be shown in the **Select a Template** list box, see Figure 1-8. If you have saved the template in any other file, choose the **Browse** button. The **Select a template file** dialog box will be displayed. You can use this dialog box to browse the directory in which the template file is saved.

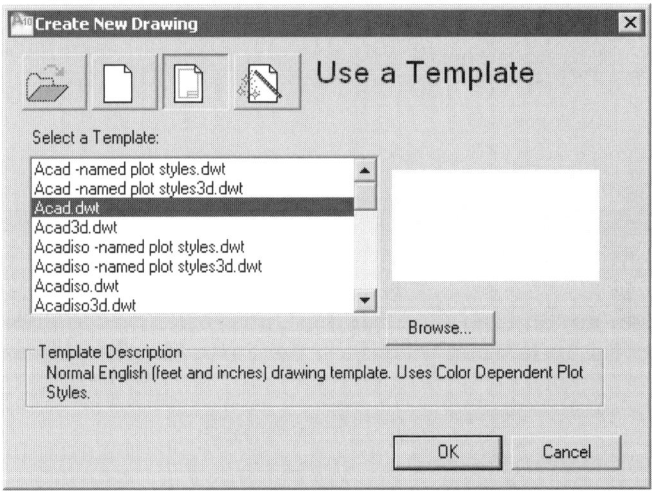

***Figure 1-8** Default templates in the **Create New Drawing** dialog box*

If you have set the **STARTUP** system variable to 0, the **Select template** dialog box appears when you choose the **New** button. This dialog box also displays the default **Template** folder and all template files saved in it, see Figure 1-9. You can use this dialog box to select the template file you that want to open.

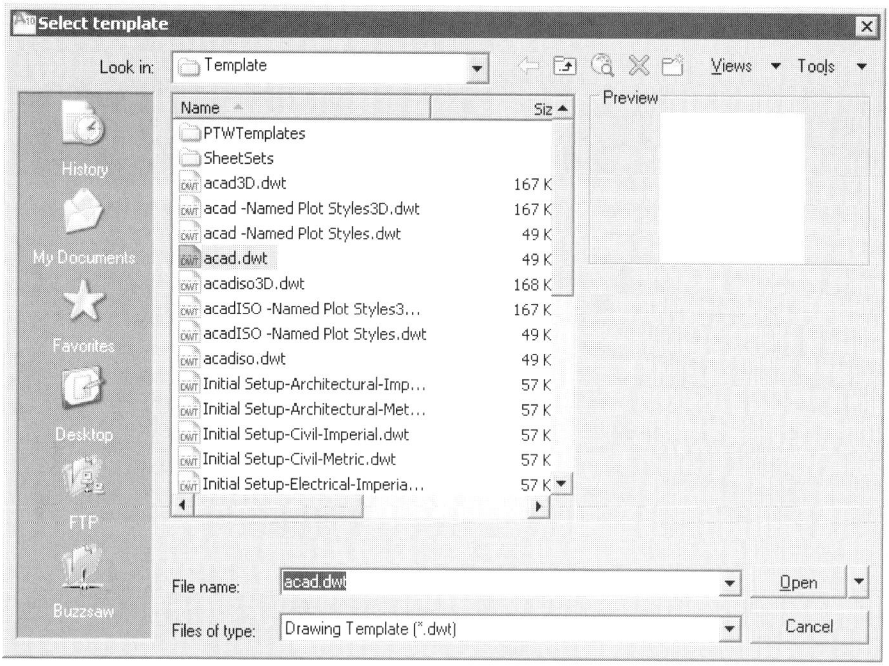

Figure 1-9 *The* ***Select template*** *dialog box that appears when the option to show the* ***Startup*** *dialog box is not selected*

Using any of the previously mentioned dialog boxes, select the *proto1.dwt* template drawing. AutoCAD will start a new drawing that will have the same setup as that of the template drawing *proto1.dwt*.

You can have several template drawings, each with a different setup. For example, **PROTOB** for a 18" by 12" drawing, **PROTOC** for a 24" by 18" drawing. Each template drawing can be created according to user-defined specifications. You can then load any of these template drawings, as discussed previously.

CUSTOMIZING DRAWINGS WITH LAYERS AND DIMENSIONING SPECIFICATIONS

Most production drawings need multiple layers for different groups of objects. In addition to layers, it is a good practice to assign different colors to different layers to control the line width at the time of plotting. You can generate a template drawing that contains the desired number of layers with linetypes and colors according to your company specifications. You can then use this template drawing to make a new drawing. The next example illustrates the procedure used for customizing a drawing with layers, linetypes, and colors.

Example 3

Create a template drawing *proto3.dwt* that has a border and the company's title block, as shown in Figure 1-10.

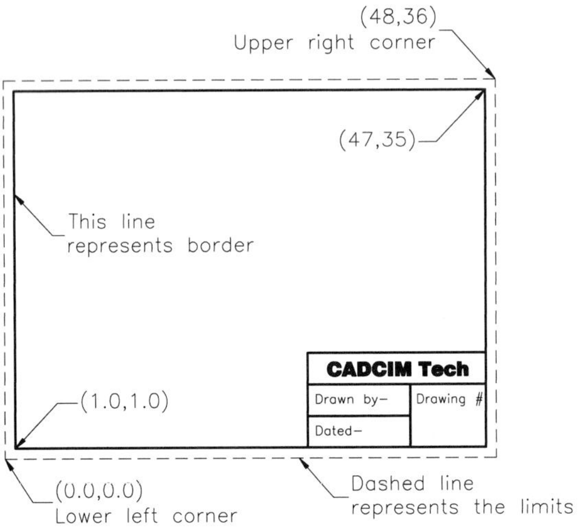

***Figure 1-10** The template drawing for Example 3*

This template drawing will have the following initial drawing setup:

Limits	48.0,36.0
Text height	0.25
Border line lineweight	0.012"
Ltscale	4.0

DIMENSIONS
Overall dimension scale factor 4.0
Dimension text above the extension line
Dimension text aligned with dimension line

LAYERS

Layer Name	Line Type	Color
0	Continuous	White
OBJ	Continuous	Red
CEN	Center	Yellow
HID	Hidden	Blue
DIM	Continuous	Green
BOR	Continuous	Magenta

Step 1
Setting limits, text size, polyline width, polyline, and linetype scaling

Start a new drawing with default parameters by selecting the **Start from Scratch** option in the **Create New Drawings** dialog box. In the new drawing file, use the AutoCAD commands to set up the values as given for this example. Also, draw a border and a title block, as shown in Figure 1-10. In this figure, the hidden lines indicate drawing limits. The border lines are 1.0 units inside the drawing limits. For border lines, increase the lineweight to a value of 0.012".

Use the following procedure to produce the prototype drawing for Example 3.

1. Invoke the **LIMITS** command by choosing **Application Menu > Format > Drawing Limits** or by entering **LIMITS** at the Command prompt. The following is the prompt sequence

   Command: **LIMITS**
   Reset Model space limits:
   Specify lower left corner or [ON/OFF] <0.0000,0.0000>: Enter
   Specify upper right corner <12.0000,9.0000>: **48,36**

2. Increase the drawing display area by choosing the **All** option of the **ZOOM** command.

3. Enter **TEXTSIZE** at the Command prompt to change the text size.

   Command: **TEXTSIZE**
   Enter new value for TEXTSIZE <0.2000>: **0.25**

4. Next, draw the border using the **RECTANG** command. The prompt sequence to draw the rectangle is:

   Command: **RECTANG**
   Specify first corner point or [Chamfer/Elevation/Fillet/Thickness/Width]: **1.0,1.0**
   Specify other corner point or [Area/Dimensions/Rotation]: **47.0,35.0**

5. Now, select the rectangle and change its lineweight to **0.012"**. Make sure that the **Show/Hide Lineweight** button is chosen in the Status Bar.

6. Enter **LTSCALE** at the Command prompt to change the linetype scale.

   Command: **LTSCALE**
   Enter new linetype scale factor<Current>: **4.0**

Step 2
Setting dimensioning parameters

You can use the **Dimension Style Manager** dialog box to set the dimension variables. Choose the inclined arrow displayed on the **Dimensions** panel title bar in the **Annotate** tab of the **Ribbon**; the **Dimension Style Manager** dialog box will be displayed, as shown in Figure 1-11.

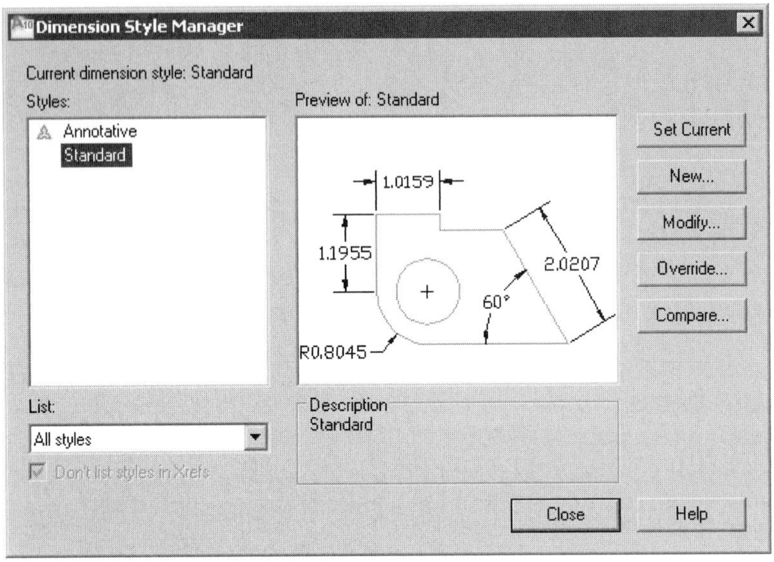

Figure 1-11 *The* ***Dimension Style Manager*** *dialog box*

You can also invoke this dialog box by entering **DIMSTYLE** at the Command prompt. Choose the **New** button from the **Dimension Style Manager** dialog box; the **Create New Dimension Style** dialog box will be displayed. Specify the new style name as **MYDIM1** in the **New Style Name** edit box, as shown in the Figure 1-12 and then choose the **Continue** button. The **New Dimension Style:MYDIM1** dialog box is displayed.

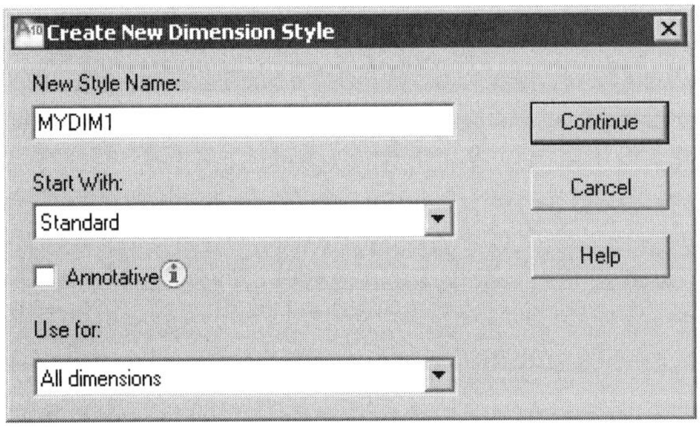

Figure 1-12 *The* ***Create New Dimension Style*** *dialog box*

Overall dimension scale factor

To specify a dimension scale factor, choose the **Fit** tab of the **New Dimension Style: MYDIM1** dialog box. Set the value in the **Use overall scale of** as **4** in the **Scale for dimension features** area, as shown in Figure 1-13.

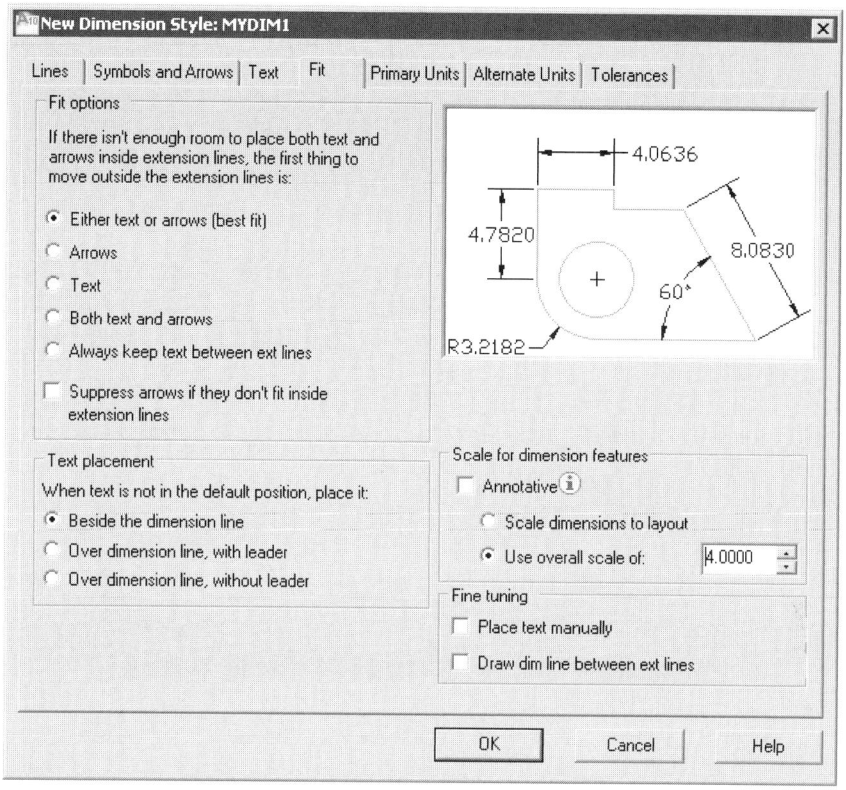

Figure 1-13 *The **Fit** tab of the **New Dimension Style: MYDIM1** dialog box*

Dimension text over the dimension line
In the **Fit** tab of the **New Dimension Style: MYDIM1** dialog box. Select the **Over dimension line, with leader** radio button from the **Text placement** area.

Dimension text aligned with the dimension line
In the **Text Alignment** area of the **Text** tab of the **New Dimension Style: MYDIM1** dialog box, select the **Aligned with the dimension line** radio button and then choose **OK** button.

Setting the new dimension style to current
A new dimension style with the name **MYDIM1** is shown in the **Styles** area of the **Dimension Style Manager** dialog box. Select this dimension style and then choose the **Set Current** button to make it the current dimension style. Choose the **Close** button to exit this dialog box.

Step 3
Setting layers
Choose **Home > Layers > Layer Properties** from the **Ribbon** or choose **Format > Layer** from the menu bar to invoke the **Layer Properties Manager** dialog box. You can also invoke the **Layer Properties Manager** dialog box from the **Layers** toolbar. Choose the **New Layer**

button in the **Layer Properties Manager** dialog box and rename Layer1 as OBJ. Choose the color swatch of the OBJ layer to display the **Select Color** dialog box. Select the **Red** color and choose **OK**; the red color is assigned to the OBJ layer. Again, choose the **New Layer** button in the **Layer Properties Manager** dialog box and rename the Layer1 as CEN. Choose the linetype swatch to display the **Select Linetype** dialog box.

If different linetypes are not already loaded, choose the **Load** button to display the **Load or Reload Linetypes** dialog box. Select the **CENTER** linetype from the **Available Linetypes** area and choose **OK**; the **Select Linetype** dialog box will reappear. Select the **CENTER** linetype from the **Loaded linetypes** area and choose **OK**. Choose the color swatch to display the **Select Color** dialog box. Select the **Yellow** color and choose **OK**; the color yellow and linetype center will be assigned to the layer CEN.

Similarly, different linetypes and different colors can be set for different layers mentioned in the statement of this example, as shown in Figure 1-14.

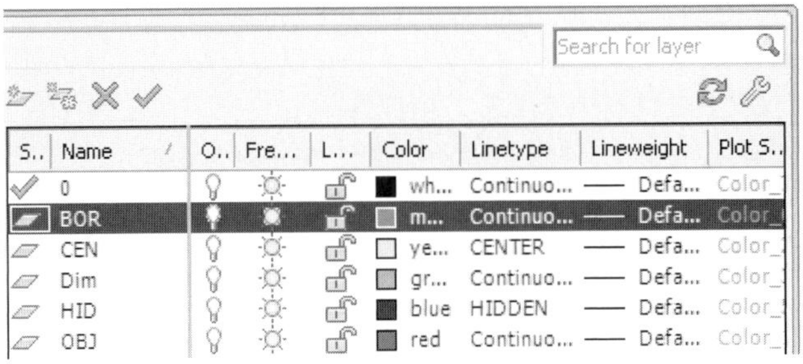

***Figure 1-14** Partial display of the **Layer Properties Manager** dialog box*

Step 4
Adding title block

Next, add the title block and the text, as shown in Figure 1-10. After completing the drawing, save it as *proto3.dwt*. You have created a template drawing (PROTO3) that contains all the information given in Example 3.

CUSTOMIZING A DRAWING WITH LAYOUT

The Layout (paper space) provides a convenient way to plot multiple views of a three-dimensional (3D) drawing or multiple views of a regular two-dimensional (2D) drawing. It takes quite some time to set up the viewports in the model space with different vpoints and scale factors. You can create prototype drawings that contain predefined viewport settings, with vpoint and the other desired information. If you create a new drawing or insert a drawing, the views are automatically generated. The following example illustrates the procedure for generating a prototype drawing with paper space and model space viewports:

Example 4

Create a drawing template, as shown in Figure 1-15, with four views in Layout3 (Paper space) that display front, top, side, and 3D views of the object. The plot size is 10.5 by 8 inches. The plot scale is 0.5 or 1/2" = 1". The paper space viewports should have the following vpoint settings:

Viewports	Vpoint	View
Top right	1,-1,1	3D view
Top left	0,0,1	Top view
Lower right	1,0,0	Right side view
Lower left	0,-1,0	Front view

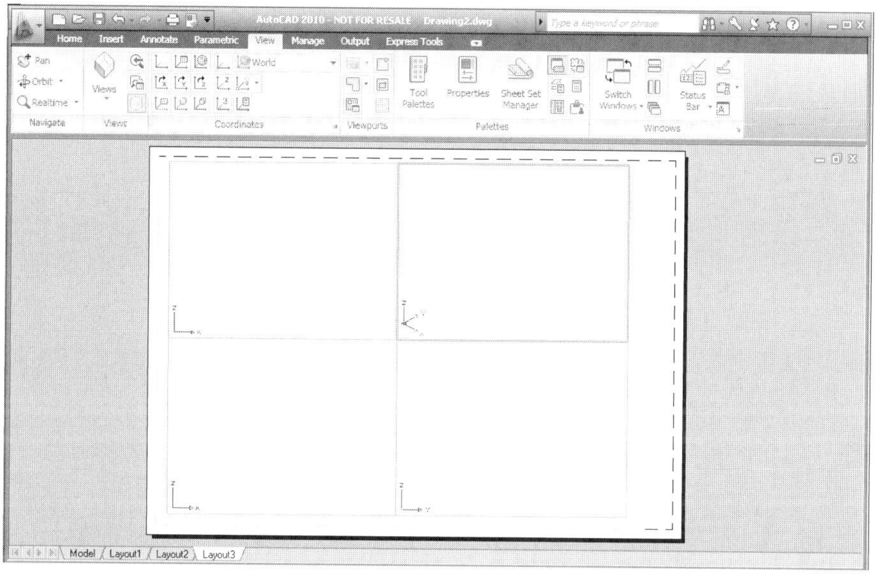

Figure 1-15 *Paper space with four viewports*

Start AutoCAD and create a new drawing. Use the following commands and options to set up various parameters:

Step 1

First, you need to create a new layout using the **LAYOUT** command. You can also choose **New Layout** button from the **Quick View Layouts** panel available on choosing the **Quick View Layouts** button on the Status Bar. A new layout is automatically created with the default name. Alternatively, you can also use the **LAYOUT** command to create a new layout. The Command prompt that is displayed is given next.

Command: **LAYOUT**
Enter layout option [Copy/Delete/New/Template/Rename/SAveas/Set/?] <set>: **N**
Enter new Layout name <Layout3>: Enter

Step 2

Next, choose the new layout (**Layout3**) tab. The new layout (**Layout3**) will be displayed with the default viewport. Use the **ERASE** command to erase this viewport. Invoke the **Page Setup Manager** dialog box from the shortcut menu that will by displayed when you right click on the **Quick View Layouts** button and choose the **Modify** button to modify the default page setup. In the **Page Setup - Layout3** dialog box, select the required printer from the **Printer/plotter** area. In this example, HP LaserJet 4000 has been used.

Step 3

From the **Paper size** area, select the paper size that is supported by the selected plotting device. In this example, the paper size is A4. Choose the **OK** button to accept the settings and return to the **Page Setup Manager** dialog box. Choose **Close** in this dialog box to close the **Page Setup Manager** dialog box.

Step 4

Next, you need to set up a layer with the name VIEW for viewports and assign it green color. Invoke the **Layer Properties Manager** dialog box. Choose the **New Layer** button and name the Layer1 as VIEW. Choose the color swatch of the VIEW layer to display the **Select Color** dialog box. Select the color **Green** and choose the **OK** button. This color will be assigned to **View** layer. Also, make the VIEW layer current and then exit the **Layer Properties Manager** dialog box.

Step 5

To create four viewports, use the **MVIEW** command. You can directly choose **View > Viewports > 4Viewports** from the menu bar or enter **MVIEW** command at the Command prompt. The following is the prompt sequence when you invoke the **MVIEW** command.

Command: **MVIEW**
Specify corner of viewport or
[ON/OFF/Fit/Shadeplot/Lock/Object/Polygonal/Restore/LAyer/2/3/4] <Fit>: **4**
Specify first corner or [Fit] <Fit>: **0.25,0.25**
Specify opposite corner: **10.25,7.75**

Step 6

Choose the **PAPER** button in the Status Bar to activate the model space or enter **MSPACE** at the Command prompt.

Command: **MSPACE** (or **MS**)

Make the lower left viewport active by clicking on it. Next, you need to change the viewpoints of different paper space viewports using the **VPOINT** command. To invoke this command, choose **View > 3D Views > Viewpoint** from the menu bar or enter **VPOINT** at the Command prompt. The viewpoint values for different viewports are given in the statement of Example 4. To set the view point for the lower left viewport the Command prompt sequence is given next.

Command: **VPOINT**
Current view direction: VIEWDIR= 0.0000,0.0000,1.0000
Specify a view point or [Rotate] <display compass and tripod>: **0,-1,0**

Similarly, use the **VPOINT** command to set the viewpoint of the other viewports.

Make the top left viewport active by selecting a point in the viewport and then use the **ZOOM** command to specify the paper space scale factor to 0.5. The **ZOOM** command can be invoked by choosing **View > Zoom > Scale** from the **Application Menu** or by entering **ZOOM** at the Command prompt. The Command prompt sequence is given next.

Command: **ZOOM**
Specify corner of window, enter a scale factor (nX or nXP), or
[All/Center/Dynamic/Extents/Previous/Scale/Window/Object] <real time>: **0.5XP**

Now, make the next viewport active and specify the zoom scale factor. Do the same for the remaining viewports.

Step 7

Use the **MODEL** button in the Status Bar to change to paper space and then set a new layer PBORDER with yellow color. Make the PBORDER layer current, draw a border, and if needed, a title block using the **PLINE** command. You can also change to paper space by entering **PSPACE** at the Command prompt.

The **PLINE** command can be invoked by choosing the **Polyline** button from the **Draw** toolbar or choosing **Home > Draw > Polyline** from the **Ribbon**. The **PLINE** command can also be invoked by entering **PLINE** at the Command prompt.

Command: **PLINE**
Specify start point: **0,0**
Current line-width is 0.0000
Specify next point or [Arc/Halfwidth/Length/Undo/Width]: **0,8.0**
Specify next point or [Arc/Close/Halfwidth/Length/Undo/Width]: **10.5,8.0**
Specify next point or [Arc/Close/Halfwidth/Length/Undo/Width]: **10.5,0**
Specify next point or [Arc/Close/Halfwidth/Length/Undo/Width]: **C**

Step 8

The last step is to choose the **Model** tab (or change the **TILEMODE** to 1), if not already active and save the prototype drawing as template. To test the layout that you just created, make the 3D drawing, as shown in Figure 1-18 or make any 3D object. Switch to **Layout 3** tab; you will find four different views of the object (Figure 1-16). If the object views do not appear in the viewports, use the **PAN** commands to position the views in the viewports. You can freeze the VIEW layer so that the viewports do not appear on the drawing. You can plot this drawing from the Layout3 with a plot scale factor of 1:1 and the size of the plot will be exactly as specified.

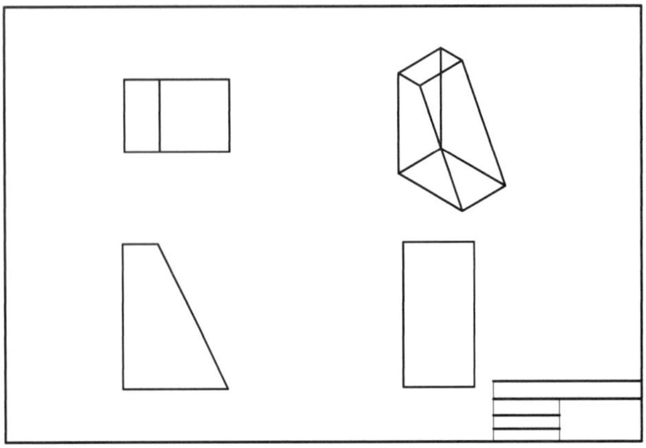

Figure 1-16 *Four views of a 3D object in paper space*

CUSTOMIZING DRAWINGS WITH VIEWPORTS

In certain applications, you may need multiple model space viewport configurations to display different views of an object. This involves setting up the desired viewports and then changing the viewpoint for different viewports. You can create a prototype drawing that contains a required number of viewports and the viewpoint information. If you insert a 3D object in one of the viewports of the prototype drawing, you will automatically get different views of the object without setting viewports or viewpoints. The following example illustrates the procedure for creating a prototype drawing with a standard number (four) of viewports and viewpoints.

Example 5

Create a prototype drawing with four viewports, as shown in Figure 1-17. The viewports should have the following viewpoints:

Viewports	Vpoint	View
Top right	1,-1,1	3D view
Top left	0,0,1	Top view
Lower right	1,0,0	Right side view
Lower left	0,-1,0	Front view

Step 1

Start AutoCAD and create a new drawing from scratch.

Step 2
Setting viewports

Viewports and corresponding viewpoints can be set with the **VPORTS** command. You can also choose the **Display Viewports Dialog** button from the **Viewports** toolbar or choose **View > Viewports > New** from the **Ribbon** to display the **Viewports** dialog box, as shown

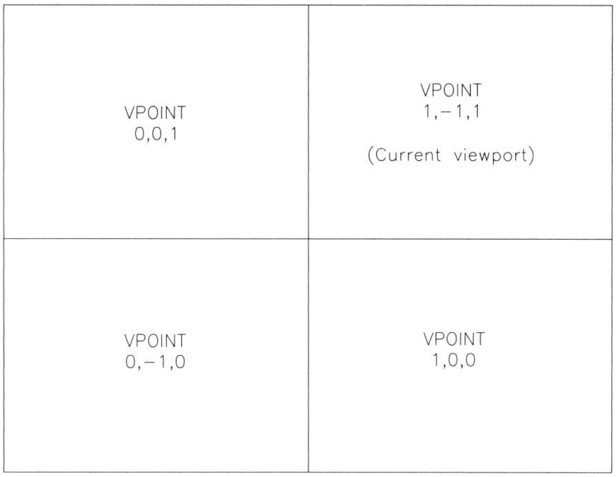

Figure 1-17 *Viewports with different viewpoints*

in Figure 1-18. Select **Four: Equal** from the **Standard viewports** area. In the **Preview** area four equal viewports are displayed. Select **3D** from the **Setup** drop-down list. The four viewports with the different viewpoints will be displayed in the **Preview** area as Top, Front, Right and SE Isometric, respectively. **Top** represents the viewpoints as (0,0,1), **Front** represents the

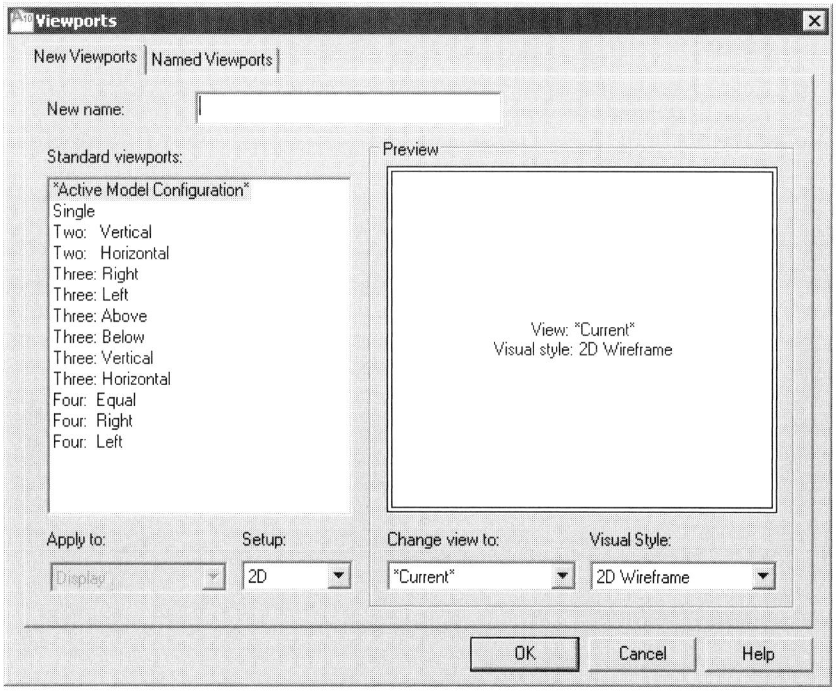

Figure 1-18 *The* ***Viewports*** *dialog box*

viewpoints as (0,-1,0), **Right** represents the viewpoints as (1,0,0) and **SE Isometric**, represents the viewpoints as (1,-1,1) respectively. Choose the **OK** button. Save the drawing as *proto5.dwt*.

Viewports and viewpoints can also be set by entering **-VPORTS** and **VPOINT** at the Command prompt, respectively.

Step 3

Start a new drawing and draw the 3D tapered block, as shown in Figure 1-19.

Step 4

Again, start a new drawing, TEST, using the prototype drawing *proto5.dwt*. Make the top right viewport current and then insert or create a drawing, as shown in Figure 1-19. Four different views will be automatically displayed on the screen, as shown in Figure 1-20.

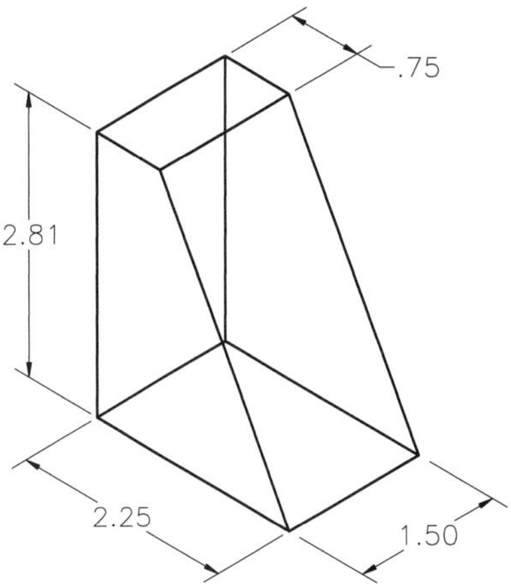

Figure 1-19 *3D tapered block*

Note

The method for creating 3D models is discussed in Chapter 25 (Getting Started with 3D) of AutoCAD 2010: A Problem-Solving Approach textbook by Proff. Sham Tickoo.

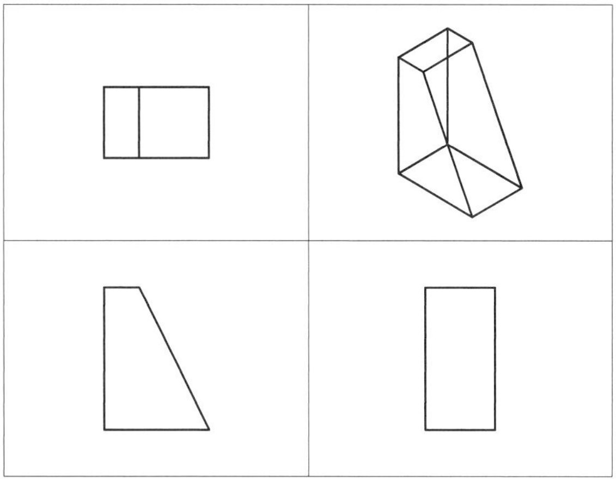

Figure 1-20 *Different views of a 3D tapered block*

CUSTOMIZING DRAWINGS ACCORDING TO PLOT SIZE AND DRAWING SCALE

For controlling the plot area, it is recommended to use layouts. You can make the drawing of any size, use the layout to specify the sheet size, and then draw the border and title block. However, you can also plot a drawing in the model space and set up the system variables so that the plotted drawing is to your specifications. You can generate a template drawing according to plot size and scale. For example, if the scale is 1/16" = 1' and the drawing is to be plotted on a 36" by 24" area, you can calculate drawing parameters like limits, **DIMSCALE**, and **LTSCALE** and save them in a template drawing. This will save considerable time in the initial drawing setup and provide uniformity in the drawings. The next example explains the procedure involved in customizing a drawing according to a certain plot size and scale.

Note

You can also use the paper space to specify the paper size and scale.

Example 6

Create a drawing template (**PROTO6**) with the following specifications:

Plotted sheet size	36" by 24" (Figure 1-20)
Scale	1/8" = 1.0'
Snap	3'
Grid	6'
Text height	1/4" on plotted drawing
Linetype scale	Calculate
Dimscale factor	Calculate
Units	Architectural

	Precision, 16-denominator of the smallest fraction
	Angle in degrees/minutes/seconds
	Precision, 0d00'
	Direction control, base angle, east
	Angle positive, if measured counterclockwise
Border	Border should be 1" inside the edges of the plotted drawing sheet, using PLINE 1/32" wide when plotted (Figure 1-21)

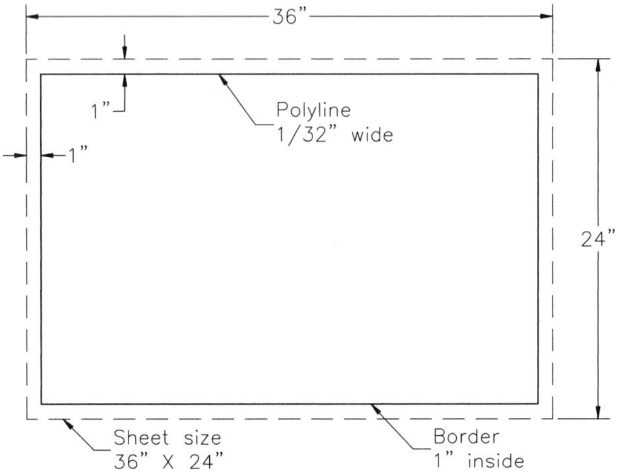

Figure 1-21 *Border of the template drawing*

Step 1

Calculating limits, text height, linetype scale, dimension scale, and polyline width

In this example, you need to calculate some values before you set the parameters. The limits of the drawing depend on the plotted size of the drawing and its scale. Similarly, **LTSCALE** and **DIMSCALE** depend on the plot scale of the drawing. The following calculations explain the procedure for finding the values of limits, ltscale, dimscale, and text height.

Limits

Given:

Sheet size 36" x 24"

Scale 1/8" = 1'

or 1" = 8'

Calculate:

X Limit

Y Limit

Since sheet size is 36" x 24" and scale is 1/8"=1'

Therefore, X Limit = 36 x 8' = 288'

Y Limit = 24 x 8' = 192'

Text height

Given:
Text height when plotted = 1/4"
Scale 1/8" = 1'

Calculate:
Text height
Since scale is 1/8" = 1'
or 1/8" = 12"
or 1" = 96"
Therefore, scale factor = 96
Text height = 1/4" x 96
= 24" = 2'

Linetype scale and dimension scale

Known:
Since scale is 1/8" = 1'
or 1/8" = 12"
or 1" = 96"

Calculate:
LTSCALE and DIMSCALE
Since scale factor = 96
Therefore, LTSCALE = Scale factor = 96
Similarly, DIMSCALE = 96
(All dimension variables, like DIMTXT and DIMASZ, will be multiplied by 96.)

Polyline Width

Given:
Scale is 1/8" = 1'

Calculate:
PLINE width
Since scale is 1/8" = 1'
or 1" = 8'
or 1" = 96"

Therefore,
PLINE width = 1/32 x 96
= 3"

After calculating the parameters, use the following AutoCAD commands to set up the drawing and save the drawing as *proto6.dwt.*

Step 2
Setting units

Start a new drawing and enter **UNITS** at the Command prompt to display the **Drawing Units** dialog box. Select **Architectural** from the **Type** drop-down list in the **Length** area. Select **0'-01/16"** from the **Precision** drop-down list. Make sure the **Clockwise** check box in the **Angle** area is not selected. Select **Deg/Min/Sec** from the **Type** drop-down list and select **0d00'** from the **Precision** drop-down list in the **Angle** area. Now, choose the **Direction** button to display the **Directional Control** dialog box. Select the **East** radio button, if it is not selected, in the **Base Angle** area and then choose **OK**.

Step 3
Setting limits, snap and grid, textsize, linetype scale, dimension scale, dimension style and pline

To set limits, enter **LIMITS** at the Command prompt.

Command: **LIMITS**
Specify lower left corner or [ON/OFF] <0'-0",0'-0">:**0,0**
Specify upper right corner <1'-0",0'-9">: **288',192'**

Invoke the **All** option of the **ZOOM** command to increase the drawing display area.

Right-click on the **Snap Mode** or **Grid Display** button in the Status Bar to invoke the shortcut menu. In the shortcut menu, choose **Settings** to display the **Drafting Settings** dialog box. You can also choose the **Osnap Settings** button from the **Object Snap** toolbar. In the dialog box, choose the **Snap and Grid** tab. Enter **3'** and **3'** in the **Snap X spacing** and **Snap Y spacing** edit boxes, respectively. Enter **6'** and **6'** in the **Grid X spacing** and **Grid Y spacing** edit boxes, respectively. Then choose **OK**.

You can also set these values by entering **SNAP** and **GRID** at the Command prompt.

The size of the text can be changed by entering **TEXTSIZE** at the Command prompt.

Command: **TEXTSIZE**
Enter new value for TEXTSIZE <current>: **2'**

To set the **LTSCALE**, choose the **Format > Linetype** from the menu bar or enter **LINETYPE** at the Command prompt to invoke the **Linetype Manager** dialog box. Choose the **Show details** button. Specify the **Global scale factor** as **96** in the **Global scale factor** edit box.

You can also change the scale of the linetype by entering **LTSCALE** at the Command prompt.

Choose the inclined arrow displayed on the **Dimensions** panel title bar from the **Annotate** tab of the **Ribbon** to invoke the **Dimension Style Manager** dialog box, as shown in Figure 1-11. Choose the **New** button from the **Dimension Style Manager** dialog box to invoke the **Create New Dimension Style** dialog box. Specify the new style name as **MYDIM2** in the **New Style Name** edit box and then choose the **Continue** button. The **New Dimension**

Style: MYDIM2 dialog box will be displayed. Choose the **Fit** tab and set the value in the **Use overall scale of** spinner to **96** in the **Scale for dimension features** area. Now, choose the **OK** button to again display the **Dimension Style Manager** dialog box. Choose **Close** to exit the dialog box.

You can invoke the **PLINE** command by choosing **Home > Draw > Polyline** button from the **Ribbon** or entering **PLINE** at the Command prompt.

Command: **PLINE**
Specify start point: **8',8'**
Current line-width is **0'-0"**
Specify next point or [Arc/Halfwidth/Length/Undo/Width]: **W**
Specify starting width<0'-00">: **3**
Specify ending width<0'-3">: **ENTER**
Specify next point or [Arc/Halfwidth/Length/Undo/Width]: **280,0**
Specify next point or [Arc/Close/Halfwidth/Length/Undo/Width]: **0,184**
Specify next point or [Arc/Close/Halfwidth/Length/Undo/Width]: **-280,0**
Specify next point or [Arc/Close/Halfwidth/Length/Undo/Width]: **C**

Now, save the drawing as *proto6.dwt*.

Self-Evaluation Test

Answer the following questions and then compare them to those given at the end of this chapter:

1. The template drawings are saved in __________ format.

2. To use a template file, select the __________ option in the **Create New Drawing** dialog box.

3. To start a drawing with the default setup, select the __________ option in the **Create New Drawing** dialog box.

4. If the plot size is 36" x 24", and the scale is 1/2" = 1', then X Limit = __________ and Y Limit = __________.

5. You can use AutoCAD's __________ command to set up a viewport in the paper space.

Review Questions

Answer the following questions:

1. The default value of **DIMSCALE** is __________.

2. The default value of **DIMTXT** is __________.

3. The default value of **SNAP** is _________.

4. Architectural units can be selected by using the _________ command or the _________ command.

5. Name three standard template drawings that come with AutoCAD software _________, _________, and _________.

6. If the plot size is 24" x 18", and the scale is 1:20, the X Limit = _________ and Y Limit = _________.

7. If the plot size is 200 x 150 and limits are (0.00,0.00) and (600.00,450.00), the **LTSCALE** factor = _________.

8. _________ provides a convenient way to plot multiple views of a 3D drawing or multiple views of a regular 2D drawing.

9. You can use the _________ command to change to paper space.

10. You can use AutoCAD's _________ command to change to model space.

11. The values that can be assigned to **TILEMODE** are _________ and _________.

12. In the model space, if you want to reduce the display size by half, the scale factor you enter in the **ZOOM > Scale** option is _________.

Exercises

Exercise 1 — *General*

Create a drawing template (*protoe1.dwt*) with the following specifications:

Units	Architectural with precision 0'-0 1/16
Angle	Decimal Degrees with precision 0.
Base angle	East.
Angle direction	Counterclockwise.
Limits	48' x 36'

Exercise 2 — *General*

Create a drawing template (*protoe2.dwt*) with the following specifications:

Limits	36.0,24.0
Snap	0.5

Grid	1.0
Text height	0.25
Units	Decimal
	Precision 0.00
	Decimal degrees
	Precision 0
	Base angle, East
	Angle positive if measured counterclockwise

Exercise 3 *General*

Create a drawing template (*protoe3.dwt*) with the following specifications:

Limits	48.0,36.0
Text height	0.25
PLINE width	0.03
LTSCALE	4.0
DIMSCALE	4.0
Plot size	10.5 x 8

LAYERS

Layer Name	Line Type	Color
0	Continuous	White
OBJECT	Continuous	Green
CENTER	Center	Magenta
HIDDEN	Hidden	Blue
DIM	Continuous	Red
BORDER	Continuous	Cyan

Exercise 4 *General*

Create a prototype drawing with the following specifications (the name of the drawing is *protoe4.dwt*):

Limits	36.0,24,0
Border	35.0,23.0
Grid	1.0
Snap	0.5
Text height	0.15
Units	Decimal (up to 2 places)
LTSCALE	1
Current layer	Object

LAYERS

Layer Name	Linetype	Color
0	Continuous	White
Object	Continuous	Red
Hidden	Hidden	Yellow
Center	Center	Green
Dim	Continuous	Blue
Border	Continuous	Magenta
Notes	Continuous	White

This prototype drawing should have a border line and a title block as shown in Figure 1-22.

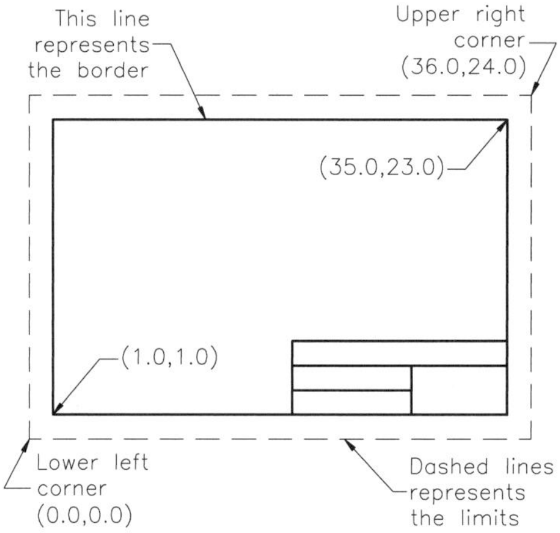

Figure 1-22 *Prototype drawing*

Exercise 5 *General*

Create a template drawing shown in Figure 1-23 with the following specifications and save it with the name *protoe5.dwt*:

Plotted sheet size	36" x 24" (Figure 1-22)
Scale	1/2" = 1.0'
Text height	1/4" on plotted drawing
LTSCALE	24
DIMSCALE	24
Units	Architectural 32-denominator of smallest fraction to display Angle in degrees/minutes/seconds Precision 0d00"00" Angle positive if measured counterclockwise
Border	Border is 1-1/2" inside the edges of the plotted drawing sheet, using the PLINE 1/32" wide when plotted.

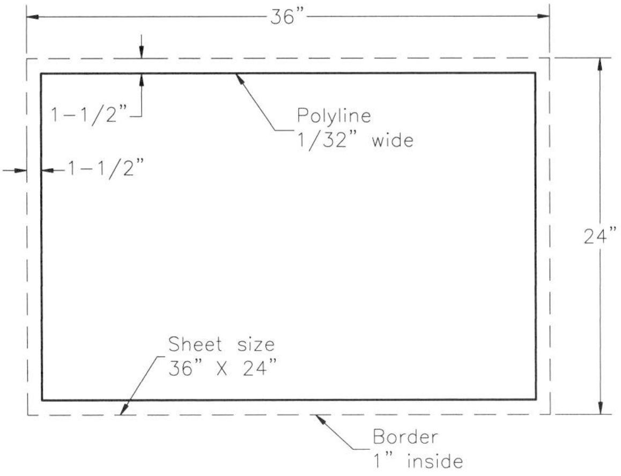

Figure 1-23 *Drawing for Exercise 5*

Exercise 6 *General*

Create a prototype drawing, as shown in Figure 1-24 with the following specifications (the name of the drawing is *protoe6.dwt*):

Plotted sheet size	24" x 18" (Figure 1-23)
Scale	1/2"=1.0'
Border	The border is 1" inside the edges of the plotted drawing sheet, using the PLINE 0.05" wide when plotted (Figure 1-24)

Dimension text over the dimension line
Dimensions aligned with the dimension line
Calculate overall dimension scale factor
Enable the display of alternate units
Dimensions to be associative.

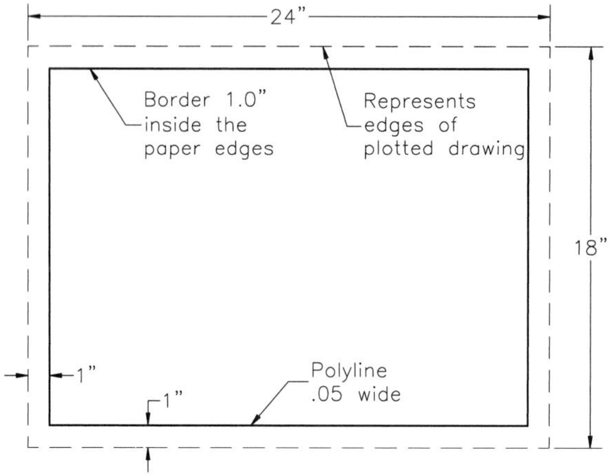

Figure 1-24 *Prototype drawing*

Exercise 7 *Mechanical*

Create a template with the necessary specifications to draw the Articulated Rod shown in Figure 1-25. Also, use this template to draw the views of the Articulated Rod.

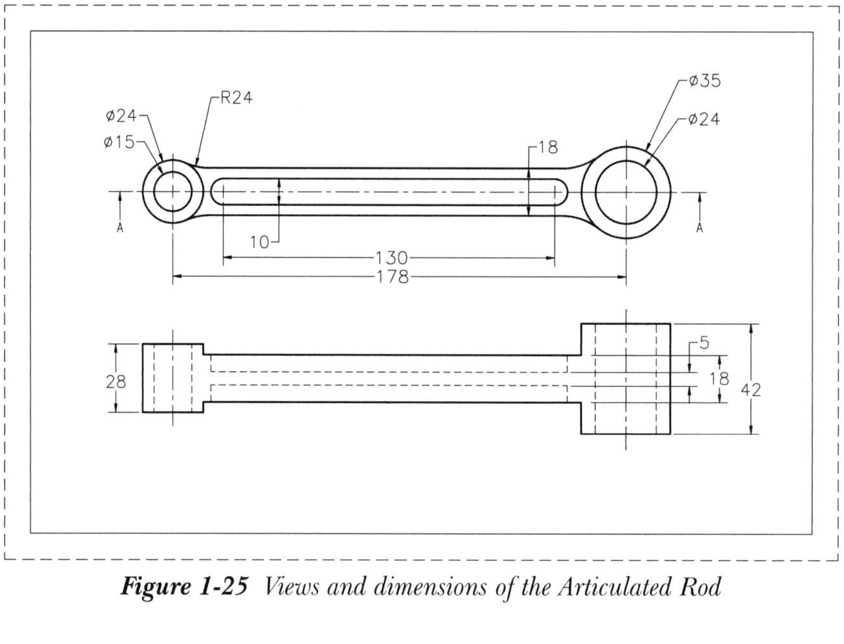

Figure 1-25 *Views and dimensions of the Articulated Rod*

Answers to Self-Evaluation Test

1. *.dwt*, **2. Use a Template**, **3. Start from Scratch**, **4.** 72",48", **5. MVIEW**.

Chapter 2

Script Files and Slide Shows

Learning Objectives

After completing this chapter, you will be able to:

- *Write script files and use the **Run Script** option to run script files.*
- *Use the **RSCRIPT** and **DELAY** commands in script files.*
- *Run script files while loading AutoCAD.*
- *Create a slide show.*
- *Preload slides when running a slide show.*

WHAT ARE SCRIPT FILES?

AutoCAD has provided a facility called script files that allow you to combine different AutoCAD commands and execute them in a predetermined sequence. The commands can be written as a text file using any text editor like Notepad. These files, generally known as script files, have an extension *.scr* (example: *plot1.scr*). A script file is executed with the AutoCAD **SCRIPT** command.

Script files can be used to generate a slide show, do the initial drawing setup, or plot a drawing to a predefined specification. They can also be used to automate certain command sequences that are used frequently in generating, editing, or viewing drawings. Remember that the script files cannot access dialog boxes or menus. When commands that open dialog boxes are issued from a script file, AutoCAD runs the command line version of the command instead of opening the dialog box.

Example 1

Write a script file that will perform the following initial setup for a drawing (file name *script1.scr*). It is assumed that the drawing will be plotted on 12x9 size paper (Scale factor for plotting = 4).

Ortho	On		Zoom	All
Grid	2.0		Text height	0.125
Snap	0.5		LTSCALE	4.0
Limits	0,0	48.0,36.0	DIMSCALE	4.0

Step 1: Understanding commands and prompt entries

Before writing a script file, you need to know the AutoCAD commands and the entries required in response to the Command prompts. To find out the sequence of the Command prompt entries, you can type the command and then respond to different Command prompts. The following is the list of AutoCAD commands and prompt entries for Example 1:

Command: **ORTHO**
Enter mode [ON/OFF] <OFF>: **ON**

Command: **GRID**
Specify grid spacing(X) or [ON/OFF/Snap/Major/aDaptive/Limits/Follow/Aspect] <0.5000>: **2.0**

Command: **SNAP**
Specify snap spacing or [ON/OFF/Aspect/Style/Type]<0.5000>: **0.5**

Command: **LIMITS**
Reset Model space limits:
Specify lower left corner or [ON/OFF] <0.0,0.0>: **0,0**
Specify upper right corner <12.0,9.0>: **48.0,36.0**

Command: **ZOOM**
Specify corner of window, enter a scale factor (nX or nXP), or
[All/Center/Dynamic/Extents/Previous/Scale/Window/Object] <real time>: **A**

Command: **TEXTSIZE**
Enter new value for TEXTSIZE <0.02>: **0.125**

Command: **LTSCALE**
Enter new linetype scale factor <1.0000>: **4.0**

Command: **DIMSCALE**
Enter new value for DIMSCALE <1.0000>: **4.0**

Step 2: Writing the script file

Once you know the commands and the required prompt entries, you can write a script file using any text editor such as the Notepad.

As you invoke the **NOTEPAD** command, AutoCAD prompts you to enter the file to be edited. Press ENTER in response to the prompt to display the **Notepad** editor. Write the script file in the **Notepad** editor. Given below is the listing of the script file for Example 1:

```
ORTHO
ON
GRID
2.0
SNAP
0.5
LIMITS
0,0
48.0,36.0
ZOOM
ALL
TEXTSIZE
0.125
LTSCALE
4.0
DIMSCALE 4.0
```

Note that the commands and the prompt entries in this file are in the same sequence as mentioned earlier. You can also combine several statements in one line, as shown in the following list:

```
;This is my first script file, SCRIPT1.SCR
ORTHO ON
GRID 2.0
```

```
SNAP 0.5
LIMITS 0,0 48.0,36.0
ZOOM
ALL
TEXTSIZE 0.125
LTSCALE 4.0
DIMSCALE 4.0
```

Save the script file as *script1.scr* on A or C drive and exit the text editor. Remember that if you do not save the file in the *.scr* format, it will not work as a script file. Notice the space between the commands and the prompt entries. For example, between **ORTHO** command and **ON**, there is a space. Similarly, there is a space between **GRID** and **2.0**.

Note

In the script file, a space is used to terminate a command or a prompt entry. Therefore, spaces are very important in these files. Make sure there are no extra spaces, unless they are required to press ENTER more than once.

*After you change the limits, it is a good practice to use the **ZOOM** command with the **All** option to increase the drawing display area.*

Keyboard shortcuts are not allowed in the script files. Therefore, make sure not to use them in the script files.

Tip

AutoCAD ignores and does not process any lines that begin with a semicolon (;). This allows you to put comments in the related file or line.

RUNNING SCRIPT FILES

The **SCRIPT** command allows you to run a script file while you are at the drawing editor. Choose **Manage > Applications > Run Script** from the **Ribbon**; the **Select Script File** dialog box will be invoked, as shown in Figure 2-1. You can enter the name of the script file or you can accept the default file name. The default script file name is the same as the drawing name. If you want to enter a new file name, type the name of the script file without the file extension (**.SCR**). (The file extension is assumed and need not be included with the file name.)

Step 3: Running the script file

To run the script file of Example 1, invoke the **SCRIPT** command, select the file **SCRIPT1**, and then choose the **Open** button in the **Select Script File** dialog box (Figure 2-1) . You will see the changes taking place on the screen as the script file commands are executed.

You can also enter the name of the script file at the Command prompt by setting **FILEDIA**=0. The sequence for invoking the script using the Command line is given next.

Command: **FILEDIA**
Enter new value for FILEDIA <1>: **0**
Command: **SCRIPT**
Enter script file name <current>: *Specify the script file name.*

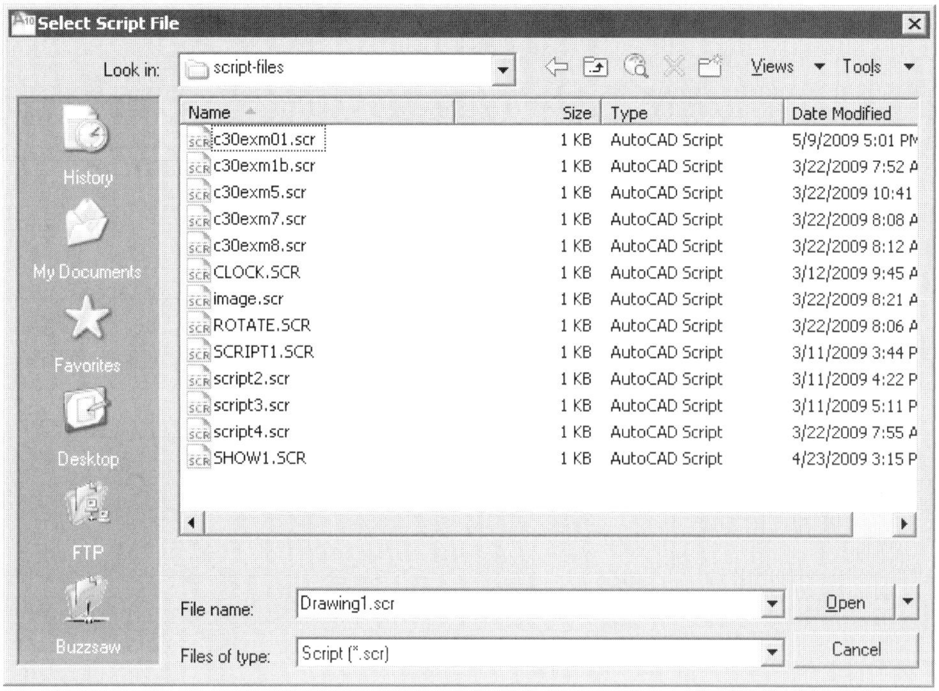

Figure 2-1 The ***Select Script File*** *dialog box*

Example 2

Write a script file that will set up the following layers with the given colors and linetypes (file name *script2.scr*).

Layer Names	Color	Linetype	Line Weight
OBJECT	Red	Continuous	default
CENTER	Yellow	Center	default
HIDDEN	Blue	Hidden	default
DIMENSION	Green	Continuous	default
BORDER	Magenta	Continuous	default
HATCH	Cyan	Continuous	0.05

Step 1: Understanding commands and prompt entries

You need to know the AutoCAD commands and the required prompt entries before writing a script file. For Example 2, you need the following commands to create the layers with the given colors and linetypes.

Command: **-LAYER**
Enter an option
[?/Make/Set/New/Rename/ON/OFF/Color/Ltype/LWeight/MATerial/Plot/Freeze/Thaw/LOck/Unlock/stAte/Description/rEconcile]: **N**
Enter name list for new layer(s): **OBJECT,CENTER,HIDDEN,DIMENSION,BORDER,HATCH**

Enter an option
[?/Make/Set/New/Rename/ON/OFF/Color/Ltype/LWeight/MATerial/Plot/Freeze/Thaw/LOck/Unlock/stAte/Description/rEconcile]: **L**
Enter loaded linetype name or [?] <Continuous>: **CENTER**
Enter name list of layer(s) for linetype "CENTER" <0>: **CENTER**

Enter an option
[?/Make/Set/New/Rename/ON/OFF/Color/Ltype/LWeight/MATerial/Plot/Freeze/Thaw/LOck/Unlock/stAte/Description/rEconcile]: **L**
Enter loaded linetype name or [?] <Continuous>: **HIDDEN**
Enter name list of layer(s) for linetype "HIDDEN" <0>: **HIDDEN**

Enter an option
[?/Make/Set/New/Rename/ON/OFF/Color/Ltype/LWeight/MATerial/Plot/Freeze/Thaw/LOck/Unlock/stAte/Description/rEconcile]: **C**
New color [Truecolor/COlorbook]: **RED**
Enter name list of layer(s) for color 1 (red) <0>: **OBJECT**

Enter an option
[?/Make/Set/New/Rename/ON/OFF/Color/Ltype/LWeight/MATerial/Plot/Freeze/Thaw/LOck/Unlock/stAte/Description/rEconcile]: **C**
New color [Truecolor/COlorbook]: **YELLOW**
Enter name list of layer(s) for color 2 (yellow) <0>: **CENTER**

Enter an option
[?/Make/Set/New/Rename/ON/OFF/Color/Ltype/LWeight/MATerial/Plot/Freeze/Thaw/LOck/Unlock/stAte/Description/rEconcile]: **C**
New color [Truecolor/COlorbook]: **BLUE**
Enter name list of layer(s) for color 5 (blue)<0>: **HIDDEN**

Enter an option
[?/Make/Set/New/Rename/ON/OFF/Color/Ltype/LWeight/MATerial/Plot/Freeze/Thaw/LOck/Unlock/stAte/Description/rEconcile]: **C**
New color [Truecolor/COlorbook]: **GREEN**
Enter name list of layer(s) for color 3 (green)<0>: **DIMENSION**

Enter an option
[?/Make/Set/New/Rename/ON/OFF/Color/Ltype/LWeight/MATerial/Plot/Freeze/Thaw/LOck/Unlock/stAte/Description/rEconcile]: **C**

New color [Truecolor/COlorbook]: **MAGENTA**
Enter name list of layer(s) for color 6 (magenta)<0>: **BORDER**

Enter an option
[?/Make/Set/New/Rename/ON/OFF/Color/Ltype/LWeight/MATerial/Plot/Freeze/Thaw/
LOck/Unlock/stAte/Description/rEconcile]: **C**
New color [Truecolor/COlorbook]: **CYAN**
Enter name list of layer(s) for color 4 (cyan)<0>: **HATCH**

Enter an option
[?/Make/Set/New/Rename/ON/OFF/Color/Ltype/LWeight/MATerial/Plot/Freeze/Thaw/
LOck/Unlock/stAte/Description/rEconcile]: **LW**
Enter lineweight (0.0mm - 2.11mm): **0.05**
Enter name list of layers(s) for lineweight 0.05mm <0>: **HATCH**
[?/Make/Set/New/Rename/ON/OFF/Color/Ltype/LWeight/MATerial/Plot/Freeze/Thaw/
LOck/Unlock/stAte/Description/rEconcile]: Enter

Step 2: Writing the script file

The following file is a listing of the script file that creates different layers and assigns the given colors and linetypes to them:

```
;This script file will create new layers and
;assign different colors and linetypes to layers
LAYER
NEW
OBJECT,CENTER,HIDDEN,DIMENSION,BORDER,HATCH
L
CENTER
CENTER
L
HIDDEN
HIDDEN
C
RED
OBJECT
C
YELLOW
CENTER
C
BLUE
HIDDEN
C
GREEN
DIMENSION
C
MAGENTA
```

BORDER
C
CYAN
HATCH
(This is a blank line to terminate the **LAYER** command. End of script file.)

Save the script file as *script2.scr*.

Step 3: Running the script file

To run the script file of Example 2, choose **Manage > Applications > Run Script** from the **Ribbon** or enter **SCRIPT** at the Command prompt to invoke the **Select Script File** dialog box. Select **SCRIPT2.SCR** and then choose **Open**. You can also enter the **SCRIPT** command and the name of the script file at the Command prompt by setting **FILEDIA**=0.

Example 3

Write a script file that will rotate the line and the circle, as shown in Figure 2-2, around the lower endpoint of the line through 45-degree increments. The script file should be able to produce a continuous rotation of the given objects with a delay of two seconds after every 45-degree rotation (file name *script3.scr*). It is assumed that the line and circle are already drawn on the screen.

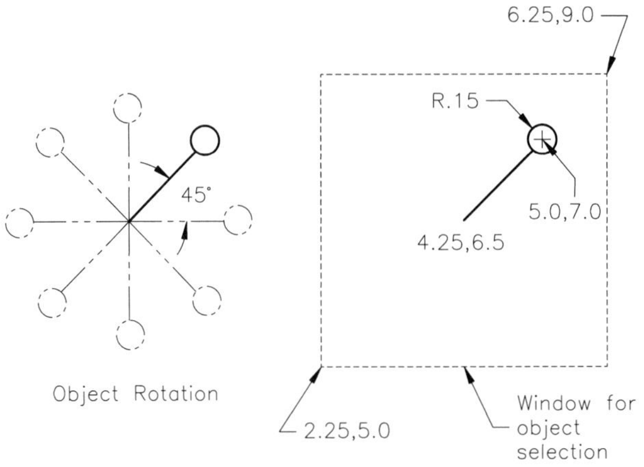

Figure 2-2 *Line and circle rotated through 45-degree increments*

Step 1: Understanding commands and prompt entries

Before writing the script file, enter the required commands and prompt entries. Write down the exact sequence of the entries in which they have been entered to perform the given operations. The following is the list of the AutoCAD command sequences needed to rotate the circle and the line around the lower endpoint of the line:

Command: **ROTATE**
Current positive angle in UCS: ANGDIR=counterclockwise ANGBASE=0
Select objects: **W** *(Window option to select object)*
Specify first corner: **2.25, 5.0**
Specify opposite corner: **6.25, 9.0**
Select objects: Enter
Specify base point: **4.25,6.5**
Specify rotation angle or [Reference]: **45**

Step 2: Writing the script file

Once the AutoCAD commands, command options, and their sequences are known, you can write a script file. You can use any text editor to write a script file. The following is a listing of the script file that will create the required rotation of the circle and line of Example 3. The line numbers and *(Blank line for Return)* are not a part of the file. They are shown here for reference only.

```
1 ROTATE
2 W
3 2.25,5.0
4 6.25,9.0
5              (Blank line for Return.)
6 4.25,6.5
7 45
```


Line 1
ROTATE
In this line, **ROTATE** is an AutoCAD command that rotates the objects.

Line 2
W
In this line, **W** is the **Window** option for selecting the objects that need to be edited.

Line 3
2.25,5.0
In this line, 2.25 defines the *X* coordinate and 5.0 defines the *Y* coordinate of the lower left corner of the object selection window.

Line 4
6.25,9.0
In this line, 6.25 defines the *X* coordinate and 9.0 defines the *Y* coordinate of the upper right corner of the object selection window.

Line 5
Line 5 is a blank line that terminates the object selection process.

Line 6
4.25,6.5
In this line, 4.25 defines the *X* coordinate and 6.5 defines the *Y* coordinate of the base point for rotation.

Line 7
45
In this line, 45 is the incremental angle of rotation.

Note

One of the limitations of the script file is that all the information has to be contained within the file. These files do not let you enter information. For instance, in Example 3, if you want to use the ***Window*** *option to select the objects, the* ***Window*** *option (W) and the two points that define this window must be contained within the script file. The same is true for the base point and all other information that goes in a script file. There is no way that a script file can prompt you to enter a particular piece of information and then resume the script file, unless you embed AutoLISP commands to prompt for user input.*

Step 3: Saving the script file
Save the script file with the name *script3.scr*.

Step 4: Running the script file
Choose **Tools > Run Script** from the menu bar, or choose **Manage > Applications > Run Script** from the **Ribbon**, or enter **SCRIPT** at the Command prompt to invoke the **Select Script File** dialog box. Select **SCRIPT3.SCR** and then choose **Open**. You will notice that the line and circle that were drawn on the screen are rotated once through an angle of 45-degree. However, there will be no continuous rotation of the sketched entities. The next section (Repeating Script Files) explains how to continue the steps mentioned in the script file. You will also learn how to add a time delay between the continuous cycles in later sections of this chapter.

REPEATING SCRIPT FILES

The **RSCRIPT** command allows the user to execute the script file indefinitely until canceled. It is a very desirable feature when the user wants to run the same file continuously. For example, in the case of a slide show for a product demonstration, the **RSCRIPT** command can be used to run the script file repeatedly until it is terminated by pressing the ESC key. Similarly, in Example 3, the rotation command needs to be repeated indefinitely to create a continuous rotation of the objects. This can be accomplished by adding **RSCRIPT** at the end of the file, as shown in the following listing of the script file:

ROTATE
W
2.25,5.0
6.25,9.0
(Blank line for Return.)
4.25,6.5

45
RSCRIPT

The **RSCRIPT** command in line 8 will repeat the commands from line 1 to line 7, and thus set the script file in an indefinite loop. If you run the *script3.scr* file now, you will notice that there is a continuous rotation of the line and circle around the specified base point. However, the speed at which the entities rotate makes it difficult to view the objects. As a result, you need to add time delay between every repetition. The script file can be stopped by pressing the ESC or the BACKSPACE key.

Note
You cannot provide conditional statements in a script file to terminate the file when a particular condition is satisfied unless you use the AutoLISP functions in the script file.

INTRODUCING TIME DELAY IN SCRIPT FILES

As mentioned earlier, some of the operations in the script files happen very quickly and make it difficult to see the operations taking place on the screen. It might be necessary to intentionally introduce a pause between certain operations in a script file. For example, in a slide show for a product demonstration, there must be a time delay between different slides so that the audience have enough time to see each slide. This is accomplished by using the **DELAY** command, which introduces a delay before the next command is executed. The general format of the **DELAY** command is given below.

Command: DELAY Time
Where **Command** ------ AutoCAD command prompt
DELAY ---------- **DELAY** command
Time ------------- Time in milliseconds

The **DELAY** command is to be followed by the delay time in milliseconds. For example, a delay of 2,000 milliseconds means that AutoCAD will pause for approximately two seconds before executing the next command. It is approximately two seconds because computer processing speeds vary. The maximum time delay you can enter is 32,767 milliseconds (about 33 seconds). In Example 3, a two-second delay can be introduced by inserting a **DELAY** command line between line 7 and line 8, as in the following file listing:

ROTATE
W
2.25,5.0
6.25,9.0
(Blank line for Return.)
4.25,6.5
45
DELAY 2000
RSCRIPT

The first seven lines of this file rotate the objects through a 45-degree angle. Before the **RSCRIPT** command on line 8 is executed, there is a delay of 2,000 milliseconds (about two seconds). The **RSCRIPT** command will repeat the script file that rotates the objects through another 45-degree angle. Thus, a slide show is created with a time delay of two seconds after every 45-degree increment.

RESUMING SCRIPT FILES

If you cancel a script file and then want to resume it, you can use the **RESUME** command.

Command: **RESUME**

The **RESUME** command can also be used if the script file has encountered an error that causes it to be suspended. The **RESUME** command will skip the command that caused the error and continue with the rest of the script file. If the error occurs when the command is in progress, use a leading apostrophe with the **RESUME** command (**'RESUME**) to invoke the **RESUME** command in the transparent mode.

Command: **'RESUME**

COMMAND LINE SWITCHES

The command line switches can be used as arguments to the *acad.exe* file that launches AutoCAD. You can also use the **Options** dialog box to set the environment or by adding a set of environment variables in the *autoexec.bat* file. The command line switches and environment variables override the values set in the **Options** dialog box for the current session only. These switches do not alter the system registry. The following is the list of the command line switches:

Switch	Function
/c	Controls where AutoCAD stores and searches for the hardware configuration file. The default file is *acad 2010.cfg*
/s	Specifies which directories to search for support files if they are not in the current directory
/b	Designates a script to run after AutoCAD starts
/t	Specifies a template to use when creating a new drawing
/nologo	Starts AutoCAD without first displaying the logo screen
/v	Designates a particular view of the drawing to be displayed on start-up of AutoCAD
/r	Reconfigures AutoCAD with the default device configuration settings
/p	Specifies the profile to use on start-up

RUNNING A SCRIPT FILE WHILE LOADING AutoCAD

The script files can also be run while loading AutoCAD, before it is actually started. The following is the format of the command for running a script file while loading AutoCAD.

"Drive**Program Files\AutoCAD 2010\acad.exe" [existing-drawing] [/t template] [/v view] /b Script-file**

In the following example, AutoCAD will open the existing drawing (MYdwg1) and then run the script file (Setup) through the **Run** dialog box, as shown in Figure 2-3.

Figure 2-3 *Invoking the script file while loading AutoCAD using the* ***Run*** *dialog box*

Example:

"C:\Program Files\AutoCAD 2010\acad.exe" MYdwg1 /b Setup

Where **AutoCAD 2010** AutoCAD 2010 subdirectory containing
-------------------- AutoCAD system files
acad.exe -------- ACAD command to start AutoCAD
MYDwg1 -------- Existing drawing file name
Setup ------------ Name of the script file

In the following example, AutoCAD will start a new drawing with the default name (Drawing), using the template file temp1, and then run the script file (Setup).

Example:

"C:\Program Files\AutoCAD 2010\acad.exe" /t temp1 /b Setup

Where **temp1** ----------- Existing template file name
Setup ------------ Name of the script file

or

"C:\ProgramFiles\AutoCAD 2010\acad.exe"/t temp1 "C:\MyFolder"/b Setup

Where C**Program Files\AutoCAD 2010\acad.exe** Path name for acad.exe
C:\MyFolder --- Path name for the Setup script file

In the following example, AutoCAD will start a new drawing with the default name (Drawing), and then run the script file (Setup).

Example:

"C:\Program Files\AutoCAD 2010\acad.exe" /b Setup

Where **Setup** ------------ Name of the script file

Here, it is assumed that the AutoCAD system files are loaded in the AutoCAD 2010 directory.

Note

*To invoke a script file while loading AutoCAD, the drawing file or the template file specified in the command must exist in the search path. You cannot start a new drawing with a given name. You can also use any template drawing file that is found in the template directory to run a script file through the **Run** dialog box.*

Tip

You should avoid abbreviations to prevent any confusion. For example, C can be used as a close option when you are drawing lines. It can also be used as a command alias for drawing a circle. If you use both of them in a script file, it might be confusing.

Example 4

Write a script file that can be invoked while loading AutoCAD and create a drawing with the following setup (filename *script4.scr*).

Grid	3.0
Snap	0.5
Limits	0,0
	36.0,24.0
Zoom	All
Text height	0.25
LTSCALE	3.0
DIMSCALE	3.0

Layers

Name	Color	Linetype
OBJ	Red	Continuous
CEN	Yellow	Center
HID	Blue	Hidden
DIM	Green	Continuous

Step 1: Writing the script file

Write a script file and save the file under the name *script4.scr*. The following is the listing of the script file that performs the initial setup for a drawing:

```
GRID 3.0
SNAP 0.5
LIMITS 0,0 36.0,24.0 ZOOM ALL
TEXTSIZE 0.25
LTSCALE 3
DIMSCALE 3.0
LAYER NEW
OBJ,CEN,HID,DIM
```

L CENTER CEN
L HIDDEN HID
C RED OBJ
C YELLOW CEN
C BLUE HID
C GREEN DIM
(Blank line for ENTER.)

Step 2: Loading the script file through the Run dialog box

After you have written and saved the file, quit the drawing editor. To run the script file, script4, select **Start > Run** and then enter the following command line.

"C:\Program Files\AutoCAD 2010\acad.exe" /t EX4 /b script4

Where **acad.exe** -------- ACAD to load AutoCAD
EX4 -------------- Template file name
SCRIPT4 ------- Name of the script file

Here, it is assumed that the template file (EX4) and the script file (script4) is on C drive. When you enter this line, AutoCAD is loaded and the file *ex4.dwt* is opened. The script file, script4, is then automatically loaded and the commands defined in the file are executed.

In the following example, AutoCAD will start a new drawing with the default name (Drawing), and then run the script file (script4) (Figure 2-4).

Example:

"C:\Program Files\AutoCAD 2010\acad.exe" /b script4

Where **script4** ----------- Name of the script file

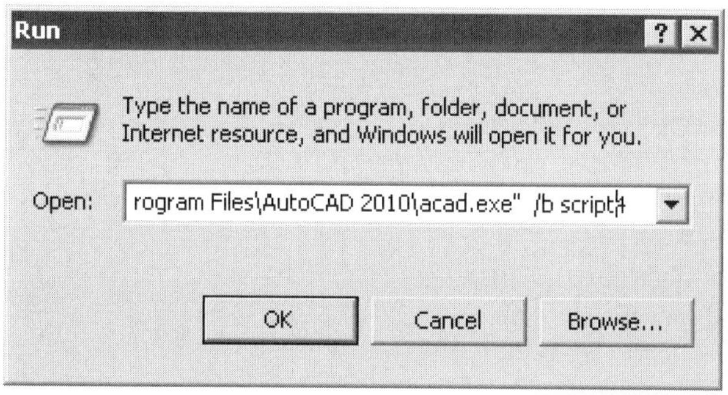

***Figure 2-4** Invoking the script file while loading AutoCAD using the **Run** dialog box*

Here, it is assumed that the AutoCAD system files are loaded in the AutoCAD 2010 directory.

Example 5

Write a script file that will plot a 36" by 24" drawing to the maximum plot size on a 8.5" by 11" paper, using your system printer/plotter. Use the **Window** option to select the drawing to be plotted.

Step 1: Understanding commands and prompt entries

Before writing a script file to plot a drawing, find out the plotter specifications that must be entered in the script file to obtain the desired output. To determine the prompt entries and their sequence to set up the plotter specifications, enter the **-PLOT** command. Note the entries you make and their sequence (the entries for your printer or plotter will probably be different). The following is a listing of the plotter specifications with the new entries:

Command: **-PLOT**
Detailed plot configuration? [Yes/No] <No>: **Y**
Enter a layout name or [?] <Model>: Enter
Enter an output device name or [?] <HP LaserJet 4000 Series PCL 6>: Enter
Enter paper size or [?] <Letter>: Enter
Enter paper units [Inches/Millimeters] <Inches>: **I**
Enter drawing orientation [Portrait/Landscape] <Landscape>: **L**
Plot upside down? [Yes/No] <No>: **N**
Enter plot area [Display/Extents/Limits/View/Window] <Display>: **W**
Enter lower left corner of window <0.000000,0.000000>: **0,0**
Enter upper right corner of window <0.000000,0.000000>: **36,24**
Enter plot scale (Plotted Inches=Drawing Units) or [Fit] <Fit>: **F**
Enter plot offset (x,y) or [Center] <0.00,0.00>: **0,0**
Plot with plot styles? [Yes/No] <Yes>: **Yes**
Enter plot style table name or [?] (enter . for none) <>: **.**
Plot with lineweights? [Yes/No] <Yes>: **Y**
Enter shade plot setting [As displayed/Wireframe/Hidden/Visual styles/Rendered] <As displayed>: Enter
Write the plot to a file [Yes/No] <N>: **N**
Save changes to page setup? Or set shade plot quality? [Yes/No/Quality] <N>: Enter
Proceed with plot [Yes/No] <Y>: **Y**

Step 2: Writing the script file

Now you can write the script file by entering the responses of these prompts in the file. The following file is a listing of the script file that will plot a 36" by 24" drawing on 8.5" by 11" paper after making the necessary changes in the plot specifications. The comments on the right are not a part of the file.

PLOT
Y
(Blank line for ENTER, selects default layout.)
(Blank line for ENTER, selects default printer.)
(Blank line for ENTER, selects the default paper size.)

```
I
L
N
w
0,0
36,24
F
0,0
Y
.              (Enter . for none)
Y
               (Blank line for ENTER, plots as displayed.)
N
N
Y
```


Saving and running the script file for this example is the same as that described in the previous examples. You can use a blank line to accept the default value for a prompt. A blank line in the script file will cause a Return. However, you must not accept the default plot specifications because the file may have been altered by another user or by another script file. Therefore, always enter the actual values in the file so that when you run a script file, it does not take the default values.

Example 6

Write a script file to animate a clock with continuous rotation of the second hand (longer needle) through 5-degree and the minutes hand (shorter needle) through 2-degree clockwise around the center of the clock, (Figure 2-5).

The specifications are given next.

Specifications for the rim made of donut.

Color of Donut	Blue
Inside diameter of Donut	8.0
Outside diameter of Donut	8.4
Center point of Donut	5,5

Specifications for the digit mark made of polyline.

Color of the digit mark	Green
Start point of Pline	5,8.5
Initial width of Pline	0.5
Final width of Pline	0.5
Height of Pline	0.5

Specifications for second hand (long needle) made of polyline.

Color of the second hand	Red

Figure 2-5 *Drawing for Example 6*

Start point of Pline	5,5
Initial width of Pline	0.5
Final width of Pline	0.0
Length of Pline	3.5
Rotation of the second hand	5 degree clockwise

Specifications for minute hand (shorter needle) made of polyline.

Color of the minute hand	Cyan
Start point of Pline	5,5
Initial width of Pline	0.35
Final width of Pline	0.0
Length of Pline	3.0
Rotation of the minute hand	2 degree clockwise

Step 1: Understanding the commands and prompt entries for creation of the clock

For this example, you can create two script files and then link them. The first script file will demonstrate the creation of the clock on the screen. The next script file will demonstrate the rotation of the needles of the clock.

First write a script file to create the clock as follows and save the file under the name *clock.scr*. The following is the listing of this file:

Command: **-COLOR**
Enter default object color [Truecolor/COlorbook] <BYLAYER>: **Blue**
Command: **DONUT**
Specify inside diameter of donut<0.5>: **8.0**
Specify outside diameter of donut<0.5>: **8.4**
Specify center of donut or <exit>: **5,5**

Specify center of donut or <exit>: Enter
Command: **-COLOR**
Enter default object color [Truecolor/COlorbook] <BYLAYER>: **Green**
Command: **PLINE**
Specify start point: **5,8.5**
Specify next point or [Arc/Halfwidth/Length/Undo/Width]: **Width**
Specify starting width<0.00>: **0.25**
Specify ending width<0.25>: **0.25**
Specify next point or [Arc/Halfwidth/Length/Undo/Width]: **@0.25<270**
Specify next point or [Arc/Halfwidth/Length/Undo/Width]: Enter
Command: **-ARRAY**
Select objects: **Last**
Select objects: Enter
Enter the type of array[Rectangular/Polar]<R>: **Polar**
Specify center point of array or [Base]: **5,5**
Enter the number of items in the array: **12**
Specify the angle to fill(+= ccw, -=cw)<360>: **360**
Rotate arrayed objects ? [Yes/No]<Y>: **Y**
Command: **-COLOR**
Enter default object color [Truecolor/COlorbook] <BYLAYER>: **RED**
Command: **PLINE**
Specify start point: **5,5**
Specify next point or [Arc/Close/Halfwidth/Length/Undo/Width]: **Width**
Specify starting width<.25>: **0.5**
Specify ending width<0.5>: **0**
Specify next point or [Arc/Halfwidth/Length/Undo/Width]: **@3.5<0**
Specify next point or [Arc/Close/Halfwidth/Length/Undo/Width]: Enter
Command: **-COLOR**
Enter default object color [Truecolor/COlorbook] <BYLAYER>: **Cyan**
Command: **PLINE**
Specify start point: **5,5**
Specify next point or [Arc/Halfwidth/Length/Undo/Width]: **Width**
Specify starting width<0.0000>: **0.35**
Specify ending width<0.35>: **0**
Specify next point or [Arc/Halfwidth/Length/Undo/Width]: **@3<90**
Specify next point or [Arc/Close/Halfwidth/Length/Undo/Width]: Enter
Command: **SCRIPT**
ROTATE.SCR

Now you can write the script file by entering the responses to these prompts in the file *color.scr*. Next, you will write the script to rotate the clock hands and save the file with the name *rotate.scr*. Remember that while entering the commands in the script files, you do not need to add a hyphen (-) as a prefix to the command name to execute them from the command line. When a command is entered using the script file, the dialog box is not displayed and it is executed using the command line. For example, in this script file, the **COLOR** and the **ARRAY** command will be executed using the command line. Listing of the script file is given next.

```
Color
Blue
Donut
8.0
8.4
5,5
                    Blank line for ENTER
Color
Green
Pline
5,8.5
W
0.25
0.25
@0.25<270
                    Blank line for ENTER
Array
L
                    Blank line for ENTER
P
5,5
12
360
Y
Color
Red
Pline
5,5
W
0.5
0
@3.5<0
                    Blank line for ENTER
Color
Cyan
Pline
5,5
w
0.35
0
@3<90
                    Blank line for ENTER
Script
ROTATE.SCR          (Name of the script file that will cause rotation)
```

Save this file as *clock.scr* in a directory that is specified in the AutoCAD support file search path. It is recommended that the *rotate.scr* file should also be saved in the same directory. Remember that if the files are not saved in the directory that is specified in the AutoCAD support file search path using the **Options** dialog box, the linked script file (*rotate.scr*) may not run.

Step 2: Understanding the commands and sequences for rotation of the needles

The last line in the above script file is *rotate.scr*. This is the name of the script file that will rotate the clock hands. Before writing the script file, enter the ROTATE command and respond to the command prompts that will cause the desired rotation. The following is the listing of the AutoCAD command sequences needed to rotate the objects:

Command: **ROTATE**
Select objects: **L**
Select objects: Enter
Specify base point: 5,5
Specify rotation angle or [Copy/Reference] <0>: -2
Command: **ROTATE**
Select objects: **C**
Specify first corner: 3,3
Specify other corner: 7,7
Select objects: **Remove**
Remove objects: **L**
Remove objects: Enter
Specify base point: 5,5
Specify rotation angle or [Copy/Reference]: -5

Now you can write the script file by entering the responses to these prompts in the file *rotate.scr*. The following is the listing of the script file that will rotate the clock hands:

```
Rotate
L
                                    Blank line for ENTER
5,5
-2
Rotate
c
3,3
7,7
R
L
                                    Blank line for ENTER
5,5
-5
Rscript
```

Save the above script file as *rotate.scr*. Now, run the script file *clock.scr*. Since this file is linked with *rotate.scr*, it will automatically run *rotate.scr* after running *clock.scr*. Note that if the linked file is not saved in a directory specified in the AutoCAD support file search path, the last line of the *clock.scr* must include a fully-resolved path to *rotate.scr*, or AutoCAD would not be able to locate the file.

Exercise 1 *General*

Write a script file that will plot a 288' by 192' drawing on a 36" x 24" sheet of paper. The drawing scale is 1/8" = 1'. (The filename is *script9.scr*. In this exercise, assume that AutoCAD is configured for the HPGL plotter and the plotter description is HPGL-Plotter.)

WHAT IS A SLIDE SHOW?

AutoCAD provides a facility of using script files to combine the slides in a text file and display them in a predetermined sequence. In this way, you can generate a slide show for a slide presentation. You can also introduce a time delay in the display so that the viewer has enough time to view each slide.

A drawing or parts of a drawing can also be displayed using the AutoCAD display commands. For example, you can use **ZOOM**, **PAN**, or other commands to display the details you want to show. If the drawing is very complicated, it takes quite some time to display the desired information and it may not be possible to get the desired views in the right sequence. However, with slide shows you can arrange the slides in any order and present them in a definite sequence. In addition to saving time, this also helps to minimize the distraction that might be caused by constantly changing the drawing display. Also, some drawings are confidential in nature and you may not want to display some portions or views of them. You can send a slide show to a client without losing control of the drawings and the information that is contained in them.

WHAT ARE SLIDES?

A slide is the snapshot of a screen display; it is like taking a picture of a display with a camera. The slides do not contain any vector information like AutoCAD drawings, which means that the entities do not have any information associated with them. For example, the slides do not retain any information about the layers, colors, linetypes, start point, or endpoint of a line or viewpoint. Therefore, slides cannot be edited like drawings. If you want to make any changes in the slide, you need to edit the drawing and then make a new slide from the edited drawing.

CREATING SLIDES

In AutoCAD, slides are created using the **MSLIDE** command. If **FILEDIA** is set to 1, the **MSLIDE** command displays the **Create Slide File** dialog box (Figure 2-6) on the screen. You can enter the slide file name in this dialog box. If **FILEDIA** is set to 0, the command will prompt you to enter the slide file name.

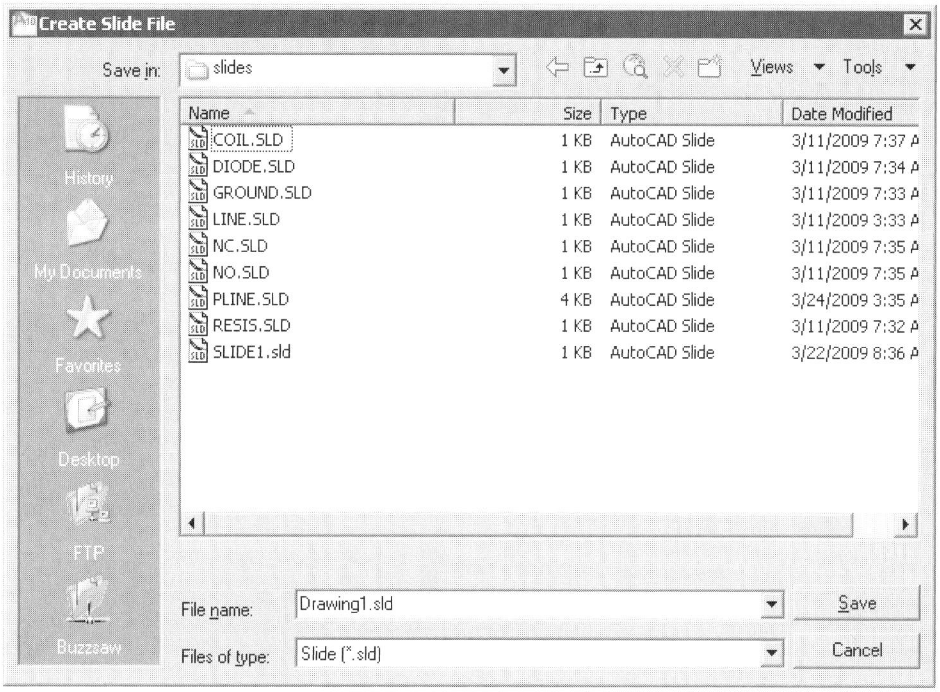

***Figure 2-6** The **Create Slide File** dialog box*

Command: **MSLIDE**
Enter name of slide file to create <Default>: *Slide file name.*

Example:
Command: **MSLIDE**
Slide File: <Drawing1> **SLIDE1**
Where **Drawing1** ------- Default slide file name
SLIDE1 --------- Slide file name

In this example, AutoCAD will save the slide file as *slide1.sld*.

Note

*In model space, you can use the **MSLIDE** command to make a slide of the existing display in the current viewport. If you are in the paper space viewport, you can make a slide of the display in the paper space that includes any floating viewports.*

*When the viewports are not active, the **MSLIDE** command will make a slide of the current screen display.*

VIEWING SLIDES

To view a slide, use the **VSLIDE** command at the Command prompt; the **Select Slide File** dialog box will be displayed, as shown in the Figure 2-7. Choose the file that you want to view and then choose **OK**. The corresponding slide will be displayed on the screen. If **FILEDIA** is 0, the slide that you want to view can be directly entered at the Command prompt.

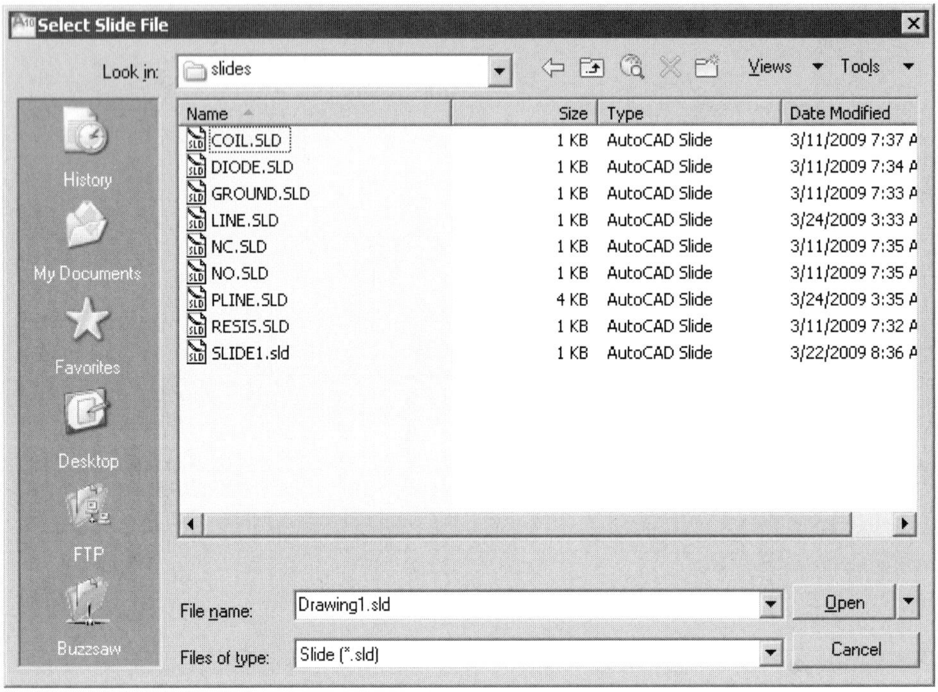

***Figure 2-7** The **Select Slide File** dialog box*

Command: **VSLIDE**
Enter name of slide file to view<Default>: *Name*.

Example:
Command: **VSLIDE**
Slide file <Drawing1>: SLIDE1
Where **Drawing1** ------- Default slide file name
SLIDE1 --------- Name of slide file

Note

After viewing a slide, you can use the ***REDRAW*** *command, roll the wheel, or pan with a mouse to remove the slide display and return to the existing drawing on the screen.*

Any command that is automatically followed by a ***REDRAW*** *command will also display the existing drawing. For example, AutoCAD* ***GRID***, ***ZOOM ALL***, *and* ***REGEN*** *commands will automatically return to the existing drawing on the screen.*

You can view the slides on high-resolution or low-resolution monitors. Depending on the resolution of the monitor, AutoCAD automatically adjusts the image. However, if you are using a high-resolution monitor, it is better to make the slides using the same monitor to take full advantage of that monitor.

Example 7

Write a script file that will create a slide show of the following slide files, with a time delay of 15 seconds after every slide (Figure 2-8).

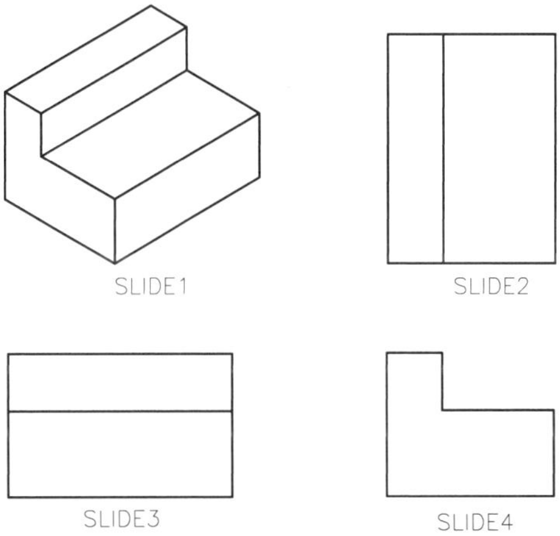

Figure 2-8 *Slides for slide show*

Step 1: Creating the slides

The first step in a slide show is to create the slides using the **MSLIDE** command. The **MSLIDE** command will invoke the **Create Slide File** dialog box. Enter the name of the slide as **SLIDE1** and choose the **Save** button to exit the dialog box. Similarly, other slides can be created and saved. Figure 2-8 shows the drawings that have been saved as slide files **SLIDE1**, **SLIDE2**, **SLIDE3**, and **SLIDE4**. The slides must be saved to a directory in AutoCAD's search path otherwise, the script would not find the slides.

Step 2: Writing the script file

The second step is to find out the sequence, in which you want these slides to be displayed, with the necessary time delay, if any, between slides. Then you can use any text editor or the AutoCAD **EDIT** command (provided the *acad.pgp* file is present and **EDIT** is defined in the file) to write the script file with the extension *.scr*.

The following is the listing of the script file that will create a slide show of the slides in Figure 2-8. The name of the script file is **SLDSHOW1**.

```
VSLIDE SLIDE1
DELAY 15000
VSLIDE SLIDE2
DELAY 15000
VSLIDE SLIDE3
DELAY 15000
VSLIDE SLIDE4
DELAY 15000
```

Step 3: Running the script file

To run this script file, choose **Manage > Applications > Run Script** from the **Ribbon** or enter **SCRIPT** at the Command prompt to invoke the **Select Script File** dialog box. Choose **SLDSHOW1** and then choose the **Open** button from the **Select Script File** dialog box. You can see the changes taking place on the screen.

PRELOADING SLIDES

In the script file of Example 7, VSLIDE SLIDE1 in line 1 loads the slide file, **SLIDE1**, and displays it on the screen. After a pause of 15,000 milliseconds, it starts loading the second slide file, **SLIDE2**. Depending on the computer and the disk access time, you will notice that it takes some time to load the second slide file. The same is true for the other slides. To avoid the delay in loading the slide files, AutoCAD has provided a facility to preload a slide while viewing the previous slide. This is accomplished by placing an asterisk (*) in front of the slide file name.

```
VSLIDE SLIDE1           (View slide, SLIDE1.)
VSLIDE *SLIDE2          (Preload slide, SLIDE2.)
DELAY 15000             (Delay of 15 seconds.)
VSLIDE                  (Display slide, SLIDE2.)
VSLIDE *SLIDE3          (Preload slide, SLIDE3.)
DELAY 15000             (Delay of 15 seconds.)
VSLIDE                  (Display slide, SLIDE3.)
VSLIDE *SLIDE4
DELAY 15000
VSLIDE
DELAY 15000
RSCRIPT                 (Restart the script file.)
```

Example 8

Write a script file to generate a continuous slide show of the following slide files, with a time delay of two seconds between slides SLD1, SLD2, and SLD3.

The slide files are located in different subdirectories, as shown in Figure 2-9.

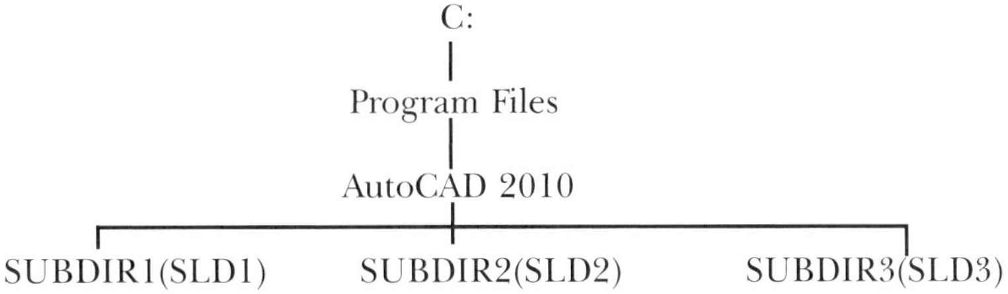

Figure 2-9 *Subdirectories of the C drive*

Where
C: -------------------- *Root directory.*
Program Files ------------- *Root directory.*
AutoCAD 2010 ---------- *Subdirectory where the AutoCAD files are loaded.*
SUBDIR1----------------- *Drawing subdirectory.*
SUBDIR2 ----------------- *Drawing subdirectory.*
SUBDIR3 ----------------- *Drawing subdirectory.*
SLD1 -------------------- *Slide file in SUBDIR1 subdirectory.*
SLD2 -------------------- *Slide file in SUBDIR2 subdirectory.*
SLD3 -------------------- *Slide file in SUBDIR3 subdirectory.*

The following is the listing of the script files that will generate a slide show for the slides in Example 8:

```
1 VSLIDE "C:/Program Files/AutoCAD 2010/SUBDIR1/SLD1.SLD"
2 DELAY  2000
3 VSLIDE "C:/Program Files/AutoCAD 2010/SUBDIR2/SLD2.SLD"
4 DELAY 2000
5 VSLIDE "C:/Program Files/AutoCAD 2010/SUBDIR3/SLD3.SLD"
6 DELAY 2000
7 RSCRIPT
```

Line 1
VSLIDE "C:/Program Files/AutoCAD 2010/SUBDIR1/SLD1.SLD"
In this line, the AutoCAD command **VSLIDE** loads the slide file **SLD1**. The path name is mentioned along with the command **VSLIDE**. If the path name directory contains spaces, then the path name must be enclosed in quotes.

Line 2
DELAY 2000
This line uses the AutoCAD **DELAY** command to create a pause of approximately two seconds before the next slide is loaded.

Line 3
VSLIDE "C:/Program Files/AutoCAD 2010/SUBDIR2/SLD2.SLD"
In this line, the AutoCAD command **VSLIDE** loads the slide file **SLD2**, located in the subdirectory **SUBDIR2**. If the slide file is located in a different subdirectory, you need to define the path with the slide file.

Line 5
VSLIDE "C:/Program Files/AutoCAD 2010/SUBDIR3/SLD3.SLD"
In this line, the **VSLIDE** command loads the slide file SLD3, located in the subdirectory **SUBDIR3**.

Line 7
RSCRIPT
In this line, the **RSCRIPT** command executes the script file again and displays the slides on the screen. This process continues indefinitely until the script file is canceled by pressing the ESC key or the BACKSPACE key.

SLIDE LIBRARIES

AutoCAD provides a utility, SLIDELIB, which constructs a library of the slide files. The format of the **SLIDELIB** utility command is as follows:

SLIDELIB (Library filename) <(Slide list filename)

Example:
SLIDELIB SLDLIB <SLDLIST
Where **SLIDELIB**------ AutoCAD's SLIDELIB utility
SLDLIB --------- Slide library filename
SLDLIST ------- List of slide filenames

The **SLIDELIB** utility is supplied with the AutoCAD software package. You can find this utility (SLIDELIB.EXE) in the *C:\Program Files\AutoCAD 2010* subdirectory. The slide file list is a list of the slide file names that you want in a slide show. It is a text file that can be written by using any text editor like Notepad. The slide files in the slide file list should not contain any file extension (*.sld*). However, if you want to add a file extension, it should be *.sld*.

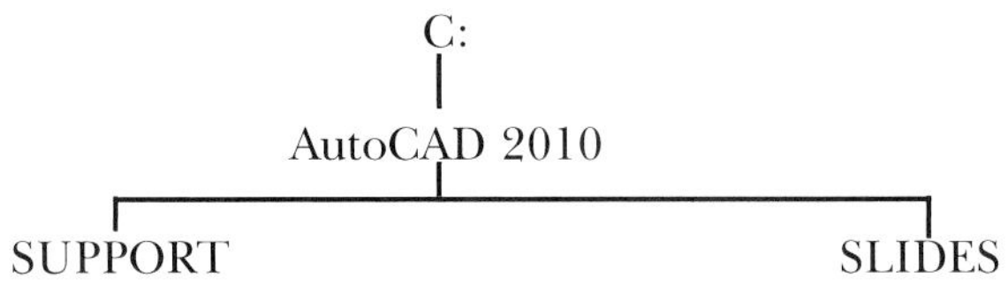

The slide file list can also be created by using the following command, if you have a DOS version 5.0 or above. You can use the make directory (md) or change directory (cd) commands in the DOS mode while making or changing directories.

C:\AutoCAD 2010\SLIDES>**DIR *.SLD/B>SLDLIST**

In this example, assume that the name of the slide file list is **SLDLIST** and all slide files are in the SLIDES subdirectory. To use this command to create a slide file list, all slide files must be in the same directory.

When you use the SLIDELIB utility, it reads the slide file names from the file that is specified in the slide list and the file is then written to the file specified by the library. In Example 9, the SLIDELIB utility reads the slide filenames from the file SLDLIST and writes them to the library file SLDSHOW1:

C:\>**SLIDELIB SLDSHOW1 <SLDLIST**

Note

You cannot edit a slide library file. If you want to change anything, you have to create a new list of the slide files and then use the SLIDELIB utility to create a new slide library.

*If you edit a slide while it is being displayed on the screen, the slide will not be edited. Instead, the current drawing that is behind the slide gets edited. Therefore, do not use any editing commands while you are viewing a slide. Use the **VSLIDE** and **DELAY** commands only while viewing a slide.*

The path name is not saved in the slide library. This is the reason if you have more than one slide with the same name, even though they are in different subdirectories, only one slide will be saved in the slide library.

Example 9

Use AutoCAD's SLIDELIB utility to generate a continuous slide show of the following slide files with a time delay of 2.5 seconds between the slides. (The filenames are: SLDLIST for slide list file, SLDSHOW1 for slide library, SHOW1 for script file.)

front, top, rside, 3dview, isoview

The slide files are located in different subdirectories, as shown in Figure 2-10.

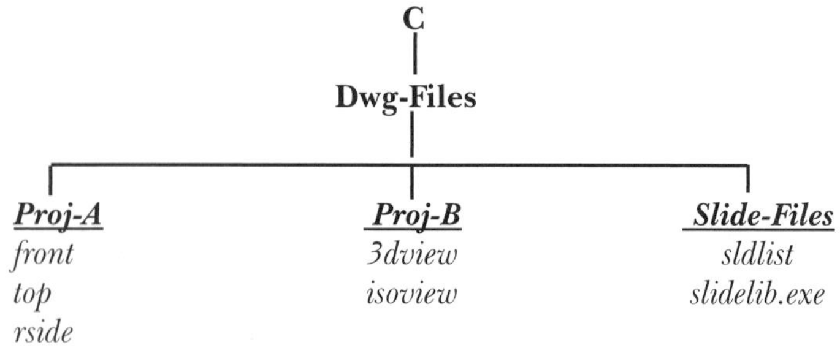

Where **C** ----------------- *(C drive.)*
Dwg-Files ------- *(Subdirectory where drawing files are located)*
Proj-A ------------ *(Drawing subdirectory.)*
Proj-B ------------ *(Drawing subdirectory.)*
Slide-Files ------ *(Directory where Slidelib.exe and sldlist are copied)*

Figure 2-10 *Subdirectories to be created in C drive*

Step 1

The first step is to create a list of the slide file names with the drive and the directory information. Assume that you are in the **Slide-Files** subdirectory. You can use a text editor like Notepad to create a list of the slide files that you want to include in the slide show. These files do not need a file extension. However, if you choose to give them a file extension, it should be *.SLD*. The following file is a listing of the file SLDLIST for Example 9:

```
c:\Dwg-Files\Proj-A\front
c:\Dwg-Files\Proj-A\top
c:\Dwg-Files\Proj-A\rside
c:\Dwg-Files\Proj-B\3dview
c:\Dwg-Files\Proj-B\isoview
```

Step 2

The second step is to use AutoCAD's SLIDELIB utility program to create the slide library. The name of the slide library is assumed to be **sldshow1** for this example. Before creating the slide library, copy the slide list file (SLDLIST) and the SLIDELIB utility from the support directory to the Slide-Files directory. This ensures that all the required files are in one directory. Enter **SHELL** command at AutoCAD Command prompt and then press the ENTER key at the OS Command prompt. The **AutoCAD Shell Active** Command window will be displayed on the screen, see Figure 2-11. You can also use Windows DOS box by choosing **All Programs > Accessories > Command Prompt**.

Command: **SHELL**
OS Command: Enter

Now, enter the following to run the SLIDELIB utility to create the slide library. Here it is assumed that **Slide-Files** directory is the current directory.

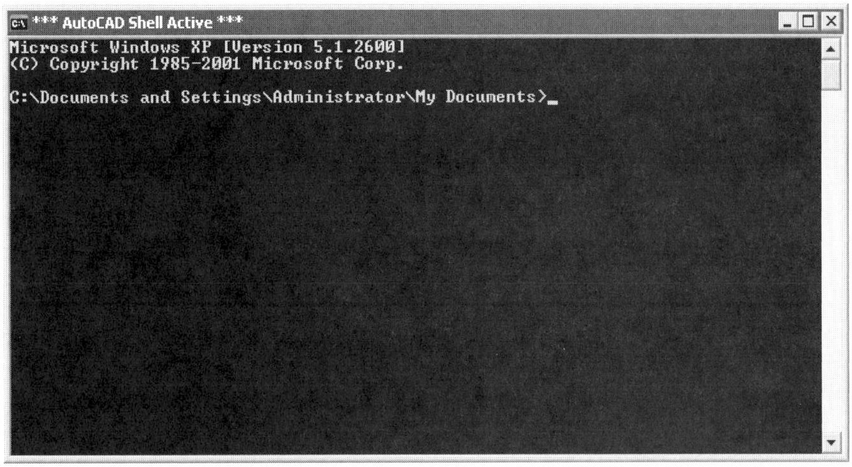

Figure 2-11 *The* ***AutoCAD Shell Active*** *Command window*

C:\Dwg-Files\Slide-Files>SLIDELIB sldshow1 <sldlist

Where **SLIDELIB** ------ AutoCAD's SLIDELIB utility
sldshow1 -------- Slide library
sldlist ------------ Slide file list

Step 3

Now, you can write a script file for the slide show that will use the slides in the slide library. The name of the script file for this example is assumed to be SHOW1.

```
VSLIDE sldshow1(front)
DELAY 2500
VSLIDE sldshow1(top)
DELAY 2500
VSLIDE sldshow1(rside)
DELAY 2500
VSLIDE sldshow1(3dview)
DELAY 2500
VSLIDE sldshow1(isoview)
DELAY 2500
RSCRIPT
```

Step 4

Invoke the **Select Script File** dialog box, as shown in Figure 2-12 by choosing **Manage > Applications > Run Script** from the **Ribbon** or enter the **SCRIPT** command at the Command prompt. You can also enter the **SCRIPT** command at the Command prompt after setting the system variable FILEDIA to 0.

Command: **SCRIPT**
Enter script file name<default>: **SHOW1**

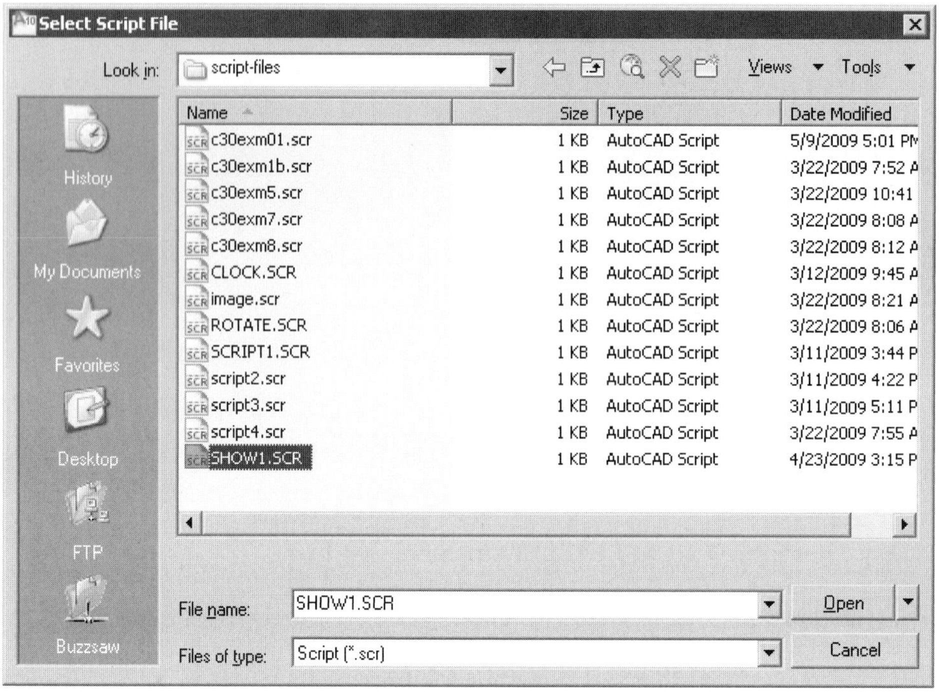

***Figure 2-12** Selecting the script file from the **Select Script File** dialog box*

Self-Evaluation Test

Answer the following questions and then compare them to those given at the end of this chapter:

SCRIPT FILES

1. AutoCAD has provided a facility of __________ that allows you to combine different AutoCAD commands and execute them in a predetermined sequence.

2. Before writing a script file, you need to know the AutoCAD __________ and the __________ required in response to the command prompts.

3. The AutoCAD __________ command is used to run a script file.

4. In a script file, the __________ is used to terminate a command or a prompt entry.

5. The **DELAY** command is to be followed by __________ in milliseconds.

SLIDE SHOW

6. Slides do not contain any _________ information, which means that the entities do not have any information associated with them.

7. Slides _________ edited like a drawing.

8. Slides can be created using the AutoCAD _________ command.

9. To view a slide, use the AutoCAD _________ command.

10. AutoCAD provides a utility that constructs a library of the slide files. This is done with AutoCAD's utility program called _________.

Review Questions

Answer the following questions:

SCRIPT FILES

1. The _________ files can be used to generate a slide show, do the initial drawing setup, or plot a drawing to a predefined specification.

2. In a script file, you can _________ several statements in one line.

3. When you run a script file, the default script file name is the same as the _________ name.

4. Type the name of the script file without the _________ file, when you run a script file.

5. One of the limitations of script files is that all the information has to be contained _________ the file.

6. The AutoCAD _________ command allows you to re-execute a script file indefinitely until the command is canceled.

7. You cannot provide a _________ statement in a script file to terminate the file when a particular condition is satisfied.

8. The AutoCAD _________ command schedules a delay before the next command is executed.

9. If the script file is canceled and you want to resume the script file, you need to use the ___________________ command.

SLIDE SHOW

10. AutoCAD provides a facility through _________ files to combine the slides in a text file and display them in a predetermined sequence.

11. A _________ can also be introduced in the script file so that the viewer has enough time to view a slide.

12. Slides are the _________ of a screen display.

13. In model space, you can use the **MSLIDE** command to make a slide of the _________ display in the _________ viewport.

14. If you are in paper space, you can make a slide of the display in paper space that _________ any floating viewports.

15. If you want to make any changes in the slide, you need to _________ the drawing, then make a new slide from the edited drawing.

16. If the slide is in the slide library and you want to view it, the slide library name has to be _________ with the slide filename.

17. You cannot _________ a slide library file. If you want to change anything, you have to create a new list of the slide files and then use the _________ utility to create a new slide library.

18. The path name _________ be saved in the slide library. Therefore, if you have more than one slide with the same name, although with different subdirectories, only one slide will be saved in the slide library.

Exercises

SCRIPT FILES

Exercise 2 *General*

Write a script file that will perform the following initial setup for a drawing.

Grid	2.0
Snap	0.5
Limits	0,0
	18.0,12.0
Zoom	All
Text height	0.25
LTSCALE	2.0
Overall dimension scale factor is 2	

Aligned dimension text with the dimension line
Dimension text above the dimension line
Size of the center mark is 0.75

Exercise 3 *General*

Write a script file that will set up the following layers with the given colors and linetypes.

Contour	Red	Continuous
SPipes	Yellow	Center
WPipes	Blue	Hidden
Power	Green	Continuous
Manholes	Magenta	Continuous
Trees	Cyan	Continuous

Exercise 4 *General*

Write a script file that will perform the following initial setup for a new drawing.

Limits	0,0 24,18
Grid	1.0
Snap	0.25
Ortho	On
Snap	On
Zoom	All
Pline width	0.02
PLine	0,0 24,0 24,18 0,18 0,0
Ltscale	1.5
Units	Decimal units
	Precision 0.00
	Decimal degrees
	Precision 0
	Base angle East (0.00)
	Angle measured counterclockwise

Layers

Name	**Color**	**Linetype**
Obj	Red	Continuous
Cen	Yellow	Center
Hid	Blue	Hidden
Dim	Green	Continuous

Exercise 5 *General*

Write a script file that will plot a given drawing according to the following specifications. (Use the plotter for which your system is configured and adjust the values accordingly.)

Plot, using the Window option
Window size (0,0 24,18)
Do not write the plot to file
Size in inch units
Plot origin (0.0,0.0)
Maximum plot size (8.5,11 or the smallest size available on your printer/plotter)
90-degree plot rotation
No removal of hidden lines
Plotting scale (Fit)

Exercise 6 *General*

Write a script file that will continuously rotate a line in 10-degree increments around its midpoint (Figure 2-13). The time delay between increments is one second.

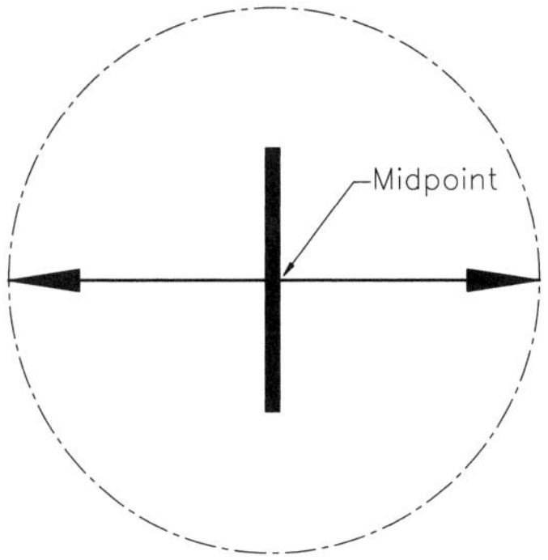

Figure 2-13 *Drawing for Exercise 6*

Exercise 7 *General*

Write a script file that will continuously rotate the following arrangements (Figure 2-14). One set of two circles and one line should rotate clockwise, while the other set of two circles and one line should rotate counterclockwise. Assume the rotation to be 5-degree around the intersection of the lines for both sets of arrangements.

Specifications are given below:

Start point of the horizontal line	2,4
End point of the horizontal line	8,4
Center point of circle at the start point of horizontal line	2,4
Diameter of the circle	1.0
Center point of circle at the end point of horizontal line	8,4
Diameter of circle	1.0
Start point of the vertical line	5,1
End point of the vertical line	5,7
Center point of circle at the start point of the vertical line	5,1
Diameter of the circle	1.0
Center point of circle at the end point of the vertical line	5,7
Diameter of the circle	1.0

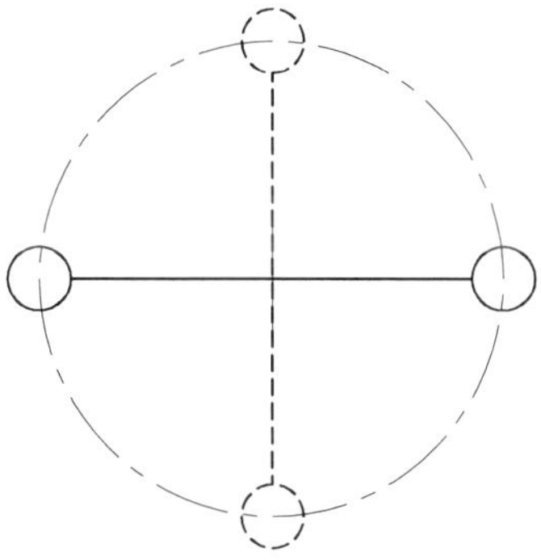

Figure 2-14 *Drawing for Exercise 7*

Hint: *Select one set of two circles and one line and then create one group. Similarly, select another set of two circles and one line and then create another group. Rotate one group clockwise and another group counterclockwise.*

SLIDE SHOW

Exercise 8 *General*

Make the slides shown in Figure 2-15 and write a script file for a continuous slide show. Provide a time delay of 5 seconds after every slide. (You are not restricted to use only the slides shown in Figure 2-15. You can use any slides of your choice.)

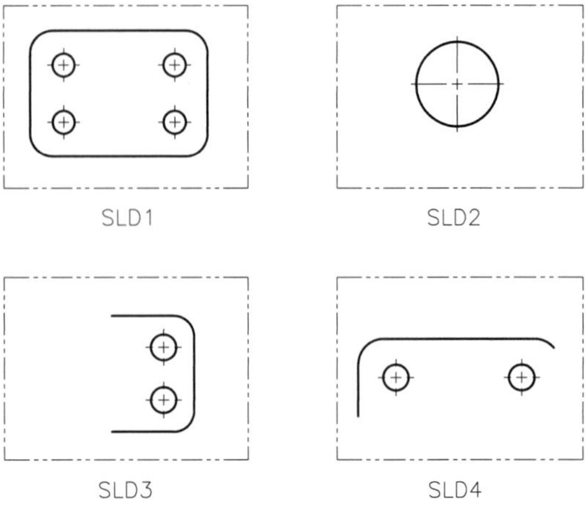

Figure 2-15 *Slides for slide show*

Exercise 9 *General*

List the slides in Exercise 8 in a file SLDLIST2 and create a slide library file SLDLIB2. Then write a script file SHOW2 using the slide library with a time delay of 5 seconds after every slide.

Answers to Self-Evaluation Test

1. SCRIPT files, **2.** Commands, options, **3. SCRIPT**, **4.** blank space, **5.** time, **6.** vector, **7.** cannot be, **8. MSLIDE**, **9. VSLIDE**, **10. SLIDELIB**

Chapter 3

Creating Linetypes and Hatch Patterns

Learning Objectives

After completing this chapter, you will be able to:

Create Linetypes:
- *Write linetype definitions.*
- *Create different linetypes.*
- *Create linetype files.*
- *Determine **LTSCALE** for plotting the drawing to given specifications.*
- *Define alternate linetypes and modify existing linetypes.*
- *Create string and shape complex linetypes.*

Create Hatch Patterns:
- *Understand hatch pattern definition.*
- *Create new hatch patterns.*
- *Determine the effect of angle and scale factor on hatch.*
- *Create hatch patterns with multiple descriptors.*
- *Save hatch patterns in a separate file.*
- *Define custom hatch pattern file.*

STANDARD LINETYPES

The AutoCAD software package comes with a library of standard linetypes that has 38 different standard linetypes and seven complex linetypes, including ISO linetypes. These linetypes are saved in the *acad.lin* and *acadiso.lin* file. You can modify the existing linetypes or create new ones.

LINETYPE DEFINITIONS

All linetype definitions consist of two parts: **header line** and **pattern line**.

Header Line

The **header line** consists of an asterisk (*) followed by the name of the linetype and the linetype description. The name and the linetype description should be separated by a comma. If there is no description, the comma that separates the linetype name and the description is not required.

The format of the header line is:

*** Linetype Name, Description**

Example:
***HIDDENS,__ __ __ __ __ __**

Where * ------------------- Asterisk sign
HIDDENS ------ Linetype name
, -------------------- Comma
__ __ __ __ ------ Linetype description

All linetype definitions require a linetype name. When you want to load a linetype or assign a linetype to an object, AutoCAD recognizes the linetype by the name you have assigned to the linetype definition. The names of the linetype definition should be selected to help the user recognize the linetype by its name. For example, the linetype name LINEFCX does not give the user any idea about the type of line. However, a linetype name like DASHDOT gives a better idea about the type of line that a user can expect.

The linetype description is a textual representation of the line. This representation can be generated by using dashes, dots, and spaces at the keyboard. The graphic is used to display the linetypes on the screen using the **LINETYPE** command with the ? option or using the dialog box. The linetype description cannot exceed 47 characters.

Pattern Line

The **pattern line** contains the definition of the line pattern consisting of the alignment field specification and the linetype specification, separated by a comma.

The format of the pattern line is:

Alignment Field Specification, Linetype Specification

Example:

A,.75,-.25,.75

Where **A** ----------------- Alignment field specification
, -------------------- Comma
.75,-.25,.75 ----- Linetype specification

The letter used for alignment field specification is A. This is the only alignment field supported by AutoCAD; therefore, the pattern line will always start with the letter A. The linetype specification defines the configuration of the dash-dot pattern to generate a line. The maximum number for dash length specification in the linetype is 12, provided the linetype pattern definition fits on one 80-character line.

ELEMENTS OF LINETYPE SPECIFICATION

All linetypes are created by combining the basic elements in a desired configuration. There are three basic elements that can be used to define a linetype specification.

Dash	(Pen down)
Dot	(Pen down, 0 length)
Space	(Pen up)

Example:

_______ . _______ . _______ . _______

Where **.** -------------------- Dot (pen down with 0 length)
Blank space -------------- Space (pen up)
_______ -------------------- Dash (pen down with specified length)

The dashes are generated by defining a positive number. For example, .5 will generate a dash 0.5 units long. Similarly, spaces are generated by defining a negative number. For example, -.2 will generate a space 0.2 units long. The dot is generated by defining 0 length.

Example:

A,.5,-.2,0,-.2,.5

Where **0** ------------------- Dot (zero length)
-.2 ----------------- Length of space (pen up)
.5 ------------------ Length of dash (pen down)

CREATING LINETYPES

Before creating a linetype, you need to decide the type of line that you want to generate. Draw the line on a piece of paper and measure the length of each element that constitutes the line. You need to define only one segment of the line because the pattern is repeated when you draw a line. Linetypes can be created or modified by any one of the following methods:

Using a text editor like Notepad
Adding a new linetype in the *acad.lin* or *acadiso.lin* file
Using the **-LINETYPE** command

The following example, Example 1, explains how to create a new linetype using the three methods mentioned above.

Example 1

Create linetype DASH3DOT (Figure 3-1) with the following specifications:

Length of the first dash 0.5
Blank space 0.125
Dot
Blank space 0.125
Dot
Blank space 0.125
Dot
Blank space 0.125

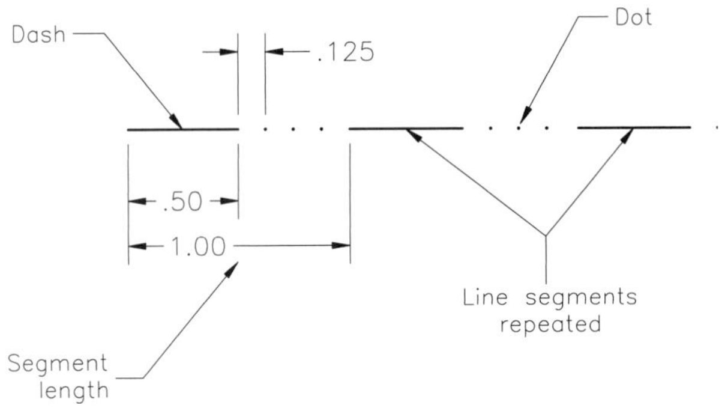

Figure 3-1 *Linetype specifications of DASH3DOT*

Using a Text Editor

Step 1: Writing definition of linetype

You can start a new linetype file and then add the line definitions to this file. To do this, use any text editor like Notepad to start a new file (*newlt.lin*) and then add the linetype definition of the DASH3DOT linetype. The name and the description must be separated by a comma (,). The description is optional. If you decide not to give one, omit the comma after the linetype name DASH3DOT.

***DASH3DOT,____ . . . ____ . . . ____**
A,.5,-.125,0,-.125,0,-.125,0,-.125

Save it as *newlt.lin* in AutoCAD's Support directory.

Step 2: Loading the linetype

To load this linetype, choose **Home > Properties > Linetype > Other** from the **Ribbon** to display the **Linetype Manager** dialog box. Next, choose the **Load** button in the **Linetype Manager** dialog box to display the **Load or Reload Linetypes** dialog box. Choose the **File** button in the **Load or Reload Linetypes** dialog box to display the **Select Linetype File** dialog box, as shown in Figure 3-2. Choose the *newlt.lin* file in the **Select Linetype File** dialog box and then choose **Open**. Again the **Load or Reload Linetypes** dialog box is displayed. Choose the **DASH3DOT** linetype in the **Available Linetypes** area and then choose **OK**. The **Linetype Manager** dialog box is displayed. Select the **DASH3DOT** linetype and then choose the **Current** button to make the selected linetype current. Then, choose **OK**.

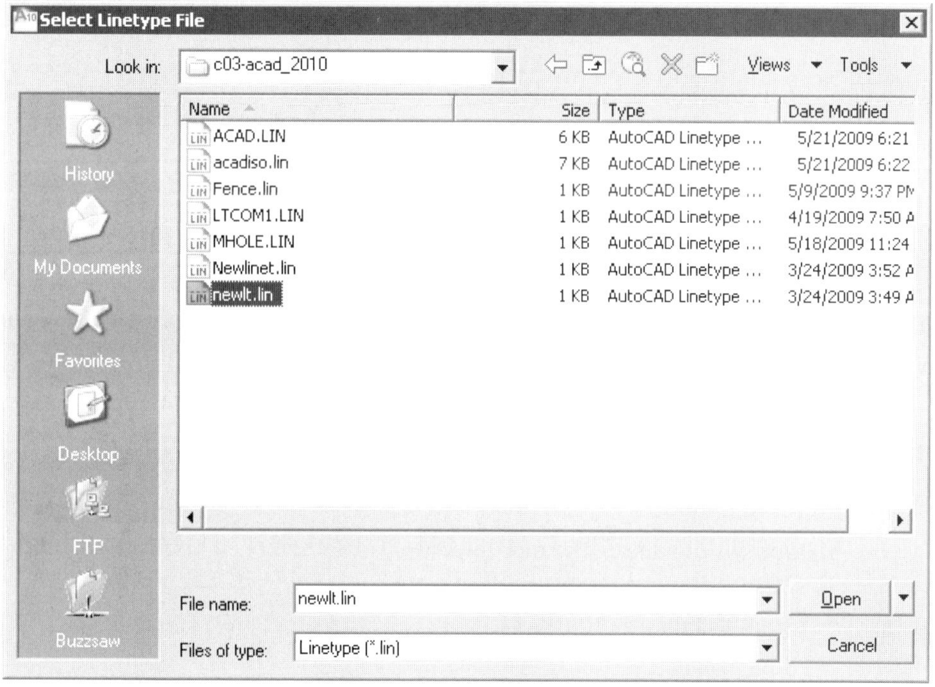

Figure 3-2 *The* ***Select Linetype File*** *dialog box*

Adding a New Linetype in the *acad.lin* File

Step 1: Adding a new linetype in the *acad.lin*

You can also use a text editor (like Notepad) to create a new linetype. Using the text editor, load the file and insert the lines that define the new linetype. The following file is a partial listing of the *acad.lin* file after adding a new linetype to the file:

```
*BORDER,Border __ __ . __ __ . __ __ . __ __ . __ __ .
A,.5,-.25,.5,-.25,0,-.25
*BORDER2,Border (.5x) __.__.__.__.__.__.__.__.__.__.__.
A,.25,-.125,.25,-.125,0,-.125
*BORDERX2,Border (2x) ____  ____  .  ____  ____  .  ___
A,1.0,-.5,1.0,-.5,0,-.5

*CENTER,Center ____ _ ____ _ ____ _ ____ _ ____ _ ____
A,1.25,-.25,.25,-.25
*CENTER2,Center (.5x) ___ _ ___ _ ___ _ ___ _ ___ _ ___
A,.75,-.125,.125,-.125
*CENTERX2,Center (2x) ________  __  ________  __  _____
A,2.5,-.5,.5,-.5

*DASHDOT,Dash dot __ . __ . __ . __ . __ . __ . __
A,.5,-.25,0,-.25
*DASHDOT2,Dash dot (.5x) _._._._._._._._._._._._._._.
A,.25,-.125,0,-.125
*DASHDOTX2,Dash dot (2x) ____  .  ____  .  ____  .  ___
A,1.0,-.5,0,-.5

|
|

*GAS_LINE,Gas line ----GAS----GAS----GAS----GAS----GAS----GAS--
A,.5,-.2,["GAS",STANDARD,S=.1,R=0.0,X=-0.1,Y=-.05],-.25
*ZIGZAG,Zig zag /\/\/\/\/\/\/\/\/\/\/\/\/\/\/\/\/\/\/\/\/\/\/\
A,.0001,-.2,[ZIG,ltypeshp.shx,x=-.2,s=.2],-.4,[ZIG,ltypeshp.shx,r=180,x=.2,s=.2],-.2
*DASH3DOT,___ . . . ___ . . . ___
A,.5,-.125,0,-.125,0,-.125,0,-.125
```

The last two lines of this file define the new linetype, DASH3DOT. The first line contains the name DASH3DOT and the description of the line (___ . . .___). The second line contains the alignment and the pattern definition.

Step 2: Loading the linetype

Save the file and then load the linetype using the **LINETYPE** command. The procedure of loading the linetype is the same as described earlier in this example. The lines and polylines that this linetype will generate are shown in Figure 3-3.

Figure 3-3 *Lines created by linetype DASH3DOT*

Note

If you change the LTSCALE factor, all lines in the drawing are affected by the new ratio.

Using the LINETYPE Command

Step 1: Creating a linetype

To create a linetype using the **LINETYPE** command, first make sure that you are in the drawing editor. Then enter the **-LINETYPE** command and choose the **Create** option to create a linetype.

Command: **-LINETYPE**
Enter an option [?/Create/Load/Set]: **C**

Enter the name of the linetype and the name of the library file in which you want to store the definition of the new linetype.

Enter name of linetype to create: **DASH3DOT**

If **FILEDIA**=1, the **Create or Append Linetype File** dialog box (Figure 3-4) will appear on the screen. If **FILEDIA=**0, you are prompted to enter the name of the file.

Enter linetype file name for new linetype definition <default>: **Acad**

If the linetype already exists, the following message will be displayed on the screen:

Wait, checking if linetype already defined...
"Linetype" already exists in this file. Current definition is:
alignment, dash-1, dash-2, _____.
Overwrite?<N>

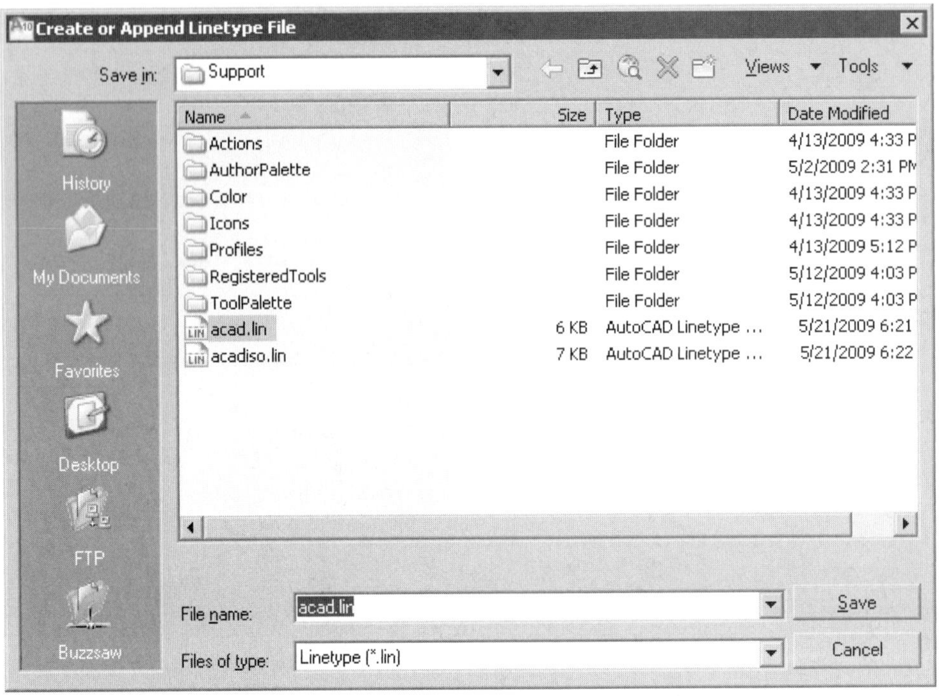

***Figure 3-4** The **Create or Append Linetype File** dialog box*

If you want to redefine the existing line style, enter **Y**. Otherwise, type **N** or press ENTER to choose the default value of N. You can then repeat the process with a different name of the linetype. After entering the name of the linetype and the library file name, you are prompted to enter the descriptive text and the pattern of the line.

Descriptive text: ***DASH3DOT,____ . . . ____ . . . ____**
Enter linetype pattern (on next line):
A,.5,-.125,0,-.125,0,-.125,0,-.125

Descriptive Text

***DASH3DOT,____ . . . ____ . . . ____**

For the descriptive text, you have to type an asterisk (*) followed by the name of the linetype. For Example 1, the name of the linetype is DASH3DOT. The name *DASH3DOT can be followed by the description of the linetype; the length of this description cannot exceed 47 characters. In this example, the description is dashes and dots ____ . . . ____. It could be any text or alphanumeric string. The description is displayed on the screen when you list the linetypes.

Pattern

A,.5,-.125,0,-.125,0,-.125,0,-.125

The line pattern should start with an alignment definition. By default, AutoCAD supports only one type of alignment—A. Therefore, it is displayed on the screen when you select the **LINETYPE** command with the **Create** option. After entering **A** for the pattern alignment, define the pen position. A positive number (.5 or 0.5) indicates a "pen-down" position, and a negative number (-.25 or -0.25) indicates a "pen-up" position. The length of the dash or the space is designated by the magnitude of the number. For example, 0.5 will draw a dash 0.5 units long, and -0.25 will leave a blank space of 0.25 units. A dash length of 0 will draw a dot (.). The following are the pattern definition elements for Example 1:

.5	**pen down**	**0.5 units long dash**
-.125	**pen up**	**.125 units blank space**
0	**pen down**	**dot**
-.125	**pen up**	**.125 units blank space**
0	**pen down**	**dot**
-.125	**pen up**	**.125 units blank space**
0	**pen down**	**dot**
-.125	**pen up**	**.125 units blank space**

After you enter the pattern definition, the linetype (DASH3DOT) is automatically saved in the *acad.lin* file

Step 2: Loading the linetype

You can use the **LINETYPE** command to load the linetype or choose **Home > Properties > Linetype > Other** from the **Ribbon**. The linetype (DASH3DOT) can also be loaded using the **-LINETYPE** command and selecting the **Load** option.

ALIGNMENT SPECIFICATION

As the name suggests, the alignment specifies the pattern alignment at the start and the end of the line, circle, or arc. In other words, the line always starts and ends with the dash (___). The alignment definition "A" requires the first element be a dash or dot (pen down), followed by a negative (pen up) segment. The minimum number of dash segments for alignment A is two. If there is not enough space for the line, a continuous line is drawn.

For example, in the linetype DASH3DOT of Example 1, the length of each line segment is 1.0 (.5 + .125 + .125 + .125 + .125 = 1.0). If the length of the line drawn is less than 1.00, a single line is drawn that looks like a continuous line, see Figure 3-5. If the length of the line is 1.00 or greater, the line will be drawn according to DASH3DOT linetype. AutoCAD automatically adjusts the length of the dashes and the line always starts and ends with a dash. The length of the starting and ending dashes is at least half the length of the dash as specified in the file. If the length of the dash as specified in the file is 0.5, the length of the starting and ending dashes is at least 0.25. To fit a line that starts and ends with a dash, the length of these dashes can also increase, as shown in Figure 3-5.

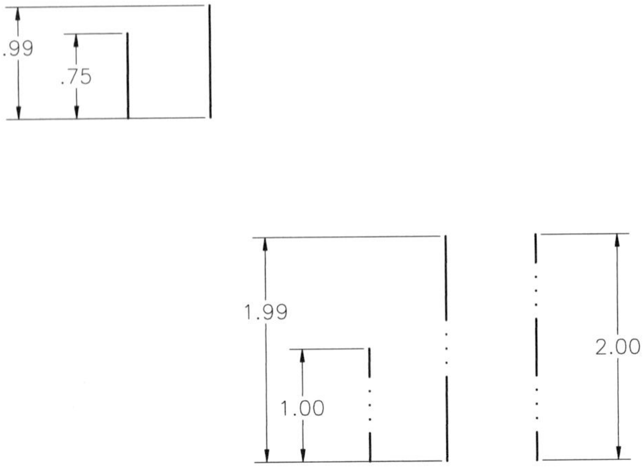

Figure 3-5 *Alignment of linetype DASH3DOT*

LTSCALE COMMAND

As mentioned earlier, the length of each line segment in the DASH3DOT linetype is 1.0 (.5 + .125 + .125 + .125 + .125 = 1.0). If you draw a line that is less than 1.0 units long, a single dash is drawn that looks like a continuous line, see Figure 3-6. This problem can be rectified by changing the linetype scale factor variable **LTSCALE** to a smaller value. This can be accomplished using the **LTSCALE** command.

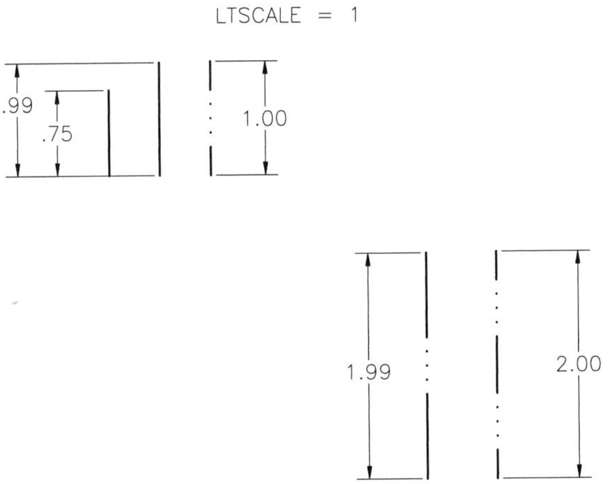

Figure 3-6 *Alignment when LTSCALE = 1*

Command: **LTSCALE**
Enter new linetype scale factor <default>: *New value.*

The default value of the **LTSCALE** variable is 1.0. If the LTSCALE is changed to 0.75, the length of each segment is reduced by 0.75 (1.0 x 0.75 = 0.75). Then, if you draw a line 0.75 units or longer, it will be drawn according to the definition of DASH3DOT (___ . . . ___) (Figures 3-7 and 3-8).

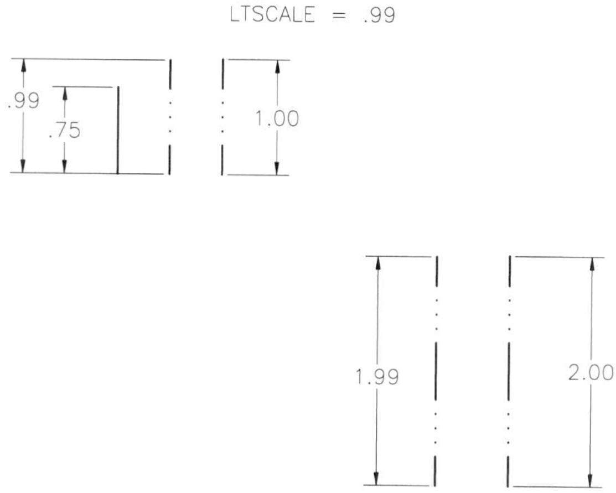

Figure 3-7 *Alignment when LTSCALE = 0.99*

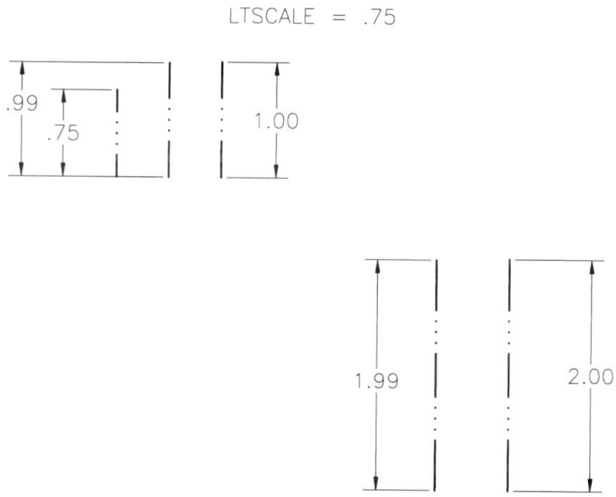

Figure 3-8 *Alignment when LTSCALE = 0.75*

The appearance of the lines is also affected by the limits of the drawing. Most of the AutoCAD linetypes work fine for drawings that have the limits 12,9. Figure 3-9 shows a line of linetype DASH3DOT that is four units long and the limits of the drawing are 12,9. If you increase the limits to 48,36 the lines will appear as continuous lines. If you want the line to appear the same as before on the screen, the LTSCALE needs to be changed. Since the limits of the drawing have increased four times, the LTSCALE should also be increased by the same amount. If you change the scale factor to four, the line segments will also increase by a factor of four. As shown in Figure 3-9, the length of the starting and the ending dash has increased to one unit.

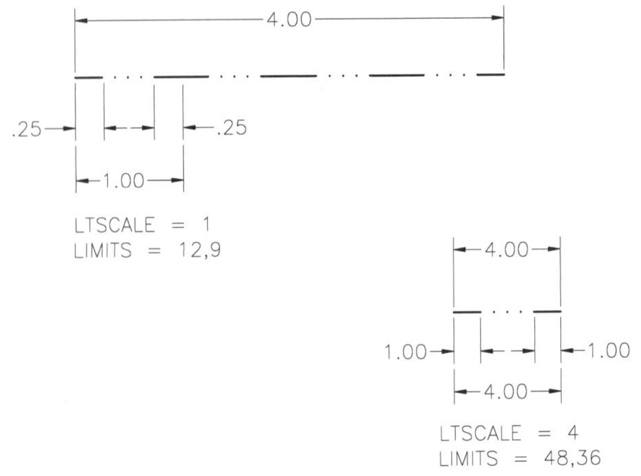

Figure 3-9 *Linetype DASH3DOT before and after changing the LTSCALE factor*

In general, the approximate LTSCALE factor for screen display can be obtained by dividing the X -limit of the drawing by the default X -limit (12.00). However, it is recommended that the linetype scale must be set according to plot scale discussed in the next section.

LTSCALE factor for SCREEN DISPLAY = X -limits of the drawing/12.00

Example:
Drawing limits are 48,36
LTSCALE factor for screen display= 48/12 = 4

Drawing sheet size is 36,24 and scale is 1/4" = 1'
LTSCALE factor for screen display = 12 x 4 x (36 / 12) = 144

LTSCALE FACTOR FOR PLOTTING

The LTSCALE factor for plotting depends on the size of the sheet used to plot the drawing. For example, if the limits are 48 by 36, the drawing scale is 1:1, and you want to plot the drawing on a 48" by 36" size sheet, the LTSCALE factor is 1. If you check the specification of a hidden line in the *acad.lin* file, the length of each dash is 0.25. Therefore, when you plot a drawing with 1:1 scale, the length of each dash in a hidden line is 0.25.

However, if the drawing scale is 1/8" = 1' and you want to plot the drawing on a 48" by 36" paper, the LTSCALE factor must be 96 (8 x 12 = 96). If you increase the LTSCALE factor to 96, the length of each dash in the hidden line will increase by a factor of 96. As a result, the length of each dash will be 24 units (0.25 x 96 = 24). At the time of plotting, the scale factor must be 1:96 to plot the 384' by 288' drawing on a 48" by 36" size paper. Each dash of the hidden line that was 24" long on the drawing will be 0.25 (24/96 = 0.25) inch long when plotted. Similarly, if the desired text size on the paper is 1/8", the text height in the drawing must be 12" (1/8 x 96 = 12").

LTSCALE Factor for PLOTTING = Drawing Scale

Sometimes your plotter may not be able to plot a 48" by 36" drawing or you may like to decrease the size of the plot so that the drawing fits within the specified area. To get the correct dash lengths for hidden, center, or other lines, you must adjust the LTSCALE factor. For example, if you want to plot the previously mentioned drawing in a 45" by 34" area, the correction factor is:

Correction factor	= 48/45
	= 1.0666
New LTSCALE factor	= LTSCALE factor x Correction factor
	= 96 x 1.0666
	= 102.4

New LTSCALE Factor for PLOTTING = Drawing Scale x Correction Factor

Note

If you change the LTSCALE factor, all lines in the drawing are affected by the new ratio.

CURRENT LINETYPE SCALING (CELTSCALE)

Like **LTSCALE**, the **CELTSCALE** system variable controls the linetype scaling. The difference is that **CELTSCALE** determines the current linetype scaling. For example, if you set the **CELTSCALE** to 0.5, all lines drawn after setting the new value for **CELTSCALE** will have the linetype scaling factor of 0.5. The value is retained in the **CELTSCALE** system variable. The first line (a) in Figure 3-10 is drawn with the **CELTSCALE** factor of 1 and the second line (b) is drawn with the **CELTSCALE** factor of 0.5. The length of the dashes is reduced by a factor of 0.5 when the **CELTSCALE** is 0.5.

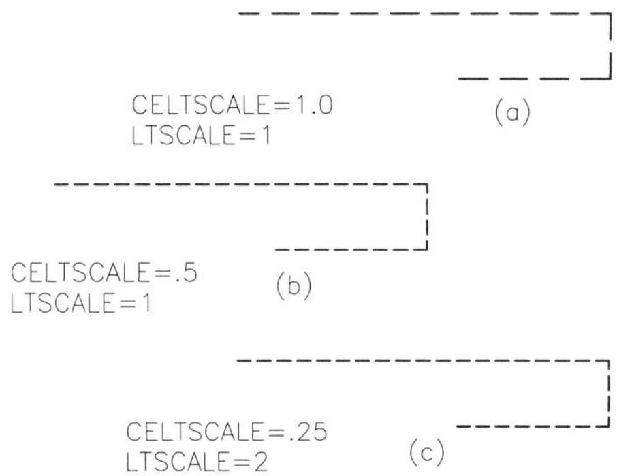

Figure 3-10 *Using* ***CELTSCALE*** *to control current linetype scaling*

The **LTSCALE** system variable controls the global scale factor. For example, if **LTSCALE** is set to 2, all lines in the drawing will be affected by a factor of 2. The net scale factor is equal to the product of **CELTSCALE** and **LTSCALE**. Figure 3-10(c) shows a line that is drawn with **LTSCALE** of 2 and **CELTSCALE** of 0.25. The net scale factor is = **LTSCALE** x **CELTSCALE** = 2 x 0.25 = 0.5.

Note

You can change the current linetype scale factor of a line by using the ***Properties*** *palette that can be invoked by choosing the* ***Properties*** *button from the* ***Quick Access*** *Toolbar (Customize to add). You can also use the* ***CHANGE*** *command and then choose the* ***ltScale*** *option.*

ALTERNATE LINETYPES

One of the problems with the LTSCALE factor is that it affects all the lines in the drawing. As shown in Figure 3-11(a), the length of each segment in all DASH3DOT type lines is approximately equal, no matter how long the lines. You may want to have a small segment length if the lines are small and a longer segment length if the lines are long. You can accomplish this by using CELTSCALE (discussed later in this chapter) or by defining an alternate linetype with a different segment length. For example, you can define a linetype DASH3DOT and DASH3DOTX with different line pattern specifications.

```
*DASH3DOT,____ . . . ____ . . . ____ . . . ____
A,0.5,-.125,0,-.125,0,-.125,0,-.125
*DASH3DOTX,________ . . . __________
A,1.0,-.25,0,-.25,0,-.25,0,-.25
```

In the DASH3DOT linetype, the segment length is one unit. Whereas, in the DASH3DOTX linetype the segment length is two units. You can have several alternate linetypes to produce the lines with different segment lengths. Figure 3-11(b) shows the lines generated by DASH3DOT and DASH3DOTX.

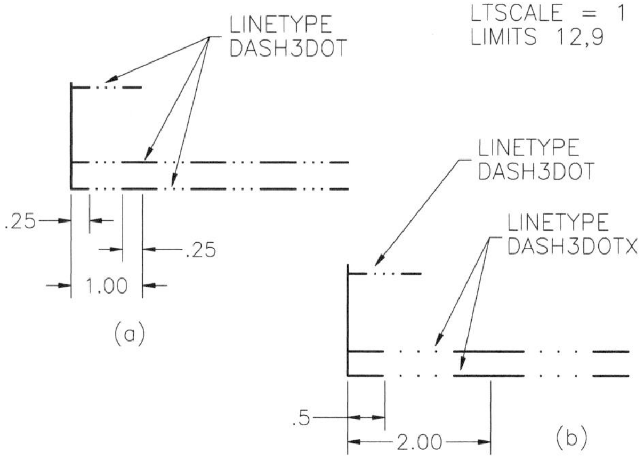

Figure 3-11 *Linetypes generated by DASH3DOT and DASH3DOTX*

Note

Although you may have used various linetypes with different segment lengths, the lines will be affected equally when you change the LTSCALE factor. For example, if the LTSCALE factor is 0.5, the segment length of DASH3DOT line will be 0.5 units and the segment length of DASH3DOTX will be 1.0 units.

MODIFYING LINETYPES

You can also modify the linetypes that are defined in the *acad.lin* or *acadiso.lin* file. You must save a copy of the original *acad.lin* or *acadiso.lin* file before making any changes to it. You need a text editor, such as Notepad, to modify the linetype. You can also use the **EDIT** function of DOS, or the **EDIT** command (provided the *acad.pgp* file is present and **EDIT** is defined in the file). For example, if you want to change the dash length of the border linetype from 0.5 to 0.75, load the file, then edit the pattern line of the border linetype. The following file is a partial listing of the *acad.lin* file after changing the border and centerx2 linetypes.

```
;; AutoCAD Linetype Definition file
;; Version 3.0
;; Copyright (C) 1991-2009 by Autodesk, Inc.  All Rights Reserved.


*BORDER,Border __ __ . __ __ . __ __ . __ __ . __ __ .
A,.5,-.25,.5,-.25,0,-.25
*BORDER2,Border (.5x) __.__.__.__.__.__.__.__.__.__.__.
A,.25,-.125,.25,-.125,0,-.125
*BORDERX2,Border (2x) ____  ____  .  ____  ____  .  ___
A,1.0,-.5,1.0,-.5,0,-.5

*CENTER,Center ____ _ ____ _ ____ _ ____ _ ____ _ ____
A,1.25,-.25,.25,-.25
*CENTER2,Center (.5x) ___ _ ___ _ ___ _ ___ _ ___ _ ___
A,.75,-.125,.125,-.125
*CENTERX2,Center (2x) ________  __  ________  __  _____
A,2.5,-.5,.5,-.5

*DASHDOT,Dash dot __ . __ . __ . __ . __ . __ . __
A,.5,-.25,0,-.25
*DASHDOT2,Dash dot (.5x) _._._._._._._._._._._._._._.
A,.25,-.125,0,-.125
*DASHDOTX2,Dash dot (2x) ____  .  ____  .  ____  .  ___
A,1.0,-.5,0,-.5

*DASHED,Dashed __ __ __ __ __ __ __ __ __ __ __ __ __ _
A,.5,-.25
*DASHED2,Dashed (.5x) _ _ _ _ _ _ _ _ _ _ _ _ _ _ _ _ _ _
A,.25,-.125
*DASHEDX2,Dashed (2x) ____  ____  ____  ____  ____  ___
A,1.0,-.5

*DIVIDE,Divide ____ . . ____ . . ____ . . ____ . . ____
A,.5,-.25,0,-.25,0,-.25
```

```
*DIVIDE2,Divide (.5x) __..__..__..__..__..__..__..__
A,.25,-.125,0,-.125,0,-.125
*DIVIDEX2,Divide (2x) ________  . .  ________  . .  _
A,1.0,-.5,0,-.5,0,-.5

*DOT,Dot . . . . . . . . . . . . . . . . . . . . . . . . .
A,0,-.25
*DOT2,Dot (.5x) .......................................
A,0,-.125
*DOTX2,Dot (2x) .  .  .  .  .  .  .  .  .  .  .  .  .  .
A,0,-.5

*HIDDEN,Hidden __ __ __ __ __ __ __ __ __ __ __ __ __ __
A,.25,-.125
*HIDDEN2,Hidden (.5x) _ _ _ _ _ _ _ _ _ _ _ _ _ _ _ _ _ _
A,.125,-.0625
*HIDDENX2,Hidden (2x) ____ ____ ____ ____ ____ ____ ____
A,.5,-.25

*PHANTOM,Phantom ______  __  __  ______  __  __  ______
A,1.25,-.25,.25,-.25,.25,-.25
*PHANTOM2,Phantom (.5x) ___ _ _ ___ _ _ ___ _ _ ___ _ _
A,.625,-.125,.125,-.125,.125,-.125
*PHANTOMX2,Phantom (2x) ____________    ____    ____   _
A,2.5,-.5,.5,-.5,.5,-.5

;;
;;  ISO 128 (ISO/DIS 12011) linetypes
;;
;;  The size of the line segments for each defined ISO line is
;;  defined for an usage with a pen width of 1 mm. To use them with
;;  the other ISO predefined pen widths, the line has to be scaled
;;  with the appropriate value (e.g. pen width 0,5 mm -> ltscale 0.5).
;;
*ACAD_ISO02W100,ISO dash __ __ __ __ __ __ __ __ __ __ __ __ __
A,12,-3
|
|
*ACAD_ISO15W100,ISO double-dash triple-dot __ __ . . . __ __ . .
A,12,-3,12,-3,0,-3,0,-3,0,-3

;;  Complex linetypes
;;  Complex linetypes have been added to this file.
;;  These linetypes were defined in LTYPESHP.LIN in
;;  Release 13, and are incorporated in ACAD.LIN in
```

```
;; Release 14.
;;
;; These linetype definitions use LTYPESHP.SHX.
;;
*FENCELINE1,Fenceline circle ----0-----0----0-----0----0-----0--
A,.25,-.1,[CIRC1,ltypeshp.shx,x=-.1,s=.1],-.1,1
*FENCELINE2,Fenceline square ----[]-----[]----[]-----[]----[]---
A,.25,-.1,[BOX,ltypeshp.shx,x=-.1,s=.1],-.1,1
*TRACKS,Tracks -|-|-|-|-|-|-|-|-|-|-|-|-|-|-|-|-|-|-|-|-|-|-|-|-
A,.15,[TRACK1,ltypeshp.shx,s=.25],.15
*BATTING,Batting SSSSSSSSSSSSSSSSSSSSSSSSSSSSSSSSSSSSSSSSSSSSSSS
A,.0001,-.1,[BAT,ltypeshp.shx,x=-.1,s=.1],-.2,[BAT,ltypeshp.shx,r=180,x=.1,s=.1],-.1
*HOT_WATER_SUPPLY,Hot water supply ---- HW ---- HW ---- HW ----
A,.5,-.2,["HW",STANDARD,S=.1,R=0.0,X=-0.1,Y=-.05],-.2
*GAS_LINE,Gas line ----GAS----GAS----GAS----GAS----GAS----GAS--
A,.5,-.2,["GAS",STANDARD,S=.1,R=0.0,X=-0.1,Y=-.05],-.25
*ZIGZAG,Zig zag /\/\/\/\/\/\/\/\/\/\/\/\/\/\/\/\/\/\/\/\/\/\/\/
A,.0001,-.2,[ZIG,ltypeshp.shx,x=-.2,s=.2],-.4,[ZIG,ltypeshp.shx,r=180,x=.2,s=.2],-.2
```

Example 2

Create a new file, *newlint.lin*, and define a linetype VARDASH with the following specifications:

Length of first dash 1.0
Blank space 0.25
Length of second dash 0.75
Blank space 0.25
Length of third dash 0.5
Blank space 0.25
Dot
Blank space 0.25
Length of next dash 0.5
Blank space 0.25
Length of next dash 0.75

Step 1: Writing definition of linetype

Use a text editor and insert the following lines that define the new linetype **VARDASH**.

```
*VARDASH,———— ——— —— . —— ——— ————
A,1,-.25,.75,-.25,.5,-.25,0,-.25,.5,-.25,.75,-.25
```

Step 2: Loading the linetype

You can use the **LINETYPE** command or choose **Home > Properties > Linetype > Other** from the **Ribbon** to load the linetype. The type of lines that this linetype will generate are shown in Figure 3-12.

Figure 3-12 Lines generated by linetype VARDASH

COMPLEX LINETYPES

AutoCAD has provided a facility to create complex linetypes. The complex linetypes can be classified into two groups: string complex linetype and shape complex linetype. The difference between the two is that the string complex linetype has a text string inserted in the line, whereas the shape complex linetype has a shape inserted in the line. The facility of creating complex linetypes increases the functionality of lines. For example, if you want to draw a line around a building that indicates the fence line, you can do it by defining a complex linetype that will automatically give you the desired line with the text string (Fence). Similarly, you can define a complex linetype that will insert a shape (symbol) at predefined distances along the line.

CREATING A STRING COMPLEX LINETYPE

While writing the definition of a string complex linetype, the actual text and its attributes must be included in the linetype definition, refer to Figure 3-13. The format of the string complex linetype is:

["String", Text Style, Text Height, Rotation, X-Offset, Y-Offset]

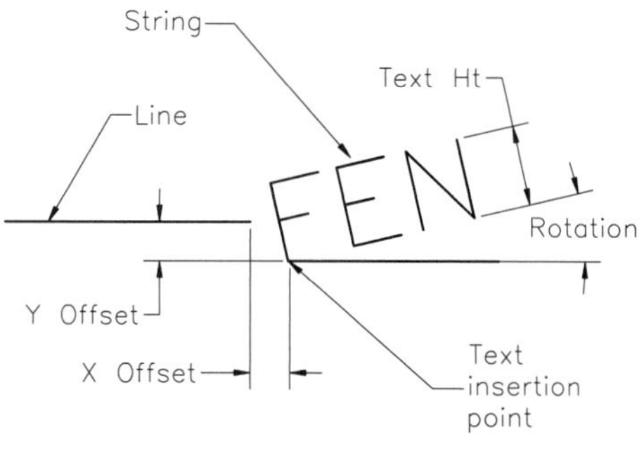

Figure 3-13 Attributes of a string complex linetype

String. It is the actual text that you want to insert along the line. The text string must be enclosed in quotation marks (" ").

Text Style. This is the name of the text style file that you want to use for generating the text string. The text style must be predefined.

Text Height. This is the actual height of the text, if the text height defined in the text style is 0. Otherwise, it acts as a scale factor for the text height specified in the text style. In Figure 3-13, the height of the text is 0.1 units.

Rotation. The rotation can be specified as an absolute or relative angle. In the absolute rotation the angle is always measured with respect to the positive *X* axis, no matter what AutoCAD's direction setting. The absolute angle is represented by letter "a".

In relative rotation the angle is always measured with respect to orientation of dashes in the linetype. The relative angle is represented by the letter "r". The angle can be specified in radians (r), grads (g), or degrees (d). The default is degrees.

X-Offset. This is the distance of the lower left corner of the text string from the endpoint of the line segment measured along the line. If the line is horizontal, then the X-Offset distance is measured along the *X* axis. In Figure 3-13, the X-Offset distance is 0.05.

Y-Offset. This is the distance of the lower left corner of the text string from the endpoint of the line segment measured perpendicular to the line. If the line is horizontal, then the Y-Offset distance is measured along the *Y* axis. In Figure 3-13, the Y-Offset distance is -0.05. The distance is negative because the start point of the text string is 0.05 units below the endpoint of the first line segment.

Example 3

In the following example, you will write the definition of a string complex linetype that consists of the text string "Fence" and line segments. The length of each line segment is 0.75. The height of the text string is 0.1 units, and the space between the end of the text string and the following line segment is 0.05, see Figure 3-14.

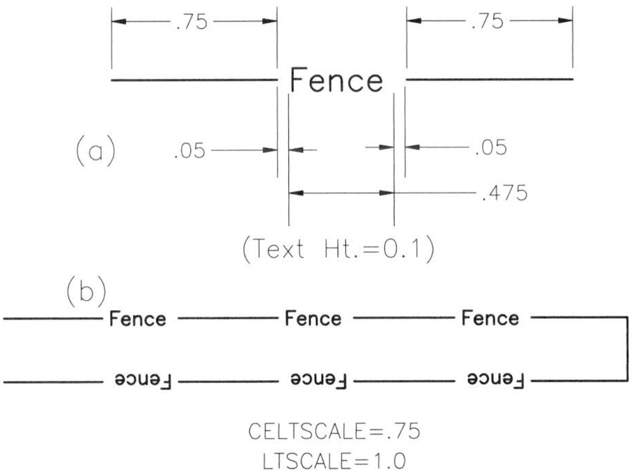

Figure 3-14 *The attributes of a string complex linetype and line specifications for Example 3*

Step 1: Determining the line specifications

Before writing the definition of a new linetype, it is important to determine the line specification. One of the ways to do this is to actually draw the lines and the text the way you want them to appear in the drawing. Once you have drawn the line and the text to your satisfaction, measure the distances needed to define the string complex linetype. The values are given as follows:

Text string=	Fence
Text style=	Standard
Text height=	0.1
Text rotation=	0
X-Offset=	0.05
Y-Offset=	-0.05
Length of the first line segment=	0.75
Distance between the line segments=	0.575

Step 2: Writing the definition of string complex linetype

Use a text editor to write the definition of the string complex linetype. You can add the definition to the *acad.lin* or *acadiso.lin* file or create a separate file. The extension of the file must be *.lin*. The following file is the listing of the *fence.lin* file for Example 3. The name of the linetype is NEWFence1.

```
*NEWFence1,New fence boundary line
A,0.75,["Fence",Standard,S=0.1,A=0,X=0.05,Y=-0.05],-0.575
or
A,0.75,-0.05,["Fence",Standard,S=0.1,A=0,X=0,Y=-0.05],-0.525
```

Step 3: Loading the linetype

You can use the **LINETYPE** command to load the linetype or choose **Home > Properties > Linetype > Other** from the **Ribbon**. Draw a line or any object to check if the line is drawn to the given specifications as shown in Figure 3-15. Notice that the text is always drawn along the *X* axis. Also, when you draw a line at an angle, polyline, circle, or spline, the text string does not align with the object (Figure 3-15).

Step 4: Aligning the text with the line

In the NEWFence linetype definition, the specified angle is 0-degree (Absolute angle A = 0). Therefore, when you use the NEWFence linetype to draw a line, circle, polyline, or spline, the text string (Fence) will be at zero degrees. If you want the text string (Fence) to align with the polyline (Figure 3-16), spline, or circle, specify the angle as relative angle (R = 0) in the NEWFence linetype definition. The following is the linetype definition for NEWFence linetype with relative angle R = 0:

```
*NEWFence2,New fence boundary line
A,0.75,["Fence",Standard,S=0.1,R=0,X=0.05,Y=-0.05],-0.575
```

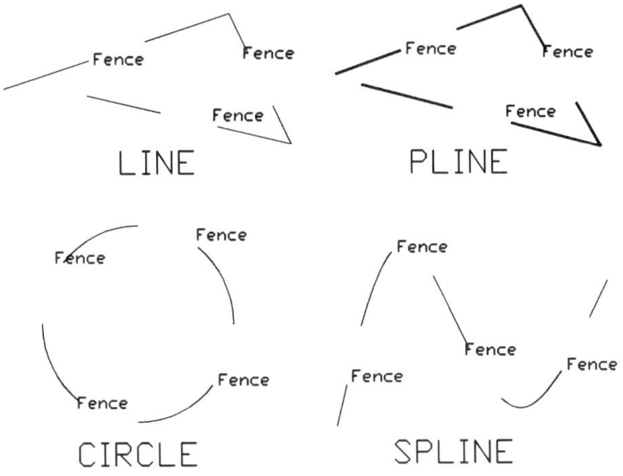

Figure 3-15 *Using string complex linetype with angle A=0*

Step 5: Aligning the midpoint of text with the line

In Figure 3-16, you will notice that the text string is not properly aligned with the circumference of the circle. This is because AutoCAD draws the text string in a direction that is tangent to the circle at the text insertion point.

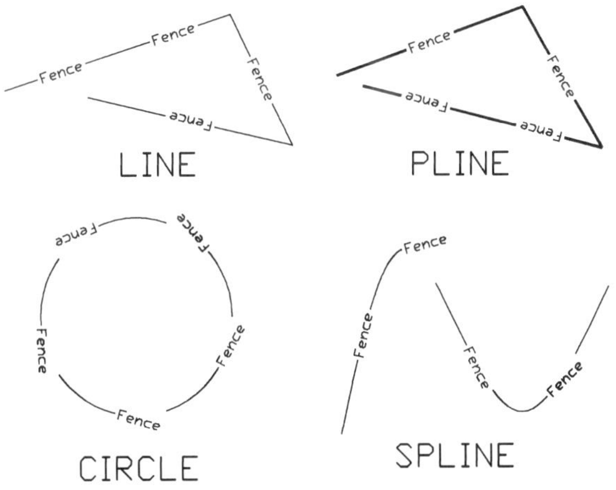

Figure 3-16 *Using a string complex linetype with angle R = 0*

To resolve this problem, you must define the middle point of the text string as the insertion point. Also, the line specifications should be measured accordingly. Figure 3-17 gives the measurements of the NEWFence linetype with the middle point of the text as the insertion point and Figure 3-18 shows the entities sketched with the selected linetype.

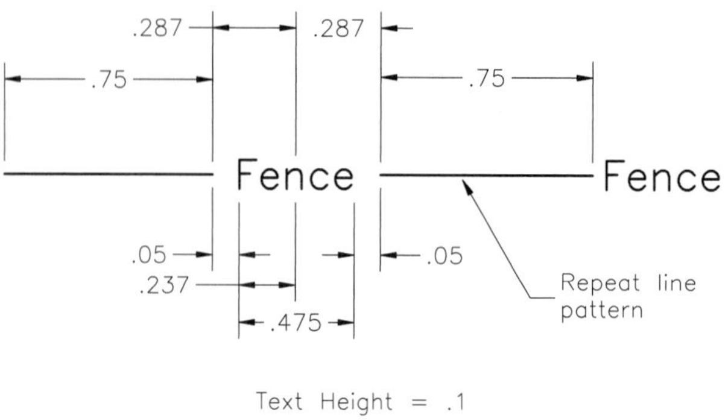

Figure 3-17 *Specifications of a string complex linetype with the middle point of the text string as the text insertion point*

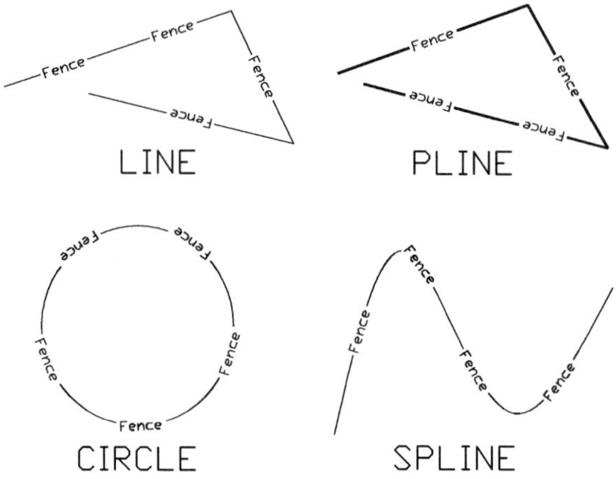

Figure 3-18 *Using a string complex linetype with the middle point of the text string as the text insertion point*

The following is the linetype definition for NEWFence linetype:

*NEWFence3,New fence boundary line
A,0.75,-0.287,["Fence",Standard,S=0.1,X=-0.237,Y=-0.05],-0.287

Note

If no angle is defined in the line definition, it acquires the default value of angle R = 0. Also, the text does not automatically insert to its midpoint like the regular text with MID justification.

Creating a Shape Complex Linetype

As with the string complex linetype, when you write the definition of a shape complex linetype, the name of the shape, the name of the shape file, and other shape attributes, like rotation, scale, X-Offset, and Y-Offset, must be included in the linetype definition. The format of the shape complex linetype is:

[Shape Name, Shape File, Scale, Rotation, X-Offset, Y-Offset]

The following is the description of the attributes of Shape Complex Linetype (Figure 3-19).

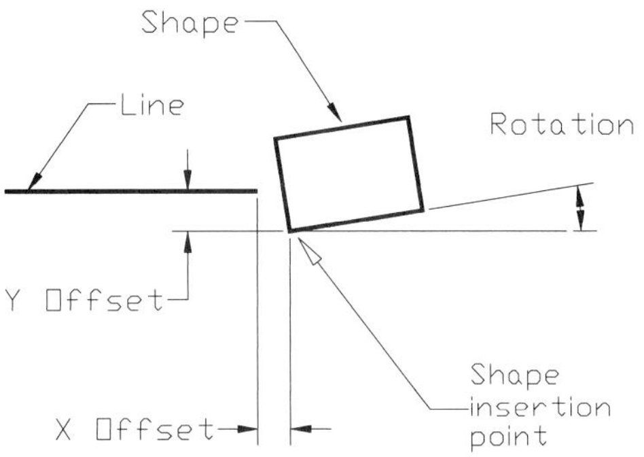

Figure 3-19 *The attributes of a shape complex linetype*

Shape Name. This is the name of the shape that you want to insert along the line. The shape name must exist; otherwise, no shape will be generated along the line.

Shape File. This is the name of the compiled shape file (*.shx*) that contains the definition of the shape being inserted in the line. The name of the subdirectory where the shape file is located must be in the AutoCAD search path. The shape files (*.shp*) must be compiled before using the **SHAPE** command to load the shape.

Note
To know more about the shape files (.shp) refer to Chapter 11.

Scale. This is the scale factor by which the defined shape size is to be scaled. If the scale is 1, the size of the shape will be the same as defined in the shape definition (*.shp* file).

Rotation. The rotation can be specified as an absolute or relative angle. In absolute rotation, the angle is always measured with respect to the positive *X* axis, no matter what AutoCAD's direction setting. The absolute angle is represented by letter "a." In relative rotation, the angle is always measured with respect to the orientation of dashes in the linetype. The relative angle is represented by the letter "r." The angle can be specified in radians (r), grads (g), or degrees (d). The default is degrees.

X-Offset. This is the distance of the shape insertion point from the endpoint of the line segment measured along the line. If the line is horizontal, then the X-Offset distance is measured along the *X* axis. In Figure 3-20, the X-Offset distance is 0.2.

Y-Offset. This is the distance of the shape insertion point from the endpoint of the line segment measured perpendicular to the line. If the line is horizontal, then the Y-Offset distance is measured along the *Y* axis. In Figure 3-20, the Y-Offset distance is 0.

Example 4

Write the definition of a shape complex linetype that consists of the shape (Manhole; the name of the shape is MH) and a line. The scale of the shape is 0.1, the length of each line segment is 0.75, and the space between line segments is 0.2.

Step 1: Determining the line specifications

Before writing the definition of a new linetype, it is important to determine the line specifications. One of the ways this can be done is to actually draw the lines and the shape the way you want them to appear in the drawing (Figure 3-20). Once you have drawn the line and the shape to your satisfaction, measure the distances needed to define the shape complex linetype. In this example, the values are as follows:

Shape name MH
Shape file name *mhole.shx* (Name of the compiled shape file.)
Scale 0.1
Rotation 0
X-Offset 0.2
Y-Offset 0
Length of the first line segment = 0.75
Distance between the line segments = 0.2

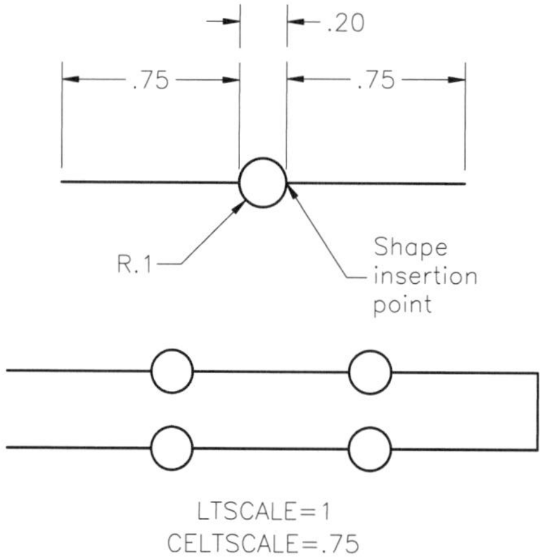

Figure 3-20 *The attributes of the shape complex linetype and line specifications for Example 4*

Step 2: Writing the definition of the shape

Use a text editor to write the definition of the shape file. The extension of the file must be *.shp*. The following file is the listing of the *mhole.shp* file for Example 4. The name of the shape is MH.

```
*215,9,MH
001,10,(1,007),
001,10,(1,071),0
```

Step 3: Compiling the shape

Use the **COMPILE** command to compile the shape file (*.shp* file). Remember that the path of the folder in which you save the shape file should be specified in the AutoCAD support file search path. When you use this command, the **Select Shape or Font File** dialog box will be displayed (Figure 3-21). If **FILEDIA** = 0, this command will be executed using the command line. The following is the command sequence for compiling the shape file:

Command: **COMPILE**
Enter shape (.SHP) or PostScript font (.PFB) file name: **MHOLE**

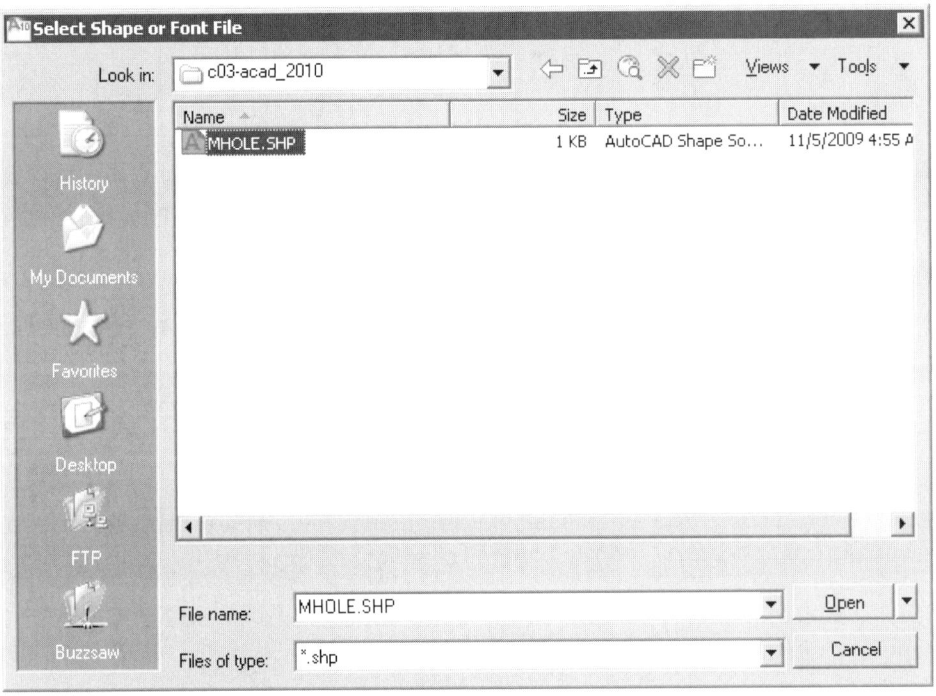

Figure 3-21 *The **Select Shape or Font File** dialog box*

Step 4: Writing the definition of the shape complex linetype

Use a text editor to write the definition of the shape complex linetype. You can add the definition to the *acad.lin* file or create a separate file. The extension of the file must be *.lin*. The following file is the listing of the *mhole.lin* file for Example 4. The name of the linetype is MHOLE.

```
*MHOLE,Line with Manholes
A,0.75,[MH,MHOLE.SHX,S=0.10,X=0.2,Y=0],-0.2
```

Step 5: Loading the linetype

To test the linetype, load the linetype using the **LINETYPE** command. Assign the linetype to a layer. Draw a line or any object to check if the line is drawn to the given specifications. The shape is drawn upside down when you draw a line from right to left. Figure 3-22 shows the execution of the linetype *mhole.lin* using line, pline, and spline. Figure 3-23 displays the region hatched using the string complex and shape complex line types respectively.

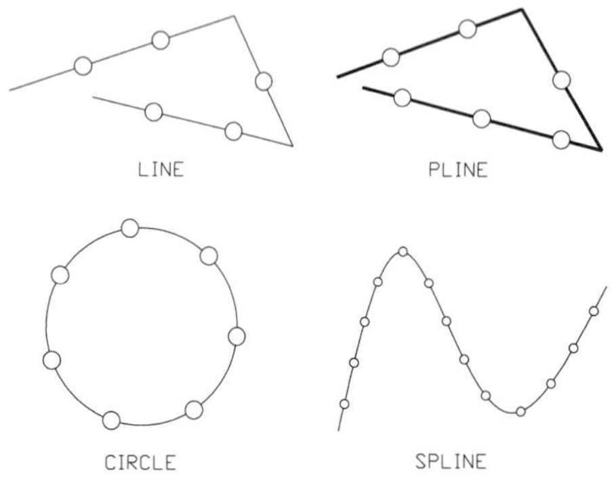

Figure 3-22 *Using a shape complex linetype*

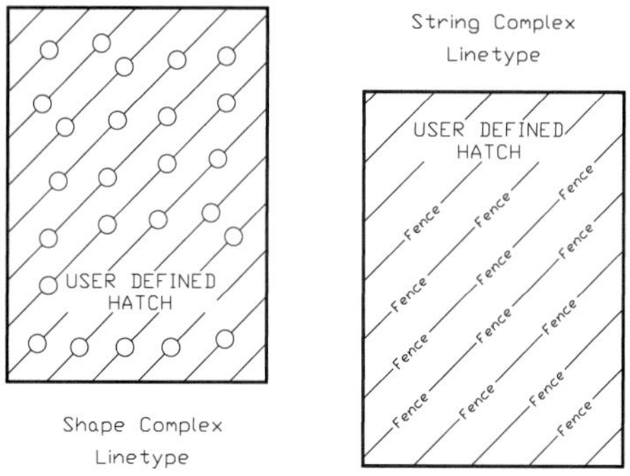

Figure 3-23 *Using shape and string complex linetypes to create custom hatch*

HATCH PATTERN DEFINITION

AutoCAD has a default hatch pattern library file, *acad.pat and acadiso.pat*, that contains 72 hatch patterns. Generally, you can hatch all the drawings using these default hatch patterns. However, if you need a different hatch pattern, AutoCAD lets you create your own hatch patterns. There is no limit to the number of hatch patterns you can define.

The hatch patterns you define can be added to the hatch pattern library file, *acad.pat or acadiso.pat*. You can also create a new hatch pattern library file, provided the file contains only one hatch pattern definition, and the name of the hatch is the same as the name of the file.

The hatch pattern definition consists of the following two parts: **Header Line** and **Hatch Descriptors**.

Header Line

The **header line** consists of an asterisk (*) followed by the name of the hatch pattern. The hatch name is the name used in the hatch command to hatch an area. After the hatch name comes the hatch description. Both are separated from each other by a comma (,). The general format of the header line is:

***HATCH Name [, Hatch Description]**

Where * ----------------------------- Asterisk
HATCH Name ----------- Name of hatch pattern
Hatch Description ------ Description of hatch pattern

The description can be any text that describes the hatch pattern. It can also be omitted, in which case, a comma should not follow the hatch pattern name.

Example:
***DASH45, Dashed lines at 45-degree**
Where **DASH45** --------- Hatch name
Dashed lines at 45-degree ------- Hatch description

Hatch Descriptors

The **hatch descriptors** consist of one or more lines that contain the definition of the hatch lines. The general format of the hatch descriptor is:

Angle, X-origin, Y-origin, D1, D2 [,Dash Length.....]
Where **Angle** ------------ Angle of hatch lines
X-origin --------- *X* coordinate of hatch line
Y-origin --------- *Y* coordinate of hatch line
D1 ---------------- Displacement of second line (Delta-X)
D2 ---------------- Distance between hatch lines (Delta-Y)
Length ----------- Length of dashes and spaces (Pattern line definition)

Example:
45,0,0,0,0.5,0.5,-0.125,0,-0.125
Where **45** ----------------- Angle of hatch line
0 ------------------- X-Origin
0 ------------------- Y-Origin
0 ------------------- Delta-X
0.5 ---------------- Delta-Y
0.5 ---------------- Dash (pen down)
-0.125 ------------ Space (pen up)
0 ------------------- Dot (pen down)
-0.125 ------------ Space (pen up)
0.5,-0.125,0,-0.125 Pattern line definition

Hatch Angle

X-origin and Y-origin. The hatch angle is the angle that the hatch lines make with the positive *X* axis. The angle is positive if measured counterclockwise (Figure 3-24), and negative if the angle is measured clockwise. When you draw a hatch pattern, the first hatch line starts from the point defined by X-origin and Y-origin.

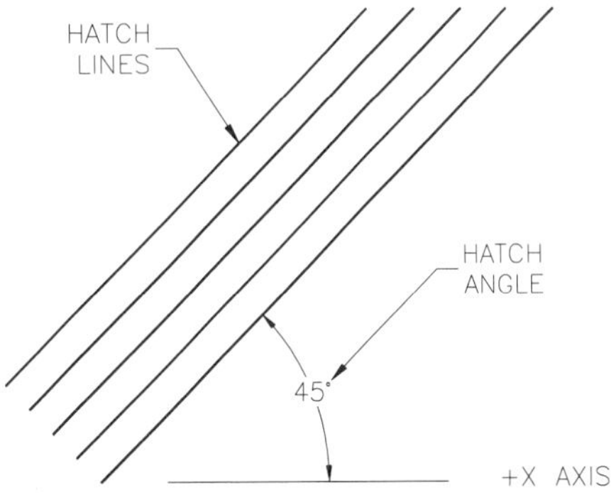

Figure 3-24 *Hatch angle*

The remaining lines are generated by offsetting the first hatch line by a distance specified by delta-X and delta-Y. In Figure 3-25(a), the first hatch line starts from the point with the coordinates X = 0 and Y = 0. In Figure 3-25(b), the first line of hatch starts from a point with the coordinates X = 0 and Y = 0.25.

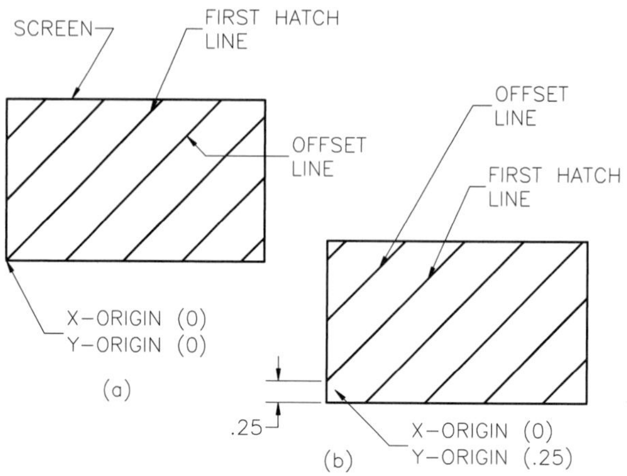

***Figure 3-25** X-origin and Y-origin of hatch lines*

Delta-X and Delta-Y. Delta-X is the displacement of the offset line in the direction in which the hatch lines are generated. For example, if the lines are drawn at a 0-degree angle and delta-X = 0.5, the offset line will be displaced by a distance delta-X (0.5) along the 0-angle direction. Similarly, if the hatch lines are drawn at a 45-degree angle, the offset line will be displaced by a distance delta-X (0.5) along a 45-degree direction (Figure 3-26). Delta-Y is the displacement of the offset lines measured perpendicular to the hatch lines. For example, if delta-Y = 1.0, the space between any two hatch lines will be 1.0 (Figure 3-26).

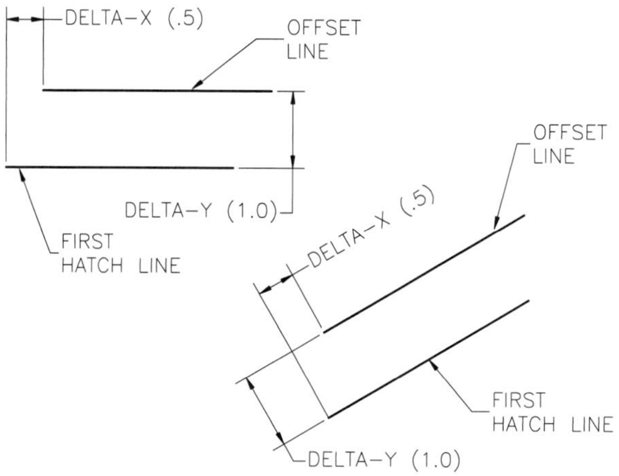

***Figure 3-26** Delta-X and Delta-Y of hatch lines*

HOW HATCH WORKS?

When you hatch an area, infinite number of hatch lines of infinite length are generated. The first hatch line always passes through the point specified by the X-origin and Y-origin. The remaining lines are generated by offsetting the first hatch line in both directions. The offset distance is determined by delta-X and delta-Y, as shown in Figure 3-26. All selected entities that form the boundary of the hatch area are then checked for intersection with these lines. Any hatch lines found within the defined hatch boundaries are turned on, and the hatch lines outside the hatch boundary are turned off, as shown in Figure 3-27. Since the hatch lines are generated by offsetting, the hatch lines in all the areas of the drawing are automatically aligned relative to the drawing's snap origin. Figure 3-27(a) shows the hatch lines as computed by AutoCAD. These lines are not drawn on the screen; they are shown here for illustration only. Figure 3-27(b) shows the hatch lines generated in the circle that was defined as the hatch boundary.

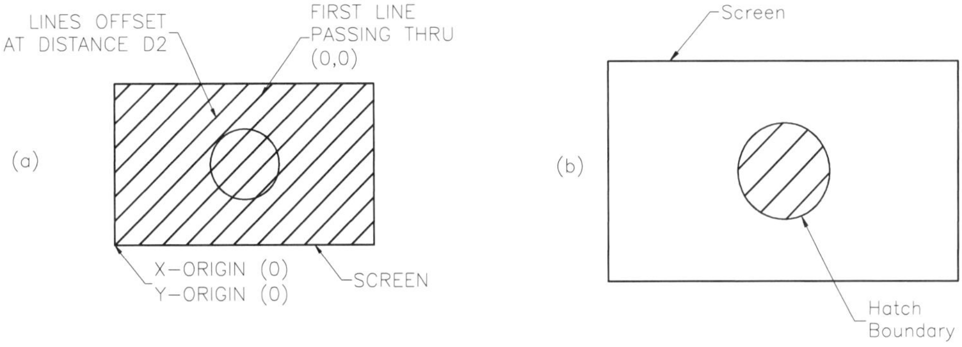

Figure 3-27 Hatch lines outside the hatch boundary are turned off

SIMPLE HATCH PATTERN

It is good practice to develop the hatch pattern specification before writing a hatch pattern definition. For simple hatch patterns it may not be that important, but for more complicated hatch patterns you should know the detailed specifications. Example 5 illustrates the procedure for developing a simple hatch pattern.

Example 5

Write a hatch pattern definition for the hatch pattern shown in Figure 3-28, with the following specifications:

Name of the hatch pattern =	HATCH1
X-Origin =	0
Y-Origin =	0
Distance between hatch lines =	0.5
Displacement of hatch lines =	0
Hatch line pattern =	Continuous

Step 1: Creating the hatch pattern file

This hatch pattern definition can be added to the existing *acad.pat* hatch file. You can use any text editor (like Notepad) to write the file. Load the *acad.pat* file that is located in the *C:\Documents and Settings\User\Application Data\Autodesk\AutoCAD 2010\R18\enu\Support* directory and insert the following two lines at the end of the file.

*HATCH1,Hatch Pattern for Example 5
45,0,0,0,.5

Where **45** ----------------- Hatch angle
0 ------------------- X-origin
0 ------------------- Y-origin
0 ------------------- Displacement of second hatch line
.5 ------------------ Distance between hatch lines

The first field of hatch descriptors contains the angle of the hatch lines and the value of this angle in this particular case is 45-degree with respect to the positive *X* axis. The second and third fields describe the *X* and *Y* coordinates of the first hatch line origin. The first line of the hatch pattern will pass through this point. If the values of the X-origin and Y-origin were 0.5 and 1.0, respectively, then the first line would pass through the point with the *X* coordinate of 0.5 and the *Y* coordinate of 1.0, with respect to the drawing origin 0,0. The remaining lines are generated by offsetting the first line, as shown in Figure 3-28.

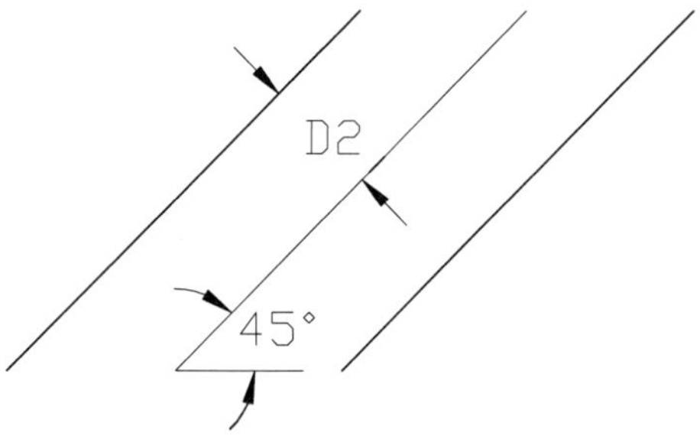

***Figure 3-28** Hatch pattern angle and offset distance*

Step 2: Loading the hatch pattern

Choose **Home > Draw > Hatch** button from the **Ribbon** to display the **Hatch and Gradient** dialog box. Make sure **Predefined** is selected in the **Type** edit box. Select the hatch pattern name from the drop-down list or choose the **[...]** button adjacent to the **Pattern** drop-down list to display the **Hatch Pattern Palette** dialog box. Choose the **Other Predefined** tab and select the hatch pattern file displayed there. Then, choose **OK** to display the **Hatch and Gradient** dialog box again. Change the **Scale** and **Angle**, if needed, and then hatch a circle to test the hatch pattern.

The **Hatch and Gradient** dialog box can also be invoked by entering **BHATCH** at the Command prompt.

EFFECT OF ANGLE AND SCALE FACTOR ON HATCH

When you hatch an area, you can alter the angle and displacement of hatch lines you have specified in the hatch pattern definition to get a desired hatch spacing. You can do this by entering an appropriate value for angle and scale factor in the **HATCH** command.

To understand how the angle and the displacement can be changed, hatch an area with the hatch pattern HATCH1 in Example 5. You will notice that the hatch lines have been generated according to the definition of hatch pattern HATCH1. Notice the effect of hatch angle and scale factor on the hatch. Figure 3-29(a) shows a hatch with a 0-degree angle and a scale factor of 1.0. If the angle is 0, the hatch will be generated with the same angle as defined in the hatch pattern definition (45-degree in Example 5). Similarly, if the scale factor is 1.0, the distance between the hatch lines will be the same as defined in the hatch pattern definition. Figure 3-29(b) shows a hatch that is generated when the hatch scale factor is 0.5. If you measure the distance between the successive hatch lines, it will be 0.5 x 0.5 = 0.25. Figures 3-29(c) and (d) show the hatch when the angle is 45-degree and the scale factors are 1.0 and 0.5, respectively.

Scale and Angle can also be set by entering **-HATCH** at the Command prompt.

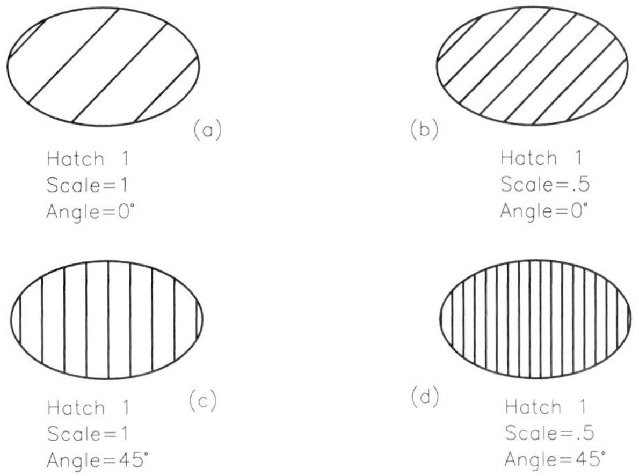

Figure 3-26 *Delta-X and Delta-Y of hatch lines*

HATCH PATTERN WITH DASHES AND DOTS

The lines you can use in a hatch pattern definition are not restricted to continuous lines. You can define any line pattern to generate a hatch pattern. The lines can be a combination of dashes, dots, and spaces in any configuration. However, the maximum number of dashes you can specify in the line pattern definition of a hatch pattern is six. Example 6 uses a dash-dot linetype to create a hatch pattern.

Example 6

Write a hatch pattern definition for the hatch pattern shown in Figure 3-30, with the following specifications. Define a new path say *C:\Program Files\Hatch1* and save the hatch pattern in that path.

Name of the hatch pattern	HATCH2
Hatch angle =	0

Chapter 3

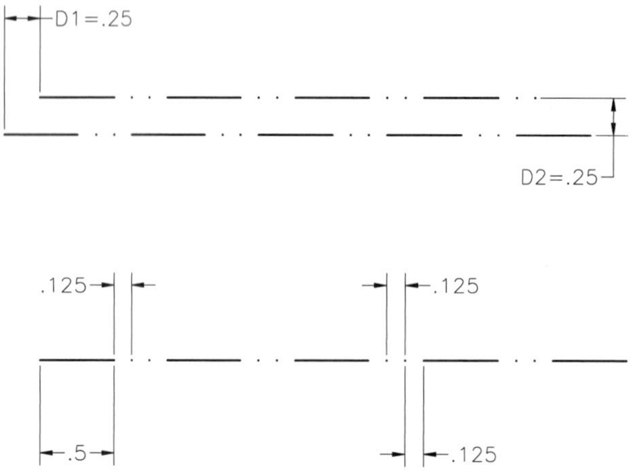

Figure 3-30 *Hatch lines made of dashes and dots*

X-origin =	0
Y-origin =	0
Displacement of lines (D1) =	0.25
Distance between lines (D2) =	0.25
Length of each dash =	0.5
Space between dashes and dots =	0.125
Space between dots =	0.125

Writing the definition of a hatch pattern

You can use the **EDIT** command or any text editor (Notepad) to edit the *acad.pat* file. The general format of the header line and the hatch descriptors is:

***HATCH NAME, Hatch Description**
Angle, X-Origin, Y-Origin, D1, D2 [,Dash Length.....]

Substitute the values from Example 6 in the corresponding fields of the header line and field descriptor:

*HATCH2,Hatch with dashes and dots
0,0,0,0.25,0.25,0.5,-0.125,0,-0.125,0,-0.125

Where **0** ------------------- Angle
0 ------------------- X-origin
0 ------------------- Y-origin
0.25 --------------- Delta-X
0.25 --------------- Delta-Y
0.5 ---------------- Length of dash
-0.125 ------------ Space (pen up)
0 ------------------- Dot (pen down)
-0.125 ------------ Space (pen up)

0 ------------------- Dot
-0.125 ------------ Space

Specifying a New Path for Hatch Pattern Files

When you enter a hatch pattern name for hatching, AutoCAD looks for that file name in the **Support** directory or the directory paths specified in the support file search path. You can specify a new path and directory to store your hatch files.

Create a new folder *Hatch1* in *C* drive under the *Program Files* folder. Save the *acad.pat* file with hatch pattern **HATCH2** definition in the same subdirectory, Hatch1. Right-click in the drawing area to activate the shortcut menu. Choose **Options** from the shortcut menu to display the **Options** dialog box. The **Options** dialog box can also be invoked by choosing the **Options** button from the **Application Menu** or by directly entering **OPTIONS** at the command prompt. Choose the **Files** tab in the **Options** dialog box to display the **Search paths, file names and file locations** area. Click on the **plus** sign of the **Support File Search Path** to display the different subdirectories of the **Support File Search Path**, as shown in Figure 3-31. Now, choose the **Add** button to display the space to add a new subdirectory. Enter the location of the new subdirectory, **C:\Program Files\Hatch1** or click on the **Browse** button to specify

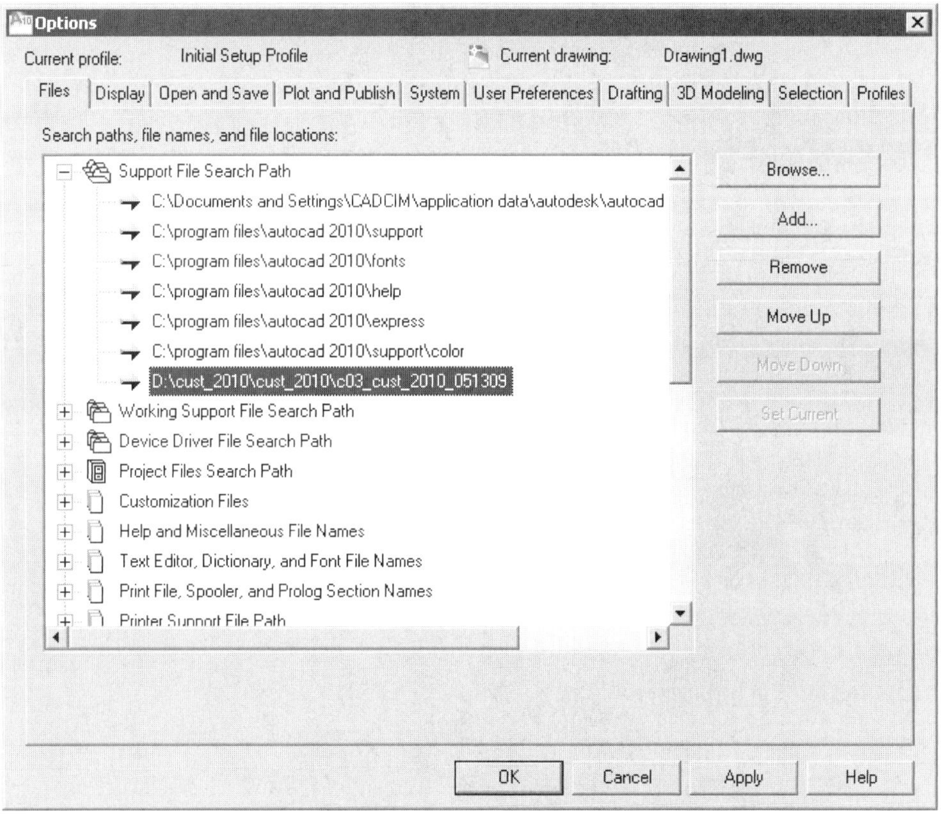

Figure 3-31 *The **Options** dialog box*

the path. Choose the **Apply** button and then choose **OK** to exit the dialog box. You have created a subdirectory and specified the search path for the hatch files.

Follow the procedure as described in Example 5 to activate the hatch pattern. The hatch thus generated is shown in Figure 3-32. Figure 3-32(a) shows the hatch with a 0-degree angle and a scale factor of 1.0. Figure 3-32(b) shows the hatch with a 45-degree angle and a scale factor of 0.5.

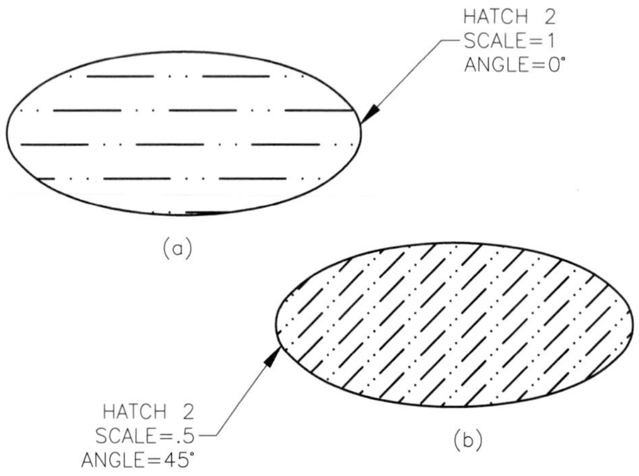

Figure 3-32 *Hatch pattern at different angles and scales*

HATCH WITH MULTIPLE DESCRIPTORS

Some hatch patterns require multiple lines to generate a shape. For example, if you want to create a hatch pattern of a brick wall, you need a hatch pattern that has four hatch descriptors to generate a rectangular shape. You can have any number of hatch descriptor lines in a hatch pattern definition. It is up to the user to combine them in any conceivable order. However, there are some shapes you cannot generate. A shape that has a nonlinear element, like an arc cannot be generated by hatch pattern definition. However, you can simulate an arc by defining short line segments because you can use only straight lines to generate a hatch pattern. Example 7 uses three lines to define a triangular hatch pattern.

Example 7

Write a hatch pattern definition for the hatch pattern shown in Figure 3-33, with the following specifications:

Name of the hatch pattern = HATCH3
Vertical height of the triangle = 0.5
Horizontal length of the triangle = 0.5
Vertical distance between the triangles = 0.5
Horizontal distance between the triangles = 0.5

Each triangle in this hatch pattern consists of the following three elements: a vertical line, a horizontal line, and a line inclined at 45-degree.

Step 1: Defining specifications for vertical line

For the vertical line, the specifications are (Figure 3-34):

Hatch angle =	90-degree
X-origin =	0
Y-origin =	0
Delta-X (D1) =	0
Delta-Y (D2) =	1.0
Dash length =	0.5
Space =	0.5

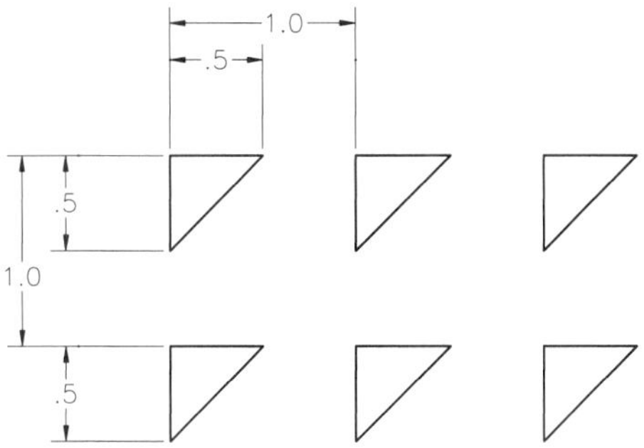

Figure 3-33 *Triangle hatch pattern*

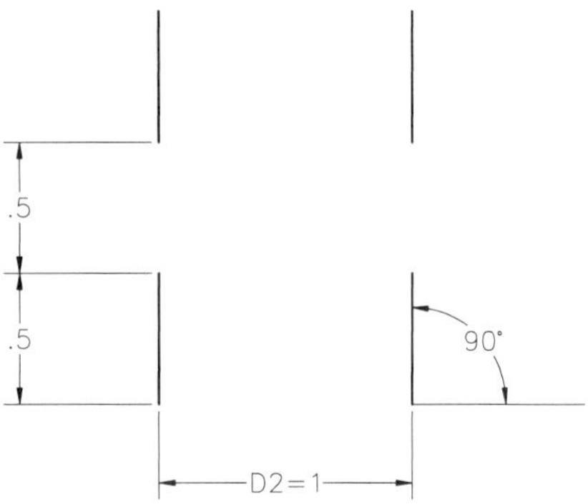

Figure 3-34 *Vertical line*

Substitute the values from the vertical line specification in various fields of the hatch descriptor to get the following line:

90,0,0,0,1,.5,-.5

Where **90** ----------------- Hatch angle
0 ------------------- X-origin
0 ------------------- Y-origin
0 ------------------- Delta-X
1 ------------------- Delta-Y
.5 ------------------ Dash (pen down)
-.5 ----------------- Space (pen up)

Step 2: Defining specifications of horizontal line

For the horizontal line (Figure 3-35), the specifications are:

Hatch angle =	0-degree
X-origin =	0
Y-origin =	0.5
Delta-X (D1) =	0
Delta-Y (D2) =	1.0
Dash length =	0.5
Space =	0.5

The only difference between the vertical line and the horizontal line is the angle. For the horizontal line, the angle is 0-degree, whereas for the vertical line, the angle is 90-degree. Substitute the values from the vertical line specification to obtain the following line:

0,0,0.5,0,1,.5,-.5

Where **0** ------------------- Hatch angle
0 ------------------- X-origin
0.5 ---------------- Y-origin
0 ------------------- Delta-X
1 ------------------- Delta-Y
.5 ------------------ Dash (pen down)
-.5 ----------------- Space (pen up)

Step 3: Defining specifications of the inclined line

This line is at an angle; therefore, you need to calculate the distances delta-X (D1) and delta-Y (D2), the length of the dashed line, and the length of the blank space. Figure 3-36 shows the calculations to find these values.

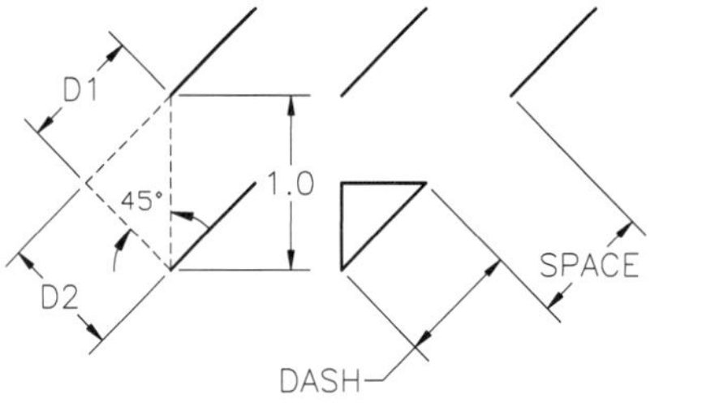

D1 = 1.0 x COS 45 D2 = 1.0 x SIN 45
D1 = .7071 D2 = .7071

DASH = SQRT(0.5^2 +0.5^2)
=.7071
SPACE = DASH =.7071

Figure 3-36 *Line inclined at 45-degree*

Hatch angle =	45-degree
X-Origin =	0
Y-Origin =	0
Delta-X (D1) =	0.7071
Delta-Y (D2) =	0.7071
Dash length =	0.7071
Space =	0.7071

After substituting the values in the general format of the hatch descriptor, you will obtain the following line:

45,0,0,.7071,.7071,.7071,-.7071

Where **45** ----------------- Hatch angle
0 ------------------- X-origin
0 ------------------- Y-origin
.7071 ------------- Delta-X
.7071 ------------- Delta-Y
.7071 ------------- Dash (pen down)
-.7071 ------------ Space (pen up)

Step 4: Loading the hatch pattern

Now you can combine the three lines and insert them at the end of the *acad.pat* file or you can enter the values in a separate hatch file and save it. You can also use the **EDIT** command to edit the file and insert the lines.

The following file is a partial listing of the *acad.pat* file, after adding the hatch pattern definitions from Examples 5, 6, and 7.

```
*SOLID, Solid fill
45, 0,0, 0,.125
*Angle,Angle steel
0, 0,0, 0,.275, .2,-.075
90, 0,0, 0,.275, .2,-.075
*Ansi31,ANSI Iron, Brick, Stone masonry
45, 0,0, 0,.125
*Ansi32,ANSI Steel
45, 0,0, 0,.375
45, .176776695,0, 0,.375
*Ansi33,ANSI Bronze, Brass, Copper
45, 0,0, 0,.25
45, .176776695,0, 0,.25, .125,-.0625
*Ansi34,ANSI Plastic, Rubber
45, 0,0, 0,.75
45, .176776695,0, 0,.75
45, .353553391,0, 0,.75
45, .530330086,0, 0,.75
```

```
*Ansi35,ANSI Fire brick, Refractory material
45, 0,0, 0,.25
45, .176776695,0, 0,.25, .3125,-.0625,0,-.0625
*Ansi36,ANSI Marble, Slate, Glass
45, 0,0, .21875,.125, .3125,-.0625,0,-.0625


*Swamp,Swampy area
0, 0,0, .5,.866025403, .125,-.875
90, .0625,0, .866025403,.5, .0625,-1.669550806
90, .078125,0, .866025403,.5, .05,-1.682050806
90, .046875,0, .866025403,.5, .05,-1.682050806
|
|
0, 0,0, .125,.125, .125,-.125
90, .125,0, .125,.125, .125,-.125
*HATCH1,Hatch at 45 Degree Angle
45,0,0,0,.5
*HATCH2,Hatch with Dashes & Dots:
0,0,0,.25,.25,0.5,-.125,0,-.125,0,-.125
*HATCH3,Triangle Hatch:
90,0,0,0,1,.5,-.5
0,0,0.5,0,1,.5,-.5
45,0,0,.7071,.7071,.7071,-.7071
```

Load the hatch pattern as described in Example 5 (Hatch3.pat) and test the hatch. Figure 3-37 shows the hatch pattern that will be generated by this hatch pattern (HATCH3). In Figure 3-37(a) the hatch pattern is at a 0-degree angle and the scale factor is 0.5. In Figure 3-37(b) the hatch pattern is at a -45-degree angle and the scale factor is 0.5.

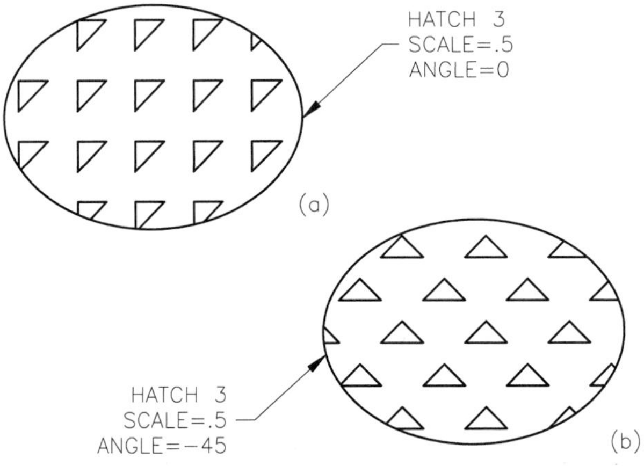

Figure 3-37 *Hatch generated by HATCH3 pattern*

SAVING HATCH PATTERNS IN A SEPARATE FILE

When you load a certain hatch pattern, AutoCAD looks for that definition in the *acad.pat* or *acadiso.pat* file. This is the reason, the hatch pattern definitions must be in that file. However, you can add the new pattern definition to a different file and then copy that file to *acad.pat* or *acadiso.pat*. Be sure to make a copy of the original *acad.pat* or *acadiso.pat* file so that you can copy that file back when needed. Assume the name of the file that contains your custom hatch pattern definitions is *customh.pat*.

1. Copy *acad.pat* or *acadiso.pat* file to *acadorg.pat*
2. Copy *customh.pat* to *acad.pat* or *acadiso.pat*

If you want to use the original hatch pattern file, copy the *acadorg.pat* file to *acad.pat* or *acadiso.pat*

CUSTOM HATCH PATTERN FILE

As mentioned earlier, you can add the new hatch pattern definitions to the *acad.pat* or *acadiso.pat* file. There is no limit to the number of hatch pattern definitions you can add to this file. However, if you have only one hatch pattern definition, you can define a separate file. It has the following requirements:

1. The name of the file has to be the same as the hatch pattern name.
2. The file can contain only one hatch pattern definition.
3. The hatch pattern name and the hatch file name should be unique.
4. If you quite often use the hatch patterns saved on the removable drive to hatch the drawings, you can add removable drive to the AutoCAD search path using the **Options** dialog box. AutoCAD will automatically search the file on the removable drive and will display it in the **Hatch and Gradient** dialog box.

```
*HATCH3,Triangle Hatch:
90,0,0,0,1,.5,-.5
0,0,0.5,0,1,.5,-.5
45,0,0,.7071,.7071,.7071,-.7071
```


Note

The hatch lines can be edited after exploding the hatch with the ***EXPLODE*** *command. After exploding, each hatch line becomes a separate object.*

However, it is recommended not to explode a hatch because it increases the size of the drawing database. For example, if a hatch consists of 100 lines, save it as a single object. However, after you explode the hatch, every line becomes a separate object and you have 99 additional objects in the drawing.

Keep the hatch lines in a separate layer to facilitate editing of the hatch lines.

Assign a unique color to hatch lines so that you can control the width of the hatch lines at the time of plotting.

Tip

1. The file or the subdirectory in which hatch patterns have been saved must be defined in the ***Support File Search Path*** *in the* ***File*** *tab of the* ***Options*** *dialog box.*

2. The hatch patterns that you create automatically get added to AutoCAD's slide library as an integral part of AutoCAD 2010 and are displayed in the ***Preview Area*** *in the* ***Hatch Pattern Palette*** *dialog box under the* ***Hatch and Gradient*** *dialog box. Hence there is no need to create a slide library.*

Self-Evaluation Test

Answer the following questions and then compare them to those given at the end of this chapter:

CREATING LINETYPES

1. The _________ command can be used to change the linetype scale factor.
2. The linetype description should not be more than _________ characters long.
3. A positive number denotes a pen _________ segment.
4. The segment length _________ generates a dot.
5. A negative number denotes a pen _________ segment.
6. The option_________ of the **LINETYPE** Command is used to generate a new linetype.
7. The description in the case of header line is _________. (optional/necessary)
8. The standard linetypes are stored in the _________ file.
9. The_________ determines the current linetype scaling.

CREATING HATCH PATTERNS

10. The header line consists of an asterisk, the pattern name, and _________.
11. The *acad.pat* or *acadiso.pat* file contains _________ number of hatch pattern definitions.
12. The standard hatch patterns are stored in the _________ file.

13. The first hatch line passes through a point whose coordinates are specified by _________ and _________.

Review Questions

CREATING LINETYPES

1. The _________ command can be used to create a new linetype.
2. The _________ command can be used to load a linetype.
3. In AutoCAD, the linetypes are saved in the _________ file.
4. AutoCAD supports only _________ alignment field specification.
5. A line pattern definition always starts with _________.
6. A header line definition always starts with _________.

CREATING HATCH PATTERNS

7. The perpendicular distance between the hatch lines in a hatch pattern definition is specified by _________.
8. The displacement of the second hatch line in a hatch pattern definition is specified by _________.
9. The maximum number of dash lengths that can be specified in the line pattern definition of a hatch pattern is _________.
10. The hatch lines in different areas of the drawing will automatically _________ because the hatch lines are generated by offsetting.
11. The hatch angle as defined in the hatch pattern definition can be changed further when you use the AutoCAD _________ command.
12. When you load a hatch pattern, AutoCAD looks for that hatch pattern in the _________ file.
13. The hatch lines can be edited after _________ the hatch by using the _________ command.

Exercises

CREATING LINETYPES

Exercise 1 — *General*

Using the **LINETYPE** command, create a new linetype "DASH3DASH" with the following specifications:

Length of the first dash 0.75
Blank space 0.125
Dash length 0.25
Blank space 0.125
Dash length 0.25
Blank space 0.125
Dash length 0.25
Blank space 0.125

Exercise 2 — *General*

Use a text editor to create a new file, *newlt2.lin*, and a new linetype, DASH2DASH, with the following specifications:

Length of the first dash 0.5
Blank space 0.1
Dash length 0.2
Blank space 0.1
Dash length 0.2
Blank space 0.1

Exercise 3 — *General*

a. Write the definition of a string complex linetype (hot water line), as shown in Figure 3-38(a). To determine the length of the HW text string, you should first draw the text (HW) using any text command and then measure its length.

b. Write the definition of a string complex linetype (gas line), as shown in Figure 3-38(b). Determine the length of the text string as mentioned in part a.

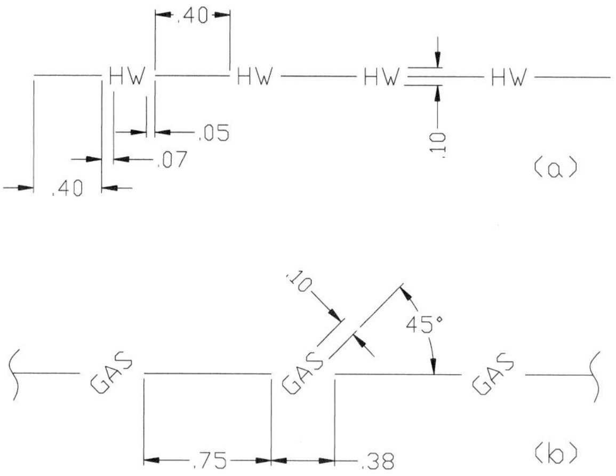

Figure 3-38 *Specifications for a string complex linetype*

Creating Hatch Patterns

Exercise 4 *General*

Determine the hatch pattern specifications and write a hatch pattern definition for the hatch pattern in Figure 3-39.

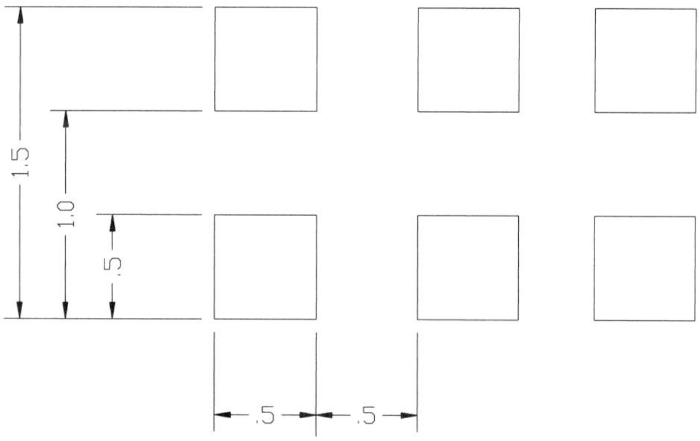

Figure 3-39 *Drawing for Exercise 4*

Exercise 5 — *General*

Determine the hatch pattern specifications and write a hatch pattern definition for the hatch pattern in Figure 3-40.

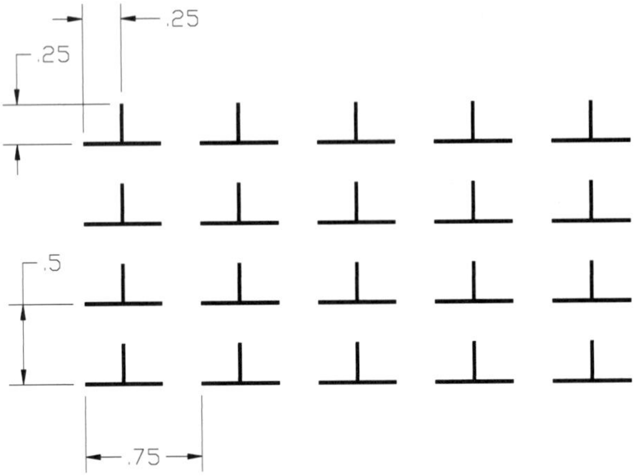

Figure 3-40 *Hatch pattern for Exercise 5*

Answers to Self-Evaluation Test

1. LTSCALE, **2.** 47, **3.** down, **4.** zero, **5.** up, **6. Create**, **7.** optional, **8.** *acad.lin* or *acadiso.lin*, **9. CELTSCALE**, **10.** Pattern Description, **11.** 67, **12.** *acad.pat* or *acadiso.pat*, **13.** X-origin, Y-origin.

Chapter 4

Customizing the ACAD.PGP File

Learning Objectives

After completing this chapter, you will be able to:

- *Customize the **ACAD.PGP** file.*
- *Edit different sections of the **ACAD.PGP** file.*
- *Abbreviate commands by defining command aliases.*
- *Use the **REINIT** command to reinitialize the PGP file.*

WHAT IS THE ACAD.PGP FILE?

AutoCAD software comes with the program parameters file *acad.pgp*, which defines aliases for the operating system commands and some of the AutoCAD commands. When you install AutoCAD on a computer that runs on Windows 2000, XP, or Vista operating system, this file is automatically copied on the *C:\Documents and Settings\User\Application Data\ Autodesk\AutoCAD 2010\R18.0\enu\Support* subdirectory of the hard drive. However, if the computer runs on Windows 98 operating system, the file will be copied on the *C:\Windows\Local Settings\Application Data\Autodesk\AutoCAD 2010\R18.0\enu\Support* directory. The *acad.pgp* file lets you access the operating system commands from the drawing editor. For example, if you want to delete a file, all you need to do is enter DEL at the Command prompt **(Command: DEL)**, and AutoCAD will prompt you to enter the name of the file that you want to delete.

The file also contains command aliases of some frequently used AutoCAD commands. For example, the command alias for the **LINE** command is L. This is the reason if you enter **L** at the Command prompt **(Command: L)**, AutoCAD will treat it as the **LINE** command. The *acad.pgp* file also contain comment lines that give you some information about different sections of the file. The *acad.pgp* file can be opened by choosing **Manage > Customization > Edit Aliases** button from the **Ribbon**. You can also open the *acad.pgp* file by choosing **Tools > Customize > Edit Program Parameters (acad.pgp)** option from the menu bar. The following file is a listing of the standard *acad.pgp* file. Some of the lines have been deleted to make the file shorter.

```
;
;
; Program Parameters File For AutoCAD 2010
; External Command and Command Alias Definitions

; Copyright (C) 1997-2009 by Autodesk, Inc.  All Rights Reserved.

; Each time you open a new or existing drawing, AutoCAD searches
; the support path and reads the first acad.pgp file that it finds.

; -- External Commands --
; While AutoCAD is running, you can invoke other programs or utilities
; such Windows system commands, utilities, and applications.
; You define external commands by specifying a command name to be used
; from the AutoCAD command prompt and an executable command string
; that is passed to the operating system.

; -- Command Aliases --
; The Command Aliases section of this file provides default settings for
; AutoCAD command shortcuts.  Note: It is not recommended that  you directly
; modify this section of the PGP file., as any changes you make to this section of the
; file will not migrate successfully if you upgrade your AutoCAD to a
; newer version.  Instead, make changes to the new
; User Defined Command Aliases
```

```
; section towards the end of this file.

; -- User Defined Command Aliases --
; You can abbreviate frequently used AutoCAD commands by defining
; aliases for them in the User Defined Command Aliases section of acad.pgp.
; You can create a command alias for any AutoCAD command,
; device driver command, or external command.

; Recommendation: back up this file before editing it. To ensure that
; any changes you make to PGP settings can successfully be migrated
; when you upgrade to the next version of AutoCAD, it is suggested that
; you make any changes to the default settings in the User Defined Command
; Aliases section at the end of this file.

; External command format:
; <Command name>,[<Shell request>],<Bit flag>,[*]<Prompt>,

; The bits of the bit flag have the following meanings:
; Bit 1: if set, don't wait for the application to finish
; Bit 2: if set, run the application minimized
; Bit 4: if set, run the application "hidden"
; Bit 8: if set, put the argument string in quotes
;
; Fill the "bit flag" field with the sum of the desired bits.
; Bits 2 and 4 are mutually exclusive; if both are specified, only
; the 2 bit is used. The most useful values are likely to be 0
; (start the application and wait for it to finish), 1 (start the
; application and don't wait), 3 (minimize and don't wait), and 5
; (hide and don't wait). Values of 2 and 4 should normally be avoided,
; as they make AutoCAD unavailable until the application has completed.
;
; Bit 8 allows commands like DEL to work properly with filenames that
; have spaces such as "long filename.dwg". Note that this will interfere
; with passing space delimited lists of file names to these same commands.
; If you prefer multiplefile support to using long file names, turn off
; the "8" bit in those commands.

; Examples of external commands for command windows

DEL,      DEL,      8,File to delete: ,
DIR,      DIR,      8,File specification: ,
SH,       ,         1,*OS Command: ,
SHELL,    ,         1,*OS Command: ,
START,    START,    1,*Application to start: ,
TYPE,     TYPE,     8,File to list: ,
```

External commands area

Chapter 4

```
; Examples of external commands for Windows
; See also the (STARTAPP) AutoLISP function for an alternative method.

EXPLORER,  START EXPLORER, 1,,
NOTEPAD,  START NOTEPAD,  1,*File to edit: ,
PBRUSH,   START PBRUSH,   1,,

; Command alias format:
;   <Alias>,*<Full command name>

;  The following are guidelines for creating new command aliases.
;  1. An alias should reduce a command by at least two characters.
;      Commands with a control key equivalent, status bar button,
;      or function key do not require a command alias.
;      Examples: Control N, O, P, and S for New, Open, Print, Save.
;  2. Try the first character of the command, then try the first two,
;      then the first three.
;  3. Once an alias is defined, add suffixes for related aliases:
;      Examples: R for Redraw, RA for Redrawall, L for Line, LT for
;      Linetype.
;  4. Use a hyphen to differentiate between command line and dialog
;      box commands.
;      Example: B for Block, -B for -Block.
;
; Exceptions to the rules include AA for Area, T for Mtext, X for Explode.

;  -- Sample aliases for AutoCAD commands --
;  These examples include most frequently used commands.  NOTE: It is recommended
;  that you not make any changes to this section of the PGP file to ensure the
;  proper migration of your customizations when you upgrade to the next version of
;  AutoCAD.  The aliases listed in this section are repeated in the User Custom
;  Settings section at the end of this file, which can safely be edited while
;  ensuring your changes will successfully migrate.

3A,              *3DARRAY
3DMIRROR,        *MIRROR3D
3DNavigate,      *3DWALK
3DO,             *3DORBIT
3DP,             *3DPRINT
3DPLOT,          *3DPRINT
3DW,             *3DWALK
3F,              *3DFACE
3M,              *3DMOVE
3P,              *3DPOLY
3R,              *3DROTATE
3S,              *3DSCALE
```

Command alias section

```
A,            *ARC
AC,           *BACTION
ADC,          *ADCENTER
AECTOACAD,    *-ExportToAutoCAD
AA,           *AREA
AL,           *ALIGN
3AL,          *3DALIGN
AP,           *APPLOAD
APLAY,        *ALLPLAY
AR,           *ARRAY
-AR,          *-ARRAY
ARR,          *ACTRECORD
ARM,          *ACTUSERMESSAGE
-ARM,         *-ACTUSERMESSAGE
ARU,          *ACTUSERINPUT
ARS,          *ACTSTOP
-ARS,         *-ACTSTOP
ATI,          *ATTIPEDIT
ATT,          *ATTDEF
-ATT,         *-ATTDEF
ATE,          *ATTEDIT
-ATE,         *-ATTEDIT
ATTE,         *-ATTEDIT
B,            *BLOCK
|
|
HE,           *HATCHEDIT
HI,           *HIDE
I,            *INSERT
-I,           *-INSERT
IAD,          *IMAGEADJUST
IAT,          *IMAGEATTACH
ICL,          *IMAGECLIP
IM,           *IMAGE
-IM,          *-IMAGE
IMP,          *IMPORT
IN,           *INTERSECT
INF,          *INTERFERE
IO,           *INSERTOBJ
QVD,          *QVDRAWING
QVDC,         *QVDRAWINGCLOSE
QVL,          *QVLAYOUT
QVLC,         *QVLAYOUTCLOSE
J,            *JOIN
```

```
L,             *LINE
LA,            *LAYER
-LA,           *-LAYER
LAS,           *LAYERSTATE
LE,            *QLEADER
LEN,           *LENGTHEN
LI,            *LIST
LINEWEIGHT,    *LWEIGHT
LMAN,          *LAYERSTATE
LO,            *-LAYOUT
LS,            *LIST
LT,            *LINETYPE
-LT,           *-LINETYPE
LTYPE,         *LINETYPE
-LTYPE,        *-LINETYPE
LTS,           *LTSCALE
LW,            *LWEIGHT
M,             *MOVE
MA,            *MATCHPROP
MAT,           *MATERIALS
ME,            *MEASURE
MI,            *MIRROR
ML,            *MLINE
MLA,           *MLEADERALIGN
MLC,           *MLEADERCOLLECT
MLD,           *MLEADER
|
|
R,             *REDRAW
RA,            *REDRAWALL
RC,            *RENDERCROP
RE,            *REGEN
REA,           *REGENALL
REC,           *RECTANG
REG,           *REGION
REN,           *RENAME
-REN,          *-RENAME
REV,           *REVOLVE
RO,            *ROTATE
|
|
; The following are alternative aliases and aliases as supplied
;  in AutoCAD Release 13.
```

```
AV,                     *DSVIEWER
CP,                     *COPY
DIMALI,                 *DIMALIGNED
DIMANG,                 *DIMANGULAR
DIMBASE,                *DIMBASELINE
DIMCONT,                *DIMCONTINUE
DIMDIA,                 *DIMDIAMETER
DIMED,                  *DIMEDIT
DIMTED,                 *DIMTEDIT
DIMLIN,                 *DIMLINEAR
DIMORD,                 *DIMORDINATE
DIMRAD,                 *DIMRADIUS
DIMSTY,                 *DIMSTYLE
DIMOVER,                *DIMOVERRIDE
LEAD,                   *LEADER
TM,                     *TILEMODE

; Aliases for Hyperlink/URL Release 14 compatibility
SAVEURL,                *SAVE
OPENURL,                *OPEN
INSERTURL,              *INSERT

; Aliases for commands discontinued in AutoCAD 2000:
AAD,                    *DBCONNECT
AEX,                    *DBCONNECT
ALI,                    *DBCONNECT
ASQ,                    *DBCONNECT
ARO,                    *DBCONNECT
ASE,                    *DBCONNECT
DDATTDEF,               *ATTDEF
DDATTEXT,               *ATTEXT
DDCHPROP,               *PROPERTIES
DDCOLOR,                *COLOR
DDLMODES,               *LAYER
DDLTYPE,                *LINETYPE
DDMODIFY,               *PROPERTIES
DDOSNAP,                *OSNAP
DDUCS,                  *UCS

; Aliases for commands discontinued in AutoCAD 2004:
ACADBLOCKDIALOG,        *BLOCK
ACADWBLOCKDIALOG,       *WBLOCK
ADCENTER,               *ADCENTER
BMAKE,                  *BLOCK
BMOD,                   *BLOCK
BPOLY,                  *BOUNDARY
```

```
PAINTER,            *MATCHPROP
PREFERENCES,        *OPTIONS
RECTANGLE,          *RECTANG
SHADE,              *SHADEMODE
VIEWPORTS,          *VPORTS

; Aliases for commands discontinued in AutoCAD 2007:
RMAT,               *MATERIALS
FOG,                *RENDERENVIRONMENT
FINISH,             *MATERIALS
SETUV,              *MATERIALMAP
SHOWMAT,            *LIST
RFILEOPT,           *RENDERPRESETS
RENDSCR,            *RENDERWIN

; Aliases for commands discontinued in AutoCAD 2009:
DASHBOARD,          *RIBBON
DASHBOARDCLOSE,     *RIBBONCLOSE

; -- User Defined Command Aliases --
; Make any changes or additions to the default AutoCAD command aliases in
; this section to ensure successful migration of these settings when you
; upgrade to the next version of AutoCAD. If a command alias appears more
; than once in this file, items in the User Defined Command Alias take
; precedence over duplicates that appear earlier in the file.
; **********----------**********  ; No xlate ; DO NOT REMOVE
```

SECTIONS OF THE ACAD.PGP FILE

The contents of the AutoCAD program parameters file (*acad.pgp*) can be categorized into three sections based on the information that is defined in the *acad.pgp* file. They do not appear in any definite order in the file, and have no section headings. For example, the comment lines can be entered anywhere in the file; the same is true with external commands and AutoCAD command aliases. The *acad.pgp* file can be divided into three sections: comments, external commands, and command aliases.

Comments

The comments of *acad.pgp* file can contain any number of comment lines and can occur anywhere in the file. Every comment line must start with a semicolon (;) (This is a comment line). Any line that is preceded by a semicolon is ignored by AutoCAD. You should use the comment line to give some relevant information about the file that will help other AutoCAD users to understand, edit, or update the file.

External Command

In the external command section, you can define any valid external command that is supported by your system. The information must be entered in the following format:

<Command name>, [OS Command name],<Bit flag>, [*]<Command prompt>,

Command Name. This is the name you want to use to activate the external command from the AutoCAD drawing editor. For example, you can use **GOWORD** as a command name to load the Word program (Command: **GOWORD**). The command name must not be an AutoCAD command name or an AutoCAD system variable name. If the name is an AutoCAD command name, the command name in the **PGP** file will be ignored. Also, if the name is an AutoCAD system variable name, the system variable will be ignored. You should use the command names that reflect the expected result of the external commands. (For example, **HELLO** is not a good command name for a directory file.) The command names can be uppercase or lowercase.

OS Command Name. The OS Command name is the name of a valid system command that is supported by an operating system. For example, in DOS, the command to delete files is DEL and therefore, the OS Command name used in the *acad.pgp* file must be DEL. The following is a list of the type of commands that can be used in the PGP file:

OS Commands (DEL, DIR, TYPE, COPY, RENAME, EDLIN, etc.)
Commands for starting a word processor, or text editors (WORD, SHELL, etc.)
Name of the user-defined programs and batch files

Bit Flag. This field must contain a number, preferably 8 or 1. The following are the bit flag values and their meaning:

Bit flag set to	Meaning
0	Start the application and wait for it to finish
1	Do not wait for the application to finish
2	Runs the application minimized
4	Runs the application hidden
8	Puts the argument string in quotes

Command Prompt. The Command prompt field of the command line contains the prompt you want to display on the screen. It is an optional field that must be replaced by a comma if there is no prompt. If the operating system (OS) command that you want to use contains spaces, the prompt must be preceded by an asterisk (*). For example, the DOS command **EDIT NEW.PGP** contains a space between EDIT and NEW; therefore, the prompt used in this command line must be preceded by an asterisk. The command can be terminated by pressing ENTER. If the OS command consists of a single word (DIR, DEL, TYPE), the preceding asterisk must be omitted. In this case, you can terminate the command by pressing the SPACEBAR or the ENTER key.

Command Aliases

It is time-consuming to enter AutoCAD commands at the keyboard because it requires typing the complete command name before pressing ENTER. AutoCAD provides a facility that can be used to abbreviate the commands by defining aliases for the commands. This is made possible by the AutoCAD program parameters file (*acad.pgp* file). Each command alias line consists of two fields **(L, *LINE)**. The first field **(L)** defines the alias of the command; the second field **(*LINE)** consists of the AutoCAD command. The command must be preceded by an asterisk for AutoCAD to recognize the command line as a command alias. The two fields must be separated by a comma. The blank lines and the spaces between the two fields are ignored. In addition to AutoCAD commands, you can also use aliases for AutoLISP command names, provided the programs that contain the definition of these commands are loaded.

Example 1

Add the following external commands and AutoCAD command aliases to the AutoCAD program parameters file (*acad.pgp*).

External Commands

Abbreviation	Command Description
GOWORD	This command loads the word processor (Winword) program from the C:\Program Files\Winword.
RN	This command executes the rename command of DOS.
COP	This command executes the copy command of DOS.

Command Aliases Section

Abbreviation	Command	Abbreviation	Command
EL	Ellipse	T	Trim
CO	Copy	CH	Chamfer
O	Offset	ST	Stretch
S	Scale	MI	Mirror

The *acad.pgp* file is an ASCII text file. To edit this file, choose **Manage > Customization > Edit Aliases** button from the **Ribbon**. You can also use text editor like notepad to edit the *acad.pgp* file. The following is a partial listing of the *acad.pgp* file after insertion of the lines for the command aliases of Example 1. **The line numbers are not a part of the file; they are shown here for reference only.** The lines that have been added to the file are highlighted in bold.

```
DEL,DEL,              8,File to delete: ,
DIR,DIR,              8,File specification ,
EDIT, START EDIT,     8,File to edit: ,
SH,,                  1,*OS Command: ,
SHELL,,               1,*OS Command: ,
```

```
6  START,START,          1,Application to start: ,
7  TYPE,TYPE,            8,File to list: ,
8
9  GOWORD, START WINWORD,1,,
10 RN, RENAME,8,File to rename:,
11 COP,COPY,8,File to copy:,
12 DIMLIN        *DIMLINEAR
13 DIMORD,       *DIMORDINATE
14 DIMRAD,       *DIMRADIUS
15 DIMSTY,       *DIMSTYLE
16 DIMOVER,      *DIMOVERRIDE
17 LEAD,         *LEADER
18 TM,           *TILEMODE
19 EL,           *ELLIPSE
20 CO,           *COPY
21 O,            *OFFSET
22 S,            *SCALE
23 MI,           *MIRROR
24 ST,           *STRETCH
```

Explanation

Line 9
GOWORD, START WINWORD,1,,
In line 9, **GOWORD** loads the word processor program **(WINWORD)**. The **winword.exe** program is located in the winword directory under Program Files.

Lines 10 and 11
RN, RENAME,8,File to rename:,
COP,COPY,8,File to copy:,
Line 10 defines the alias for the DOS command **RENAME,** and the next line defines the alias for the DOS command **COPY**. The **8** is a bit flag, and the Command prompt **File to rename and File to copy** is automatically displayed to let you know the format and the type of information that is expected.

Lines 19 and 20
EL, *ELLIPSE
CO, *COPY
Line 19 defines the alias **(EL)** for the **ELLIPSE** command, and the next line defines the alias **(CO)** for the **COPY** command. The commands must be preceded by an asterisk. You can put any number of spaces between the alias abbreviation and the command.

Note

*If a command alias definition duplicates an existing one then the one that is lower down in the file is given preference and is allowed to work. For example, in standard file, if you add S, *SCALE to the end of the file then your definition works and the one higher up in the file is ignored.*

REINITIALIZING THE ACAD.PGP FILE

When you make any changes in the *acad.pgp* file, there are two ways to reinitialize the *acad.pgp* file. One is to quit AutoCAD and then restart the AutoCAD program. When you start AutoCAD, the *acad.pgp* file is automatically loaded. You can also reinitialize the *acad.pgp* file by using the **REINIT** command. The **REINIT** command lets you reinitialize the I/O ports, digitizer, and AutoCAD program parameters file, *acad.pgp*. When you enter the **REINIT** command, AutoCAD will display the **Re-initialization** dialog box (Figure 4-1). To reinitialize the *acad.pgp* file, select the corresponding toggle box, and then choose **OK**. AutoCAD will reinitialize the program parameters file (*acad.pgp*), and then you can use the command aliases defined in the file.

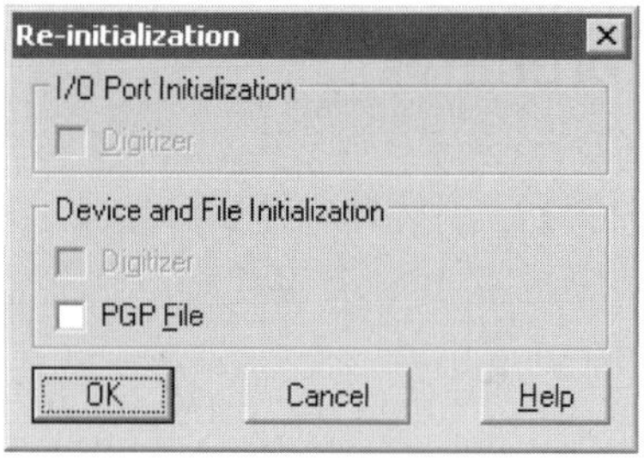

Figure 4-1 *The **Re-initialization** dialog box*

Tip

Copy and save the original acad.pgp file at some other location, so that, after you have made changes in the acad.pgp file and used it, you can restore the default settings by copying and pasting the original acad.pgp file. This lets other users use the original, unedited file.

Self-Evaluation Test

Answer the following questions and then compare them to those given at the end of this chapter:

1. One way of reinitializing the *acad.pgp* file is to ______________ AutoCAD and then ______________ it.
2. The command used to reinitialize the *acad.pgp* file is ______________ .
3. In the command alias section, the AutoCAD command must be preceded by an__________ symbol.
4. Every comment line must start with a ______________ .
5. The command alias and the AutoCAD command must be separated by a___________.

Review Questions

Indicate whether the following statements are true or false.

1. The comment section can contain any number of lines. (T/F)
2. AutoCAD ignores any line that is preceded by a semicolon. (T/F)
3. The command alias must not be an AutoCAD command. (T/F)
4. The bit flag field must contain 8. (T/F)
5. In the command alias section, the command alias must be preceded by a semicolon. (T/F)
6. You cannot use aliases for AutoLISP commands. (T/F)
7. The *acad.pgp* file does not come with AutoCAD software. (T/F)
8. The *acad.pgp* file is an ASCII file. (T/F)

Exercise

Exercise 1 *General*

Add the following external commands and AutoCAD command aliases to the AutoCAD program parameters file (*acad.pgp*).

External Command Section

Abbreviation	Command Description
MYWORDPAD	This command loads the WORDPAD program that resides in the **Program Files\Accessories** directory.
MYEXCEL	This command loads the EXCEL program that resides in the **Program Fies\Microsoft Office** directory.
CD	This command executes the CHKDSK command of DOS.
FORMAT	This command executes the FORMAT command of DOS.

Command Aliases Section

Abbreviation	Command	Abbreviation	Command
BL	**BLOCK**	LTS	**LTSCALE**
INS	**INSERT**	EXP	**EXPLODE**
DIS	**DISTANCE**	GR	**GRID**
TE	**TIME**		

Answers to the Self-Evaluation Test

1. quit, restart, **2. REINIT**, **3.** asterisk, **4.** semicolon, **5.** comma.

Chapter 5

Customizing Menus and Toolbars

Learning Objectives

After completing this chapter, you will be able to:

- *Create new pull-down menus.*
- *Load menus.*
- *Create cascading submenus in pull-down menus.*
- *Write customization files for creating menus.*
- *Create customized toolbars.*
- *Write menus to create toolbars.*
- *Create customized shortcut menus.*

CUSTOMIZING USER INTERFACE

AutoCAD allows you to customize its user interface to your requirements, thus allowing you to efficiently use this program and reduce the design time. For example, you may be using AutoCAD to perform certain type of drawing and editing actions. To do this more efficiently, you can customize the AutoCAD user interface such that all those commands are placed together for easy access. This saves the design time and also makes the entire process more user-friendly.

Under customizing user interface, you can customize menus, toolbars, shortcut menus, **Ribbon**, image tile menus, and tablet menus. These are also termed as user interface elements. In this chapter, you will learn to customize pull-down menus, shortcut menus, and toolbars.

The AutoCAD menu provides a powerful tool to customize AutoCAD. The AutoCAD software package comes with a standard user interface file named *acad.CUI*. When you start AutoCAD, the *acad.CUI* file is automatically loaded. This file contains AutoCAD commands, separated under various user interface elements for easy identification. For example, all drawing commands are under the **Draw** panel of the **Ribbon** or under the **Draw** menu and toolbar, while all editing commands are under the **Modify** panel of the **Ribbon** or under the **Modify** menu and toolbar. The **Ribbon**, menu, and toolbar headings are named and arranged to make it easier for you to locate and access the commands, as shown in Figure 5-1. However, there are some commands that you may never use. Also, some users might like to regroup and rearrange the commands so that it is easier to access the most frequently used commands. This can be easily done by customizing the user interface.

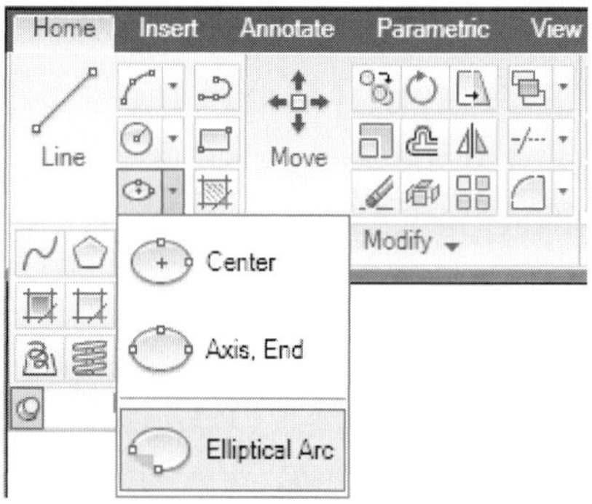

Figure 5-1 *Tabs, panels, and options in the* ***Ribbon***

Note

In AutoCAD 2010, the menu bar and toolbars are not displayed by default. To display the menu bar, left-click on the down arrow displayed on the extreme right of the ***Quick Access*** *Toolbar; a flyout will be displayed. Choose the* ***Show Menu Bar*** *option from this flyout; the menu bar will be displayed above the* ***Ribbon*** *in the AutoCAD user interface. To invoke the desired toolbar, choose the respective option from* ***Tools > Toolbars > AutoCAD*** *in the menu bar. However, if you want to work only with the menu bar and toolbars, you can switch your workspace to* ***AutoCAD Classic*** *workspace.*

CHANGES FROM THE EARLIER RELEASES

In earlier releases of AutoCAD, you had to manually write each additional menu file manually using a text editor and save it with *.mnu/mns* file formats. This was a very tedious and time consuming process and required a very high level of understanding of the customization process. However, from few recent releases, the AutoCAD customization has been made extremely simple and a person with no experience of customization can also create customized menus and toolbars. The customization of menus and other user interface elements is done using the **Customize User Interface** dialog box, shown in Figure 5-2. This dialog box can be invoked by choosing **Manage > Customization > User Interface** from the **Ribbon**. You can also invoke the **Customize User Interface** dialog box by right-clicking on any toolbar and choosing **Customize** from the shortcut menu. Alternatively, use the **CUI** command to invoke this dialog box.

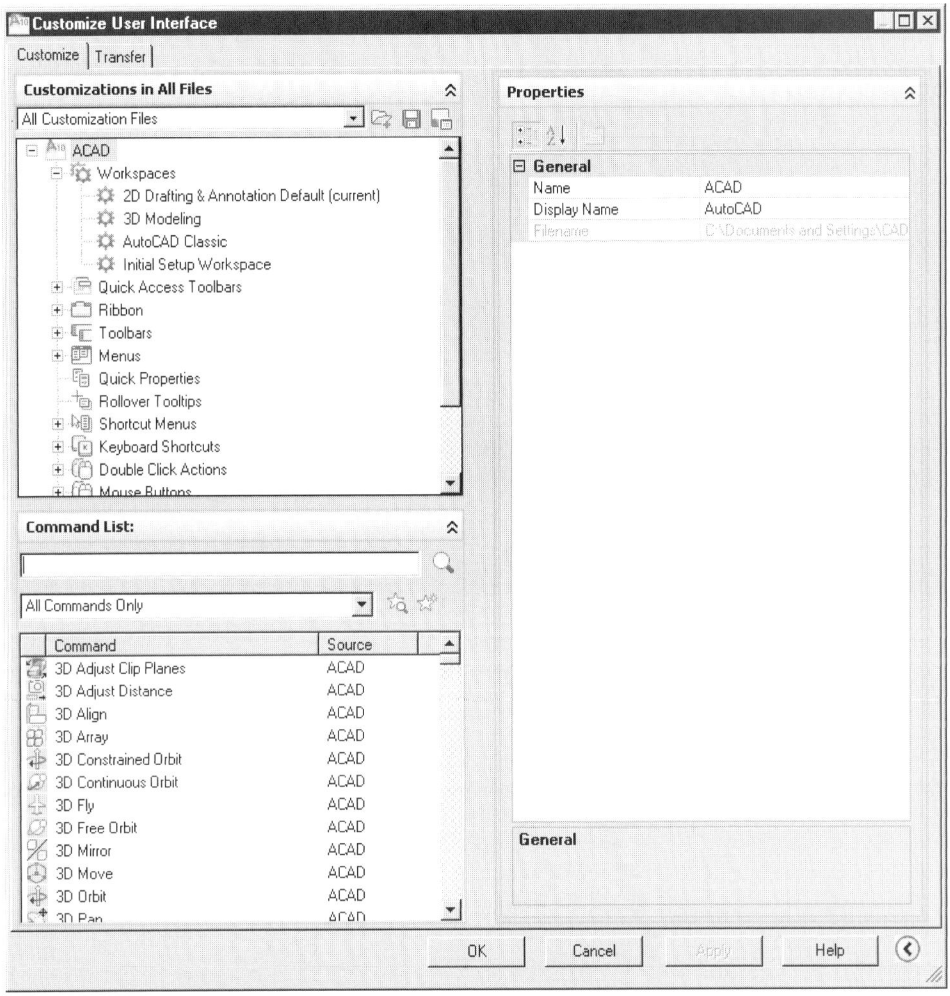

*Figure 5-2 The **Customize User Interface** dialog box*

CREATING A CUSTOMIZED MENU

Before you create a menu, you need to design it and arrange the commands the way you want them to appear on the screen. To design a menu, you should select and arrange the commands in a way that provides easy access to the most frequently used commands. A careful design saves a lot of time in the long run. Therefore, consider several possible designs with different command combinations, and then select the one best suited for the job. Suggestions from other CAD operators can prove very valuable.

To understand the process of developing a pull-down menu, consider the following example.

Example 1

Create a pull-down menu for the following AutoCAD commands and load them with the default menu items:

MYDRAW	**MYEDIT**	**MYDISPLAY**	**MYUTILITY**
LINE	**ERASE**	**REDRAW**	**SAVE**
PLINE	**MOVE**	**REGEN**	**EXIT**
CIRCLE C,R	**COPY**	**ZOOM ALL**	**PLOT**
CIRCLE C,D	**STRETCH**	**ZOOM WIN**	
CIRCLE 2P	**EXTEND**	**ZOOM PRE**	
CIRCLE 3P	**OFFSET**		

Step 1: Designing the menu

The first step in creating any menu is to design it in such a way that the commands are arranged in the desired configuration. Figure 5-3 shows one of the possible designs of the given menu.

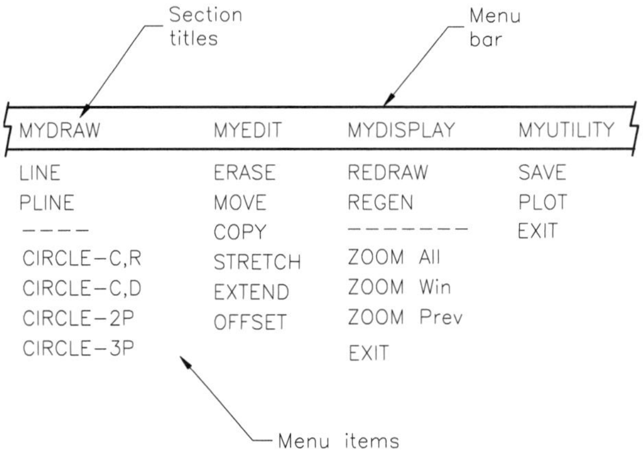

***Figure 5-3** Design of menu for Example 1*

You will notice that the **MYDRAW** and **MYDISPLAY** menus have dashed lines. These lines designate the separators that you need to insert between the menu items to separate the groups.

Step 2: Creating the menu

The customized menus are created using the **Customize User Interface** dialog box. The following steps explain the procedure of creating this menu.

1. Invoke the **Customize User Interface** dialog box by choosing **Manage > Customization >**

User Interface from the **Ribbon**. Alternatively, choose **Tools > Customize > Interface** from the menu bar to invoke this dialog box. By default, this dialog box has two tabs and each tab has two panes: the left pane and the right pane. The left pane of the **Customize** tab has two rollouts: **Customizations in All Files** and **Command List**.

2. The **Customizations in All Files** rollout displays the list of all user interface elements in a tree view. Click on the (+) sign located on the left of **Menus** to expand the tree view.

3. To add a new menu, right-click on the **Menus** element in this rollout and choose **New Menu** from the shortcut menu; a new menu with the name **Menu1** is added to the list and its properties are displayed in the right pane. Rename the menu to **MYDRAW**.

   Next, you need to add the draw commands to this menu. This is done by dragging the commands from the **Command List** rollout in the left pane.

4. Select **Draw** from the **Filter the command list by category** drop-down list in the **Command List** rollout; all draw commands are displayed in the list box.

5. Scroll to the **Line** item in the list box available in the **Command List** rollout. Drag this item and drop it on the **MYDRAW** menu, as shown in Figure 5-4; the **Line** command is added to the menu.

6. Next, drag **Polyline** from the **Command List** rollout and drop it on **MYDRAW**. Remember that if you drop it just below the menu name, it will be placed above **Line**. In that case, you will have to drag **Polyline** below **Line** to maintain the sequence of commands.

   As per the menu design shown in Figure 5-3, there is a separator below **Pline** in **MYDRAW** menu. Therefore, you need to insert a separator next.

7. Right-click on **MYDRAW** and choose **Insert Separator** from the shortcut menu, as shown in Figure 5-5.

8. Next, drag **Circle, Center, Radius** from the list box in the **Command List** rollout and drop it on **MYDRAW**; the command is added below the separator in the menu.

9. Similarly, drag and drop **Circle, Center, Diameter**; **Circle, 2 Points**; and **Circle, 3 Points** from the list box in the **Command List** rollout and drop it on **MYDRAW** to add these commands to the menu.

10. Again, right-click on **Menus** in the **Customizations in All Files** rollout and choose **New Menu** from the shortcut menu.

11. Rename the new menu as **MYEDIT**.

12. Select **Edit** from the **Filter the command list by category** drop-down list in the **Command List** rollout and drag and drop **Erase** on **MYEDIT**; this tool will be added to the menu.

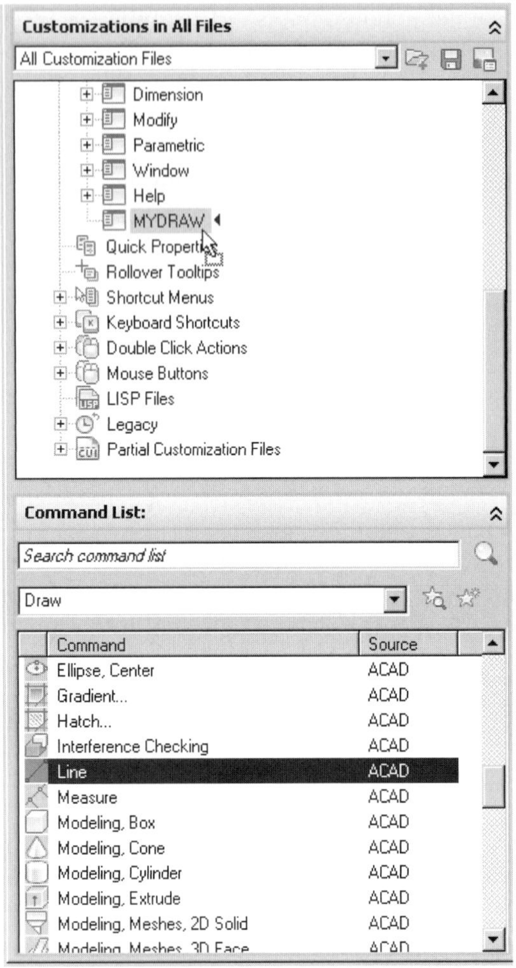

__Figure 5-4__ Dragging and adding a command to the menu

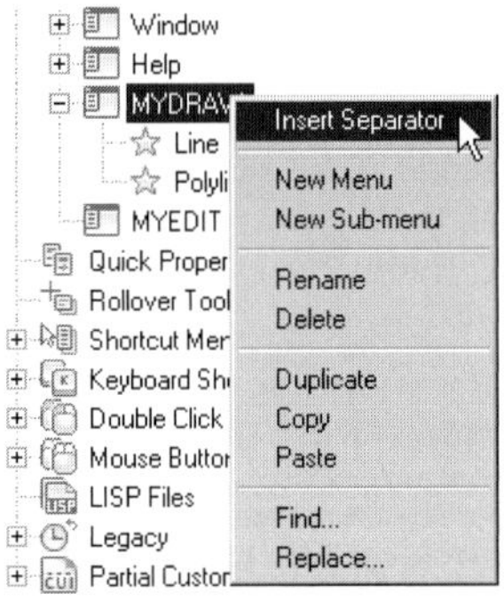

__Figure 5-5__ Choosing option from shortcut menu to insert a separator

13. Select **Modify** from the **Filter the command list by category** drop-down list in the **Command List** rollout.

14. Drag the required commands from the list box and drop them on **MYEDIT**. Make sure you drag and drop the commands in the sequence in which they are listed in Figure 5-3.

15. Similarly, create the remaining menus and add the required commands to them. You can select **All Commands and Controls** from the **Filter the command list by category** drop-down list. This will list all commands and you can easily select all the required commands for the menus.

16. Choose **Apply** and then choose **OK**. The menus will be listed at the end of the default menus in the menu bar, as shown in Figures 5-6 and 5-7.

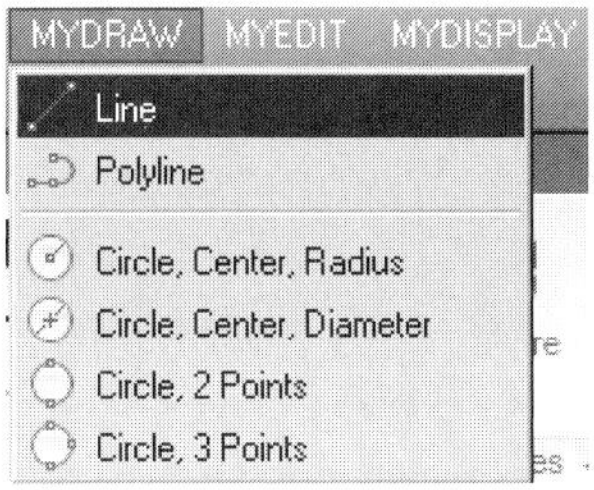

***Figure 5-6** Customized menus for Example 1*

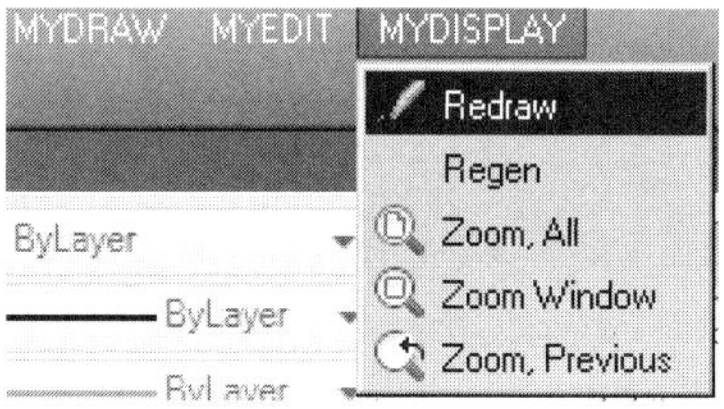

***Figure 5-7** Customized menus for Example 1*

ADDING NONSTANDARD COMMANDS TO MENUS

Sometimes, you may need to add nonstandard commands to menus. The nonstandard commands may include the commands that are not listed by default in the **Command List** rollout of the **Customize User Interface** dialog box. The example of such commands include the **TEXT** command with top-center justification. For such commands, you need to create a new command and then write its macros in the **Macro** field of the **Properties** rollout that is displayed in the right pane of the **Customize** tab of the **Customize User Interface** dialog box when you select a command from the left pane. For writing menus, you need to know the exact prompt sequence associated with the command for which you want to write macros. The following are the prompt sequence of some of the commands:

AutoCAD Command

Command: **TEXT**
Current text style: "Standard" Text height: 0.0000 Annotative: No
Specify start point of text or [Justify/Style]: J
Enter an option [Align/Fit/Center/Middle/Right/TL/TC/TR/ML/MC/MR/BL/BC/BR]: **TC**
Specify top-center point of text:

Corresponding Macro

^C^CText J TC

Where -------------------- **^C^C** Cancels the existing command twice
TEXT ------------ AutoCAD **TEXT** command
J TC ------------ Space J Space TC to invoke the **Justification** and **Top Center** options, respectively.

Example 2

In this example, you will add the **TEXT** command to write a top-centered justified text to the **MYUTILITY** menu created in Example 1. The menu item should be added at the top and a separator should be inserted after the item.

Creating the menu

1. Invoke the **Customize User Interface** dialog box by choosing **Manage > Customization > User Interface** from the **Ribbon**. Alternatively, choose **Tools > Customize > Interface** from the menu bar to invoke this dialog box.

2. Expand the **Menus** tree view and then expand the **MYUTILITY** tree view.

3. Choose the **Create a new command** button on the upper right corner of the **Command List** rollout; a command with the name **Command1** is added to the list of commands. The **Properties** rollout in the right pane displays the properties of the new command.

4. Modify the name of the new command in the **Properties** rollout to **Text, TC**.

5. In the **Description** field, enter **Writes top-center justified text**.

6. Now, in the **Macro** field, enter **^C^CText J TC**.

7. Drag and drop the new command on top of the **Save** item in the **MYUTILITY** menu. Right-click on **Text, TC** item in the menu and choose **Insert Separator** to insert a separator below this item.

8. Choose **Apply** and then choose **OK** to add this item to the menu. Figure 5-8 shows the menu after adding this command. When you invoke this command, you will be prompted to specify the top-center point of the text. Specify the top center point and enter the text.

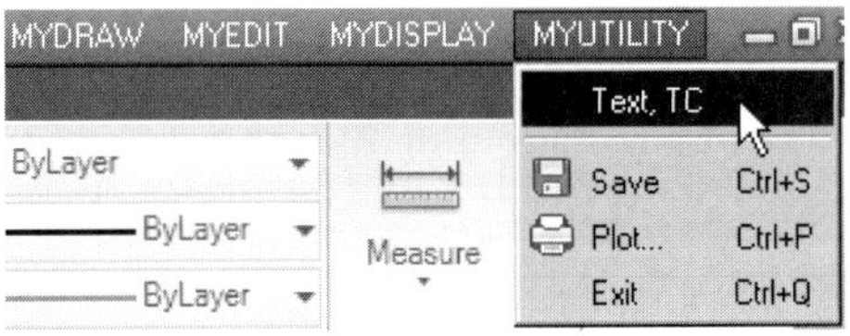

Figure 5-8 *Menu after adding a customized command*

PARTIAL CUSTOMIZATION FILES

The customized menus by default are added to the default customization file, which is *acad.CUI* file. As a result, the customized menus will be displayed in the AutoCAD environment every time, even when they are not required. To avoid this, you can export the customized menus as a partial customization file and load them only when needed. After you load the partial customization file, AutoCAD lets you use the menu with the standard AutoCAD menu. These features make it convenient to use the menus that have been developed by AutoCAD users and developers.

Transferring User Interface Elements

You can import or export some customization elements from one customization file to a another using the **Transfer** tab of the **Customize User Interface** dialog box. In the following example, you will learn to export the menus as a partial customization file.

Example 3

In this example, you will export the menus created in Examples 1 and 2 to a new *.CUI* file. Name the new CUI file as *example3.CUI*.

Exporting the menu

1. Invoke the **Customize User Interface** dialog box by choosing **Manage > Customization > User Interface** from the **Ribbon**. Alternatively, choose **Tools > Customize > Interface** from the menu bar to invoke this dialog box.

2. Choose the **Transfer** tab. This tab also has two panes. The left pane shows the user interface elements of the default CUI file and the right pane shows the headings under a new CUI file.

3. Click on the (+) sign located on the left of **Menus** in the new customization file in the right pane. You will notice that there are no items under this heading.

4. Click on the (+) sign located on the left of the **Menus** in the main customization file in the left pane. Select the **MYDRAW** menu. Now, hold the SHIFT key and select the **MYEDIT**, **MYDISPLAY**, and **MYUTILITY** menus.

5. Release the SHIFT key. Next, drag the selected menus to the **Menus** item heading in the right pane; the selected menus are displayed in the right pane also.

6. Right-click on the selected menus in the left pane and delete them to make sure they are removed from the default customization file.

7. Choose the **Save the current customization file** button, which is the right most button in the right pane; the **Save As** dialog box will be displayed.

8. Save the customization file with the name *example3.cui* in the */c05* folder. This saves the customized menus in a separate customization file.

9. Choose **Apply** and then choose the **OK** button from the **Customize User Interface** dialog box.

Loading the Partial Customization File

You can load partial customization files using the **Customize User Interface** dialog box or the **CUILOAD** command. Both these methods of loading the partial customization files are discussed next.

Loading Partial Customization Files Using the Customize User Interface Dialog Box

To load the partial customization file using this method, invoke the **Customize User Interface** dialog box. Choose the **Load partial customization file** button in the **Customizations in All Files** rollout in the left pane of the **Customize** tab; the **Open** dialog box is displayed. Select the customization file and then choose **Open**; the partial customization file is loaded and displayed in the list box of the upper rollout in the left pane of the **Customize User Interface** dialog box. When you choose **OK**, the menus available in the partial customization file are automatically loaded in the menu bar and displayed along with the default AutoCAD menu.

Loading Partial Customization Files Using the CUILOAD Command

To load the partial customization file using this method, invoke the **CUILOAD** command; the **Load/Unload Customizations** dialog box will be displayed, as shown in Figure 5-9.

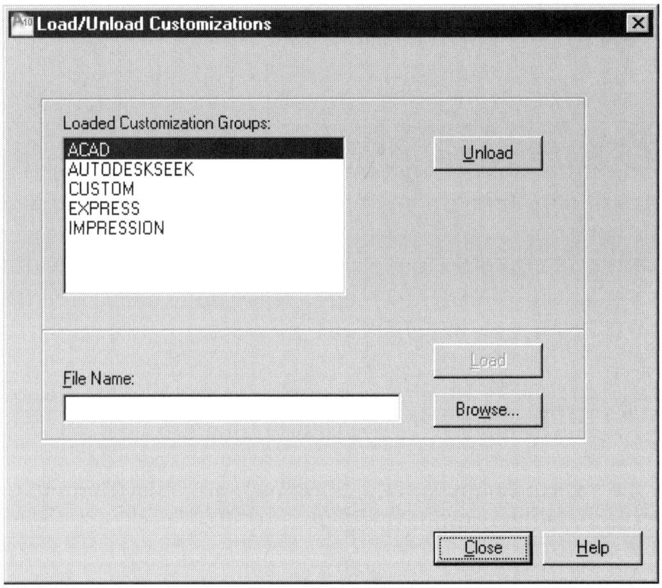

*Figure 5-9 The **Load/Unload Customizations** dialog box*

The default menu customization groups will be displayed in the **Loaded Customization Groups** area. Choose the **Browse** button; the **Select Customization File** dialog box will be displayed. Select the customization file and then choose **Open**; the name and path of the file will be displayed in the **File Name** edit box. Choose the **Load** button; the menu group will now be displayed in the **Loaded Customization Groups** area and the menus in the selected file will be displayed in the menu bar.

Example 4

In this example, you will load the partial customization file created in Example 3. You will use the **Customize User Interface** dialog box to load the file.

Loading the Partial CUI file

1. Invoke the **Customize User Interface** dialog box and choose the **Load partial customization file** button from the **Customizations in All Files** rollout in the left pane of the **Customize** tab; the **Open** dialog box is displayed.

2. Browse to the */c05* folder and select the *example3.cui* file. Next, choose the **Open** button.

3. The partial customization file is loaded and displayed in the list box of the upper rollout in the left pane of the **Customize** dialog box, as shown in Figure 5-10.

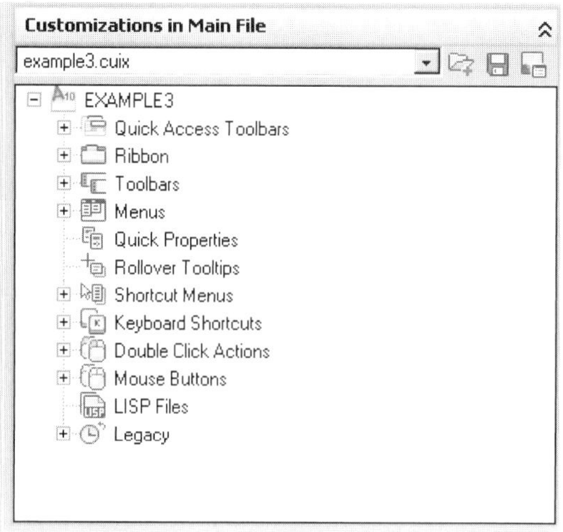

***Figure 5-10** The **Customizations in Main File** rollout displaying the partial cui file*

4. Choose the **Apply** button; the menus available in the partial customization file are automatically loaded in the menu bar and are displayed along with the default AutoCAD menu.

Exercise 1 *General*

Create a menu for the following AutoCAD commands. The menu design is shown in Figure 5-11. After creating the menu, save it as a partial customization file with the name **Exercise1.CUI**. Next, load this file using the **CUILOAD** command.

<u>DRAW</u>	<u>EDIT</u>	<u>DISP/TEXT</u>	<u>UTILITY</u>
LINE	MOVE	TEXT, MC	SAVE
PLINE	FILLET	TEXT,ML	EXIT
ELLIPSE	CHAMFER	TEXT, MR	PLOT
POLYGON	STRETCH	ZOOM WIN	
DONUT	EXTEND	ZOOM PRE	

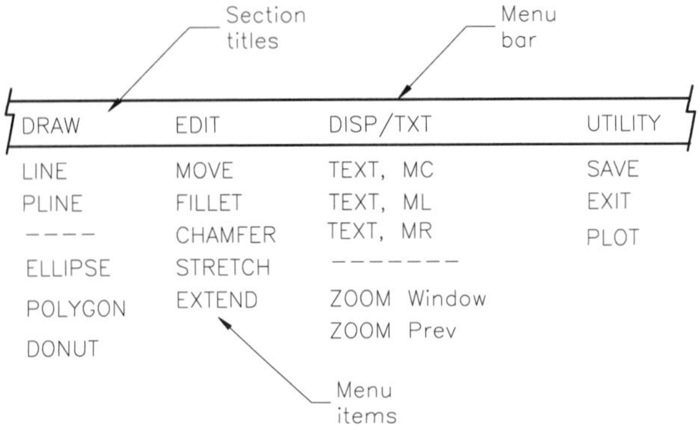

Figure 5-11 *Menu display for Exercise 1*

CASCADING SUBMENUS IN MENUS

The number of items in a menu or cursor menu can be very large, and sometimes they cannot all be accommodated on one screen. For example, consider a display device that can display a maximum of 21 menu items. If the pull-down menu or the cursor menu has more items than can be displayed, the excess menu items are not displayed on the screen and cannot be accessed. You can overcome this problem by using cascading menus that let you define smaller groups of items within a menu section. When an item is selected, it loads the cascading menu and displays the items, defined in the cascading menu, on the screen.

The cascading feature of AutoCAD allows pull-down and cursor menus to be displayed in a hierarchical order that makes it easier to select submenus.

The method of creating cascading submenus is similar to that of creating menus. The only difference is that to add a cascading submenu, you need to right-click on the new menu and choose **New Sub-menu** from the menu bar, as shown in Figure 5-12. Now, add commands to the cascading menu using the method similar to that used to add items to the menu.

Remember that if you are adding commands to a partial customization file, all AutoCAD commands will not be available in the **Commands List** rollout. Therefore, it is recommended that you create the menu and cascading submenu in the main AutoCAD menu and then export them as partial customization files.

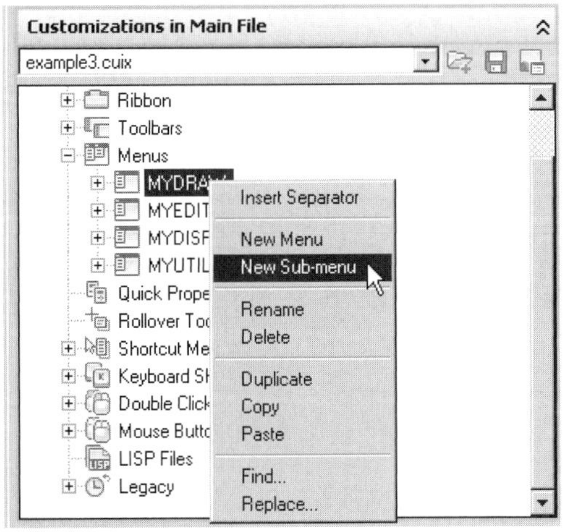

Figure 5-12 *Inserting a cascading submenu in a customized menu*

WRITING A MENU FILE

AutoCAD also allows you to manually write a menu file and then import it into the AutoCAD environment using the **Customize User Interface** dialog box. The menu files are written in a text editor such as Notepad and are saved with the *.mns* extension. The menu can have 499 sections, named as POP1, POP2, POP3, . . ., POP499. One of the most important things in developing a menu is to know the exact sequence of the commands and the prompts associated with each command. To better determine the prompt entries required in a command, you should enter all the commands and the prompt entries.

The following is a description of some of the main AutoCAD commands and the prompt entries required:

<u>LINE Command</u>

Command: **LINE**

Specify first point:

LINE
<RETURN>

<u>CIRCLE (C,R) Command</u>

Command: **CIRCLE**
Specify center point for circle or [3P/2P/Ttr (tan tan radius)]: *Specify center point.*
Specify radius of circle or [Diameter]: *Enter radius.*

Notice the command and input sequence:

CIRCLE
<RETURN>
Center point
<RETURN>
Radius
<RETURN>

<u>CIRCLE (C,D) Command</u>

Command: **CIRCLE**
Specify center point for circle or [3P/2P/Ttr (tan tan radius)]: *Specify center point.*
Specify radius of circle or [Diameter]: **D**
Specify diameter of circle: *Enter diameter.*

Notice the command and input sequence:

CIRCLE
<RETURN>
Center Point
<RETURN>
D
<RETURN>
Diameter
<RETURN>

CIRCLE (2P) Command

Command: **CIRCLE**
Specify center point for circle or [3P/2P/Ttr (tan tan radius)]: **2P**
Specify first end point of circle's diameter: *Specify first point.*
Specify second end point of circle's diameter: *Specify second point.*

Notice the command and input sequence:

CIRCLE
<RETURN>
2P
<RETURN>
Select first point on diameter
<RETURN>
Select second point on diameter
<RETURN>

ERASE Command

Command: **ERASE**

Notice the command and input sequence:

ERASE
<RETURN>

MOVE Command

Command: **MOVE**

Notice the command and prompt entry sequence:

MOVE
<RETURN>

The difference between the **Center-Radius** and **Center-Diameter** options of the **CIRCLE** command is that in the first one, you are prompted to enter the radius value by default. But in the second one, you need to enter **D** to use the **Diameter** option. This difference, although minor, is very important when writing a menu file. Similarly, the 2P (two-point) option of the **CIRCLE** command is different from the other two options. Therefore, it is important to know both the correct sequence of the AutoCAD commands and the entries made in response to the prompts associated with those commands.

As mentioned earlier, you can use any text editor (like Notepad) to write the file. You can also use the **NOTEPAD** command to write the menu file. If you use the **NOTEPAD** command, AutoCAD will prompt you to enter the file name you want to edit. If the file name exists, it will automatically be loaded; otherwise a new file will be created. After writing the file, you need to save it in the *.mnu* format and then create a *.mns* file using the *.mnu* file.

Before creating the cascading submenus, you need to understand some special characters that can be used in the name of the menu items. These characters are given below:

Character	Character Description
$(	This label character can be used with the pull-down menus to evaluate a DIESEL expression. The character must precede the label item. Example: **$(**if,**$(**getvar,orthomode),Ortho)
~	This item indicates that the label item is not available (displayed shaded); the character must precede the item. Example: [~Application not available]
!.	When used as a prefix, it displays the item with a check mark.
&	When placed directly before a character, the character is displayed underscored. For example, [W&Block] is displayed as W<u>B</u>lock. It also specifies the character as a menu accelerator key in the pull-down or Shortcut menu.
\t	This label character is used to push the text written after it to the right of the menu. For example, Line\tL will display Line on the left of the menu and L on the right of the menu. This is generally used to list the shortcut key to invoke a particular command.
->	This label character defines a cascaded submenu; it must precede the name of the submenu. Example: [**->Draw**]

<- This label character designates the last item of the cascaded pull-down menu. The character must precede the label item. Example: **[<-CIRCLE 3P]^C^CCIRCLE;3P**

The following example explains the process of writing *.mnu* file and then creating *.mns* file using it. Also, you will learn how to import these files in AutoCAD 2010.

Example 5

In this example, you will write a pull-down menu for the commands shown in Figure 5-13. After creating the menu, you will import it using the **Customize User Interface** dialog box.

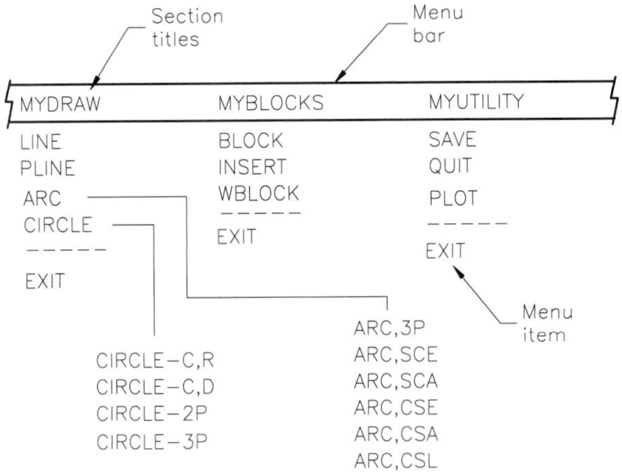

Figure 5-13 *Menu structure for Example 5*

Step 1: Writing the menu file

The following file is a listing of the menu for Example 5. **The line numbers are not a part of the menu; they are shown here for reference only.**

```
***POP1
[MYDRAW]
[LINE]^C^CLINE
[PLINE]^C^CPLINE
[->ARC]
 [ARC,3P]^C^CARC
 [ARC,SCE]^C^CARC;\C
 [ARC,SCA]^C^CARC;\C;\A
 [ARC,CSE]^C^CARC;C
 [ARC,CSA]^C^CARC;C;\\A
 [<-ARC,CSL]^C^CARC;C;\\L
[->CIRCLE]
```

```
13  [CIRCLE C,R]^C^CCIRCLE
14  [CIRCLE C,D]^C^CCIRCLE;\D
15  [CIRCLE 2P]^C^CCIRCLE;2P
16  [<-CIRCLE 3P]^C^CCIRCLE;3P
17  [--]
18  [Exit]^C
19  ***POP2
20  [MYBLOCKS]
21  [BLOCK]^C^CBLOCK
22  [INSERT]^C^CINSERT
23  [WBLOCK]^C^CWBLOCK
24  [--]
25  [Exit]^C
26  ***POP3
27  [MYUTILITY]
28  [SAVE]^C^CSAVE
29  [QUIT]^C^CQUIT
30  [PLOT]^C^CPLOT
31  [--]
32  [Exit]^C
```

Explanation

Line 5

[->ARC]

In this menu item, **ARC** is the menu item label that is preceded by the special label characters, **->**. These special characters indicate that the menu item has a submenu. The menu items that follow it (Lines 6-11) are the submenu items.

Line 11

[<-ARC,CSL]^C^CARC;C;\\L

In this line, the menu item label **ARC,CSL** is preceded by another special label characters, **<-**, which indicates the end of the submenu. The item that contains these characters must be the last menu item of the submenu.

Lines 12 and 16

[->CIRCLE]

[<-CIRCLE 3P]^C^CCIRCLE;3P

The special characters -> in front of **CIRCLE** indicate that the menu item has a submenu; the characters <- in front of **CIRCLE 3P** indicate that this item is the last menu item in the submenu. When you select the menu item **CIRCLE** from the menu, it will automatically display the submenu on the side.

Step 2: Saving the menu file

Save the menu file with the name *example5.mnu*.

Next, you need to create the *.mns* file using this *.mnu* file. The *.mns* file is similar to the *.mnu* file, with the only difference is that in the *.mns* file, you need to add the name and location of the *.mnu* file in the following format above the menu program.

```
//
//    AutoCAD menu file - Path where the menu file is saved, along with the name of file
//

***MENUGROUP=Path where the menu file is saved, along with the name of file
```

You can use the **Save As** option to save the *.mnu* file as *.mns* file and then add the above-mentioned lines at the top of the program. For the convenience of locating the file, it is recommended that the name of the *.mns* file should be the same as that of the *.mnu* file. The following would be the content of the *.mns* file for this example.

```
//
//    AutoCAD menu file - C:\customizing-2010\c05\example5.mnu
//

***MENUGROUP=C:\customizing-2010\c05\example5.mnu

***POP1
[MYDRAW]
[LINE]^C^CLINE
[PLINE]^C^CPLINE
[->ARC]
  [ARC,3P]^C^CARC;\\
  [ARC,SCE]^C^CARC;\C
  [ARC,SCA]^C^CARC;\C;\A
  [ARC,CSE]^C^CARC;C;\\
  [ARC,CSA]^C^CARC;C;\\A
  [<-ARC,CSL]^C^CARC;C;\\L
[->CIRCLE]
  [CIRCLE C,R]^C^CCIRCLE
  [CIRCLE C,D]^C^CCIRCLE;\D
  [CIRCLE 2P]^C^CCIRCLE;2P
  [<-CIRCLE 3P]^C^CCIRCLE;3P
[--]
[Exit]^C
***POP2
[BLOCKS]
[BLOCK]^C^CBLOCK
[INSERT]*^C^CINSERT
[WBLOCK]^C^CWBLOCK
[--]
[Exit]^C
```

```
***POP3
[UTILITY]
[SAVE]^C^CSAVE
[QUIT]^C^CQUIT
[PLOT]^C^CPLOT
[--]
[Exit]^C
```

Here, C:\customizing-2010\c05\example5.mnu is the path where the menu file was saved.

Save this file with the name *example5.mns* in the same folder where the *.mnu* file was saved.

Step 3: Loading the menu file as a partial customization file

In the earlier releases of AutoCAD, the *.mnu* or *.mns* files were used as it is. But from AutoCAD 2006 onward, both the *.mnu* files and *.mns* files were converted to *.cui* file. The *.cui* files are XML based files that can be edited easily using the **Customize User Interface** dialog box. While loading a *.mns* file in AutoCAD, it is automatically converted to *.cui* file. The process of loading the *.mns* files is discussed next.

1. Invoke the **Customize User Interface** dialog box.

2. In the **Customizations in All Files** rollout of the **Customize** tab, choose the **Load partial customization file** button on the right of the drop-down list.

3. In the **Open** dialog box, select **Legacy menu files (*.mnu,*.mnu)** from the **Files of type** drop-down list and double-click on the *example5.mns* file.

4. Choose **OK** from the **Warning** dialog box if displayed; the *.mns* file is loaded and is automatically converted into a *.cuix* file. Its name will be displayed as *example5.cuix* in the drop-down list of the **Customization in Main File** rollout. Choose the **Apply** button and then choose **OK**. The menus will be added to the menu bar. Alternatively, you can use the **CUILOAD** command to load the menus using the *.mns* files.

   Figure 5-14 shows the cascading menu created for Example 5.

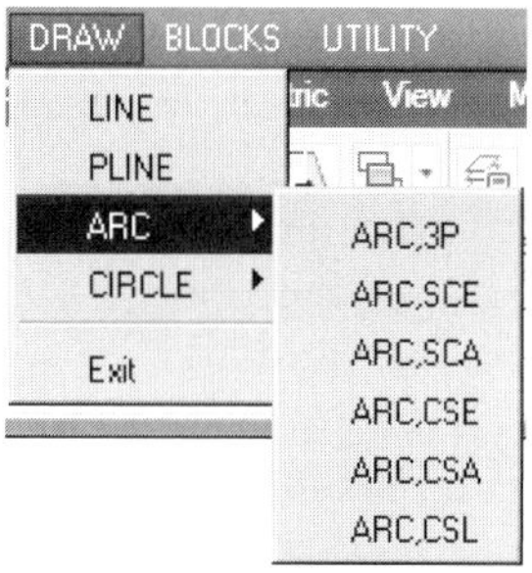

***Figure 5-14** The **ARC** cascading submenu for Example 5*

Example 6

Write a menu that has the cascading submenus for the commands shown in Figure 5-15.

Writing the menu file

The following file is a listing of the menu for Example 6. **The line numbers are not a part of the menu; they are shown here for reference only.**

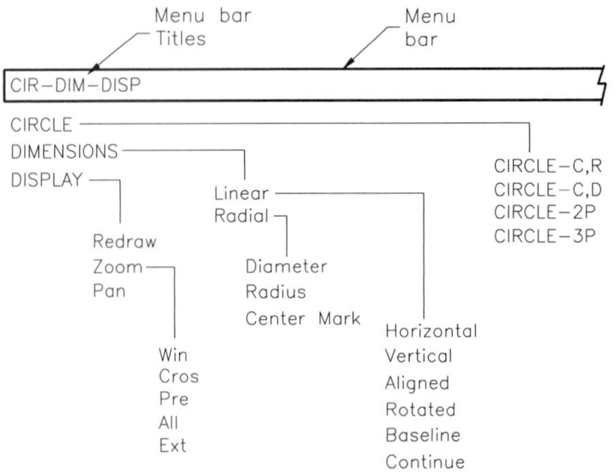

Figure 5-15 *Menu structure for Example 6*

```
***POP1
[CIR-DIM-DISP]
[->CIRCLE]
  [CIRCLE C,R]^C^C_CIRCLE
  [CIRCLE C,D]^C^C_CIRCLE;\_D
  [CIRCLE 2P]^C^C_CIRCLE;_2P
  [<-CIRCLE 3P]^C^C_CIRCLE;_3P
[->DIMENSIONS]
   [->Linear]
   [Horizontal]^C^C_dim;_horizontal
   [Vertical]^C^C_dim;_vertical
    [Aligned]^C^C_dim;_aligned
   [Rotated]^C^C_dim;_rotated
   [Baseline]^C^C_dim;_baseline
   [<-Continue]^C^C_dim;_continue
    [->Radial]
  [Diameter]^C^C_dimdiameter
   [Radius]^C^C_dimradius
   [<-<-Center Mark]^C^C_dimcenter
[->DISPLAY]
    [REDRAW]^C^CREDRAW
    [->ZOOM]
    [Win]^C^C_ZOOM;_W
[Cros]^C^C_ZOOM;_C
    [Pre]^C^C_ZOOM;_P
    [All]^C^C_ZOOM;_A
    [<-Ext]^C^C_ZOOM;_E
  [<-PAN]^C^C_Pan
```

Explanation

Lines 8 and 9

[->DIMENSIONS]

[->Linear]

The special label characters **->** in front of the menu item **DIMENSIONS** indicate that it has a submenu, and the characters -> in front of **Linear** indicate that there is another submenu. The second submenu **Linear** is within the first submenu **DIMENSIONS**, as shown in Figure 5-16. The menu items on lines 10 to 15 are defined in the **Linear** submenu, and the menu items **Linear** and **Radial** are defined in the submenu Dimensions.

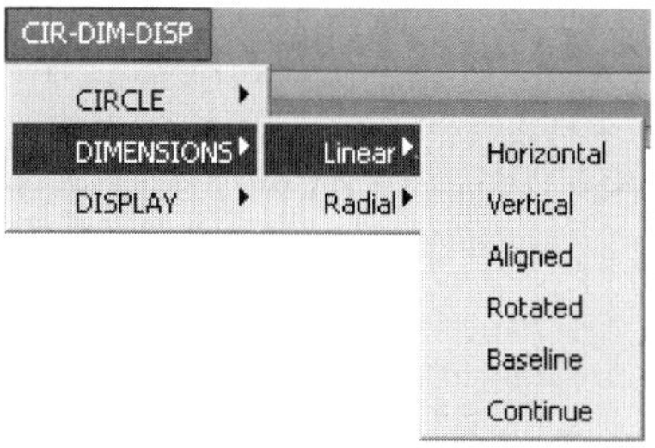

Figure 5-16 *The* ***Linear*** *cascading submenu in the* ***DIMENSIONS*** *submenu for Example 6*

Line 16

[->Radial]

This menu item defines another submenu; the menu items on line numbers 17, 18, and 19 are part of this submenu.

Line 19

[<-<-Center Mark]^C^C_dim;_center

In this menu item, the special label characters **<-<-** terminate the **Radial** and **Linear** dimensions (parent submenu) submenus.

Lines 27 and 28

[<-Ext]^C^C_ZOOM;_E

[<-PAN]^C^C_Pan

The special characters **<-** in front of the menu item **Ext** terminate the **ZOOM** submenu; the special character in front of the menu item **PAN** terminates the **DISPLAY** submenu.

Saving the menu file

Save the menu file with the name *example6.mnu*.

The following is the *.mns* file for this example.

```
//
//    AutoCAD menu file - C:\customizing-2010\c05\example6.mnu
//

***MENUGROUP=C:\customizing-2010\c05\example6.mnu

***POP1
[CIR-DIM-DISP]
[->CIRCLE]
  [CIRCLE-C,R]^C^C_CIRCLE
  [CIRCLE-C,D]^C^C_CIRCLE;\_D
  [CIRCLE-2P]^C^C_CIRCLE;_2P
```

```
  [<-CIRCLE-3P]^C^C_CIRCLE;_3P
[->DIMENSIONS]
  [->Linear]
    [Horizontal]^C^C_dim;_horizontal
    [Vertical]^C^C_dim;_vertical
    [Aligned]^C^C_dim;_aligned
    [Rotated]^C^C_dim;_rotated
    [Baseline]^C^C_dim;_baseline
    [<-Continue]^C^C_dim;_continue
  [->Radial]
   [Diameter]^C^C_dim;_diameter
   [Radius]^C^C_dim;_radius
   [<-<-Center Mark]^C^C_dim;_center
[->DISPLAY]
  [REDRAW]^C^_REDRAW
  [->ZOOM]
    [Win]^C^C_ZOOM;_W
   [Cros]^C^C_ZOOM;_C
   [Pre]^C^C_ZOOM;_P
    [All]^C^C_ZOOM;_A
   [<-Ext]^C^C_ZOOM;_E
  [<-PAN]^C^C_Pan
```

Save this file with the name *example6.mns* and then open it using the **CUILOAD** command.

CUSTOMIZING TOOLBARS

AutoCAD has provided several toolbars that should be sufficient for general use. However, sometimes you may need to customize the toolbars so that the commands that you use frequently are grouped in one toolbar. This saves time in selecting commands. It also saves the drawing space because you do not need to have several toolbars on the screen. The following example explains the process involved in creating and editing the toolbars.

Example 7

In this example, you will create a new toolbar (**MyToolbar1**) that has **Line**, **Polyline**, **Circle** (**Center Radius** option), **Arc** (**Center, Start, End** option), **Spline**, and **MTEXT** commands.

Creating the toolbar

1. Invoke the **Customize User Interface** dialog box. In the **Customizations in All Files** rollout, right-click on **Toolbars** node and choose **New Toolbar** from the shortcut menu.

2. Rename the new toolbar **MyToolbar1** and then drag the commands mentioned above to the new toolbar. You may have to drag the items in the toolbar to get the sequence mentioned in the example statement. Figure 5-17 shows the partial view of the **Customize User Interface** dialog box in the new toolbar.

3. Choose **Apply** and then choose **OK** to display this toolbar. Figure 5-18 shows this toolbar.

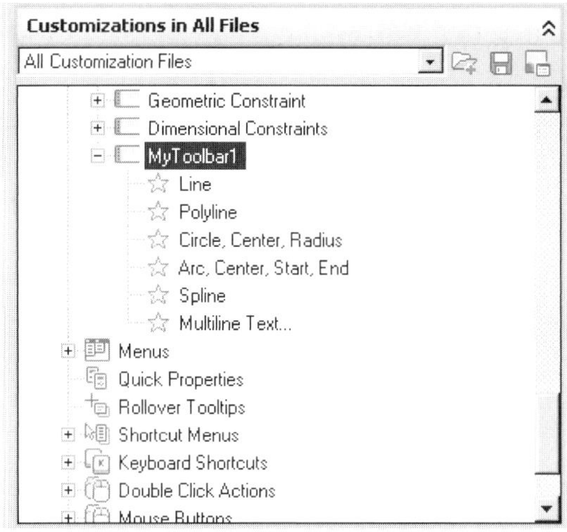

***Figure 5-17** Partial view of the **Customize User Interface** dialog box in the new toolbar*

***Figure 5-18** The **MyToolbar1** toolbar*

Adding Flyout in Toolbars

If the number of tools in a toolbar is very large, then you can group the similar type of tools in a flyout. The buttons that are grouped in a flyout can be identified by a triangular mark at their lower right corner. You can right-click on this triangular mark to access the flyout.

The method of creating a flyout is similar to that of creating a toolbar. The only difference is that to add a flyout, you need to right-click on any command added under that toolbar and choose the **New Flyout** option from the shortcut menu; a new flyout will be added under the selected command. Now, add commands to the new flyout using the method similar to the one used for adding items to the toolbar.

Note

When you add a flyout to the toolbar, a new toolbar with the same name will also be added automatically to the toolbar list. All items added to the new flyout will automatically get added to the new toolbar. Now you can also use this flyout as a separate toolbar.

Example 8

In this example, you will create a new toolbar (**MyToolbar2**) that has **Line**, **Circle** (**Center Radius** option), **Arc** (**3 Points** option), and **MTEXT** commands. Also, add a flyout under the **Arc** (**3 Points** option) that has **Arc** (**Center, Start, End** option), **Arc** (**Center, Start, Angle** option), **Arc** (**Center, Start, Length** option), and **Arc** (**Continue** option) commands.

Creating the toolbar

1. Invoke the **Customize User Interface** dialog box. In the **Customizations in All Files** rollout, right-click on the **Toolbars** node and choose **New Toolbar** from the shortcut menu.

2. Rename the new toolbar to **MyToolbar2** and then drag the **Line**, **Circle** (**Center Radius** option), **Arc** (**3 Points** option), and **Multiline Text** commands to the new toolbar. You may have to drag the items in the toolbar to get the sequence mentioned in the example statement.

Adding flyout in the toolbar

1. Right-click on the **Circle** (**Center Radius** option) command under the **MyToolbar 2** item and choose the **New Flyout** option from the shortcut menu.

2. Rename the new flyout as **Arc** and drag the **Arc** (**3 Points** option), **Arc** (**Center, Start, End** option), **Arc** (**Center, Start, Angle** option), **Arc** (**Center, Start, Length** option), and **Arc** (**Continue** option) commands to the new flyout. Drag the items in the flyout to get the sequence mentioned in the example statement. Figure 5-19 shows the partial view of the **Customize User Interface** dialog box after adding the new flyout.

3. Choose **Apply** and then choose the **OK** button from the **Customize User Interface** dialog box to display this toolbar. Figures 5-20 and 5-21 show the **MyToolbar2** toolbar with the **Arc** flyout, and the **Arc** flyout as a separate toolbar, respectively.

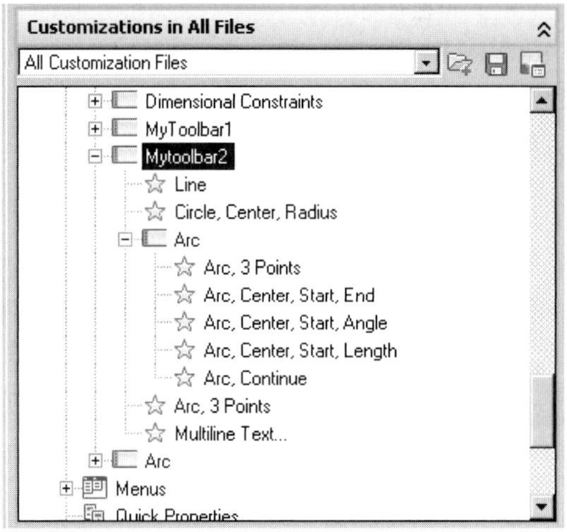

Figure 5-19 *Partial view of the* ***Customize User Interface*** *dialog box with the* ***MyToolbar2*** *toolbar*

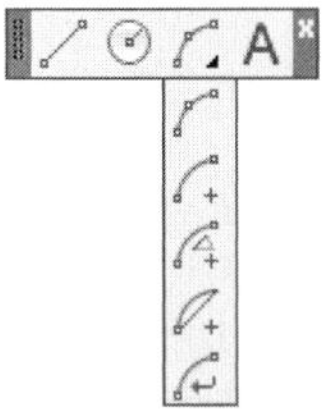

Figure 5-20 *The* ***MyToolbar2*** *toolbar with* ***Arc*** *flyout*

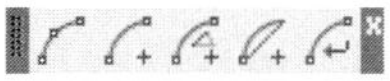

Figure 5-21 *The* ***Arc*** *flyout as separate toolbar*

Writing Menu Files for Toolbars

AutoCAD also allows you to write menu files for creating toolbars. After creating the *.mnu* and *.mns* files, you can load them in AutoCAD to display the toolbars.

Toolbar Definition. The following is the general format of the toolbar definition:

```
***TOOLBARS
**MYTOOLS1
TAG1 [Toolbar ("tbarname", orient, visible, xval, yval, rows)]
TAG2 [Button ("btnname", id_small, id_large)]macro
TAG3 [Flyout ("flyname", id_small, id_large, icon, alias)]macro
TAG4 [control (element)]
[—]
```

*****TOOLBARS** is the section label of the toolbar, and **MYTOOLS1** is the name of the submenu that contains the definition of a toolbar. Each toolbar can have five distinct items that control different elements of the toolbar: TAG1, TAG2, TAG3, TAG4, and separator ([—]). The first line of the toolbar (TAG1) defines the characteristics of the toolbar. In this line, **Toolbar** is the keyword, and it is followed by a series of options enclosed in parentheses. The available options are described as follows:

tbarname This is a text string that names the toolbar. The tbarname text string must consist of alphanumeric characters with no punctuation other than a dash (-) or an underscore (_).

orient This determines the orientation of the toolbar. The acceptable values are Floating, Top, Bottom, Left, and Right. These values are not case-sensitive.

visible This determines the visibility of the toolbar. The acceptable values are Show and Hide. These values are not case-sensitive.

xval This is a numeric value that specifies the *X* ordinate in pixels. The *X* ordinate is measured from the left edge of the screen to the left side of the toolbar.

yval This is a numeric value that specifies the *Y* ordinate in pixels. The *Y* ordinate is measured from the top edge of the screen to the top of the toolbar.

rows This is a numeric value that specifies the number of rows.

The second line of the toolbar (TAG2) defines the button. In this line, the **Button** is the key word and it is followed by a series of options enclosed in parentheses. The following is the description of the available options.

btnname This is a text string that names the button. The text string must consist of alphanumeric characters with no punctuation other than a dash (-) or an underscore (_). This text string is displayed as a tooltip when you place the cursor over the button.

id_small This is a text string that names the ID string of the small image resource (16 by 16 bitmap). The text string must consist of alphanumeric characters with no punctuation other than a dash (-) or an underscore (_). The id_small text string can also specify a user-defined bitmap (Example: RCDATA_16_CIRCLE). The **bit map images must exist or you just get a question mark**.

id_large This is a text string that names the ID string of the large image resource (32 by 32 bitmap). The text string must consist of alphanumeric characters with no punctuation other than a dash (-) or an underscore (_). The id_big text string can also specify a user-defined bitmap (Example: RCDATA_32_CIRCLE).

macro The second line (TAG2) that defines a button, is followed by a command string (macro). For example, the macro can consist of ^C^CLine. It follows the same syntax as that of a standard menu item definition.

The third line of the toolbar (TAG3) defines the flyout control. In this line, the **Flyout** is the key word, and it is followed by a series of options enclosed in parentheses. The available options are described as follows:

flyname This is a text string that names the flyout. The text string must consist of alphanumeric characters with no punctuation other than a dash (-) or an underscore (_). This text string is displayed as a tooltip when you place the cursor over the flyout button.

id_small This is a text string that names the ID string of the small image resource (16 by 16 bitmap). The text string must consist of alphanumeric characters with no punctuation other than a dash (-) or an underscore (_). The id_small text string can also specify a user-defined bitmap.

id_large This is a text string that names the ID string of the large image resource (32 by 32 bitmap). The text string must consist of alphanumeric characters with no punctuation other than a dash (-) or an underscore (_). The id_big text string can also specify a user-defined bitmap.

icon This is a Boolean key word that determines whether the button displays its own icon or the last icon selected. The acceptable values are **ownicon** and **othericon**. These values are not case-sensitive.

alias The alias specifies the name of the toolbar submenu that is defined with the standard ****aliasname** syntax.

macro The third line (TAG3), which defines a flyout control, is followed by a command string (macro). For example, the macro can consist of ^C^CCircle. It follows the same syntax as that of any standard menu item definition.

The fourth line of the toolbar (TAG4) defines a special control element. In this line, **control** is the keyword, and it is followed by the type of control element enclosed in parentheses. The following describes the available control element types.

element This parameter can have one of the following three values:

Layer:	This specifies the layer control element.
Linetype:	This specifies the linetype control element.
Color:	This specifies the color control element.

The fifth line ([**--**]) defines a separator.

After writing the *.mnu* file, you need to write the *.mns* file, which will have the following statement at the start of the file:

//
// AutoCAD menu file - *Path of the .mnu file*
//

This will be followed by the content of the *.mnu* file. The file should end with the following statement:

//
// End of AutoCAD menu file - *Path of the .mnu file*
//

The following examples explain the process of writing menus for toolbars in more detail.

Example 9

In this example, you will write a menu file for a toolbar for the **LINE**, **PLINE**, **CIRCLE**, **ELLIPSE**, and **ARC** commands. The name of the toolbar is **MyDraw1**.

Step 1: Writing program for toolbars

Use any text editor to list the toolbars. The following is the listing of the menu file containing toolbars. In this file listing, ID specifies the name tag.

```
***MENUGROUP=M1
***TOOLBARS
**TB_MyDraw1
ID_MyDraw1[_Toolbar("MyDraw1", _Floating, _Hide, 10, 200, 1)]
ID_Line   [_Button("Line", RCDATA_16_LINE, RCDATA_32_LINE)]^C^C_line
```

```
ID_Pline [_Button("Pline", RCDATA_16_PLine, RCDATA_32_PLine)]^C^C_PLine
ID_Circle[_Button("Circle", RCDATA_16_CirRAD, RCDATA_32_CirRAD)]^C^C_Circle
ID_ELLIPSE[_Button("Ellipse",RCDATA_16_EllCEN, RCDATA_32_EllCEN)]^C^C_ELLIPSE
ID_Arc[_Button("Arc 3Point", RCDATA_16_Arc3Pt, RCDATA_32_Arc3Pt)]^C^C_Arc
```

Save the file with the name *example9.mnu*.

Step 2: Writing the *.mns* file

Use the **Save As** option to save the file with the name *example9.mnu*. Modify the content of the file to add the path of the *.mnu* file above and below the program, as shown below. Save the file before closing it.

```
//
//    AutoCAD menu file - C:\customizing-2010\c05\example9.mnu
//

***Menugroup=M1
***TOOLBARS
**TB_MyDraw1
ID_MyDraw1[_Toolbar("MyDraw1", _Floating, _Hide, 10, 200, 1)]
ID_Line  [_Button("Line", RCDATA_16_LINE, RCDATA_32_LINE)]^C^C_line
ID_Pline [_Button("Pline", RCDATA_16_PLine, RCDATA_32_PLine)]^C^C_PLine
ID_Circle[_Button("Circle", RCDATA_16_CirRAD, RCDATA_32_CirRAD)]^C^C_Circle
ID_ELLIPSE[_Button("Ellipse", RCDATA_16_EllCEN,RCDATA_32_EllCEN)]^C^C_ELLIPSE
ID_Arc[_Button("Arc 3Point", RCDATA_16_Arc3Pt, RCDATA_32_Arc3Pt)]^C^C_Arc

//
//    End of AutoCAD menu file - C:\customizing-2010\c05\example9.mnu
//
```

Step 3: Loading the Toolbar

Invoke the **Customize User Interface** dialog box and then invoke the **Transfer** tab. In the right pane, choose the **Open customization file** button to display the **Open** dialog box. Select **Legacy menu files (*.mns, *.mnu)** from the **Files of type** drop-down list and select the *example9.mns* file and choose **Open**. Expand the **Toolbars** section in the right pane; **MyDraw1** toolbar will be displayed. Drag this toolbar to the **Toolbars** section in the left pane. Choose **OK** from the dialog box to close it. If the toolbar is not displayed automatically, right-click on any of the toolbar and choose **MyDraw1** from the shortcut menu; the toolbar will be displayed, as shown in Figure 5-22.

Figure 5-22 *The* ***MyDraw1*** *toolbar*

Example 10

In this example, you will write a menu file for a toolbar with a flyout. The name of the toolbar is **MyDraw2**, and it contains two buttons, **Circle** and **Arc**. When you select the **Circle** button, it should display a flyout with the **Radius**, **Diameter**, **2P**, and **3P** buttons. Similarly, when you select the **Arc** button, it should display the **3Point**, **SCE**, and **SCA** buttons.

Step 1: Writing the *.mnu* file

Use any text editor to write the menu for the toolbars. The following is the menu file listing:

```
***Menugroup=M2
***TOOLBARS
**TB_MyDraw2
ID_MyDraw2[_Toolbar("MyDraw2", _Floating, _Show, 10, 100, 1)]
ID_TbCircle[_Flyout("Circle", RCDATA_16_Circle, RCDATA_32_Circle, _OtherIcon,
M2.TB_Circle)]
ID_TbArc[_Flyout("Arc", RCDATA_16_Arc, RCDATA_32_Arc, _OtherIcon, M2.TB_Arc)]

**TB_Circle
ID_TbCircle[_Toolbar("Circle", _Floating, _Hide, 10, 150, 1)]
ID_CirRAD[_Button("Circle C,R", RCDATA_16_CirRAD, RCDATA_32_CirRAD)]^C^C_Circle
ID_CirDIA[_Button("Circle C,D", RCDATA_16_CirDIA, RCDATA_32_CirDIA)]^C^C_Circle;\D
ID_Cir2Pt[_Button("Circle 2Pts", RCDATA_16_Cir2Pt, RCDATA_32_Cir2Pt)]^C^C_Circle;2P
ID_Cir3Pt[_Button("Circle 3Pts", RCDATA_16_Cir3Pt, RCDATA_32_Cir3Pt)]^C^C_Circle;3P

**TB_Arc
ID_TbArc[_Toolbar("Arc", _Floating, _Hide, 10, 150, 1)]
ID_Arc3PT[_Button("Arc,3Pts", RCDATA_16_Arc3PT, RCDATA_32_Arc3PT)]^C^C_Arc
ID_ArcSCE[_Button("Arc,SCE", RCDATA_16_ArcSCE, RCDATA_32_ArcSCE)]^C^C_Arc;\C
ID_ArcSCA[_Button("Arc,SCA", RCDATA_16_ArcSCA, RCDATA_32_ArcSCA)]^C^C_Arc;\C;\A
```

Save the file with the name *example10.mnu*.

Explanation

ID_TbCircle[_Flyout("Circle", RCDATA_16_Circle, RCDATA_32_Circle, _OtherIcon, M2.TB_Circle)]

In this line, M2 is the MENUGROUP name (***MENUGROUP=M2) and TB_Circle is the name of the toolbar submenu. **M2.TB_Circle** will load the submenu TB_Circle that has been defined in the M2 menugroup. If M2 is missing, AutoCAD will not display the flyout when you select the **Circle** button.

ID_CirDIA[_Button("Circle C,D", RCDATA_16_CirDIA, RCDATA_32_CirDIA)] ^C^C_Circle;\D

CirDIA is a user-defined bitmap that displays the **Circle-diameter** button. If you use any other name, AutoCAD will not display the desired button.

Step 2: Writing the *.mns* file

The following is the content of the *.mns* file.

```
//
//     AutoCAD menu file - C:\customizing-2010\c05\example10.mnu
//

***Menugroup=M2
***TOOLBARS
**TB_MyDraw2
ID_MyDraw2[_Toolbar("MyDraw2", _Floating, _Show, 10, 100, 1)]
ID_TbCircle[_Flyout("Circle", RCDATA_16_Circle, RCDATA_32_Circle, _OtherIcon,
M2.TB_Circle)]
ID_TbArc[_Flyout("Arc", RCDATA_16_Arc, RCDATA_32_Arc, _OtherIcon, M2.TB_Arc)]

**TB_Circle
ID_TbCircle[_Toolbar("Circle", _Floating, _Hide, 10, 150, 1)]
ID_CirRAD[_Button("Circle C,R", RCDATA_16_CirRAD, RCDATA_32_CirRAD)]^C^C_Circle
ID_CirDIA[_Button("Circle C,D", RCDATA_16_CirDIA, RCDATA_32_CirDIA)]^C^C_Circle;\D
ID_Cir2Pt[_Button("Circle 2Pts", RCDATA_16_Cir2Pt, RCDATA_32_Cir2Pt)]^C^C_Circle;2P
ID_Cir3Pt[_Button("Circle 3Pts", RCDATA_16_Cir3Pt, RCDATA_32_Cir3Pt)]^C^C_Circle;3P

**TB_Arc
ID_TbArc[_Toolbar("Arc", _Floating, _Hide, 10, 150, 1)]
ID_Arc3PT[_Button("Arc,3Pts", RCDATA_16_Arc3PT, RCDATA_32_Arc3PT)]^C^C_Arc
ID_ArcSCE[_Button("Arc,SCE", RCDATA_16_ArcSCE, RCDATA_32_ArcSCE)]^C^C_Arc;\C
ID_ArcSCA[_Button("Arc,SCA", RCDATA_16_ArcSCA, RCDATA_32_ArcSCA)]^C^C_Arc;\C;\A

//
//     End of AutoCAD menu file - C:\customizing-2010\c05\example10.mnu
//
```

Save the file with the name *example10.mns*.

Step 3: Loading Toolbars

Using the method described in Example 9, load the toolbar. Figures 5-23 and 5-24 show the flyouts in this toolbar.

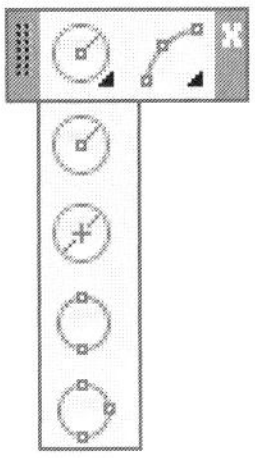

Figure 5-23 The ***Circle*** *flyout in the* ***MyDraw2*** *toolbar*

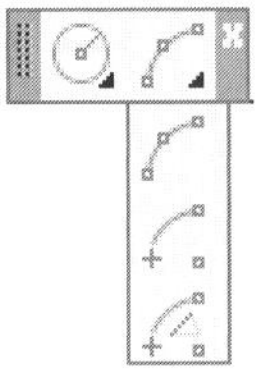

Figure 5-24 The ***Arc*** *flyout in the* ***MyDraw2*** *toolbar*

CREATING SHORTCUT MENUS

Similar to creating pull-down menus, AutoCAD also allows you to create shortcut menus that you can invoke by clicking a mouse button. The method of creating the shortcut menus is similar to that of creating the pull-down menus. However, in case of shortcut menus, you need to link it to the mouse buttons using the **Customize User Interface** dialog box to invoke them in the AutoCAD environment. The following examples show the procedure of creating shortcut menus in AutoCAD 2010.

Example 11

In this example, you will create a shortcut menu with cascading menus, as shown in Figures 5-25 and 26. The shortcut menu should be invoked by pressing the SHIFT+Right mouse button.

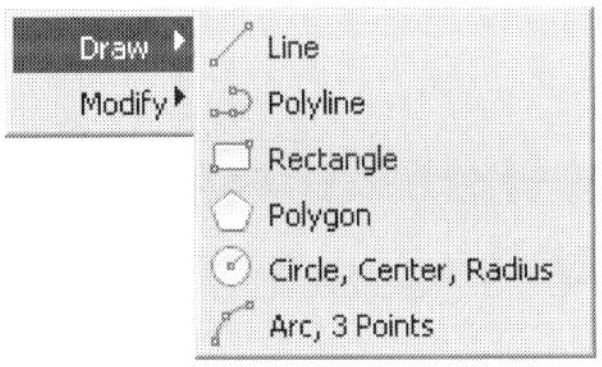

Figure 5-25 The ***Draw*** *flyout in the new shortcut menu*

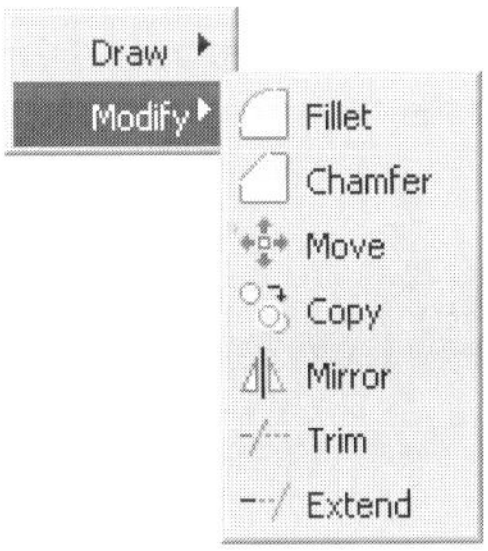

Figure 5-26 The ***Modify*** *flyout in the new shortcut menu*

Step 1: Creating the Shortcut Menu

The following steps explain the procedure of creating the shortcut menu:

1. Invoke the **Customize User Interface** dialog box.

2. Right-click on **Shortcut Menus** in the **Customizations in All Files** rollout in the left pane of the **Customize** tab and choose **New Shortcut Menu** from the shortcut menu; a new shortcut menu with the name **ShortcutMenu1** is added at the bottom of the tree view.

2. Rename the menu as **MyShortcutMenu**.

3. Click on the field that is displayed on the right of **Aliases** in the **Advanced** section of the **Properties** rollout displayed in the right pane; a swatch [...] is added on the right side of the field.

4. Choose the swatch [...]; the **Aliases** dialog box is displayed.

5. Click on the end of the default alias name of the shortcut menu, which is in form of POPnnn. Press ENTER to shift the cursor to the next line.

6. Type the alias of this menu as **My_SCT_Menu1** in the second line and then choose **OK** from the dialog box.

7. Now, right-click on **MyShortcutMenu** in the **Customizations in All Files** rollout in the left pane and choose **New Sub-menu** from the shortcut menu.

8. Rename the submenu to **Draw** and then drag the required commands to this submenu from the **Command List** rollout.

9. Similarly, create the **Modify** submenu and then drag the required commands to this submenu. The partial view of the **Customize User Interface** dialog box after creating the shortcut menu is shown in Figure 5-27.

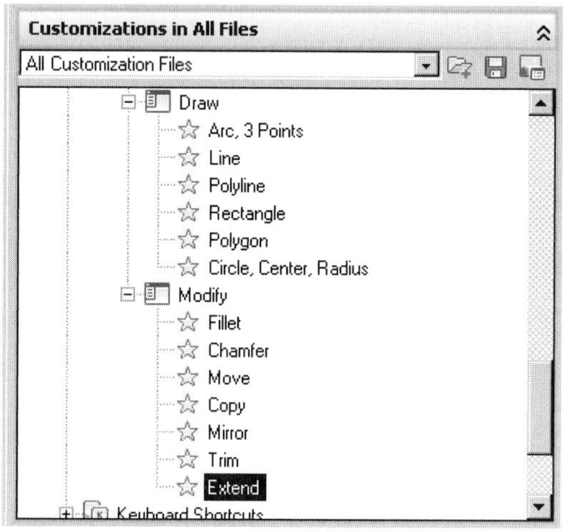

***Figure 5-27** Partial view of the **Customize User Interface** dialog box with the new shortcut menu*

Step 2: Linking the Shortcut Menu with the Mouse Button

The following steps explain the procedure of linking the shortcut menu with the mouse button:

1. Expand the **Mouse Buttons** tree view in the **Customizations in All Files** rollout; various mouse button combinations are displayed.

2. Expand the **Shift+Click** node and then click on **Button 2: Snap Menu**; the **Properties** rollout in the right pane displays the properties of this shortcut menu.

3. Click on the field on the right of **Macro** in the right pane; a swatch [...] is displayed on the right of this field.

4. Choose the swatch [...] to display the **Long String Editor** dialog box and replace the existing text with **$P0=My_SCT_Menu1 $P0=*** in this dialog box. This is the macro for

linking the new shortcut menu with the mouse button. Choose **OK** to accept and exit the dialog box.

5. Change the name in the **Command Name** edit box also to **My_SCT_Menu1**.

6. Choose **Apply** and then choose **OK**. Now if you press the SHIFT key down and right-click, the new shortcut menu will be displayed with the two cascading menus.

Note

By performing the above-mentioned steps, you will rename the default Snap menu to ***MY_SCT_MENU1****. To restore the default settings, select* ***MY_SCT_MENU1*** *from the* ***Command List*** *rollout and choose the* ***Reset Defaults*** *button in the* ***Properties*** *rollout in the right pane.*

Writing a Menu File for the Shortcut Menu

AutoCAD also allows you to write a menu file for the shortcut menu and then load it in AutoCAD using the **Customize User Interface** dialog box. Similar to writing menu files, you will have to write *.mnu* and *.mns* files for the shortcut menus.

The following is a list of some of the features of the Shortcut menu:

1. The section label of the Shortcut menu are ***POP0 and ***POP500 to ***POP999. The menu bar title defined under this section label is not displayed in the menu bar.

2. On most systems, the menu bar title is not displayed at the top of the shortcut menu. However, for compatibility reasons you should give a dummy menu bar title.

3. The POP0 menu can be accessed through the **$P0=*** menu command. The **Shortcut Menus** POP500 through POP999 must be referenced by their alias names. The reserved alias names for AutoCAD use are **GRIPS**, **CMDEFAULT**, **CMEDIT**, and **CMCOMMAND**. For example, to reference POP500 for grips, use **GRIPS command line under POP500. This command can be issued by a menu item in another menu, such as the button menu, auxiliary menu, or the screen menu. The command can also be issued from an AutoLISP or ADS program.

4. A maximum of 499 menu items can be defined in the shortcut menu and items in excess of 499 are ignored. This includes the items that are defined in the shortcut submenus.

5. The number of menu items that can be displayed on the screen depends on the system you are using. If the shortcut or pull-down menu contains more items than your screen can accommodate, the excess items are truncated. For example, if your system displays 21 menu items, the menu items in excess of 21 are automatically truncated.

6. The system variable **SHORTCUTMENU** controls the availability of **Default**, **Edit**, and **Command** mode shortcut menus. If the value is 0, it restores R2010 legacy behavior and makes the **Default**, **Edit**, and **Command** mode shortcut menus unavailable. The default value of this variable is 11.

The following example explains the process of writing the menu file for shortcut menu and then importing it in AutoCAD.

Example 12

Write a program to create a shortcut menu for the following AutoCAD commands using cascading submenus. The menu should be compatible with foreign language versions of AutoCAD. Use the SHIFT+Right mouse button to display this menu.

Osnaps	Draw	DISPLAY
Center	Line	REDRAW
Endpoint	PLINE	ZOOM
Intersection	CIR C,R	...Win
Midpoint	CIR 2P	...Cen
Nearest	ARC SCE	...Prev
Perpendicular	ARC CSE	...All
Quadrant		...Ext
Tangent		PAN
None		

Step 1: Writing the menu file

The following file is a listing of the menu file for Example 12. **The line numbers are not a part of the file; they are for reference only.**

```
***AUX1
;
$P0=*
***POP0
[O'Snaps]
[Center]_Center
[End point]_Endp
[Intersection]_Int
[Midpoint]_Mid
[Nearest]_Nea
[Perpendicular]_Per
[Quadrant]_Qua
[Tangent]_Tan
[None]_Non
[—]
[->Draw]
  [Line]^C^C_Line
  [PLINE]^C^C_Pline
  [CIR C,R]^C^C_Circle
  [CIR 2P]^C^C_Circle;_2P
  [ARC SCE]^C^C_ARC;\C
  [<-ARC CSE]^C^C_Arc;C
```

```
23 [—]
24 [->DISPLAY]
25   [REDRAW]^C^_REDRAW
26   [->ZOOM]
27     [...Win]^C^C_ZOOM;_W
28     [...Cen]^C^C_ZOOM;_C
29     [...Prev]^C^C_ZOOM;_P
30     [...All]^C^C_ZOOM;_A
31     [<-...Ext]^C^C_ZOOM;_E
32   [<-PAN]^C^C_Pan
33 ***POP1
34 [SHORTCUTMENU]
35 [SHORTCUTMENU=0]^C^CSHORTCUTMENU;0
36 [SHORTCUTMENU=1]^C^CSHORTCUTMENU;1
```

Explanation

Line 1

*****AUX1**

AUX1 is the section label for the first auxiliary menu; *** designates the menu section. The menu items that follow it, until the second section label, are a part of this buttons menu.

Lines 2 and 3

;

$P0=*

The semicolon (;) is assigned to the second button of the pointing device (the first button of the pointing device is the pick button); the special command **$P0=*** is assigned to the third button of the pointing device.

Lines 4 and 5

*****POP0**

[Osnaps]

The menu label **POP0** is the menu section label for the shortcut menu; **Osnaps** is the menu bar title. The menu bar title is not displayed, but is required. Otherwise, the first item will be interpreted as a title and will not be available.

Line 6

[Center]_Center

In this menu item, **_Center** is the center Object Snap mode. The menu files can be used with foreign language versions of AutoCAD, if AutoCAD commands and the command options are preceded by the underscore (_) character.

After loading the menu, if you press the third button of your pointing device, the shortcut menu (Figure 5-28) will be displayed at the cursor (screen crosshairs) location. If the cursor is close to the edges of the screen, the shortcut menu will be displayed at a location that is closest to the cursor position. When you select a submenu, the items contained in the submenu will be displayed, even if the shortcut menu is touching the edges of the screen display area.

Lines 33 and 34
*****POP1**
[Draw]
*****POP1** defines the first pull-down menu. If no POPn sections are defined or the status line is turned off, the Shortcut menu is automatically disabled.

Save the file with the name *example12.mnu.*

The following statement is the *.mns* file for this menu.

```
//
//     AutoCAD menu file - C:\customizing-2010\c05\example12.mnu
//

***MENUGROUP=C:\customizing-2010\c05\example12.mnu

***AUX1
;
$p0=*

***POP0
          [O'Snaps]
          [Center]_Center
          [Endpoint]_Endp
          [Intersection]_Int
          [Midpoint]_Mid
          [Nearest]_Nea
          [Perpendicular]_Per
          [Quadrant]_Qua
          [Tangent]_Tan
          [None]_Non
          [--]
          [->Draw]
            [Line]^C^C_Line
            [PLINE]^C^C_Pline
            [CIR C,R]^C^C_Circle
            [CIR 2P]^C^C_Circle;_2P
            [ARC SCE]^C^C_ARC;\C
            [<-ARC CSE]^C^C_Arc;C
          [--]
          [->DISPLAY]
            [REDRAW]^C^_REDRAW
            [->ZOOM]
              [...Win]^C^C_ZOOM;_W
              [...Cros]^C^C_ZOOM;_C
              [...Pre]^C^C_ZOOM;_P
```

```
            [...All]^C^C_ZOOM;_A
            [<-...Ext]^C^C_ZOOM;_E
          [<-PAN]^C^C_Pan

***POP1
          [SHORTCUTMENU]
          [SHORTCUTMENU=0]^C^CSHORTCUTMENU;0
          [SHORTCUTMENU=1]^C^CSHORTCUTMENU;1

//
//    End of AutoCAD menu file - C:\customizing-2010\c05\example12.mnu
//
```

Save this file with the name *example12.mns*.

Step 2: Invoking this file in AutoCAD

1. Invoke the **Customize User Interface** dialog box and choose the **Transfer** tab.

2. Choose the **Open customization file** button from the right pane. Select **Legacy menu files (*.mns, *.mnu)** from the **Files of type** drop-down list in the **Open** dialog box.

3. Double-click on the *example12.mns* file.

4. Expand the **Shortcut Menus** tree view in the right pane and then drag the **O'Snaps** menu from the right pane to the **Shortcut Menus** in the left pane; it will be added as the list item in the **Shortcut Menus** in the left pane.

5. Choose the **Customize** tab and then expand the **Shortcut Menus** tree view. Select **O'Snaps** and then add **O'Snap** as a new alias to it in the **Properties** rollout in the right pane.

6. Expand the **Mouse Buttons** tree view and then assign this shortcut menu to **Button 2: Snap Menu** in **Shift+Click**.

7. Choose **Apply** and then choose **OK**. Now, in the drawing area, press the SHIFT key down and right-click, the new shortcut menu will be displayed, as shown in Figure 5-28.

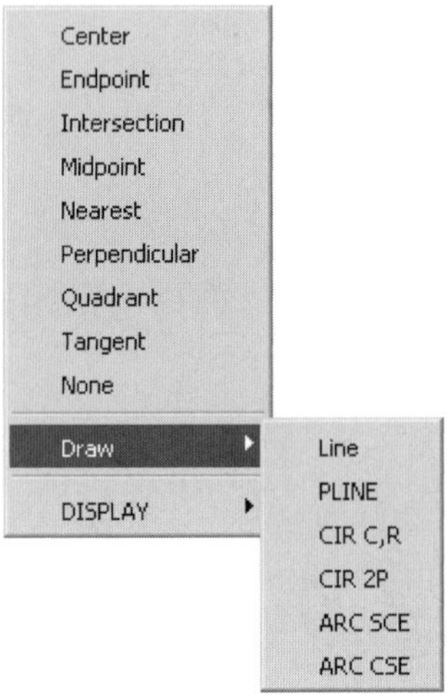

Figure 5-28 *The new shortcut menu with the* ***Draw*** *cascading menu*

Self-Evaluation Test

Answer the following questions and then compare them to those given at the end of this chapter:

1. AutoCAD allows you to customize its user interface to your requirements, thus allowing you to efficiently use this program and reduce the design time. (T/F)

2. The AutoCAD software package comes with a standard user interface file named *acad.CUIx*. (T/F)

3. By default the customized menus are added to the default customization file. (T/F)

4. AutoCAD does not allow you to manually write a menu file. (T/F)

5. In AutoCAD, you cannot create a customized toolbar. (T/F)

6. The __________ sign defines a cascaded submenu while __________ designates the last item in the menu.

7. __________ designates the last item of the pull-down or cursor menu.

8. The _________ dialog box is used to customize the user interface of AutoCAD.

9. The pop menu can have up to __________.

10. After creating the __________ files for menus and toolbars, you need to create __________ files to load the menus or toolbars.

Review Questions

Answer the following questions:

1. A pull-down menu can have ________________ sections.

2. When you start AutoCAD, the __________ file is automatically loaded.

3. AutoCAD allows you to add nonstandard commands to the menus by writing their __________.

4. The __________ label character defines a cascaded submenu.

5. The __________ label character is used to push the text written after it to the right of the menu.

6. The __________ feature of AutoCAD allows pull-down and cursor menus to be displayed in a hierarchical order that makes it easier to select submenus.

7. The __________ character when placed directly before a character displays the character as underscored.

8. The __________ in the toolbars save the drawing space because you do not need to have several toolbars on the screen.

9. The __________ text string names the toolbar.

10. The __________ is a numeric value that specifies the *X* ordinate in pixels.

Exercises

Exercise 2 *General*

Create a pull-down menu for the following AutoCAD commands. The layout of the menu is shown in Figure 5-29.

LINE	**DIMLINEAR**	**TEXT LEFT**
CIRCLE C,R	**DIMALIGNED**	**TEXT RIGHT**
CIRCLE C,D	**DIMRADIUS**	**TEXT CENTER**
ARC 3P	**DIMDIAMETER**	**TEXT ALIGNED**
ARC SCE	**DIMANGULAR**	**TEXT MIDDLE**
ARC CSE	**LEADER**	**TEXT FIT**

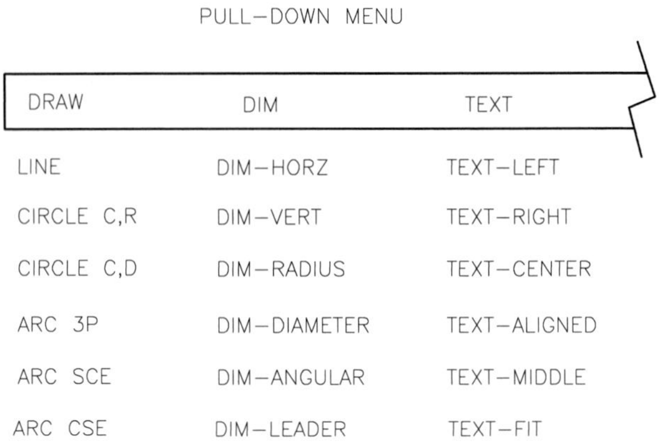

Figure 5-29 *Layout of menu for Exercise 2*

Exercise 3 *General*

Create a pull-down menu for the following AutoCAD commands.

LINE	**BLOCK**
PLINE	**WBLOCK**
CIRCLE C,R	**INSERT**
CIRCLE C,D	**BLOCK LIST**
ELLIPSE AXIS ENDPOINT	**ATTDEF**
ELLIPSE CENTER	**ATTEDIT**

Exercise 4 *General*

Using the **Customize User Interface** dialog box, create a toolbar with a flyout. The name of the toolbar is **MyDrawX1**, and it contains two buttons, **Draw** and **Modify**. When you select the **Draw** button, it should display a flyout with **all draw** buttons (**Draw** commands). Similarly, when you choose the **Modify** button, it should display a flyout with **modify** buttons (Modify commands).

Answers to Self-Evaluation Test

1. T, **2.** T, **3.** T, **4.** F, **5.** F, **6.** ->, -<, **7.** <-<-, **8. Customize User Interface**, **9.** 499, **10.** *.mnu*, *.mns*

Chapter 6

Customizing Ribbon, Workspaces, and Palettes

Learning Objectives

After completing this chapter, you will be able to:

- *Create and modify panels of* ***Ribbon****.*
- *Create and modify tabs of* ***Ribbon****.*
- *Create and modify workspaces.*
- *Customize the* ***Quick Access*** *toolbar.*
- *Customize the* ***Quick Properties*** *panel.*
- *Customize the* ***Tool Palettes****.*
- *Define and modify double-click actions.*

STRUCTURE OF RIBBON

The **Ribbon** consists of a number of tabs in which the panels are placed. The panel has two sections: upper and lower. The upper section of the panel is displayed by default, whereas the lower section is displayed by expanding the panel. To view the lower section of the panel, move the cursor on the down-arrow that is displayed at the lower right corner of the ribbon panel and click on it, see Figure 6-1. In the expanded panel, the upper section and the lower section are separated by a panel separator line. Figure 6-1 displays the collapsed view of the ribbon panel and Figure 6-2 displays the expanded view of the ribbon panel displayed in Figure 6-1.

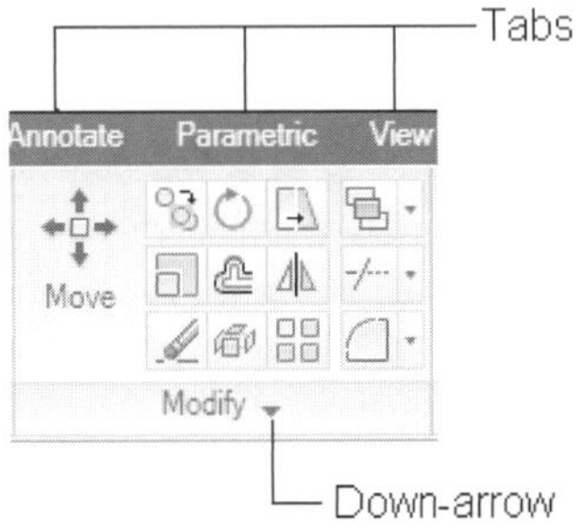

Figure 6-1 *The collapsed view of the ribbon panel*

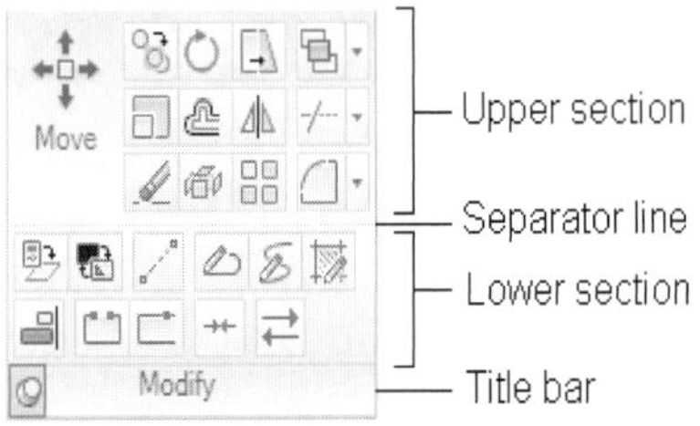

Figure 6-2 *The expanded view of the ribbon panel*

You can add or remove buttons, drop-downs, sub-panels, rows, panels, and tabs using the **Ribbon**. This in turn allows you to arrange and invoke the commands according to your requirement. The buttons in the ribbon panel are arranged in rows. If the number of buttons in a row is more, you can add drop-down in the row to accommodate those extra buttons. The down-arrow displayed at the right corner of the buttons represents a drop-down. You can access the buttons in the drop-down by clicking on this down-arrow. You can also add sub-panels to the ribbon panels which, in turn, contain rows and drop-downs, see Figure 6-3.

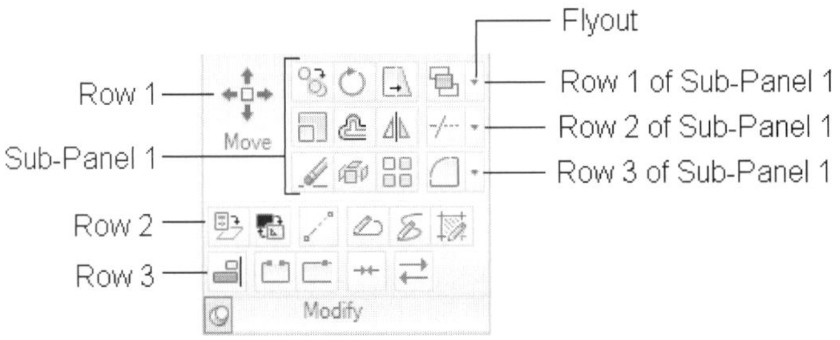

Figure 6-3 *Components of a panel in the* ***Ribbon***

CREATING A CUSTOMIZED RIBBON PANEL*

You can create a new ribbon panel using the **Customize User Interface** dialog box. To invoke this dialog box, choose **Manage > Customization > User Interface** from the **Ribbon**; the **Customize User Interface** dialog box will be displayed. In this dialog box, expand the plus sign (+) on the left of the **Ribbon** element in the **Customizations in All Files** rollout of the

Customize tab. Next, right-click on the **Panels** element under the **Ribbon** node and choose the **New Panel** option from the shortcut menu, see Figure 6-4. A new panel with the name **Panel1** will be added to the list and its properties will be displayed in the right pane. Enter the name of the panel and the name to be displayed on the title bar of the panel in the **Name** and the **Display Text** edit boxes of the **Properties** rollout, respectively.

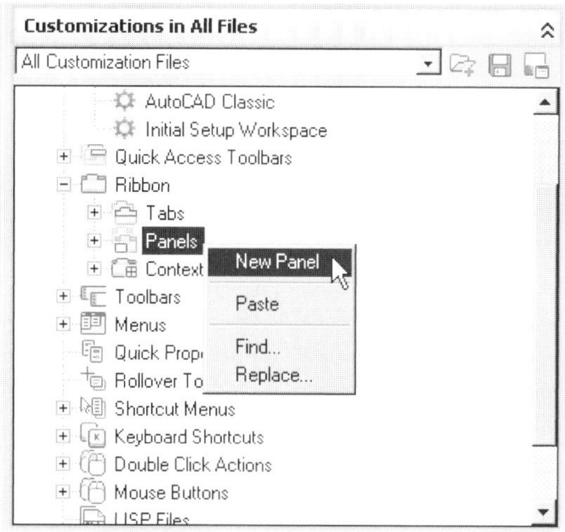

Figure 6-4** Adding a new panel to the **Ribbon

Adding a Row to the Ribbon Panel

When you create a new ribbon panel, a row (named as **Row 1**), a panel separator (named as **<SLIDEOUT>**), and a **Panel Dialog Box Launcher** are added to it by default. To add a new row in the lower section of the panel, choose the **New Row** option from the shortcut menu that will be displayed when you right-click on the panel name in the **Customizations in All Files** rollout of the **Customize User Interface** dialog box; a new row will be added below the separator line. The rows added below the panel separator can only be viewed after expanding that panel. You cannot add any row above the panel separator, so the upper section of the panel will have only one row.

The rows are displayed in the **Ribbon** according to their order of arrangement under the panel name in the **Customizations in All Files** rollout of the **Customize User Interface** dialog box. You can shuffle the positions of the rows by dragging and dropping them at the desired location. While dropping the commands in the rows, a blue colored splitter bar will be displayed to help you visually locate the exact position to drop the commands.

Adding Commands to the Rows

To add a command to the row, expand the plus sign (+) on the left of the **Panels** element under the **Ribbon** node. Next, expand the plus sign (+) on the left of the panel entry. Add a new row or expand the plus sign (+) on the left of the existing row to which you want to add the command. Drag the commands from the **Command List** rollout and drop them in front of the row to which you want to add the commands, as shown in Figure 6-5. The preview of the new ribbon panel and the command added to it is displayed in the **Panel Preview** rollout of the **Customize User Interface** dialog box.

The buttons in a row of the ribbon panel are displayed in the order in which they are arranged under the row in the **Customizations in All Files** rollout. You can shuffle the position of the buttons by dragging and dropping them at the desired location in the **Customizations in All Files** rollout.

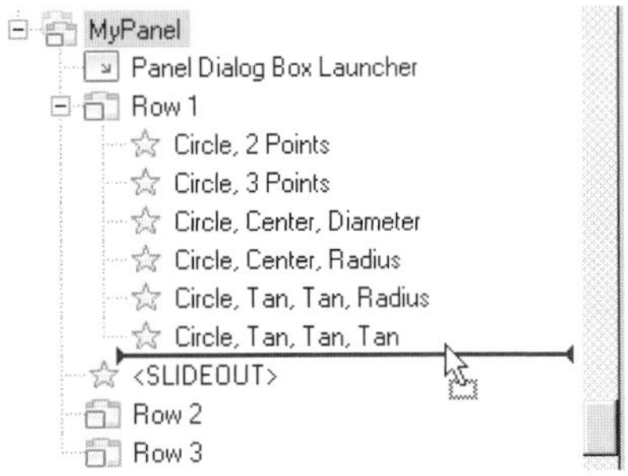

***Figure 6-5** Adding commands to a row in the ribbon panel*

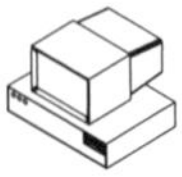

Tip

*You can control the display of the command name (command label), its display size as, well as its orientation with the button image in the ribbon panel. To do so, select a command under the **Row** node; the properties of the selected command will be displayed in the **Properties** rollout in the right pane. Choose the **Button Style** option under the **Appearance** node in the **Properties** rollout and then select the desired option from the drop-down list displayed on the right of the **Button Style** option.*

Adding Controls to the Rows

Besides command buttons, the panel also has controls like drop-down lists, sliders, text boxes, and so on. To add such type of controls, select the **Ribbon Control Elements** option from the drop-down list displayed in the **Command List** rollout of the **Customize User Interface** dialog box, as shown in Figure 6-6; the list of all control elements will be displayed. Drag the required control elements from the **Ribbon Control Elements** list and drop it in the

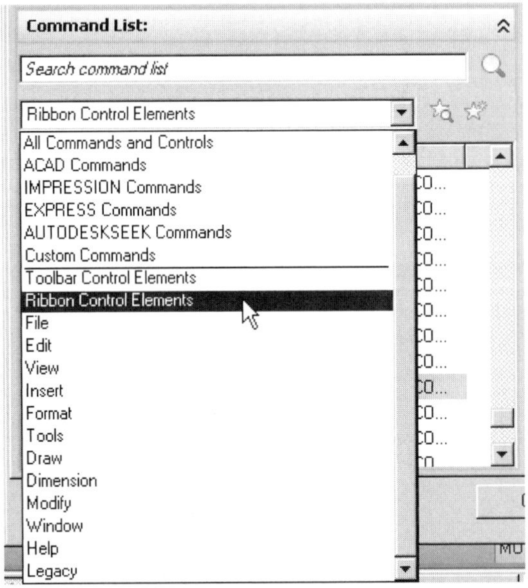

***Figure 6-6** The **Command List** rollout of the **Customize User Interface** dialog box*

Customizations in All Files rollout in front of the row name, where you want to add the control elements.

Adding a drop-down in a Row of the Panel

A drop-down can be added at any desired location in a row. To add a drop-down, right-click on that row and choose the **New Drop-down** option from the shortcut menu; a new drop-down named **New Drop-down** will be added at the end of the row. Now, to specify the location of the newly added drop-down within the row, drag and drop it at the desired location and then specify a desired name for it. You can also rename the drop-down later on by right-clicking on it and choosing the **Rename** option from the shortcut menu displayed. Now, to add the commands to the drop-down, drag them from the **Command List** rollout and drop them under the drop-down name in the **Customizations in All Files** rollout. A blue colored splitter bar will be displayed to help you visually locate the exact position for dropping the commands.

The drop-down added in a row can be displayed in two ways, either as a simple drop-down or as a split drop-down, refer to Figure 6-7. In the simple drop-down, the drop-down button and the down-arrow behave as single entity. As a result, whether you choose the button or the down-arrow, the drop-down list will be displayed in both the cases. But, in the case of split drop-down, the button and the down-arrow will behave as individual entities. As a result, if you choose the button, the assigned command will get activated and if you choose the down-arrow, the drop-down list will be displayed. To assign any one of the above mentioned behavior to the drop-down, select the particular drop-down node under the desired row; the properties of the selected drop-down element will be displayed in the **Properties** rollout in the right pane. Choose the **Behavior** option under the **Appearance** node in the **Properties** rollout and then select desired option from the drop-down list displayed.

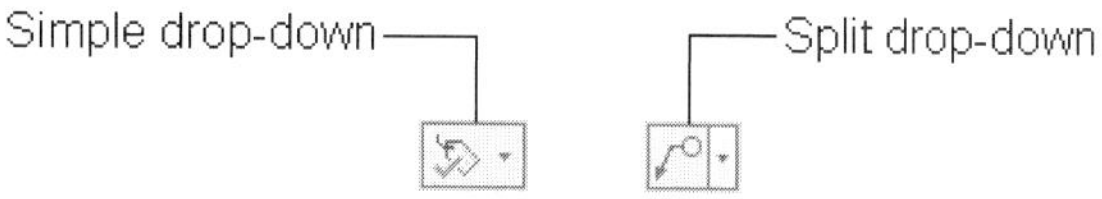

Figure 6-7 *Display modes of drop-down element*

Whenever you create a new drop-down element under any row, the **Primary Command** element will automatically be created under that element. To assign a particular command to the button of the split drop-down, drag the required command from the **Command List** rollout and drop it on the **Primary Command** element displayed under the selected drop-down element. After adding the newly created panel to the **Ribbon**, whenever you will choose the button of the split drop-down element, the assigned command will get activated.

Adding a Sub-Panel to the Row

You can also create a sub-panel within a row. To do so, right-click on any row and choose the **New Sub-Panel** option from the shortcut menu; a new sub-panel will be added under the selected row. When you add a new sub-panel, a row is also automatically added under the sub-panel. Drag the desired commands from the **Command List** rollout and drop them under the added row. You can also add new rows or drop-downs in the rows under the sub-panel. The procedure of adding rows and drop-downs is the same as discussed in the previous sections. You can also reposition the sub-panel within the row by dragging and dropping them at the desired location.

Assigning an Image to the Command Button

An existing image file can be assigned as an icon to the command button. Select the command from the panel row for which you want to change or assign an image from the **Customizations in All Files** rollout and then choose the desired image from the **Button Image** rollout, see Figure 6-8; the specified image will be assigned to the selected command.

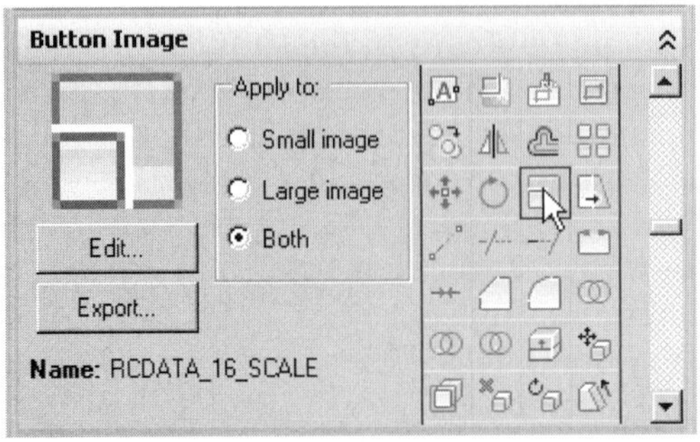

***Figure 6-8** The **Button Image** rollout of the **Customize User Interface** dialog box*

You can also edit the existing image according to your requirement and then assign it to the command button. To do so, select a command from the **Command List** rollout of the **Customize User Interface** dialog box to display the **Button Image** rollout. In the **Button Image** rollout, choose the **Edit** button; the **Button Editor** dialog box will be displayed, as shown in Figure 6-9. In this dialog box, you can edit the image using the **Pencil**, **Line**, **Circle**, and the **Erase** buttons. Select the **Grid** check box to display the grids (one small square represents one pixel) in the button that will help you to edit the image. These grids will not be displayed in the final button image. If you want to create an image starting from scratch, choose the **Clear** button from the **Button Editor** dialog box. Choose the **Save** button to save the changes. Choose the **Import/Export** button to import or export the image as BMP file. By default, the images of buttons are saved in *C:\Documents and Settings\User\ApplicationData\Autodesk\AutoCAD 2010\R18.0\enu\Support\Icons*.

You can also import the existing BMP image into the list of images in the **Button Image** rollout. To do so, right-click on any button image and choose the **Import Image** option from the shortcut menu, see Figure 6-10. The **Open** dialog box will be displayed. Locate and select the BMP image file that you want to import and choose the **Open** button; the image will be added at the end of the button image list. You can now use this image to assign it to any button or ribbon panel icon as explained earlier in this topic.

Converting the Toolbar or Dashboard to a Ribbon Panel

In AutoCAD, you can copy the commands available in any toolbar and create a ribbon panel with all those commands. To do so, expand the **Toolbars** node and select the toolbar item that you want to copy. Next, right-click and choose the **Copy To Ribbon Panels** option from

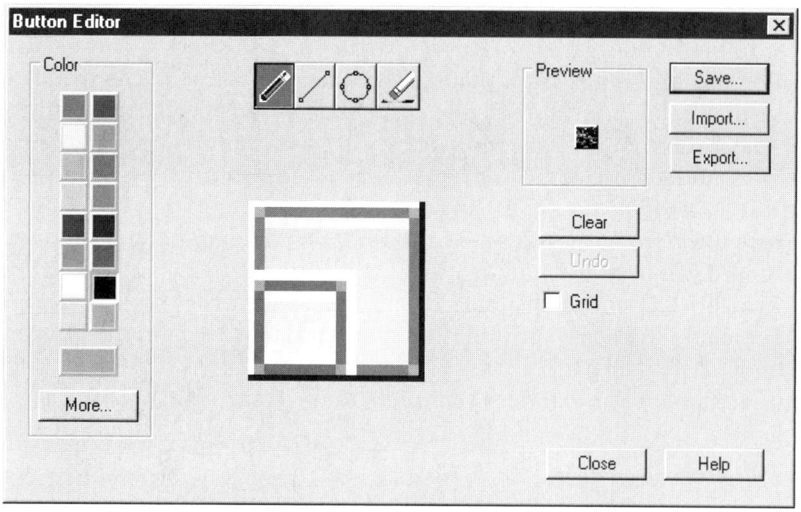

Figure 6-9 *The* ***Button Editor*** *dialog box*

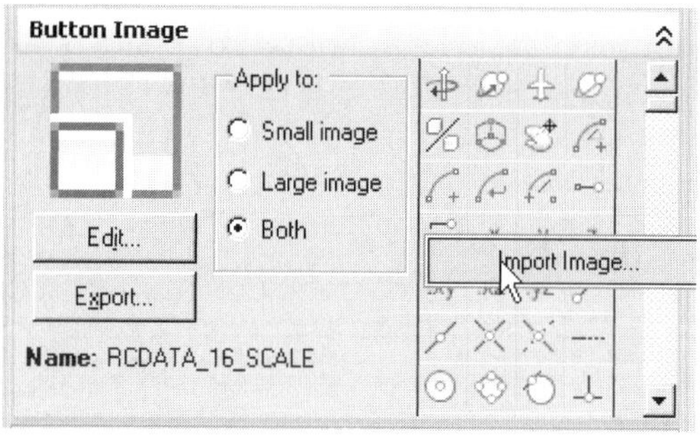

Figure 6-10 *Importing button image in the* ***Button Image*** *rollout*

the shortcut menu; the **CUI Editor - Confirm Copy to Ribbon Panels Node** message box will be displayed on the screen, see Figure 6-11. Choose the **Yes** button from this message box; a ribbon panel of the same name as that of the toolbar will be added in the **Panels** node under the **Ribbon** node. This ribbon panel will consist of all the commands that are present in the selected toolbar. Later on, you can rearrange the position of the commands in the ribbon panel by dragging and dropping them at the intended position.

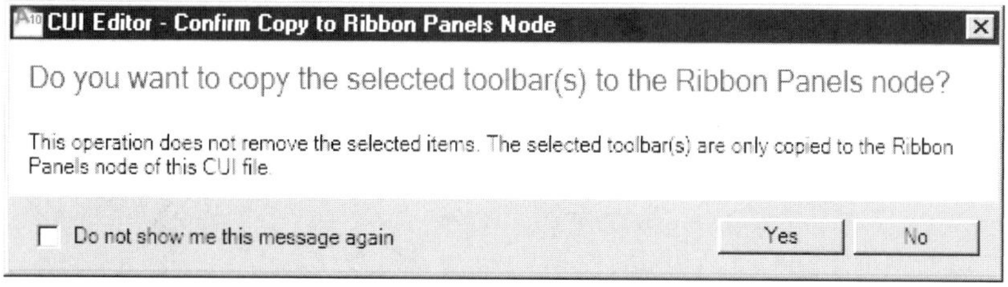

Figure 6-11 *The* ***CUI Editor - Confirm Copy to Ribbon Panels Node*** *message box*

If you have an existing toolbar or a **Dashboard** from the earlier releases of AutoCAD, you can also convert it to the ribbon panel. To do so, invoke the **Transfer** tab of the **Customize User Interface** dialog box and in the right pane of the **Transfer** tab, load the **.CUIx* file from the earlier release that is having the toolbar or the **Dashboard** to be converted to the ribbon panel. Now, select the toolbar/**Dashboard** item that you want to load in the ribbon panel. Then, right-click and choose the **Copy To Ribbon Panels** option from the shortcut menu; the **CUI Editor - Confirm Copy to Ribbon Panels Node** message box will be displayed. Choose the **Yes** button from this message box; a ribbon panel of the same name as that of the toolbar/**Dashboard** will be added under **Ribbon > Panels** in the right pane. Next, in the left pane of the **Transfer** tab of the **Customize User Interface** dialog box, load the *.cuix* file in which you want to add the newly created ribbon panel. Then, expand the **Panels** node under the **Ribbon** node in the left pane. Now, drag and drop the newly created ribbon panel from the right pane to the left pane under the **Panels** node in the current *.cuix* file.

Assigning a Command to the Panel Dialog Box Launcher

In some of the ribbon panels, an inclined arrow is displayed on the extreme right of the title bar of a panel. If you choose this arrow, a dialog box related to the settings of the options given in that particular panel will be displayed. Also, note that whenever you create a new ribbon panel, **Panel Dialog Box Launcher** is added automatically under that ribbon panel. You can customize the dialog box that will be invoked on choosing the inclined arrow by assigning the required command to the **Panel Dialog Box Launcher**. To do so, drag the command from the **Command List** rollout and drop it on the **Panel Dialog Box Launcher** element displayed under the selected ribbon panel, refer to Figure 6-12.

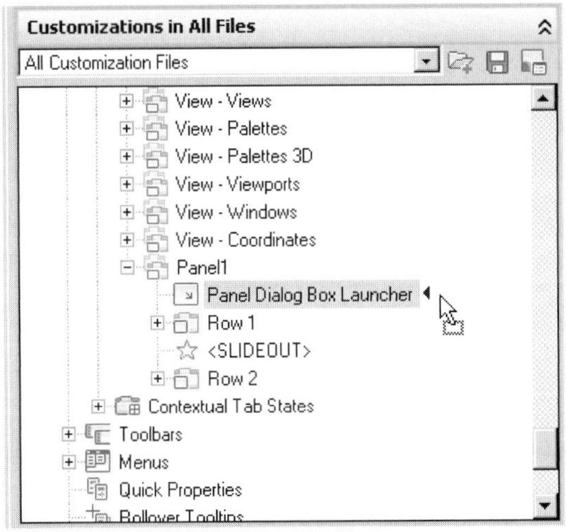

Figure 6-12 *Assigning a command to the* ***Panel Dialog Box Launcher***

Removing the Ribbon Panel/Row/Sub-panel/drop-down/ Command from the Ribbon

Select the ribbon panel/row/sub-panel/drop-down/command that you want to remove from the **Customizations in All Files** rollout and right-click; choose the **Delete/Remove** option from the shortcut menu; the corresponding option will be deleted.

Note

*The ribbon panel defined for a **Ribbon** will not vary with the change of workspace. So, the ribbon panel defined in one workspace can be used in all other workspaces.*

Tip

*The ribbon panel/row/drop-down/command once deleted cannot be undone. If you delete any entry by mistake, choose the **Cancel** button from the **Customize User Interface** dialog box to exit the dialog box without saving any changes. However, note that, this action will also undo all the changes made after the last save operation.*

Example 1

In this example, you will create a ribbon panel named, **Draw/Modify**, as shown in Figure 6-13. The structure of the ribbon panel in the **Customize User Interface** dialog box is shown in Figure 6-14.

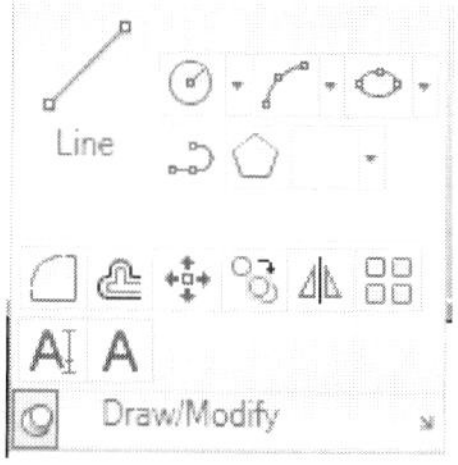

Figure 6-13 *The new **Draw/Modify** ribbon panel*

Step 1: Adding a new ribbon panel

The following steps explain the procedure of adding a ribbon panel.

1. Invoke the **Customize User Interface** dialog box by choosing **Manage > Customization > User Interface** from the **Ribbon**.

2. Expand **Ribbon** node and right-click on the **Panels** element in the **Customizations in All Files** rollout in the left pane of the **Customize** tab and choose the **New Panel** option from the shortcut menu. A new ribbon panel with the name **Panel1** will be added to the bottom of the tree view. Rename the **Panel1** to **Draw/Modify**. This name will be displayed as the title bar of the panel in the **Ribbon**.

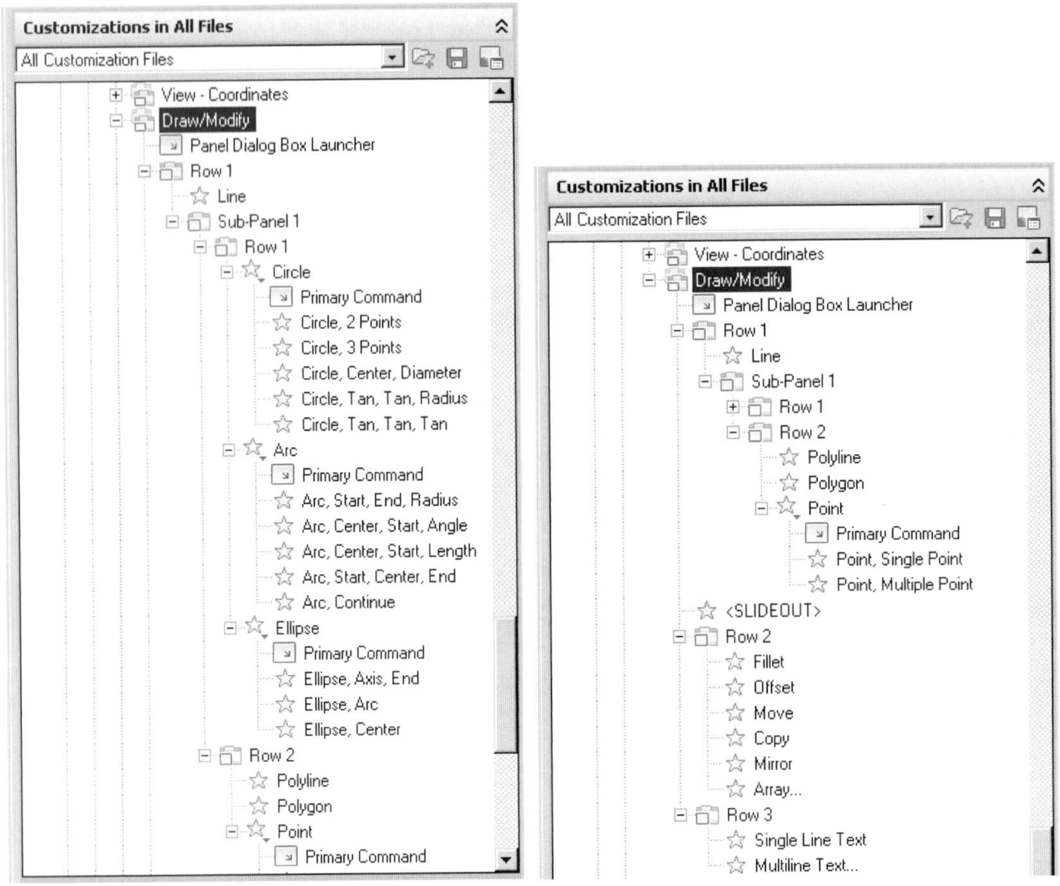

Figure 6-14 *The structure of the ribbon panel in the* ***Customize User Interface*** *dialog box*

Step 2: Adding rows to the ribbon panel

In the **Draw/Modify** ribbon panel, you need to add one row above and two rows below the panel separator. The following steps explain the procedure to add rows to the ribbon panel.

1. By default, one row is added to the upper section of the **Draw/Modify** ribbon panel. To add a new row in the lower section of the ribbon panel, right-click on the **<SLIDEOUT>** and choose the **New Row** option from the shortcut menu; a new row with the default name **Row2** is added below the **<SLIDEOUT>**.

2. Similarly, add another row below the **<SLIDEOUT>**.

Step 3: Adding command and sub-panel to the Row1

Next, you need to add a command and sub-panel to the **Row1**. The following steps explain the process of adding a command and a sub-panel.

1. Drag the **Line** command from the **Command List** rollout to the **Customizations in All Files** rollout and drop it on the **Row1**. The **Line** command is added to the **Row 1** and its preview is displayed in the **Panel Preview** rollout of the **Customize User Interface** dialog box.

2. Note that the newly added **Line** command icon should be larger than the other command icons and also, its name should be displayed below the **Line** button in the ribbon panel. To do so, select the **Line** element under the **Row1** node if it is not already selected; the properties of the **Line** element will be displayed in the **Properties** rollout in the right pane. Choose the **Button Style** option under the **Appearance** node in the **Properties** rollout and then select the **Large with Text (Vertical)** option from the drop-down list displayed.

2. Right-click on **Row1** and choose the **New Sub-Panel** option; a new sub-panel with the default name **Sub-Panel1** is added under the selected row.

Note

The rows and sub-panels added in the ribbon panel cannot be renamed.

Step 4: Adding rows, drop-downs, and commands to the Sub-Panel1

Next, you need to add rows, drop-downs, and commands to the **Sub-Panel1**. The following steps explain the process of adding commands and control elements.

1. By default, one row is added under the **Sub-Panel1**. To add one more row to it, right-click on the **Sub-Panel1** and choose the **New Row** option from the shortcut menu; a new row is added under the **Sub-Panel1**.

2. Right-click on the first row under the **Sub-Panel1** and choose the **New Drop-down** option from the shortcut menu; a drop-down with the name **New Drop-down** is added under the first row in the **Sub-Panel1**. Next, rename the **New Drop-down** to **Circle**.

3. Similarly, add two more drop-downs in the first row under the **Sub-Panel1** with the name **Arc** and **Ellipse**, respectively.

4. Drag the **Circle, 2 Points** command from the **Command List** rollout to the **Customizations in All Files** rollout and drop it on the **Circle** drop-down. The **Circle**, **2 Point** command is added to the **Circle** drop-down and its preview is displayed in the **Panel Preview** rollout of the **Customize User Interface** dialog box.

5. Similarly, drag the other commands from the **Command List** rollout and drop them under the **Circle**, **Arc**, and **Ellipse** drop-downs in a sequence, as shown in Figure 6-15.

6. Next, add the **Polyline** and **Polygon** commands to the second row under the **Sub-Panel1** in a sequence, as shown in Figure 6-15.

7. Insert a drop-down in the second row under the **Sub-Panel1** and rename the drop-down as **Point**.

8. Add the **Point, Single Point** and **Point, Multiple Point** commands to the **Point** drop-down list, see Figure 6-15.

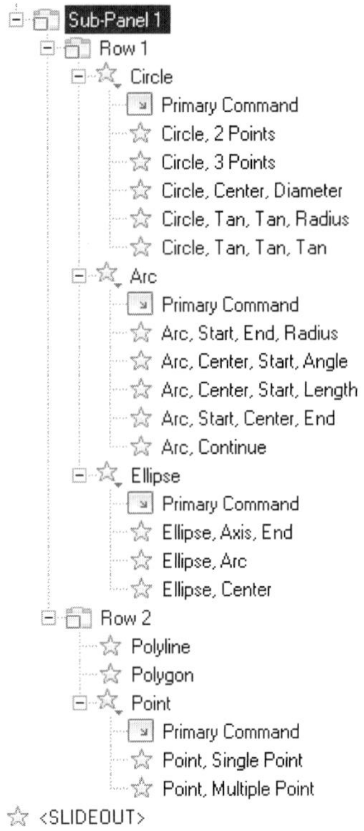

***Figure 6-15** The structure of the ribbon panel in the **Customize User Interface** dialog box*

Step 5: Defining the behavior of the drop-downs and assigning the primary commands to them

Next, you need to define the appearance and behavior of the newly added drop-downs and assign the primary commands to them.

1. Select the **Circle** drop-down node under the first row; the properties of the selected drop-down element will be displayed in the **Properties** rollout in the right pane. Choose the **Behavior** option under the **Appearance** node in the **Properties** rollout and then select the **Split** option from the drop-down list displayed.

2. Drag the **Circle, Center, Radius** command from the **Command List** rollout to the **Customizations in All Files** rollout and drop it on the **Primary Command** element displayed under the **Circle** drop-down node. Now, whenever you will choose the button of the **Circle** split drop-down element, the assigned command will get activated.

3. Similarly, define the split behavior for the **Arc** and **Ellipse** drop-down nodes under the first row and the **Point** drop-down node under the second row.

4. Follow the instructions given in the second point of Step 5 and assign the **Arcs, 3Points**; **Ellipse, Axis, End**; and **Point, Single Point** commands to the **Primary Command** element under the **Arc**, **Ellipse**, and **Point** drop-down nodes, respectively.

Step 6: Adding commands to Row2 and Row3 of the ribbon panel

Next, you need to add commands to the **Row2** and **Row 3** of the ribbon panel. The following steps explain the process of adding commands.

1. Drag the **Fillet** command from the **Command List** rollout to the **Customizations in All Files** rollout and drop it on the **Row2**; the **Fillet** command is added to the **Row2**.

2. Similarly, add other commands in **Row2** and **Row3** in a sequence, as shown in Figure 6-16.

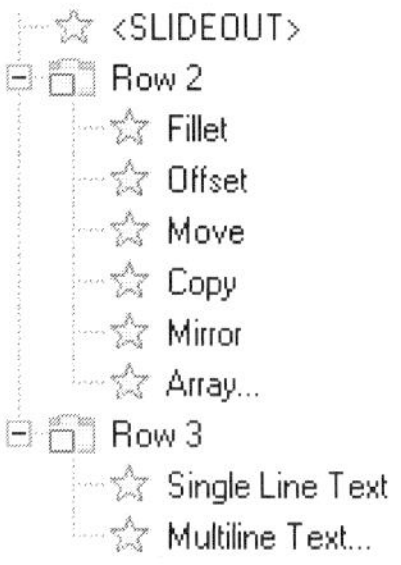

Figure 6-16 *Arrangement of commands under the* ***Row2*** *and* ***Row3***

Step 7: Assigning command to the Panel Dialog Box Launcher

Now, you need to specify the command to be invoked on choosing the inclined arrow on the title bar of the panel by assigning that command to the **Panel Dialog Box Launcher**.

1. Drag the **Text Style** command from the **Command List** rollout to the **Customizations in All Files** rollout and drop it on the **Panel Dialog Box Launcher** element displayed under the **Draw/Modify** node. Now, whenever you will choose the inclined arrow, the **Text Style** dialog box will get invoked.

Step 8: Accepting and saving the added ribbon panel

1. Choose the **Apply** button from the **Customize User Interface** dialog box to accept and apply the changes made till now, and then choose the **OK** button to save and add the **Draw/Modify** ribbon panel to the list of items under the **Panels** node.

Note

The process of adding and displaying the ribbon panel in the ***Ribbon*** *will be explained later in this chapter.*

CREATING A CUSTOMIZED RIBBON TAB

The ribbon tabs do not contain any commands or controls. It is only the means to manage the ribbon panels under a group. You can modify an existing ribbon tab or create a new ribbon tab from the **Customize User Interface** dialog box. To add a new tab, choose **Manage > Customization > User Interface** from the **Ribbon**; the **Customize User Interface** dialog box will be displayed. In this dialog box, Right-click on the **Tabs** element under the **Ribbon** node in the **Customizations in All Files** rollout of the **Customize** tab and choose the **New Tab** option from the shortcut menu, see Figure 6-17. A new tab with the name **New Tab** will be added to the list under the **Tabs** node and its properties will be displayed under the **Properties** rollout in the right pane.

Adding the Ribbon Panels to the Ribbon Tab

You can add the existing ribbon panels to the ribbon tab. To do so, invoke the **Customize User Interface** dialog box and expand the **Panels** node. Press the CTRL key and select the desired ribbon panels under the **Panels** node. Next, right-click on any panel by keeping the CTRL key pressed and then choose the **Copy** option from the shortcut menu. Now, release the CTRL key. Next, expand the **Tabs** node and right-click on the tab to which you want to add the copied panels. Choose the **Paste** option from the shortcut menu; all the copied panels will get added to the specified ribbon tab.

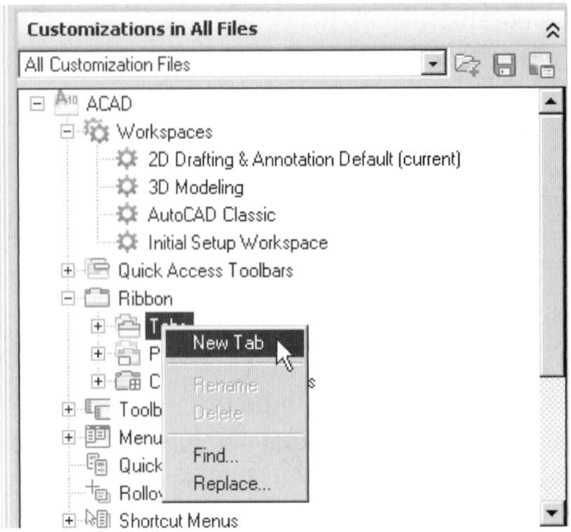

Figure 6-17 *Adding a new tab to the* ***Tabs*** *node under* ***Ribbon***

The sequence of display of the ribbon tabs in the **Ribbon** depends on the sequence of the ribbon tabs under the **Tabs** node in the **Customizations in All Files** rollout. You can change the sequence of the ribbon tabs by dragging and dropping them at the desired location under the **Tabs** node. Also, the above statement holds equally good for the ribbon panels.

Note

The process of displaying the newly created ribbon tab in the workspace is discussed later in this chapter.

CREATING CONTEXTUAL RIBBON TAB*

The name Contextual Ribbon tab itself indicates that these types of tabs are only displayed when a certain command is active or a certain object type is selected from the drawing area. For example, if you are using the **MTEXT** command to write a text, as soon as you specify the text boundary, the **Text Editor** tab will get added to the **Ribbon**. Therefore, in the referred case, the **Text Editor** tab of the **Ribbon** is the Contextual Ribbon tab. AutoCAD enables you to specify the type of ribbon tab to be displayed when a certain command is active or a certain object type is selected.

To assign an already created ribbon tab to a Contextual Ribbon tab, invoke the **Customize User Interface** dialog box. Then, expand the **Tabs** node under the **Ribbon** node in the **Customizations in All Files** rollout. Also, expand the **Contextual Tab States** node under the **Ribbon** node. Next, drag the required tab from the **Tabs** node and drop it over the desired entry listed under the **Contextual Tab States** node; the selected tab will be assigned to the entry listed under the **Contextual Tab States** node.

Note that the entries listed under the **Contextual Tab States** node are predefined on the basis of the object type selected or on the basis of the active command. These entries cannot be modified or changed. Now, if you select an object from the drawing area and a ribbon tab is assigned to that particular object type under the **Contextual Tab States** node, the assigned ribbon tab will be displayed in the **Ribbon**. After deselecting the object, the Contextual Ribbon tab will be removed from the **Ribbon**.

You can also specify how the Contextual Ribbon tab will be added and displayed in the **Ribbon**. To do so, select the ribbon tab added under any entry in the **Contextual Tab States** node; the properties of the selected ribbon tab will be displayed in the **Properties** rollout in the right pane of the **Customize User Interface** dialog box. Select the **Contextual Display Type** property under the **General** node of the **Properties** rollout; a drop-down list will be displayed on the right of the **Contextual Display Type** property. Now, you can select the required option from this drop-down list. If you select the **Full** option from this drop-down list, then the Contextual Ribbon tab will be added to the ribbon as a separate tab, and the panels of this tab will be displayed only when it is active, refer to Figure 6-18. On the other hand, if you select the **Merged** option from this drop-down list, then the Contextual Ribbon tab will not be displayed as a separate tab in the **Ribbon** and the panels of this tab will be displayed as merged with the active tab of the **Ribbon**, refer to Figure 6-19.

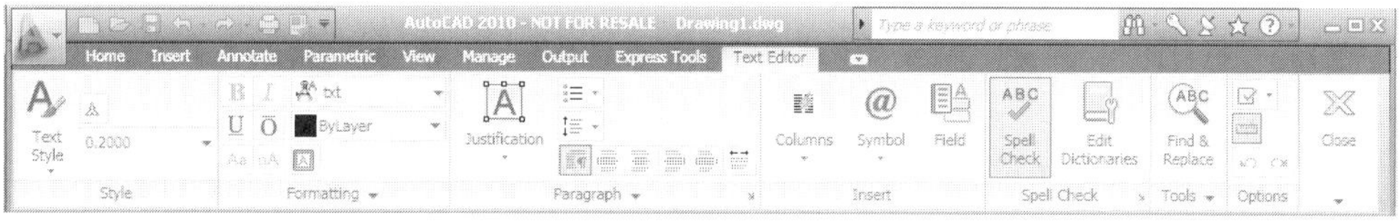

*Figure 6-18 The Contextual Ribbon tab displayed as a separate tab in the **Ribbon***

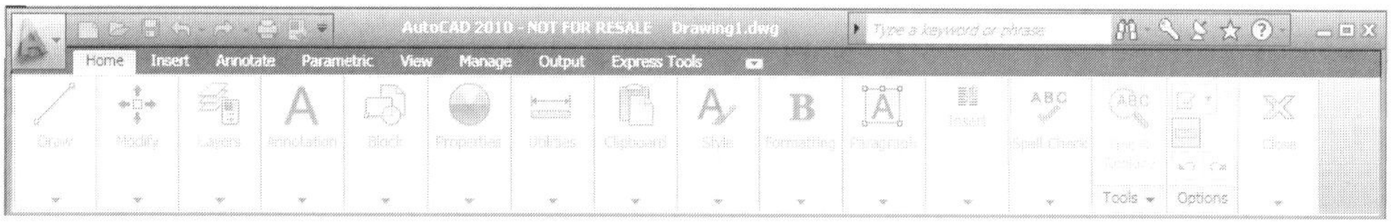

Figure 6-19** Panels of the Contextual Ribbon tab displayed as merged with the **Home** tab of the **Ribbon

CREATING AND MODIFYING WORKSPACES

A workspace is a customized arrangement of toolbars, menus, **Ribbon** and Palettes in the AutoCAD environment. Workspaces are required when you need to customize the AutoCAD environment for a specific use, where you need only a certain set of commands and tools. For such type of requirements, you can use the **Customize User Interface** dialog box to define workspaces, in which only the specified toolbars, menus, **Ribbon** and Palettes are available.

Adding a New Workspace Entry

To add a new workspace, choose the **New Workspace** option from the shortcut menu that will be displayed when you right-click on the **Workspaces** element in the **Customizations in All Files** rollout of the **Customize User Interface** dialog box, see Figure 6-20. A new workspace entry having a default name **Workspace1** will get added under the **Workspaces** node in the **Customizations in All Files** rollout. Rename the workspace in this rollout. Alternatively

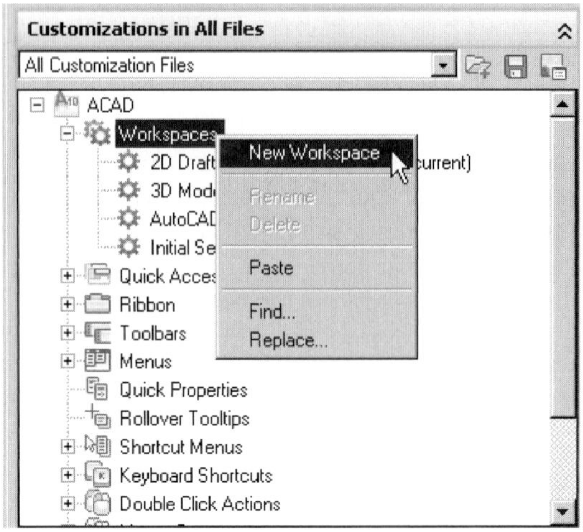

***Figure 6-20** Adding a new workspace entry under the **Workspaces** node of the **Customizations in All Files** rollout*

double-click on the **Name** edit box in the **Properties** rollout of the **Customize User Interface** dialog box and enter the desired name in place of **Workspace1**.

Adding Toolbars, Menus, and Ribbon Tabs to the Workspace

The **Workspace Contents** rollout is used to add toolbars, menus, palettes and ribbon tabs to the workspace. On selecting any workspace entry under the **Workspaces** node in the **Customizations in All Files** rollout, the **Workspace Contents** rollout is displayed in the right pane of the **Customize User Interface** dialog box, see Figure 6-21.

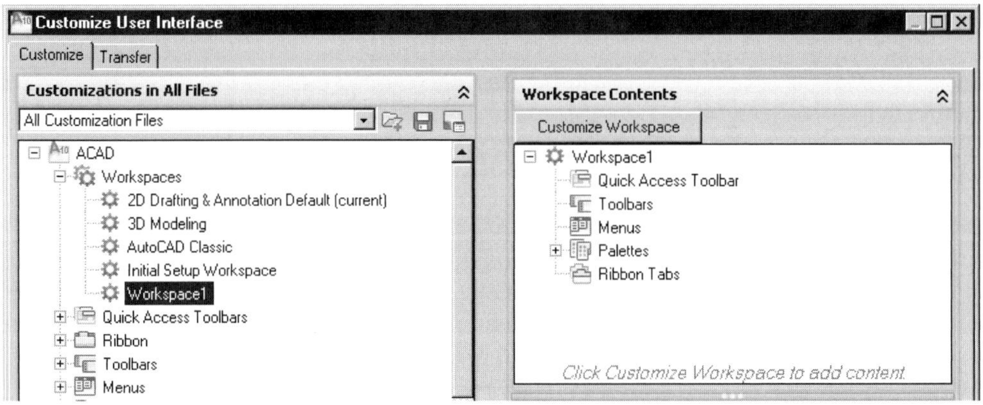

***Figure 6-21** Partial view of the **Customize User Interface** dialog box*

Choose the **Customize Workspace** button from the **Workspace Contents** rollout to facilitate adding toolbars, menus, ribbon tabs to the selected workspace, see Figure 6-22. Select the check box displayed on the left of the **Tabs** element under the **Ribbon** node in the

Customizations in All Files rollout to add all the ribbon tabs from the main *acad.cuix* file to the currently selected workspace. After selecting the check box, all ribbon tabs will get added to the **Ribbon Tabs** element in the **Workspace Contents** rollout. To add only the selected ribbon tabs to the workspace, expand the plus sign (+) on the left of the **Tabs** element in the **Customizations in All Files** rollout. All the ribbon tabs with a check box on the left of the tab names will be displayed. Select the check boxes that you want to include in the selected workspace; the selected ribbon tabs will instantaneously get added to the **Ribbon Tabs** element in the **Workspace Contents** rollout.

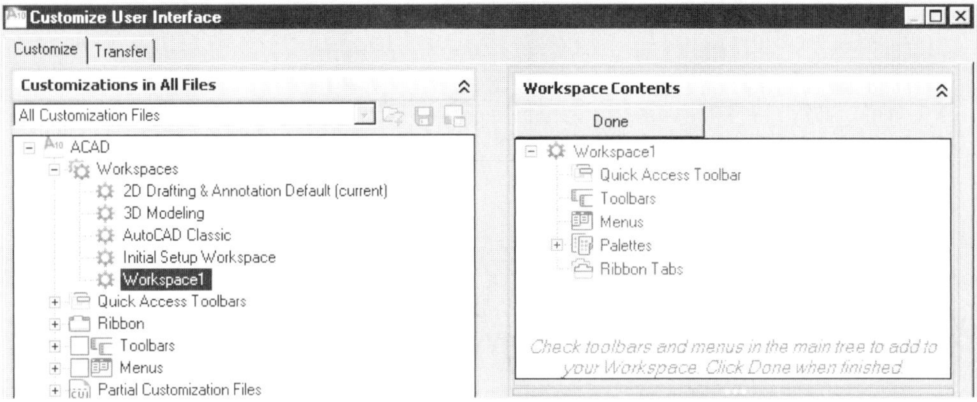

***Figure 6-22** Partial view of the **Customize User Interface** dialog box after choosing the **Customize Workspace** button*

Similarly, select the desired check boxes under the **Quick Access Toolbars**, **Menus** and the **Toolbars** elements in the **Customizations in All Files** rollout to add the selected menu and ribbon tabs to the selected workspace. You will get the preview of all toolbars, menus, and ribbon tabs added to the selected workspace in the **Workspace Contents** rollout. Next, choose the **Done** button from the **Workspace Contents** rollout to add the selected toolbars and menus to the specified workspace.

Controlling the Display of Palettes in the Workspace

You can also control the display of the palettes such as **Tool Palettes**, **Properties Palettes**, and so on, in a particular workspace. Select a workspace entry under the **Workspaces** node in the **Customizations in All Files** rollout. Next, expand the plus sign (+) on the left of the **Palettes** element in the **Workspace Contents** rollout of the **Customize User Interface** dialog box; the entries in the **Palettes** node are displayed, as shown in Figure 6-23. Now, select the desired palette under the **Palettes** node in the **Workspace Contents** rollout and click on the down arrow displayed on the right of the **Show** edit box in the **Properties** rollout. Select the desired display option from the **Show** drop-down list, see Figure 6-24.

If you select the **No** option from the **Show** drop-down list, the selected palette will not be displayed in the drawing area when you make the selected workspace current. Whereas, if

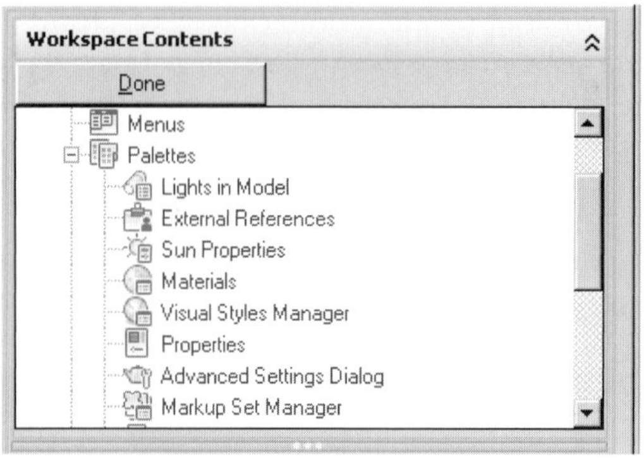

Figure 6-23 *The* ***Workspace Contents*** *rollout of the* ***Customize User Interface*** *dialog box after expanding the* ***Palettes*** *node*

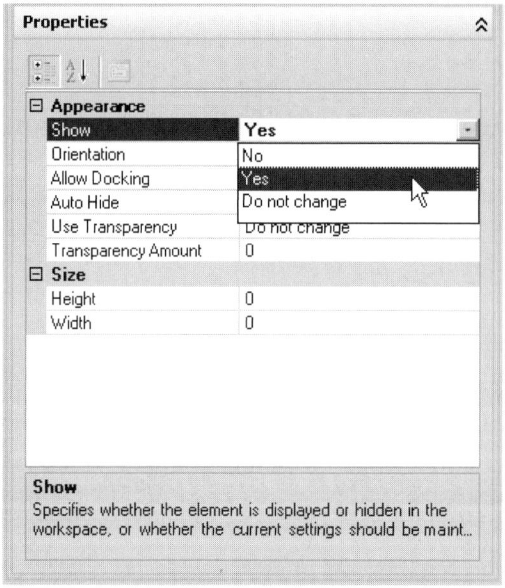

Figure 6-24 *The* ***Properties*** *rollout of the* ***Customize User Interface*** *dialog box for a selected palette entry under the* ***Palettes*** *node of the* ***Workspace Contents*** *rollout*

you select the **Yes** option from the **Show** drop-down list, the selected palette will be displayed in the drawing area by default, when you make the selected workspace current. Selecting the **Do not change** option maintains the last used display status of the palette when the selected workspace was set to current.

Similarly, you can control the other display properties such as orientation, auto-hide, size of the palette, and so on from the **Properties** rollout of the **Customize User Interface** dialog box.

Note

*You can also create a new workspace without using the **Customize User Interface** dialog box. To do so, invoke the toolbars and the window palettes that you want to display in the new workspace. Next, choose the **Workspace Switching** button from the Status Bar; a flyout will be displayed. Choose the **Save Current As** option from the flyout; the **Save Workspace** dialog box will be displayed. Enter the name of the new workspace in the **Name** edit box and choose the **Save** button; a new workspace is created with the selected tools.*

CUSTOMIZING THE QUICK ACCESS TOOLBAR*

The **Quick Access** toolbar is located on the right of the **Application Menu** and it contains a set of frequently used commands. You can customize the set of tools available in the **Quick Access** toolbar. You can also add more than one **Quick Access** toolbar in a workspace, but only one **Quick Access** Toolbar can be displayed at a time in any workspace.

Creating a New Quick Access Toolbar

To add a new **Quick Access** toolbar, right-click on the **Quick Access Toolbars** node in the **Customizations in All Files** rollout of the **Customize User Interface** dialog box; a shortcut menu will be displayed. Choose the **New Quick Access Toolbar** option from the menu; a new **Quick Access** Toolbar entry will be added under the **Quick Access Toolbars** node in the **Customizations in All Files** rollout.

Adding Commands to the Quick Access Toolbar

Whenever you create a new **Quick Access** Toolbar, some of the commonly used commands such as **New**, **Open**, **Save**, **Undo**, **Redo**, and **Plot** commands will automatically be added to it. To add more commands to the **Quick Access** Toolbar, drag the required commands from the **Command List** rollout and drop them under the desired **Quick Access Toolbar** element in the **Customizations in All Files** rollout of the **Customize User Interface** dialog box.

If you need to delete a **Quick Access Toolbar** element or the commands added in a **Quick Access** toolbar, right-click on the particular **Quick Access Toolbar** element or the command in the **Quick Access Toolbars** node of the **Customizations in All Files** rollout; a shortcut menu will be displayed. Choose the **Delete** or the **Remove** option from the shortcut menu to delete the selected **Quick Access Toolbar** element or command, respectively.

Controlling the Display of Quick Access Toolbar in the Workspace

As already discussed, you can create more than one **Quick Access** Toolbar, but only one **Quick Access** Toolbar can be displayed in a particular workspace. Therefore, you need to specify which **Quick Access** Toolbar should be displayed in that particular workspace. To do so, select a workspace entry under the **Workspaces** node in the **Customizations in All Files** rollout; the **Workspace Contents** rollout will be displayed in the right pane of the **Customize User Interface** dialog box. Choose the **Customize Workspace** button from the **Workspace Contents** rollout to specify the **Quick Access** Toolbar to be displayed. Now, select the desired check box displayed on the left of the **Quick Access Toolbar** elements under the **Quick Access Toolbars** node in the **Customization in All Files** rollout, see Figure 6-25. Note that only one check box can be selected at a time and the selected **Quick Access Toolbar** element will get added from the *acad.cuix* file to the currently selected workspace. Next, choose the **Done** button from the **Workspace Contents** rollout. Then, choose **Apply** and **OK** from the **Customize User Interface** dialog box to accept the changes and exit. Now onward, whenever the selected workspace will be set as the current workspace, the specified **Quick Access** Toolbar will be displayed on the AutoCAD Screen.

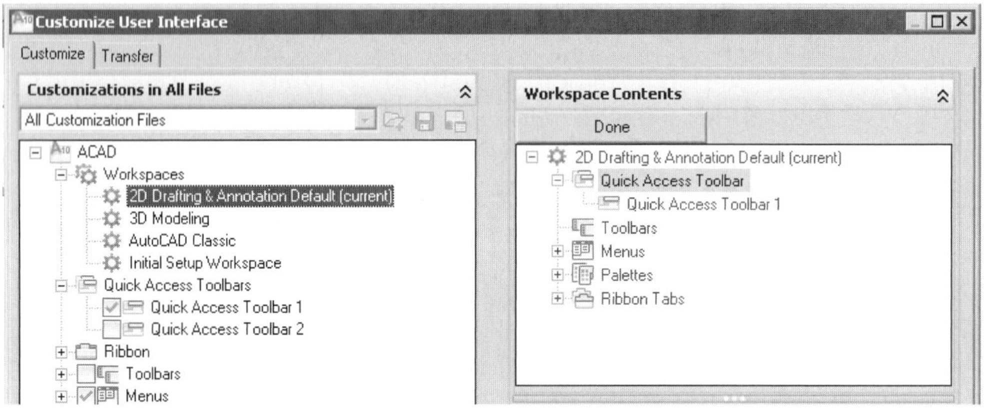

***Figure 6-25** Partial view of the **Customize User Interface** while customizing the selected workspace*

Example 2

Create a new workspace (*MyWorkspace*) such that when you set this workspace as current, the toolbars, menus, ribbon tabs, **Quick Access** toolbar, and Palettes given below are displayed in AutoCAD.

Toolbars	Menus	Ribbon Tabs	Quick Access Toolbar	Palettes
Draw	**View**	**Home-2D**	**Open**	**Properties**
Inquiry	**Edit**	**Annotate**	**Save**	**Ribbon**
Layers	**Insert**	**View-2D**	**Plot**	
Properties	**Format**	**Output**		
Zoom	**Tools**			

Step 1: Adding a new workspace entry

1. Invoke the **Customize User Interface** dialog box. Right-click on the **Workspaces** element in the **Customizations in All Files** rollout of the **Customize User Interface** dialog box; a shortcut menu will be displayed. Choose the **New Workspace** option from the shortcut menu.

2. A new workspace entry with a default name **Workspace1** will get added under the **Workspaces** node in the **Customizations in All Files** rollout. Rename **Workspace1** to **MyWorkspace**.

Step 2: Creating the Quick Access Toolbar

1. Right-click on the **Quick Access Toolbars** node in the **Customizations in All Files** rollout and then choose the **New Quick Access Toolbar** option from the shortcut menu displayed; a new **Quick Access** toolbar entry with the default name **Quick Access Toolbar 2** will get added under the **Quick Access Toolbars** node.

2. Rename the newly added **Quick Access Toolbar 2** element as **MyQAT**.

3. Expand the **MyQAT** node and then right-click on the **New** element; a shortcut menu will be displayed. Choose the **Remove** option from the shortcut menu.

4. Similarly, remove all entries under the **MyQAT** node except **Open**, **Save**, and **Plot**.

Step 3: Adding Quick Access toolbar, toolbars, menus, and ribbon tabs to the workspace

1. Select the **MyWorkspace** entry under the **Workspaces** node in the **Customizations in All Files** rollout.

2. Choose the **Customize Workspace** button from the **Workspace Contents** rollout; the elements entry in the **Customizations in All Files** rollout will change instantaneously to facilitate adding the **Quick Access** toolbar, toolbars, menus, and ribbon tabs to the selected workspace.

3. Expand the **Quick Access Toolbars** node in the **Customizations in All Files** rollout and select the check box on the left of the **MyQAT** entry; the selected **Quick Access** toolbar will get added to the new workspace.

4. Expand the **Toolbars** node; all the toolbars are displayed with a check box on the left of the toolbar name. Select the **Draw**, **Inquiry**, **Layers**, **Properties**, and **Zoom** check boxes. The selected toolbars will instantaneously get added to the **Toolbars** node in the **Workspace Contents** rollout.

5. Similarly, select the check boxes against the **Menus** and the **Ribbon > Tabs** node. The entries selected are mentioned under the **Menus** and the **Ribbon Tabs** headings in the example statement above.

Step 4: Displaying palettes in the workspace

1. Expand the **Palettes** node in the **Workspace Contents** rollout.

2. Now, select the **Properties** element under the **Palettes** node in the **Workspace Contents** rollout and select the **Show** edit box from the **Properties** rollout.

3. Click on the down arrow displayed on the right of the **Show** edit box; the **Show** drop-down list is displayed. Select the **Yes** option from the drop-down list.

4. Similarly, select the **Ribbon** element under the **Palettes** node in the **Workspace Contents** rollout and select the **Yes** option from the **Show** drop-down list.

5. Choose the **Done** button in the **Workspace Contents** rollout. Next, choose the **Apply** and **OK** buttons from the **Customize User Interface** dialog box.

Figure 6-26 shows the **Workspace Contents** rollout with the commands added under the **MyWorkspace** node.

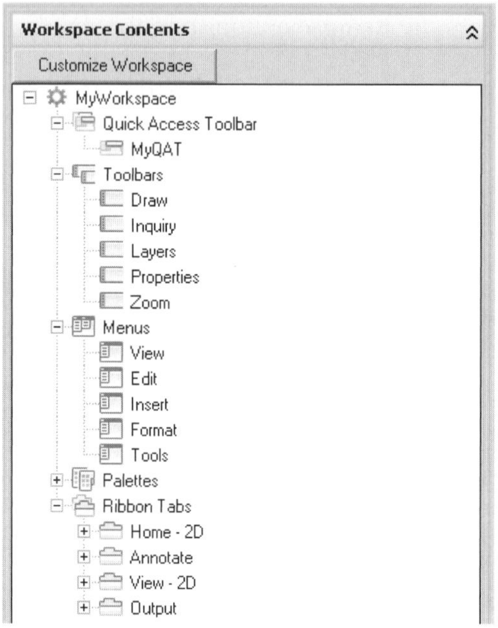

***Figure 6-26** The **Workspace Contents** rollout showing the commands added under the **MyWorkspace** node*

Set **MyWorkspace** as the current workspace. On doing so, only the toolbars, menus, ribbon tabs, **Quick Access** toolbar, and Palettes that are added under the **MyWorkspace** node in the **Customize User Interface** dialog box are displayed in the AutoCAD window.

CUSTOMIZING THE QUICK PROPERTIES PANEL

The **Quick Properties** panel enables you to see and modify some of the frequently used properties of the selected object, see Figure 6-27. The properties displayed in the **Quick Properties** panel are the subset of the properties displayed in the **Properties** palette.

Note

*If the **Quick Properties** panel is not displayed on selecting the object, choose the **Quick Properties** toggle button from the Status Bar to turn on the display of the **Quick Properties** panel.*

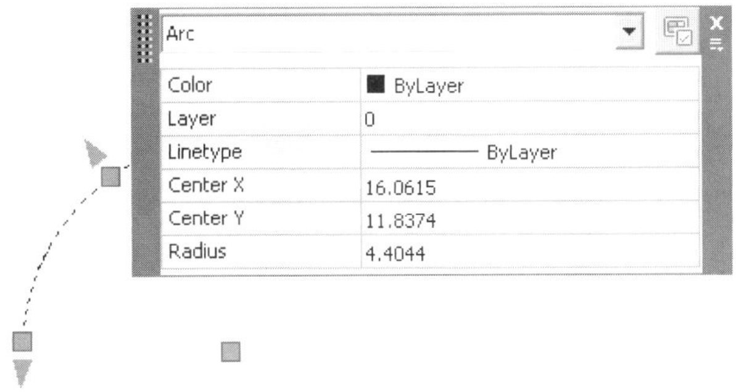

***Figure 6-27** The **Quick Properties** panel*

AutoCAD allows you to customize the **Quick Properties** panel in two ways:

1. By allowing you to specify the object type on selecting which, the **Quick Properties** panel will be displayed.

2. By allowing you to customize the properties to be displayed in the **Quick Properties** panel, when any particular object type is selected.

The above-mentioned ways of customizing the **Quick Properties** panel are discussed next.

Specifying the Object Types for Displaying the Quick Properties Panel

When you select certain object type from the drawing area, the **Quick Properties** panel will be displayed. You can specify the object types for displaying the **Quick Properties** panel by using the **Customize User Interface** dialog box. Invoke the **Customize User Interface** dialog box and select the **Quick Properties** element in the **Customizations in All Files** rollout; the Object Type and the Properties columns will be displayed in the right pane of the **Customize User Interface** dialog box, see Figure 6-28. The Object Type column lists the object types. If you select any of these objects in the drawing area, the **Quick Properties** panel will be displayed. The Properties column lists the properties that will be displayed in the **Quick Properties** panel for the object type selected in the Object Type column.

Choose the **Edit Object Type List** button from the upper right corner of the Object Type column; the **Edit Object Type List** window will be displayed on the screen, see Figure 6-29. Select the check box against the desired object type and then choose the **OK** button from the **Edit Object Type List** window. On doing so, the selected object types will be added in the Object Type column.

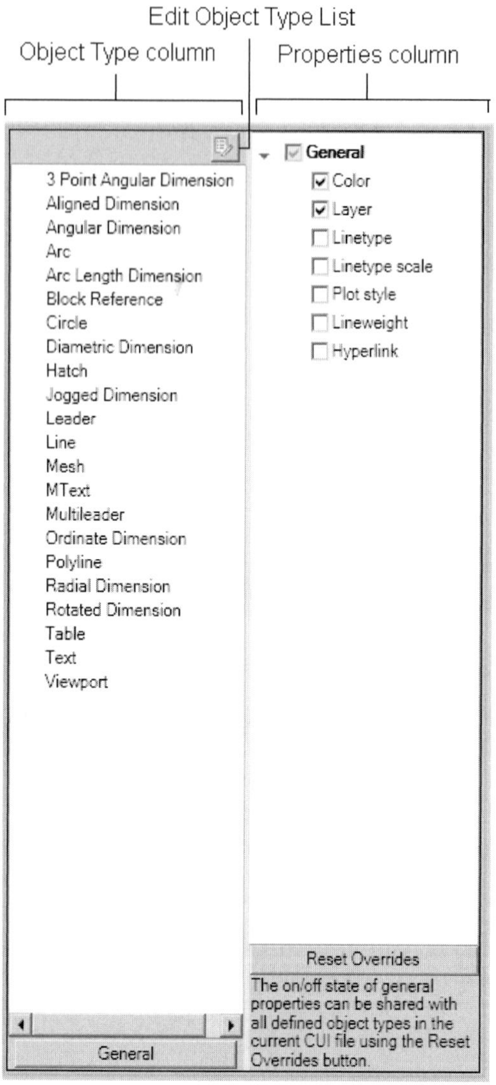

***Figure 6-28** The right pane of the **Customize User Interface** dialog box when the **Quick Properties** element is selected in the left pane*

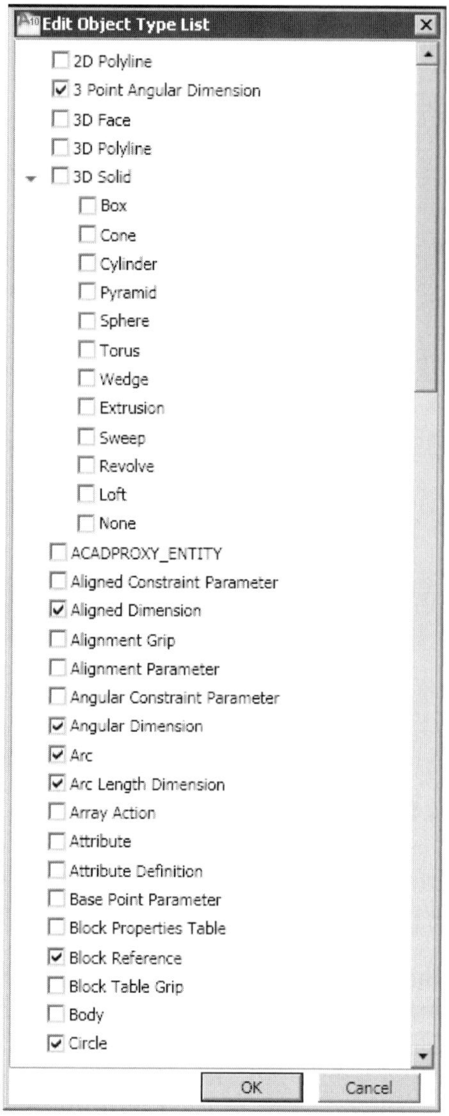

***Figure 6-29** The **Edit Object Type List** window*

Customizing the Properties to be Displayed for the Selected Object Type

To specify the object properties to be displayed for a particular object type, invoke the **Customize User Interface** dialog box and select the **Quick Properties** element in the

Customizations in All Files rollout. Now, select the required object type in the Object Type column for which you want to change the properties to be displayed; all properties related to the selected object type will be displayed in the Properties column. Select the check boxes against the desired properties.

Instead of defining the general properties (properties listed under the **General** node in the Properties column) to be displayed for all object types one by one, you can also do so in a single step. In the **Customize User Interface** dialog box, select the **Quick Properties** element in the **Customizations in All Files** rollout. If the general properties are not displayed in the Properties column, choose the **General** button at the bottom of the Object Type column; the list of general properties will be displayed in the Properties column. Select the check boxes against the desired general properties and choose the **Reset Overrides** button displayed at the bottom of the Properties column. The selected general properties will get associated with all the object types simultaneously.

Note

*You can restore the default settings of the object types and their properties to be displayed in the **Quick Properties** panel. To do so, right-click on the **Quick Properties** element in the **Customizations in All Files** rollout and choose the **Restore Default** option from the shortcut menu; the **Quick Properties-Restore Default** message box will be displayed on the screen, see Figure 6-30. Choose the **Restore the default Quick Properties settings** option from the message box.*

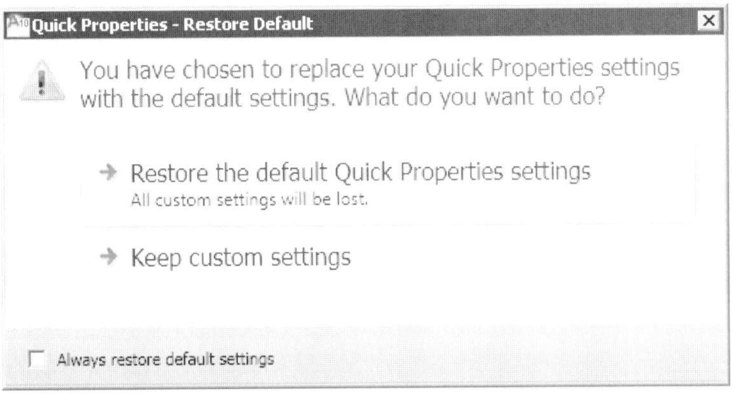

***Figure 6-30** The **Quick Properties-Restore Default** message box*

CUSTOMIZING THE ROLLOVER TOOLTIPS

Rollover tooltips enable you to preview some of the frequently used properties when you place the cursor over any object without selecting it, see Figure 6-31. Like quick properties, AutoCAD also allows you to customize the rollover tooltips. You can customize the object types and the properties to be displayed for a particular object type in the rollover tooltips. The process of customizing the rollover tooltips is similar to the one discussed for the customization of the quick properties.

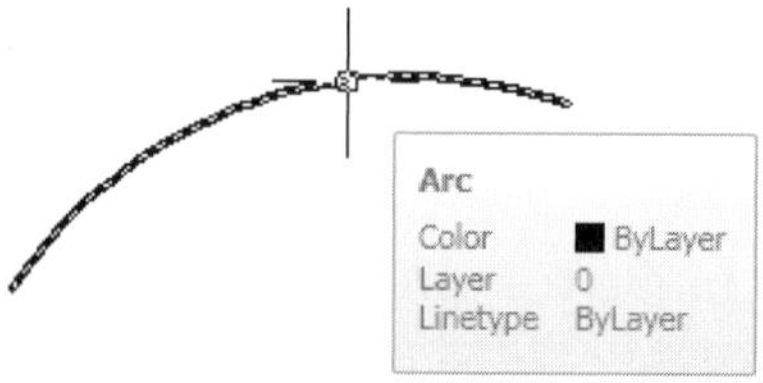

Figure 6-31 *The rollover tooltip*

You can also assign the same settings of the properties to the rollover tooltip as you did for the quick properties. In such cases, the same set of properties will be displayed in both the **Quick Properties** panel and the rollover tooltip for the same object type. To do so, right-click on the **Rollover Tooltips** element in the **Customizations in All Files** rollout and choose the **Synchronize with Quick Properties** option from the shortcut menu; the **Quick Properties-Synchronize with Rollover Tooltips** message box will be displayed, see Figure 6-32. Choose the **Apply Quick Properties settings to Rollover Tooltips** option from the message box; all object types and property settings of the **Quick Properties** panel will be applied to the **Rollover Tooltips**.

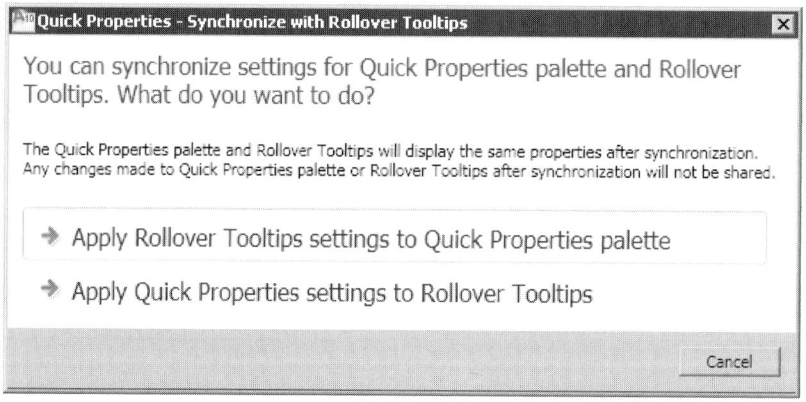

Figure 6-32 *The* ***Quick Properties-Synchronize with Rollover Tooltips*** *message box*

If you choose the **Apply Rollover Tooltips settings to Quick Properties palette** option from the **Quick Properties-Synchronize with Rollover Tooltips** message box, all object types and property settings of the **Rollover Tooltips** will be applied to the **Quick Properties** panel.

CUSTOMIZING THE TOOL PALETTES

AutoCAD provides the **Tool Palettes** as an easy and convenient way of placing and sharing the frequently used tools. In the **Tool Palettes**, the tools are organized in different groups. You can invoke only one group at a time from the **Tool Palettes**. These groups, in turn, contain the tool palettes in which the similar type of tools are organized. These tool palettes are displayed as tabs in the **Tool Palettes**.

To customize the **Tool Palettes**, choose **Manage > Customization > Tool Palettes** from the **Ribbon**; the **Customize** dialog box will be displayed, see Figure 6-33. You can also invoke the **Customize** dialog box by choosing **Tools > Customize > Tool Palettes** from the menu bar. The **Customize** dialog box has two list boxes: **Palettes** and **Palette Groups**. The **Palettes** list box lists all tool palettes available in AutoCAD, whereas the **Palette Groups** list box lists all existing groups in AutoCAD. The currently active group in the **Tool Palettes** is highlighted by bold alphabets in the **Customize** dialog box. The **Customize** dialog box helps you to create new groups in the **Tool Palettes**, as well as, add or remove tool palettes from the existing groups. It also enables you to modify the existing groups.

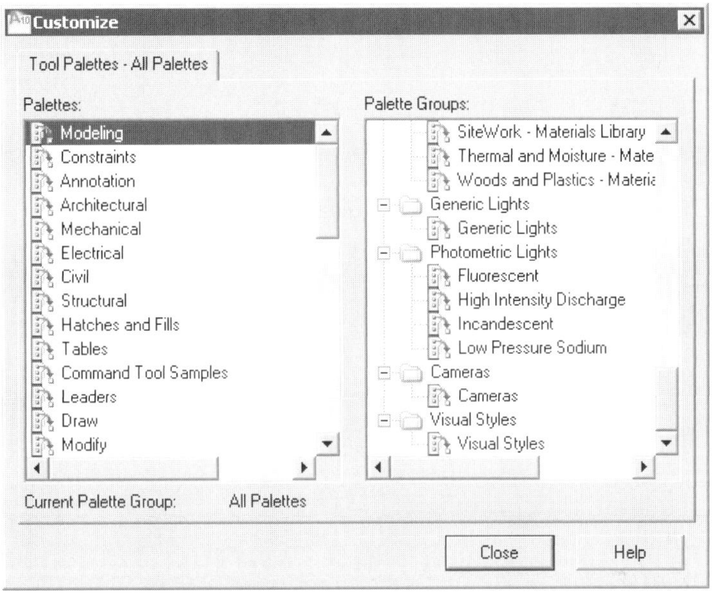

*Figure 6-33 The **Customize** dialog box*

Creating New Groups and Adding Tool Palettes to Them*

To create a new group, right-click in the blank area of the **Palette Groups** list box and then choose the **New Group** option from the shortcut menu, see Figure 6-34. A new group displaying the default name **New Group** will be added under any existing group. Rename the new group according to your requirement. Now, drag and drop the newly created group to the main tree of existing groups in the **Palettes Groups** list box, see Figure 6-35.

To add the tool palettes under the group in the **Tool Palettes**, select the desired tool palette in the **Palettes** list box, drag it to the **Palette Groups** list box and drop it under the group to which you want to add the selected tool palette, refer to Figure 6-36.

The order of display of the tool palette tabs in the **Tool Palettes** depends upon the sequence in which they are arranged under a group in the **Palette Groups** list box. You can rearrange the sequence of display of the tool palette tabs by dragging and dropping the tool palettes as per your requirement under a group in the **Palette Groups** list box.

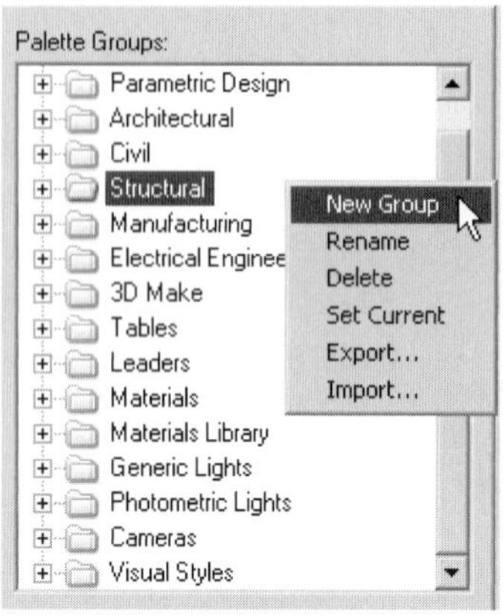

***Figure 6-34** Creating a new group in the **Palette Groups** list box*

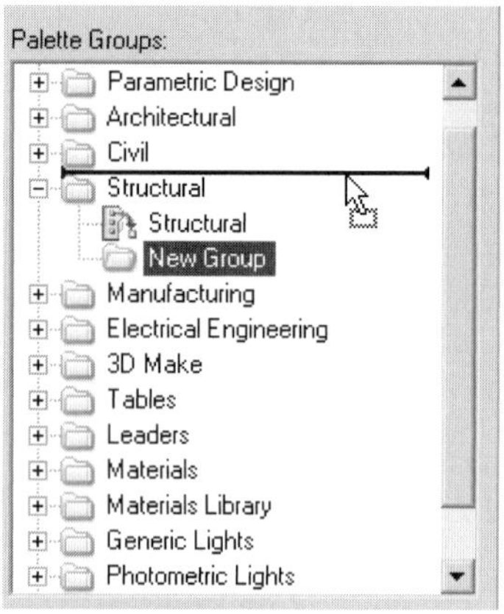

***Figure 6-35** Placing the newly created group in the **Palette Groups** list box*

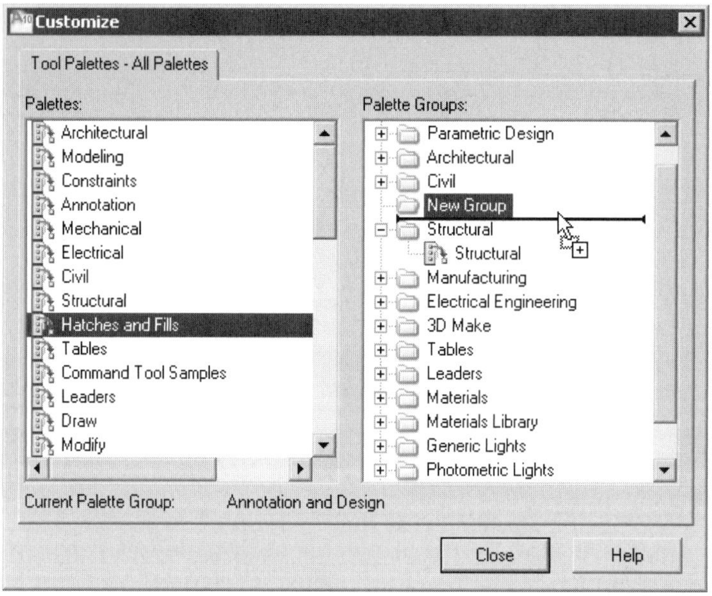

***Figure 6-36** Adding tool palette to the newly created group*

You can also add a new subgroup under any existing group. To do so, right-click on any existing group in the **Palette Groups** list box and choose the **New Group** option from the shortcut menu; a new subgroup will be added under the selected group. Next, add the tool palettes to the new subgroup one by one as discussed above.

Note

*To display any particular group of tool palettes in the **Tool Palettes**, right-click on the title bar of the **Tool Palettes** and choose the desired group from the shortcut menu.*

Exporting and Importing the Tool Palette Group

The **Customize** dialog box also enables you to export and save any tool palette group in the *.xpg* file format at the desired location. Later on, you can import the existing *.xpg* file to add the tool palette group to the **Tool Palettes**, whenever required.

To export the tool palette group, invoke the **Customize** dialog box and select the group to export from the **Palette Groups** list box. Right-click and choose the **Export** option from the shortcut menu; the **Export Group** dialog box will be displayed. Enter the name of the *.xpg* file in the **File name** edit box, specify the location where you want to export the file and then choose the **Save** button from the **Export Group** dialog box.

You can also import the existing tool palette group in the *.xpg* file format to the existing tool palette groups of the **Customize** dialog box. To do so, right-click on any of the existing group of the **Palette Groups** list box and choose the **Import** option from the shortcut menu; the **Import Group** dialog box will be displayed. Locate and select the *.xpg* file that you want to import and choose the **Open** button from the **Import Group** dialog box; the imported tool palette group will get added to the end of the existing tool palette groups list.

DEFINING DOUBLE CLICK ACTIONS

The **Double Click Action** is the command that is invoked after double-clicking on a drawn object. You can assign or modify a command to be executed when you double-click on an object. For example, when you double-click on the circle, the **PROPERTIES** command is invoked, allowing you to modify the properties of the circle. If you want any other command such as **COPY** command to be invoked on double-clicking the circle, you have to modify the double-click action assigned to the circle through the **Customize User Interface** dialog box.

To define a double-click action, invoke the **Customize User Interface** dialog box by choosing **Manage > Customization > User Interface** from the **Ribbon**. Right-click on the **Double Click Actions** in the **Customizations in All Files** rollout of the **Customize User Interface** dialog box and choose the **New Double Click Action** option from the shortcut menu; a new double-click action named **DoubleClick1** will be added at the bottom of the **Double Click Actions** tree. Rename to assign the double-click action a desired name.

After adding the double-click action in the **Customize User Interface** dialog box, you have to associate this double-click to a particular command such that when you double-click on an object created by this command, the assigned action will take place. To associate the double-click, enter the name of the command to which this double-click action will get associated in the **Object Name** edit box of the **Properties** rollout in the **Customize User Interface** dialog box. Once you exit this edit box, the values entered in the **Object Name** edit box will automatically get converted to uppercase. Next, enter the description about the double-click action in the **Description** edit box.

Note

The name entered in the ***Object Name*** *edit box must be in the valid Drawing Interchange Format (DXF). In most cases, this name is the same as the command name which is used to invoke that particular command. But in some cases, such as block and external references, the DXF names are different than the command name. For example, the* ***Object Name*** *for attributes of the block is* ***ATTRIB****.*

Next, you need to define the command that will be invoked when you double-click on an object created by the command defined in the **Object Name** edit box. Drag the desired command from the **Command List** rollout and drop it under the newly defined double-click action in the **Customizations in All Files** rollout. Choose the **OK** button from the **Customize User Interface** dialog box to accept the above defined settings. You cannot assign more than one command to a particular double-click action. If you try to drag the second command from the **Command List** rollout and drop it under the double-click action in the **Customizations in All Files** rollout, the previously added command will get replaced by the newer one.

Note

If a double-click action is not defined for a drawing object, the ***PROPERTIES*** *command will be invoked by default when you double-click on that drawing object.*

Modifying the Predefined Double Click Action

You can also modify the predefined double-click action through the **Customize User Interface** dialog box. To do so, expand the plus sign (+) on the left of the **Double Click Actions** element in the **Customizations in All Files** rollout. Choose the double-click action from the **Double Click Actions** node that you want to modify. Enter the new DXF name in the **Object Name** edit box of the **Properties** rollout to change the command with which this particular double-click is associated. To replace the current command assigned to the double-click action, drag the desired command from the **Command List** rollout and drop it under the double-click command name in the **Customizations in All Files** rollout.

Example 3

In this example, you need to define a double-click action for multileaders such that when you double-click on the multileader in the drawing, the **Multileader Style Manager** dialog box is invoked.

Step 1: Adding a new double-click action

1. Right-click on **Double Click Actions** in the **Customizations in All Files** rollout of the **Customize User Interface** dialog box and choose **New Double Click Action** from the shortcut menu.

2. A new double-click action named as **DoubleClick1** is added to the bottom of the **Double Click Actions** tree. Rename **DoubleClick1** to **Multileader**.

Step 2: Associating multileader double-click action to the MULTILEADER command

1. Enter **MULTILEADER** in the **Object Name** edit box of the **Properties** rollout in the **Customize User Interface** dialog box, see Figure 6-37.

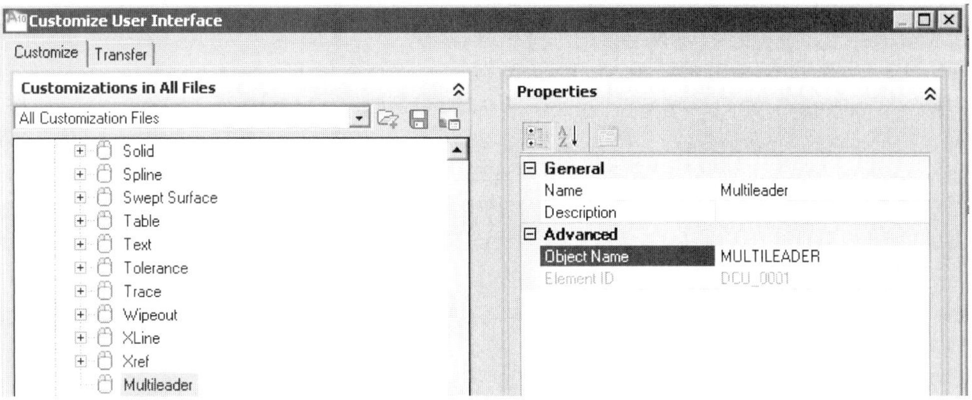

***Figure 6-37** Partial view of the **Customize User Interface** dialog box*

Step 3: Defining the command to be invoked on double-click

1. Drag **Multileader Style Manager** from the **Command List** rollout and drop it under the **Multileader** double-click action in the **Customizations in All Files** rollout.

2. Choose the **OK** button from the **Customize User Interface** dialog box to accept the above defined double-click action.

The **Multileader Style Manager** has been assigned to the Multileader double-click action. Now onwards, whenever you will double-click on a multileader in the drawing area, the **Multileader Style Manager** dialog box will be invoked.

Self-Evaluation Test

Answer the following questions and then compare them to those given at the end of this chapter:

1. The lower section of a ribbon panel is always displayed by default. (T/F)

2. The rows in the **Ribbon** are displayed in the order in which they are arranged under the ribbon panel name in the **Customizations in All Files** rollout. (T/F)

3. The command name (command label) cannot be displayed along with the button image in the ribbon panel. (T/F)

4. A sub-panel can be created within the ribbon panel. (T/F)

5. Commands available in any toolbar can be copied and transferred to a ribbon panel. (T/F)

6. You cannot specify the object type for which the **Quick Properties** panel will be displayed in the drawing area. (T/F)

7. The **All Commands** list of the **Command List** rollout displays all commands available in the selected *.cuix* file. (T/F)

8. The Contextual Ribbon tab can be defined by assigning tabs to the **Contextual Tab States** node under the **Ribbon** node of the **Customize User Interface** dialog box. (T/F)

9. The palette groups can be exported in the __________ file format.

10. The preview of the selected ribbon panel and the command added to it are displayed in the __________ rollout of the **Customize User Interface** dialog box.

Review Questions

Answer the following questions:

1. You cannot shuffle the position of rows in the ribbon panel by dragging and dropping them. (T/F)

2. You can also import an existing BMP image into the list of the images in the **Button Image** rollout. (T/F)

3. You can rename the drop-downs added under the rows, but you cannot rename the rows added under the ribbon panel in the **Customizations in All Files** rollout of the **Customize User Interface** dialog box. (T/F)

4. You can modify the predefined double-click action through the **Customize User Interface** dialog box. (T/F)

5. If a double-click action is not defined for a drawing object, the __________ command will be invoked by default while double-clicking on the drawing object.

Exercises

Exercise 1

Create the ribbon panel named *Curves/Workspaces*, as shown in Figure 6-38.

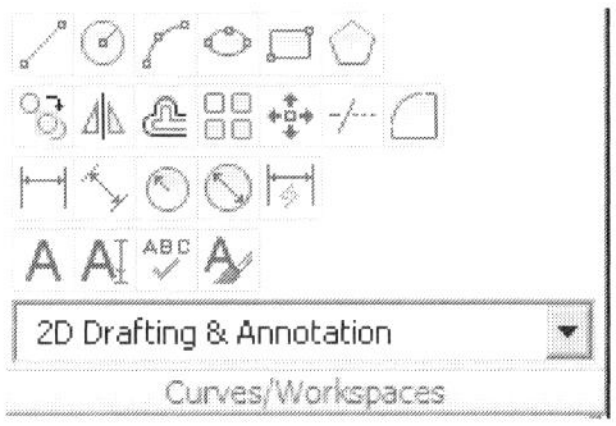

Figure 6-38 *The ribbon panel for Exercise 1*

Exercise 2

Create the ribbon tab named *MyRibbon Tab*, as shown in Figure 6-39. Add the ribbon panel created in Exercise 1 to this tab and make necessary adjustments to display the **MyRibbon Tab** tab as the first tab in the **Ribbon**.

Figure 6-39 *The ribbon tab for Exercise 2*

Exercise 3

Create a new workspace (*MyWorkspace*) such that when you set the workspace as current, the toolbars, menus, ribbon tabs, **Quick Access** toolbar, and Palettes given below are displayed in AutoCAD.

Toolbars	Menus	Ribbon Tabs	Quick Access Toolbar	Palettes
Properties	**File**	**Output-2D**	**Open**	**Ribbon**
UCS	**Edit**	**View-2D**	**New**	**Properties**
Workspaces	**Insert**	**Annotate**	**Save**	**Hatch**
Zoom	**Format**			
Multileader	**Tools**			

Exercise 4

Modify the **Quick Properties** panel that will be displayed on selecting a rectangle. It should display only the layer in which the rectangle has been created. Also, assign the same setting to the **Rollover Tooltips**.

Exercise 5

Modify the double-click action for **Tolerance** such that when you double-click on the geometric tolerance in the drawing, the **Geometric Tolerance** dialog box is invoked that allows you to create a new geometric dimension.

Exercise 6

Create a ribbon tab named **ModifyArc** with the following commands:
MOVE, **COPY**, **SCALE**, **OFFSET**, **ARRAY** and **MIRROR**.

Next, assign this ribbon tab to the **Arc selected** element under the **Contextual Tab States** node. Also, set the properties of this contextual tab such that whenever an arc is selected, this tab should be displayed as merged with the active tab of the ribbon.

Answers to Self-Evaluation Test

1. F, **2.** T, **3.** F, **4.** T, **5.** T, **6.** F, **7.** T, **8.** T, **9.** *.xtp*, **10. Panel Preview**

Chapter 7

Shapes and Text Fonts

Learning Objectives

After completing this chapter, you will be able to:

- *Write shape files.*
- *Use vector length and direction encoding to write shape files.*
- *Compile and load shape/font files.*
- *Use special codes to define a shape.*
- *Write text font files.*

SHAPE FILES

AutoCAD provides a facility to define shapes and text fonts. These files are ASCII files with the extension *.shp*. You can write these files using any text editor like Notepad.

Shape files contain information about the individual elements that constitute the shape of an object. The basic objects that are used in these files are lines and arcs. You can define any shape using these basic objects, and then insert them anywhere in a drawing. The shapes are easy to insert and take less disk space than blocks. However, there are some disadvantages of using shapes. For example, you cannot edit a shape or change it. Blocks, on the other hand, can be edited.

SHAPE DESCRIPTION

Shape description consists of the following two parts: a header and a shape specification.

Header

The header has the following format:

***SHAPE NUMBER, DEFBYTES, SHAPE NAME**
*201,21,HEXBOLT
Where ***201** --------------- Shape number
21 ----------------- Number of data bytes in shape specification
HEXBOLT ----- Shape name

Every header line starts with an asterisk (*), followed by the **SHAPE NUMBER**. The shape number is any number between 1 and 255 in a particular file and these numbers cannot be repeated within the same file. However, the numbers can be repeated in another shape file with a different name. **DEFBYTES** is the number of data bytes used by the shape specification and includes the terminating zero. **SHAPE NAME** is the name of a shape, in uppercase letters. The name is ignored if the letters are lowercase. Also, remember that the file must not contain two shapes with the same name.

Shape Specification

The shape specification contains the complete definition of the shape of an object. The shape is described with special codes, hexadecimal numbers, and decimal numbers. A hexadecimal number is designated by a leading zero (012) and a decimal number is a regular number without a leading zero (12). The data bytes are separated by a comma (,). The maximum number of data bytes is 2,000 bytes per shape, and in a particular shape file there can be more than one shape. The shape specification can have multiple lines. You should define the shape in logical blocks and enter each block on a separate line. This makes it easier to edit and debug the files. The number of characters on any line must not exceed 80. The shape specification is terminated with a zero.

VECTOR LENGTH AND DIRECTION ENCODING

Figure 7-1 shows the vector direction codes. All vectors in this figure have the same length specification. The diagonal vectors have been extended to match the closest orthogonal vector. Let us assume that the endpoint of vector 0 is two grid units from the intersection point of vectors. The endpoint of vector 1 is one grid directly above the endpoint of vector 0. Therefore, the angle of vector 1 is 26.565-degree (Tan^{-1} 1/2 = Tan^{-1} 0.5 = 26.565). Similarly, vector 2 is at 45-degree (Tan^{-1} 2/2 = Tan^{-1} 1 = 45).

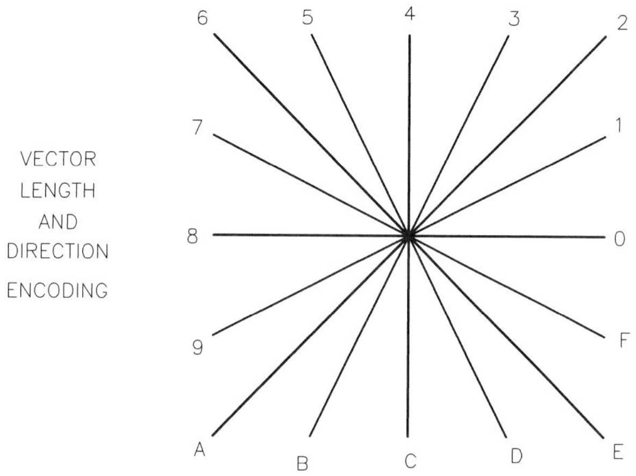

Figure 7-1 *Vector length and direction encoding*

All these vectors have the same magnitude or length specification. In other words, although their actual lengths vary, they are all considered one unit in definition. To define a vector you need its magnitude and direction. That means each shape specification byte contains vector length and a direction code. The maximum length of the vector is 15 units. Example 1 illustrates the use of vectors.

Example 1

Write a shape file for the resistor, as shown in Figure 7-2. The name of the file is *sh1.shp* and the shape name is RESIS.

Step 1: Writing the shape file

Use any text editor to write the shape file. The following two lines define the shape file for the given resistor:

***201,8,RESIS**
020,023,04D,043,04D,023,020,0

The first line is the header line and the second line is the shape specification.

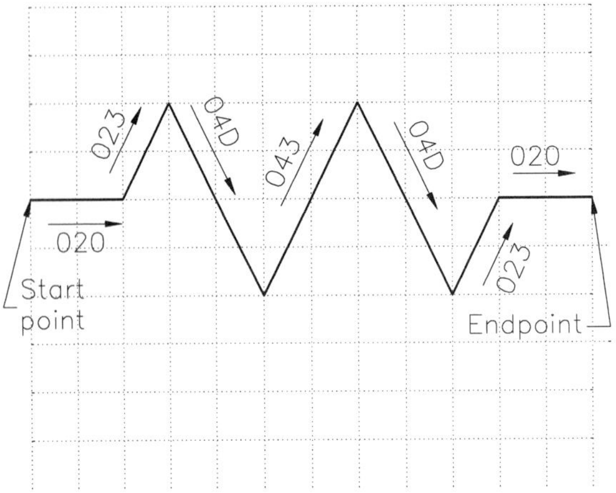

Figure 7-2 *Resistor*

Explanation

Header Line

***201,8,RESIS**

*201 is the shape number and 8 is the number of data bytes contained in the shape specification line. RESIS is the name of the shape.

Shape Specification

<u>020</u>,023,04D,043,04D,023,020,0

Where **0** ------------------- Hexadecimal notation
2 ------------------- Vector length
0 ------------------- Direction code

Each data byte in this line, except the terminating zero, has three elements. The first element (0) is the hexadecimal, the second is the length of the vector, and the third element is the direction code. For the first data byte, 020, the length of the vector is 2, and the direction is along the direction vector 0. Similarly, for the second data byte, 023, the first element, 0, is for hexadecimal; the second element, 2, is the length of the vector; and the third element, 3, is the direction code for the vector.

Step 2: Compiling the shape file

Save the file with the extension *.shp*. Now you can compile the shape file or the font file by using the **COMPILE** command. Enter the **COMPILE** command at the Command prompt to display the **Select Shape or Font File** dialog box (Figure 7-3). From this dialog box, select the shape file that you want to compile. AutoCAD will compile the file, and when the compilation is successful, the following prompt will be displayed at the command line.

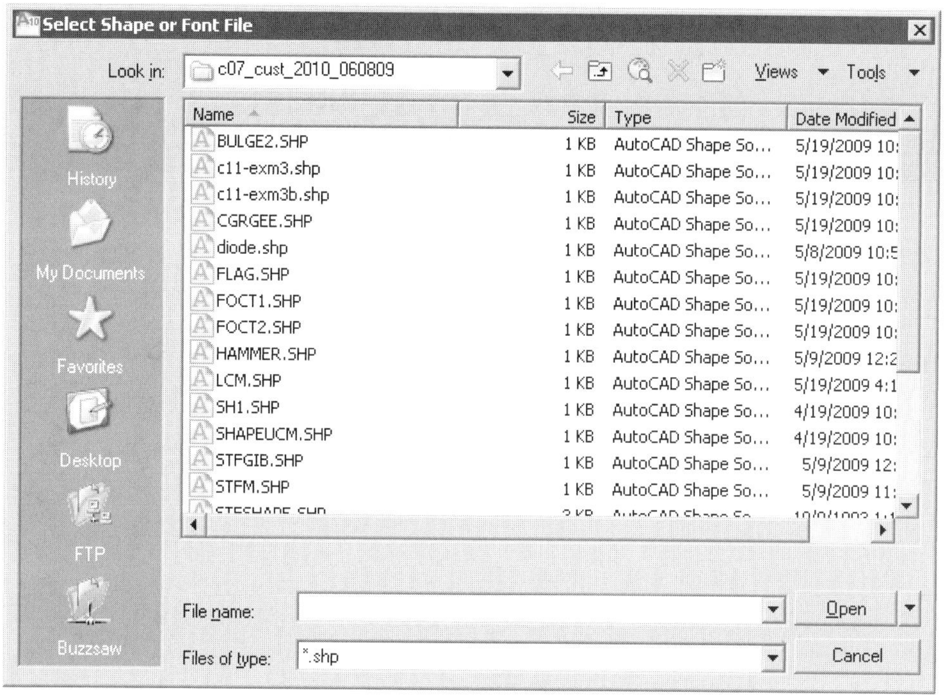

Figure 7-3 *The* ***Select Shape or Font File*** *dialog box*

Compilation successful
Output file name.shx contains nn bytes

Compilation translates a file with extension *.shp* into *.shx*. For Example 1, the name of the compiled output file is *sh1.shx* and the number of bytes is 51. If AutoCAD encounters an error in compiling a shape file, an error message will be displayed, indicating the type of error and the line number where the error occurred. You can also enter the options at the Command prompt keeping **FILEDIA** as 0.

Step 3: Loading the shape file

Enter the **LOAD** command at the Command prompt to display the **Select Shape File** dialog box. Select the file you want to load and then choose the **Open** button (Figure 7-4) or enter the command **LOAD** at the Command prompt keeping **FILEDIA** as 0.

Command: **LOAD**
Enter name of shape file to load or [?]: **SH1**

SH1 is the name of the shape file for Example 1. Do not include the extension **.SHX** with the name because AutoCAD automatically assumes the extension. If the shape file is present, AutoCAD will display the shape names that are loaded.

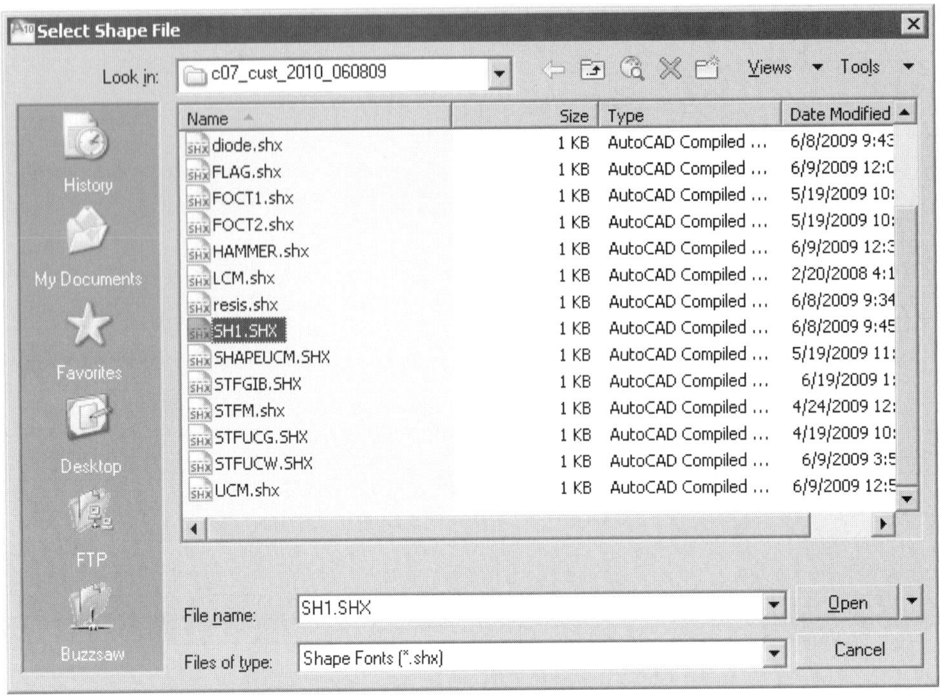

Figure 7-4 *The* ***Select Shape File*** *dialog box*

To insert the loaded shapes, use the AutoCAD **SHAPE** command:

Command: **SHAPE**
Enter shape name (or ?)<default>: *Shape name*
Specify insertion point: *Shape origin.*
Specify height<1.0>: *Number or point.*
Specify rotation angle<0.0>: *Number or point*

For Example 1, the shape name is RESIS. After you enter the information about the start point, height, and rotation, the shape will be displayed on the screen (Figure 7-5).

SPECIAL CODES

Generating shapes with the direction vectors has certain limitations. For example, you cannot draw an arc or a line that is not along the standard direction vectors. These limitations can be overcome by using special codes that add flexibility and give you better control over the shapes you want to create.

Standard Codes

There are certain standard codes that are used to denote particular action related to the shapes. The following is the listing of these codes.

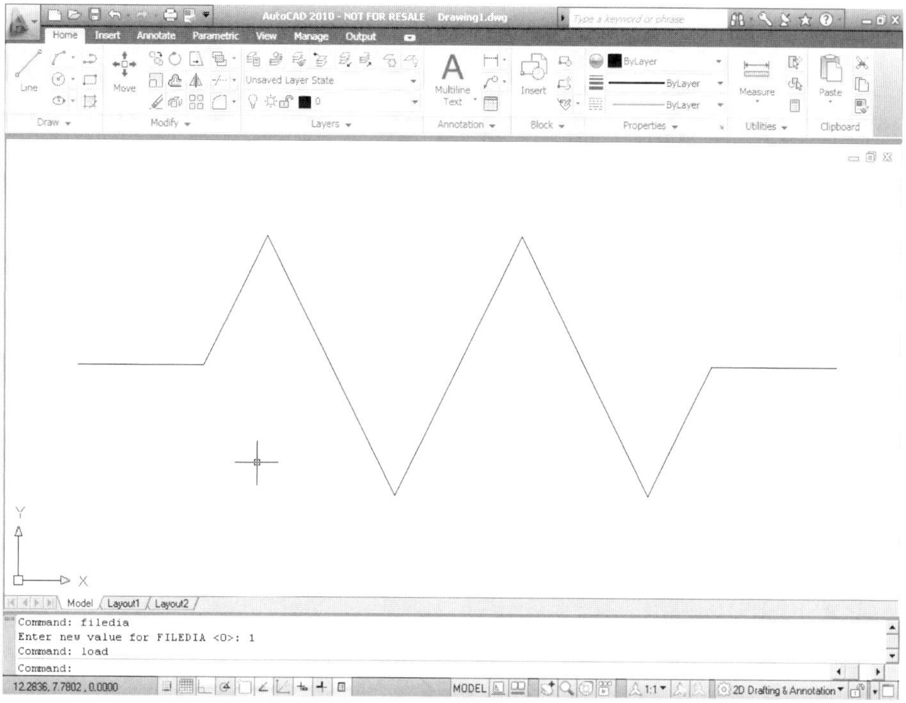

Figure 7-5 The shape (RESIS) inserted in the drawing

000	End of shape definition
001	Activate draw mode (pen down)
002	Deactivate draw mode (pen up)
003	Divide vector lengths by next byte
004	Multiply vector lengths by next byte
005	Push current location from stack (saving a location)
006	Pop current location from stack (restoring the location)
007	Draw subshape numbers given by next byte (calling a subshape)
008	X-Y displacement given by the next two bytes (non uniform lines)
009	Multiple X-Y displacement, terminated by (0,0) (continuous use of non uniform line)
00A or 10	Octant arc defined by next two bytes
00B or 11	Fractional arc defined by next five bytes
00C or 12	Arc defined by X-Y displacement and bulge
00D or 13	Multiple bulge-specified arcs (continuous use of bulge arc)
00E or 14	Process next command only if vertical text style

Code 000: End of Shape Definition

This code marks the end of a shape definition.

Code 001: Activate Draw Mode

This code turns the draw mode on. When you start a shape, the draw mode is on, so you do not need to use this code. However, if the draw mode has been turned off, you can use code 001 to turn it on.

Code 002: Deactivate Draw Mode

This code turns the draw mode off. It is used when you want to move the pen without drawing a line.

Let us say the distance from point 1 to point 2, from point 2 to point 3, and from point 3 to point 4 is 2 units each. The shape specification for this line is as follows:

020,002,020,001,020,0

The first data byte, 020, generates a line of 2 units long along the direction vector 0. The second data byte, 002, deactivates the draw mode; and the third byte, 020, generates a blank line 2 units long. The fourth data byte, 001, activates the draw mode; and the next byte, 020, generates a line that is 2 units long along direction vector 0. The last byte, 0, terminates the shape description.

Example 2

Create a shape file to generate the symbol of the diode, as shown in Figure 7-6.

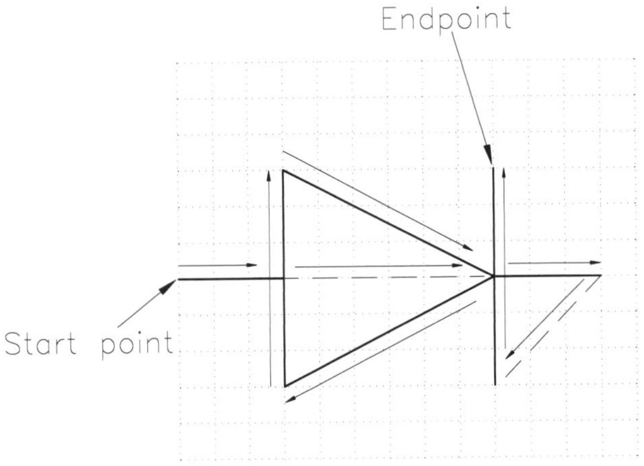

***Figure 7-6** Symbol of diode*

Step 1: Writing the shape file

You can use any text editor such as notepad to write a shape file. The name of the file is **DIODE** and the shape name is **DIOD. In the following file, the line numbers are not a part of the file; they are for reference only.**

```
1 *214,14,DIOD
2 001,030,
3 002,060,
4 001,069,064,06F,030,
5 002,03A,
6 001,064,0
7
```

Explanation

Line 1
***214,14,DIOD**
The first data byte contains an Asterisk (*) and shape number 214. The second data byte is the number of data bytes contained in the shape specification, including the terminating 0. DIOD is the name of the shape.

Line 2
001,030,
The data byte 001 activates the draw mode (pen down). In 030, the first element (0) is the hexadecimal, the second is the length of the vector and the third element is the direction code. The length of the vector is 3, and the direction is along the direction vector 0.

Line 3
002,060,
The data byte 002 deactivates the draw mode (pen up), and the next data byte defines a 6 units vector along the direction vector 0.

Line 4
001,069,064,06F,030,
The data byte 001 activates the draw mode (pen down). The data byte 069 defines a vector that is 6 units long along the direction vector 9. The data byte 064 defines a vector that is 6 units long along the direction vector 4. The data byte 06F defines a 6 units vector along the direction vector F. The data byte 030 defines a 3 units vector along the direction vector 0.

Line 5
002,03A,
The data byte 002 deactivates the draw mode (pen up). The data byte 03A defines a 3 units vector along the direction vector A.

Line 6
001,064,0
The data byte 001 activates the draw mode (pen down). The data byte 064 defines a 6 units vector along the direction vector 4. The data byte 0 terminates the shape specification.

Line 7
Blank enter

Step 2: Compiling the shape file

Save the file with the extension *.shp*. Now, you can compile the shape file or the font file by using the **COMPILE** command. Enter the **COMPILE** command at the Command prompt to display the **Select Shape or Font File** dialog box. From this dialog box, select the shape file that you want to compile. AutoCAD will compile the file, and when the compilation is successful, the following prompt will be displayed at the command line.

Compilation successful
Output file name.shx contains nn bytes

Step 3: Loading the shape file

Enter the **LOAD** command at the Command prompt to display the **Select Shape File** dialog box. Select the file you want to load and then choose the **Open** button. Alternatively, you can enter the command **LOAD** at the Command prompt keeping the **FILEDIA** as 0.

Command: **LOAD**
Enter name of shape file to load or [?]: **DIODE**

DIODE is the name of the shape file for Example 2. Do not include *.shx* with the name because AutoCAD automatically assumes the extension. If the shape file is present, AutoCAD will display the shape names that are loaded.

To insert the loaded shapes, use the AutoCAD **SHAPE** command as follows:

Command: **SHAPE**
Enter shape name (or ?)<default>: **DIOD**
Specify insertion point: *Specify the shape origin.*
Specify height<1.0: *Specify the height.*
Specify rotation angle<0.0>: *Specify the rotation angle*

Example 3

Write a shape file to generate the character "G", as shown in Figure 7-7.

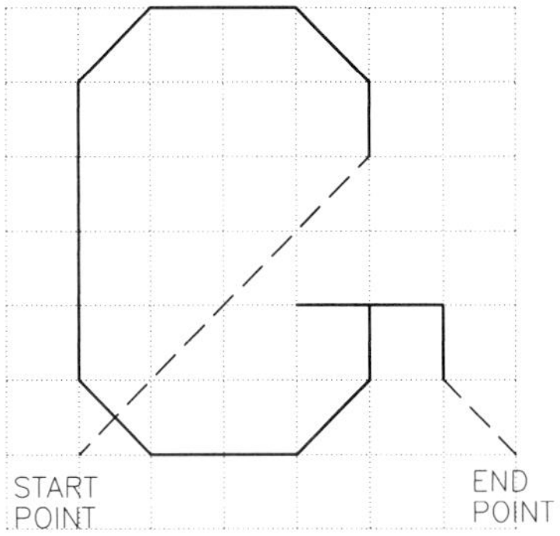

Figure 7-7 *Shape of the character "G"*

Step 1: Writing the shape file

You can use any text editor like notepad to write a shape file. The name of the file is **CHRGEE** and the shape name is GEE. **In the following file, the line numbers are not a part of the file; they are for reference only.**

```
1 *215,20,GEE
2 002,042,
3 001,014,016,028,01A,
4 04C,01E,020,012,014,
5 002,018,
6 001,020,01C,
7 002,01E,0
8
```

Explanation

Line 1
***215,20,GEE**
The first data byte contains an asterisk (*) and shape number 215. The second data byte is the number of data bytes contained in the shape specification, including the terminating 0. GEE is the name of the shape.

Line 2
002,042,
The data byte 002 deactivates the draw mode (pen up), and the next data byte defines a vector 4 units long along the direction vector 2.

Line 3
001,014,016,028,01A,
The data byte 001 activates the draw mode (pen down), and 014 defines a vector that is 1 unit

long at 90-degree (direction vector 4). The data byte 016 defines a vector that is 1 unit long along direction vector 6. The data byte 028 defines a vector that is 2 units long along direction vector 8 (180-degree). The data byte 01A defines a unit vector along the direction vector A.

Line 4
04C,01E,020,012,014,
The data byte 04C defines a vector that is 4 units long along direction vector C. The data byte 01E defines a direction vector that is 1 unit along direction vector E. The data byte 020 defines a direction vector that is 2 units long along direction vector 0 (0-degree). The data byte 012 defines a direction vector that is 1 unit long along direction vector 2. Similarly, 014 defines a vector that is 1 unit long along direction vector 4.

Line 5
002,018,
The data byte 002 deactivates the pen (pen up), and 018 defines a vector that is 1 unit long along direction vector 8.

Line 6
001,020,01C,
The data byte 001 activates the pen (pen down), and 020 defines a vector that is 2 units long along direction vector 0. The data byte 01C defines a vector that is 1 unit long along direction vector C.

Line 7
002,01E,0
The data byte 002 deactivates the pen, and the next data byte, 01E, defines a vector that is 1 unit long along direction vector E. The data byte 0 terminates the shape specification.

Line 8
Blank enter

Tip

After compiling a shape definition file in any drawing file, it need not to be recompiled in that particular drawing file. The file will reload automatically when the drawing is reopened. If you change the shape file through any process and after loading and compiling, it does not reflect the changes, then open a new drawing and compile and load the shape again. The existing drawing file does not incorporate the changes made in the shape file.

Code 003: Divide Vector Lengths by Next Byte

This code is used if you want to divide a vector by a certain number. In Example 2, if you want to divide the vectors by 2, the shape description can be written as:

003,2,020,002,020,001,020,0

The first byte, 003, is the division code, and the next byte, 2, is the number by which all the remaining vectors are divided. The length of the lines and the gap between the lines will be equal to 1 unit:

Also, the scale factors are cumulative within a shape. For example, if you insert another code, 003, in preceding shape description, the length of last vector, 020, will be divided by 4 (2 x 2):

003,2,020,002,020,001,003,2,020,0

Where **003,2** ------------ All the vectors are divided by 2
003,2 ------------ All the remaining vectors are divided by 4 (2 x 2)

Here is the output of this shape file:

Code 004: Multiply Vector Lengths by Next Byte

This code is used if you want to multiply the vectors by a certain number. It can also be used to reverse the effect of code 003. All multiplying and dividing vectors must be integers, 1-255.

003,2,020,002,020,001,004,2,020,0

Where **003,2** ------------ Divides all the vectors on the right by 2
004,2 ------------ Multiplies all the vectors on the right by 2

In this example, the code 003 divides all the vectors to the right by 2. Therefore, a vector that was 1 unit long will be 0.5 units long now. The second code, 004, multiplies the vectors to the right by 2. We know the scale factors are cumulative; therefore, the vectors that were divided by 2 earlier will be multiplied by 2 now. Because of this cumulative effect, the length of the last vector remains unchanged. This file will produce the following shape:

Codes 005 and 006: Location Save/Restore

Code 005 lets you save the current location of the pen, and code 006 restores the saved location.

Example 4

The following example illustrates the use of codes 005 and 006.

Step 1: Writing the shape file

Figure 7-8(a) shows three lines that are unit vectors and intersect at one point. After drawing the first line, the pen has to return to the origin to start a second vector. This is done using code 005, which saves the starting point (origin) of the first vector, and code 006, which restores the origin. Now, if you draw another vector, it will start from the origin. Because there are three lines, you need three code 005s and three code 006s. The following file shows the header line and the shape specification for generating three lines, as shown in Figures 7-8(a) and 7-8(b):

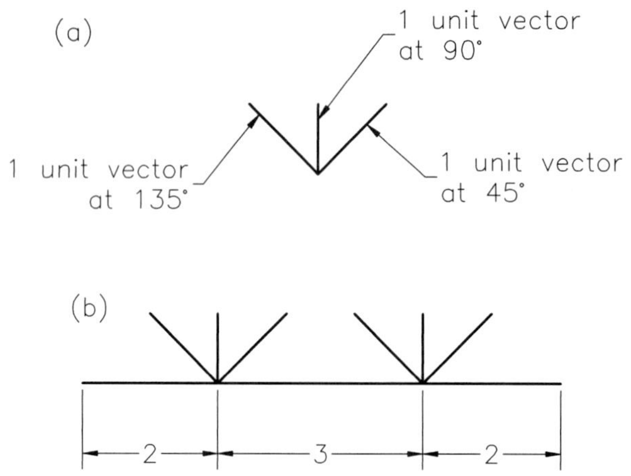

Figure 7-8 (a) Three unit vectors intersecting at a point; (b) repeating predefined subshapes

***210,10,POP1**
005,005,005,012,006,014,006,016,006,0

Where **005** --------------- Saves origin three times
012 --------------- Generates first vector
006 --------------- Restores origin
014 --------------- Generates second vector
006 --------------- Restores origin
Remember to press ENTER after the last line

The number of saves (code 005) has to equal the number of restores (code 006). If the number of saves (code 005) is more than the number of restores (code 006), AutoCAD will display the following message when the shape is drawn:

Position stack overflow in shape (shape number)

Similarly, if the number of restores (code 006) is more than the number of saves (code 005), the following message will be displayed:

Position stack underflow in shape (shape number)

The maximum number of saves and restores you can use in a particular shape definition is four.

Step 2: Compiling and loading the shape
Use the **COMPILE** and **LOAD** commands to compile and load the shape file. Next, use the **SHAPE** command to insert the shape, as described in Example 2.

Note
The name of the shape must be uppercase. Names with lowercase characters are ignored and are used to label font shape definitions. One shape file can contain 255 different shapes. Each shape is recognized by the unique name assigned to it.

Code 007: Subshape

You can define a subshape like a subroutine in a program. To reference a subshape, the subshape code, 007, has to be followed by the shape number of the subshape. The subshape has to be defined in the same shape file, and the shape number has to be from 1 to 255.

```
*210,10,POP1
005,005,005,012,006,014,006,016,006,0
*211,8,SUB1
020,007,210,030,007,210,020,0
        Where *210 -------------- Shape number
              007 --------------- Subshape reference
              210 --------------- Shape number
```

Enter SUB1 as the shape name to generate the shape, as shown in Figure 7-8(b).

Code 008: X-Y Displacement

In the previous examples, you might have noticed some limitations with the vectors. As mentioned earlier, you can draw vectors only in the 16 predefined directions, and the length of a vector cannot exceed 15 units. These restrictions make the shape files easier and more efficient, but at the same time, they are limiting. Therefore, codes 008 and 009 allow you to generate nonstandard vectors by entering the displacements along the X and Y directions. The general format for Code 008 is as follows:

008, XDISPLACEMENT, YDISPLACEMENT
or
008, (XDISPLACEMENT, YDISPLACEMENT)

X and Y displacements can range from +127 to -128. Also, a positive displacement is designated by a positive (+) number, and a negative displacement is designated by a negative (-) number. The leading positive sign (+) is optional in a positive number. The parentheses are used to improve readability, but they have no effect on the shape specification.

Code 009: Multiple X-Y Displacements

As described, code 008 allows you to generate nonstandard vectors by entering a single X and Y displacement, on the other hand, code 009 allows you to enter multiple X and Y displacements. It is terminated by a pair of 0 displacements (0,0). The general format is as follows:

009,(XDISPL, YDISPL), (XDISP, YDISPL), . . ., (0,0)

Code 00A or 10: Octant Arc

If you divide 360-degrees into eight equal parts, each angle will be 45-degrees. Each 45-degree angle segment is called an octant, and the two lines that contain an octant are called the octant boundary. The octant boundaries are numbered from 0 to 7, as shown in Figure 7-9. The general format is as follows:

10,(R,+/-0SN)

Where **R** ------------------ Radius of arc
+/- ---------------- Defines direction, + Counterclockwise, - Clockwise
0 ------------------- Hexadecimal notation
S ------------------- Starting octant boundary
N ------------------ Number of octants

10,(3,-043)

The first number, 10, is the code 00A for the octant arc. The second number, 3, is the radius of the octant arc. The negative sign indicates that the arc is to be generated in a clockwise direction. If it is positive (+) or if there is no sign, the arc will be generated in a counterclockwise direction. Zero is the hexadecimal notation, and the following number, 4, is the number of the octant boundary where the octant arc will start. The next element, 3, is the number of octants that this arc will extend. This example will generate the arc in Figure 7-10. The following is the listing of the shape file that will generate the shape, as shown in Figure 7-10. Remember to press a null ENTER at the last line.

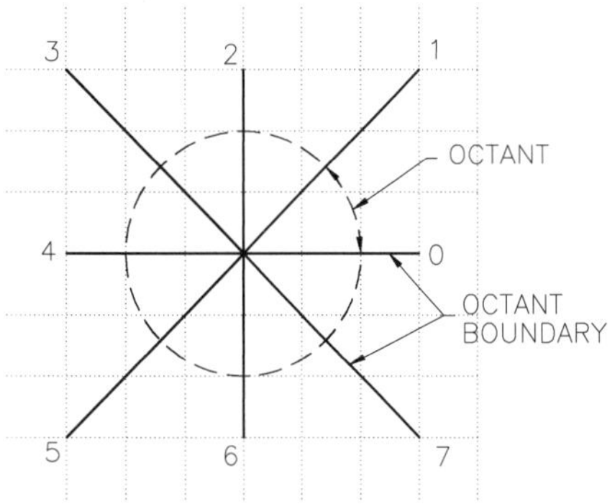

Figure 7-9 Octant boundaries

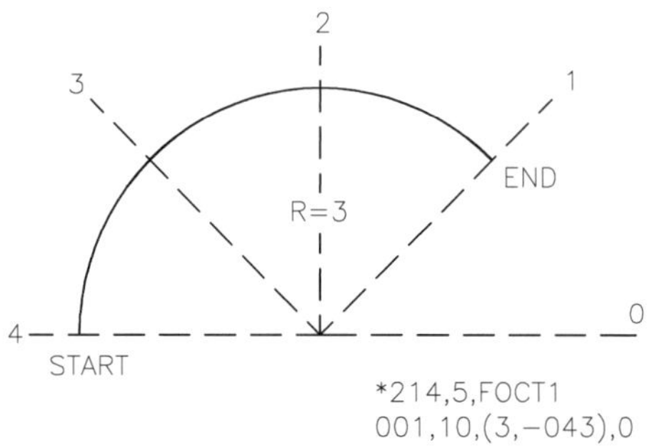

Figure 7-10 Octant arc

***214,5,FOCT1**
001,10,(3,-043),0
Null ENTER

Code 00B or 11: Fractional Arc

You can generate a nonstandard fractional arc by using code 00B or 11. This code will allow you to start and end an arc at any angle. The definition uses five bytes, and the general format is as follows:

11,(START OFFSET, END OFFSET, HIGHRADIUS, LOWRADIUS, +/-0SN)

The START OFFSET represents how far from an octant boundary the arc starts, and the END OFFSET represents how far from an octant boundary the arc ends. The HIGHRADIUS is zero if the radius is equal to or less than 255 units, and the LOWRADIUS is the radius of the arc. The positive (+) or negative (-) sign indicates, respectively, whether the arc is drawn counterclockwise or clockwise. The next element, S, is the number of the octant where the arc starts, and element N is the number of octants the arc goes through. The following example illustrates the fractional arc concept.

Start offset = (starting angle of the arc - nearest angle of the octant whose sum is less than the starting angle) x 256 / 45

End offset = (end angle of the arc - nearest angle of the octant whose sum is less than the

Example 5

end angle) x 256 / 45

Draw a fractional arc of radius 3 units that starts at a 20-degrees angle and ends at a 140-degrees angle (counterclockwise).

The solution involves the following steps:

Step 1: Calculating the parameters

Find the nearest octant boundary whose angle is less than 140-degree. The nearest octant boundary is the number 3 octant boundary, whose angle is 135-degree (3 * 45 = 135). Calculate the end offset to the nearest whole number (integer):

End offset = (140 - 135) * 256/45 = 28.44 = 28

Find the nearest octant boundary whose angle is less than 20-degree. The nearest octant boundary is 0 and its angle is 0-degree. Calculate the start offset to the nearest whole number:

Start offset = (20 - 0) * 256/45 = 113.7 = 114
Find the number of octants the arc passes through. In this example, the arc starts in the first octant and ends in the fourth octant. Therefore, the number of octants the arc passes through is four (counterclockwise).

Find the octant where the arc starts. In this example, it starts in the 0 octant.

Radius of the arc is 3. Its high radius is 0 as radius is less than 255.

Substitute the values in the general format of the fractional arc:

11,(114,28,0,3,004)

Step 2: Writing the shape file
Use any text editor or word processor to write the shape file. The following shape file will generate the fractional arc, as shown in Figure 7-11. Remember to press a null ENTER at the last line.

***221,8,FOCT2**
001,11,(114,28,0,3,004),0
blank ENTER

Step 3: Saving and loading the shape file
Save the file with the extension *.shp* Use the **COMPILE** and **LOAD** commands to compile and load the shape file. Next, use the **SHAPE** command to insert the shape in the drawing editor.

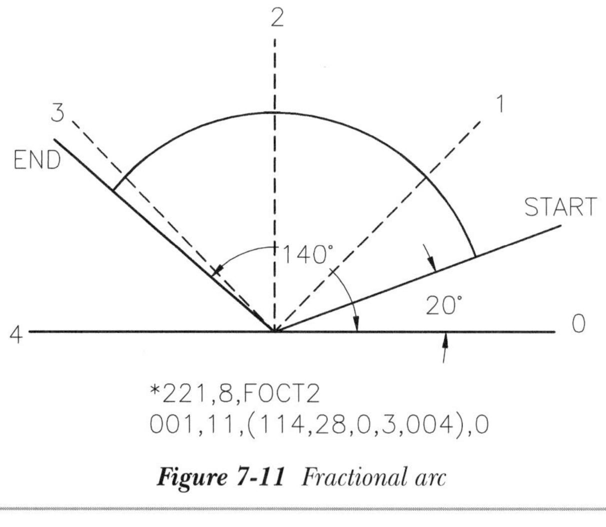

Figure 7-11 *Fractional arc*

Code 00C or 12: Arc Definition by Displacement and Bulge

Code C can be used to define an arc by specifying the displacement of the endpoint of an arc and the bulge factor. X and Y displacements may range from -127 to +127, and the bulge factor can also range from -127 to +127. A semicircle will have a bulge factor of 127, and a straight line will have a bulge factor of 0. If the bulge factor has a negative sign, the arc is drawn clockwise. The general format is as follows:

12,(X Displacement, Y Displacement, Bulge Factor)

Bulge factor = ((2 * H)/D) * 127
Where **H** ------------------ Height of arc
D ------------------ Displacement

For a semicircle, 2H = D
Therefore, bulge = (D/D) * 127 = 127
For a straight line, H = 0
Therefore, bulge = (0/D) * 127 = 0

In Figure 7-12, the distance between the start point and the endpoint of an arc is 4 units, and the height is 1 unit. Therefore, the bulge can be calculated by substituting the values in the previously mentioned relation:

Bulge = (2 * 1/4) * 127 = 63.5 = 63
(integer)

The following shape description will generate the arc, as shown in Figure 7-12.

***213,5,BULGE1**
12,(4,0,-63),0
Where **4** ------------------- X displacement
0 ------------------- Y displacement
- ------------------- Negative (-), generates clockwise arc
63 ----------------- Bulge factor

Code 00D or 13: Multiple Bulge-Specified Arcs

Code 00D or 13 can be used to generate multiple arcs with different bulge factors. It is terminated by a (0,0). The following shape description defines the arc configuration of Figure 7-13:

*214,16,BULGE2
13,(4,0,-111),
(0,4,63),
(-4,0,-111),
(0,-4,63),(0,0),0
Null ENTER

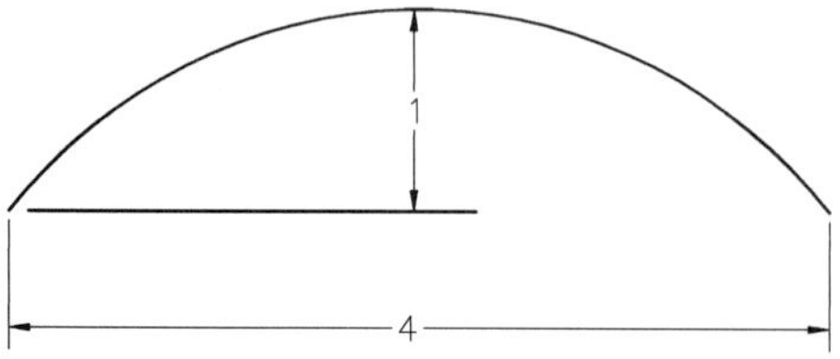

***Figure 7-12** Calculating the bulge*

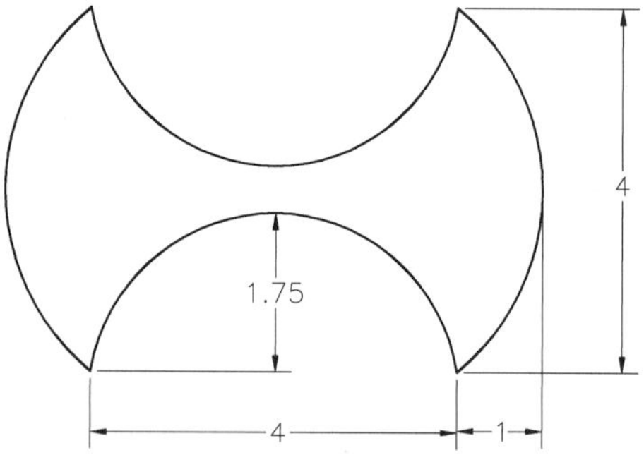

***Figure 7-13** Different arc configurations*

Note

The information about different arcs can be given in one line. For the convenience of the users the above example for ***Code 00D*** *is divided into different lines.*

Code 00E or 14: Flag Vertical Text

Code 00E or 14 is used when the same text font description is to be used in both the horizontal and vertical orientations. If the text is drawn in a horizontal direction, the vector next to code 14 is ignored. If the text is drawn in a vertical position, the vector next to code 14 is not ignored. This lets you generate text in a vertical or horizontal direction with the same shape file.

For horizontal text, the start point is the lower left point, and the endpoint is on the lower right. In vertical text, the start point is at the top center, and the endpoint is at the bottom center of the text, as shown in Figure 7-14. At first, it appears that you need two separate shape files to define the shape of a horizontal and a vertical text. However, with code 14 you can avoid the dual shape definition.

Figure 7-14 shows the pen movements for generating the text character "G." If the text is horizontal, the line that is next to code 14 is automatically ignored. However, if the text is vertical, the line is not ignored, resulting in resetting the start and endpoints appropriately for vertically aligned text.

```
*115,28,FLAG
002,14,
008,(-2,-6),
042,001,
014,016,028,01A,
04C,01E,020,012,014,
002,018,
001,020,01C,
002,01E,
```

If text is horizontal, code 14 automatically ignores next line. 008, (-2,-6),

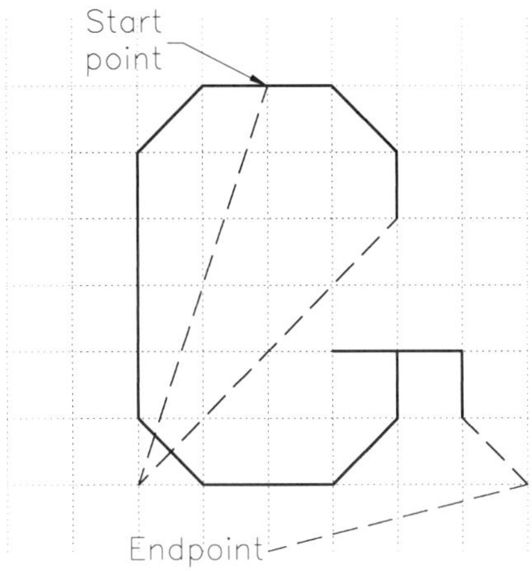

Figure 7-14 *Pen movement for generating the character "G"*

14,	If text is horizontal, code 14 automatically ignores next line. 008,(-4,-1),
008,(-4,-1),	
0	
Null ENTER	

Example 6

Write a shape file for a hammer, as shown in Figure 7-15(a). (The name of the shape file is *hmr.shp* and the shape name is HAMMER.) The following file defines the shape of a hammer. **The line numbers are not a part of the file; they are for reference only.**

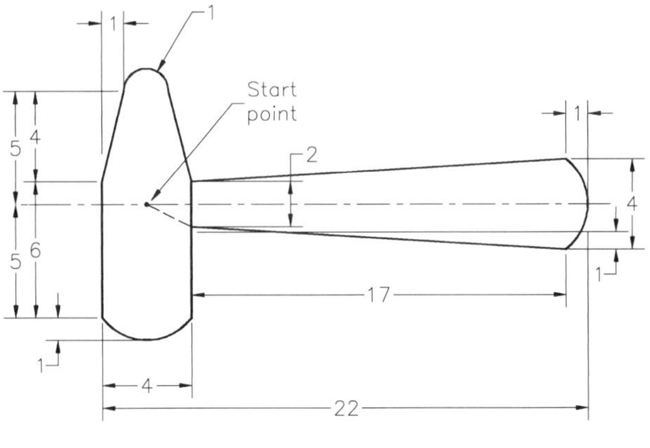

Figure 7-15(a) *Dimensions of hammer in units*

Step 1: Writing the shape file

Use any text editor or word processor to write the following shape file:

```
1 *204,34,HAMMER
2 003,22,
3 002,8,(2,-1),
4 001,024,
5 8,(-1,4),
6 00A,(1,004),
7 8,(-1,-4),06C,
8 00C,(4,0,63),
9 044,8,(17,-1),
10 00C,(0,4,63),
11 8,(-17,-1),0
12 Null Enter
```

Explanation

Line 1

***204,34,HAMMER**

This is the header line that consists of the shape number (204), the number of data bytes in the shape specification (34), and the name of the shape (HAMMER).

Line 2

003,22,

Data byte 003 has been used to divide the vectors by the next data byte, 22. This reduces the hammer shape to a unit size that facilitates the scaling operation when you insert the shape in a drawing.

Line 3

002,8,(2,-1),

Data byte 002 deactivates the pen (pen up), and the next data byte (code 008) defines a vector that has X-displacement of 2 units and Y-displacement of -1 unit. No line is drawn because the pen is deactivated.

Line 4

001,024,

Data byte 001 activates the pen (pen down) and the next data byte defines a vector that is 2 units long along the direction vector 4.

Line 5

8,(-1,4),

The first data byte, 8 (code 008), defines a vector whose X-displacement and Y-displacement is given by the next two data bytes. The X-displacement of this vector is -1 and the Y-displacement is 4 units.

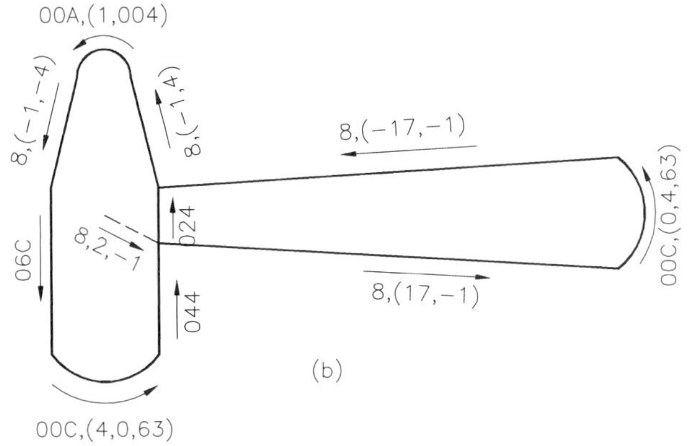

Figure 7-15(b) *Pen movements for generating the shape of a hammer*

Line 6
00A,(1,004),
Data byte 00A defines an octant arc that has a radius of 1 unit as defined by the next data byte. The first element (0) of data byte 004 is a hexadecimal notation. The second element (0) defines the starting octant of the arc, and the third element (4) defines the ending octant of the arc.

Line 7
8,(-1,-4),06C,
Data byte 8 (code 008) defines a vector that has X-displacement of -1 unit, and Y-displacement of -4 units. The next data byte defines a vector that is 6 units long along the direction vector C.

Line 8
00C,(4,0,63),
The first data byte (00C) defines an arc that has X-displacement of 4 units, Y-displacement of 0 units, and a bulge factor of 63.

$$\begin{aligned}\text{Bulge factor} &= (2 * H)/D * 127\\ &= (2 * 1)/4 * 127\\ &= 63.5\\ &= 63 \text{ (integer)}\end{aligned}$$

Line 9
044,8,(17,-1),
The first data byte defines a vector that is 4 units long along the direction vector 4. The second data byte (8) defines a vector that has X-displacement of 17 units and Y-displacement of -1 unit.

Line 10

00C,(0,4,63),

The first data byte (00C) defines an arc that has X-displacement of 0 units, Y-displacement of 4 units, and a bulge factor of 63.

$$\begin{aligned}\text{Bulge factor} &= (2 * H)/D * 127\\ &= (2 * 1)/2 * 127\\ &= 63.5\\ &= 63 \text{ (integer)}\end{aligned}$$

Line 11

8,(-17,-1),0

The first data byte (8) defines a vector that has X-displacement of -17 units and Y-displacement of -1 unit. The data byte 0 terminates the shape definition.

Step 3: Loading the shape file

Save the file with the extension *.shp* and then use **COMPILE**, **LOAD**, and **SHAPE** commands to load the shape, as described in Example 1.

TEXT FONT FILES

In addition to shape files, AutoCAD provides a facility to create new text fonts. After you have created and compiled a text font file, text can be inserted in a drawing like regular text and using the new font. These text files are regular shape files with some additional information describing the text font and the line feed. The following is the general layout of the text font file:

Text font description
Line feed
Shape definition

Text Font Description

The text font description consists of two lines:

***0,4,font name**
ABOVE, BELOW, MODES, 0

Where ***0** ----------------- Special shape number for text font
4 ------------------- Number of data bytes
font name ------ Name of font in lowercase
ABOVE---------- Upper distance
BELOW --------- Lower distance
MODES --------- 0 for horizontal text, 2 for dual orientation
0 ------------------- Terminating zero

For example, if you are writing a shape definition for an uppercase M, the text font description would be:

***0,4,ucm**
10,4,2,0

In the first line, the first data byte (0) is a special shape number for the text font, and every text font file will have this shape number. The next data byte (4) is the number of data bytes in the next line, and ucm is the shape name (name of font). The shape names in all text font files should be lowercase so that the computer does not have to save the names in memory. You can still reference the shape names for editing.

In the second line, the first data byte (10) specifies the height of an uppercase letter above the baseline. For example, in Figure 7-16 (page-27), the height of the letter M above the baseline is 10 units. The next data byte (4) specifies the distance of lowercase letters below the baseline. AutoCAD uses this information to scale the text automatically. For example, if you enter the height of text as 1 unit, the text will be 1 unit, although it was drawn 10 units high in the text font definition. The third data byte (2) defines the mode. It can have one of only two values, 0 or 2. If the text is horizontal, the mode is 0; if the text has dual orientation (horizontal and vertical), the mode is 2. The fourth data byte (0) is the zero that terminates the definition.

Line Feed

The line feed is used to space the lines so that characters do not overlap and so that a desired distance is maintained between the lines. AutoCAD has reserved the ASCII number 10 to define the line feed.

***10,5,lf**
2,8,(0,-14),0

Where ***10** ---------------- (10) ASCII number reserved for line feed
5 ------------------- (5) Number of data bytes in shape specification
lf ------------------ (lf) Shape name
2 ------------------- Deactivate pen (pen up)
(0,-14) ------------ Line feed of 14 units
0 ------------------- Terminating zero

In the first line, the first data byte (10) is the shape number reserved for line feed, and the next data byte (5) is the number of characters in the shape specification. The data byte lf is the name of the shape.

In the second line, the first data byte (2) deactivates the pen. The next data byte (8) is a special code, 008, that defines a vector by X displacement and Y displacement. The third and fourth data bytes (0,-14) are the X displacement and Y displacement of the displacement vector; they produce a line feed that is 14 units below the baseline. The fifth data byte (0) is the zero that terminates the shape definition.

Shape Definition

The shape number in the shape definition of the text font corresponds to the ASCII code for that character. For example, if you are writing a shape definition for an uppercase M, the shape number is 77.

*77,50,ucm

Where *77 ---------------- Shape number - ASCII code of uppercase "M"
50 ----------------- Number of data bytes
ucm -------------- Shape name

The ASCII codes can be obtained from the ASCII character table, which gives the ASCII codes for all characters, numbers, and punctuation marks.

32	space	56	8	80	P	104	h
33	!	57	9	81	Q	105	i
34	“	58	:	82	R	106	j
35	#	59	;	83	S	107	k
36	$	60	<	84	T	108	l
37	%	61	=	85	U	109	m
38	&	62	>	86	V	110	n
39	’	63	?	87	W	111	o
40	(	64	@	88	X	112	p
41	)	65	A	89	Y	113	q
42	*	66	B	90	Z	114	r
43	+	67	C	91	[	115	s
44	,	68	D	92	\	116	t
45	-	69	E	93	]	117	u
46	.	70	F	94	^	118	v
47	/	71	G	95	_	119	w
48	0	72	H	96	‘	120	x
49	1	73	I	97	a	121	y
50	2	74	J	98	b	122	z
51	3	75	K	99	c	123	{
52	4	76	L	100	d	124	3
53	5	77	M	101	e	125	}
54	6	78	N	102	f	126	~
55	7	79	O	103	g		

Example 7 *General*

Write a text font shape file (UCM) for an uppercase M, as shown in Figure 7-16. The font file should be able to generate horizontal and vertical text. Each grid is 1 unit, and the directions of vectors are designated with leader lines.

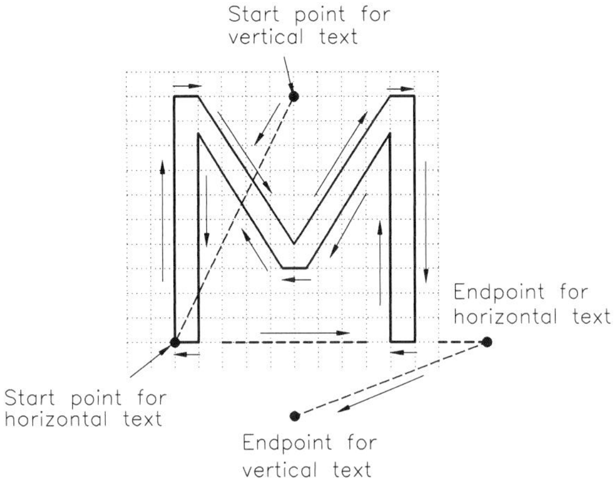

Figure 7-16 *Shape and pen movement of uppercase letter "M"*

Step 1: Writing the shape file

In the following file, the line numbers at the right are not a part of the file. They are for reference only. Use any text editor to write the following shape file.

```
1  *0,4,uppercase m
2  10,0,2,0
3  *10,13,lf
4  002,8,(0,-14),14,9,(0,14),(14,0),(0,0),0
5  *77,51,ucm
6  2,14,8,(-5,-10),
7  001,009,(0,10),(1,0),(4,-6),(4,6),(1,0),
8  (0,-10),(-1,0),(0,0),
9  003,2,
10 009,(0,17),(-7,-11),(-2,0),(-7,11),
11 (0,-17),(-2,0),(0,0),
12 002,8,(28,0),
13 004,2,
14 14,8,(-9,-4),0
```

Explanation

Line 1

***0,4,uppercase m**

The first data byte (0) is the special shape number for the text font file. The next data byte (4) is the number of data bytes, and the third data byte is the name of the shape.

Line 2

10,0,2,0

The first data byte (10) represents the total height of the character M, and the second data byte (0) represents the length of the lowercase letters that extend below the base line. Data byte 2 is the text mode for dual orientation of the text (horizontal and vertical). If the text is required in the horizontal direction only, the mode will be 0. The fourth data byte (0) is the zero that terminates the definition of this particular shape.

Line 3

***10,13,lf**

The first data byte (10) is the code reserved for line feed, and the second data byte (13) is the number of data bytes in the shape specification. The third data byte (lf) is the name of the shape.

Line 4

002,8,(0,14),14,9,(0,-14),(14,0),(0,0),0

The first data byte (002 or 2) is the code to deactivate the pen (pen up). The next three data bytes [8,(0,14)] define a displacement vector whose X displacement and Y displacement are 0 and -14 units, respectively. This will cause a carriage return that is 14 units to the right of the text insertion point of the first text line. This will work fine if the text is drawn in the horizontal direction only. However, if the text is vertical, the carriage return should produce a displacement that is below the text insertion point. This is accomplished by the next seven data bytes. Data byte 14 ignores the next code if the text is horizontal. If the text is vertical, the next code is processed. The next set of data bytes (0,-14) defines a displacement vector that is 14 units below the previous point, D1 in Figure 7-17.

Data bytes (14,0) define a displacement vector that is 14 units to the right, D2 in Figure 7-17. These four data bytes combined will result in a carriage return that is 4 units to the right of the existing line. The next set of data bytes (0,0) terminates the code 9, and the last data byte (0) terminates the shape specification.

Line 5

***77,51,ucm**

The first data byte (77) is the ASCII code of the uppercase M. The second data byte (51) is the number of data bytes in the shape specification. The next data byte (ucm) is the name of the shape file in lowercase letters.

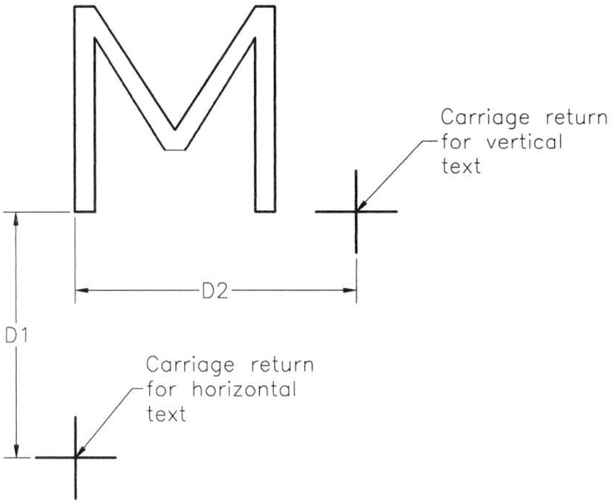

Figure 7-17 *Carriage return for vertical and horizontal texts*

Line 6
2,14,8,(-5,-10),
The first data byte code (2) deactivates the pen (pen up), and the next data byte code (14) will cause the next code to be ignored if the text is horizontal. In the horizontal text, the insertion point of the text is the starting point of that text line (Figure 7-16). However, if the text is vertical, the starting point of the text is the upper middle point of the character M. This is accomplished by the next three data bytes [8,(-5,-10)], which displace the starting point of the text 5 units to the left (width of character M is 10) and 10 units down (height of character M is 10).

Lines 7,8
001,009,(0,10),(1,0),(4,-6),(4,6),(1,0),
(0,-10),(-1,0),(0,0),
The first byte (001) activates the draw mode (pen down), and the remaining bytes define the next seven vectors.

Lines 9 to11
003,2,
009,(0,17),(-7,-11),(-2,0),(-7,11),
(0,-17),(-2,0),(0,0),
The inner vertical line of the right leg of the character M is 8.5 units long, and you cannot define a vector that is not an integer. However, you can define a vector that is 2 x 8.5 = 17 units long and then divide that vector by 2 to get a vector 8.5 units long. This is accomplished by code 003 and the next data byte, 2. All the vectors defined in the next two lines will be divided by 2.

Line 12

002,8,(28,0),

The first data byte (002) deactivates the draw mode, and the next three data bytes define a vector that is 28/2 = 14 units to the right. This means that the next character will start 14 - 10 = 4 units to the right of the existing character so that it will produce a horizontal text.

Line 13

004,2,

The code 004 multiplies the vectors that follow it by 2; therefore, it nullifies the effect of code 003,2.

Line 14

14,8,(-9,-4),0

If the text is vertical, the next letter should start below the previous letter. This is accomplished by data bytes 8,(-9,-4), which define a vector that is -9 units along the *X* axis and -4 units along the *Y* axis. The data byte 0 terminates the definition of the shape.

Step 2: Loading the shape file

Save the file as **ucm.shp**. To compile the shape file, use the C**OMPILE** command. Now, you can create text using this shape. Invoke the **Text Style** dialog box from the **Home > Annotation** panel of the **Ribbon**. Next, define a text style with the name **MYUCM1** with the settings shown in Figure 7-18 and set it as the current text style.

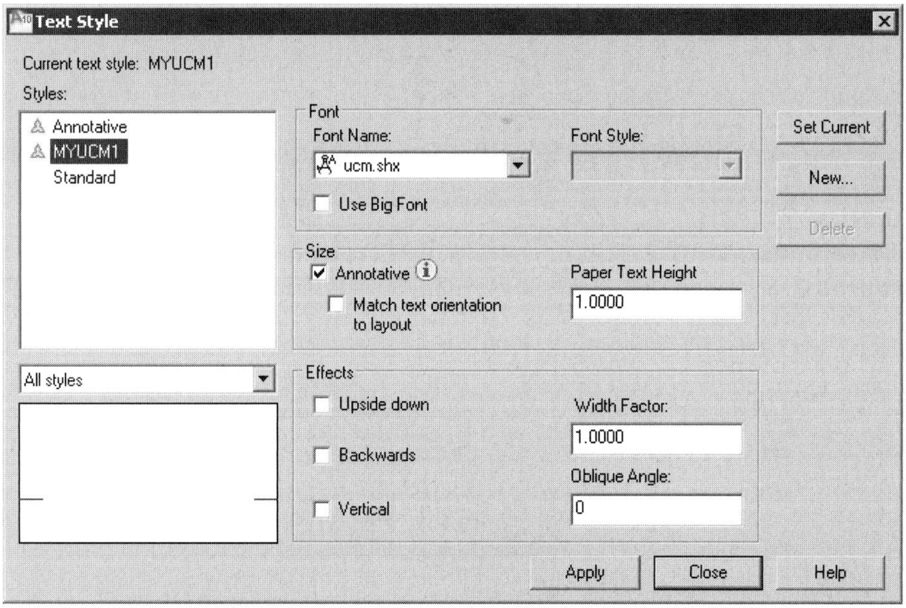

***Figure 7-18** The **Text Style** dialog box with the specified settings for the new text style*

Now use the **TEXT** command to write the text **M** with the text style **MYUCM1**, as shown in Figure 7-19.

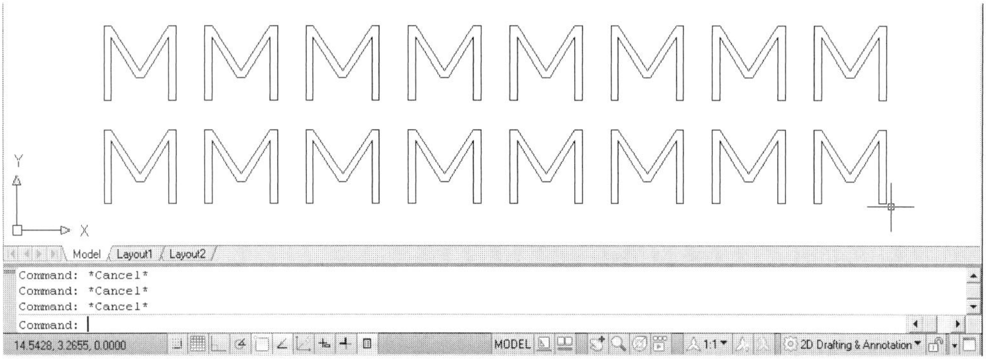

Figure 7-19 *Using the defined text font*

Example 8

Write a text font shape file for the lowercase m, as shown in Figure 7-20. The font file should be able to generate horizontal and vertical text. Each grid is 1 unit, and the direction of the vectors is designated with the leader lines.

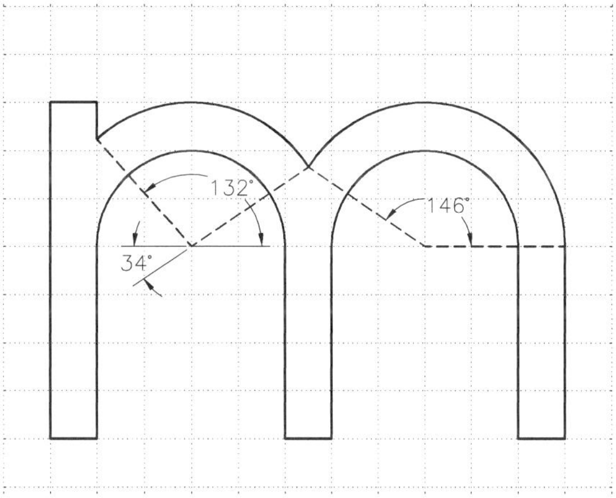

Figure 7-20 *Shape of lowercase character, m*

Step 1: Writing the shape file

The following file is a listing of the text font shape file for Example 8. **The line numbers are not a part of the file; they are shown here for reference only**. Use any text editor to write down the following shape file.

```
1 *0,4,lower-case m
2 14,3,2,0
3 *10,13,lf
4 002,8,(0,-18),14,9,(0,18),(27,0),(0,0),0
5 *109,57,lcm
6 2,14,8,(-11,-14),
7 005,005,001,020,084,
8 00A,(4,-044),
9 08C,020,084,
10 00A,(4,-044),
11 08C,020,084,
12 00B,(0,62,0,6,004),
13 00B,(193,239,0,6,003),
14 006,9,(0,14),(2,0),(0,0),
15 003,5,07C,004,5
16 006,2,8,(27,0),
17 14,8,(-16,-5),0
18 Null ENTER
```

Explanation

The format of most of the lines is the same as in the previous example, except for the following lines, which use save/restore origin, octant, and fractional arcs.

Line 7
005,005,001,020,084,
The first and second data bytes (005) are used to save the location of the point twice. The remaining data bytes activate the draw mode and define the vectors.

Line 8
00A,(4,-044),
The first data byte code (00A) is the code for the octant arc and the second data byte (4) defines the radius of the arc. The negative sign (-) in the third data byte generates an arc in the clockwise direction. The first element (0) is a hexadecimal notation. The second element defines the starting octant and the third element (4) defines the number of octants that the arc passes through.

Line 12
00B,(0,62,0,6,004),
The first data byte (00B) is the code for the fractional arc that is defined by the next five data bytes. The second data byte (0) is the starting offset of the first arc, as shown in the calculations given next.

<u>**1st Arc**</u>

Starting angle = 0	Ending angle = 146
Starting octant = 0	Ending octant= 4

Starting offset = (0-0)*256/45
= 0
Ending offset = (146-135)*256/45
= 62.57
= 62 (integer)

The third data byte (62) is the ending offset of the arc and the fourth data byte (0) is the high radius. The fifth data byte (6) defines the radius of the arc. The second element (0) of the next data byte is the starting octant and the third element (4) is the number of octants the arc goes through.

Line 13
00B,(193,239,0,6,003),
The first data byte (00B) is the code for the fractional arc. The remaining data bytes define various parameters of the fractional arc as explained earlier. The offset angles have been obtained from the following calculations:

2nd Arc
Starting angle = 34 Ending angle = 132
Starting octant = 0 Ending octant = 3
Starting offset = (34-0)*256/45
= 193.4
= 193 (integer)
Ending offset = (132-90)*256/45
= 238.9
= 239 (integer)

Note
Because the offset values have been rounded, it is not possible to describe an arc that is very accurate. Therefore, in this example the origin has been restored after two arcs were drawn. This origin was then used to draw the remaining lines.

Line 14
006,9,(0,14),(2,0),(0,0),
The first data byte (006) restores the previously saved point, and the remaining data bytes define the vectors using code 009.

Step 2: loading the shape file

Follow the procedure described in Example 7 to load the shape file and then use this shape to write the text.

Note
The compiled font files (.shx), like shape files, must be available whenever the drawing is opened. This is the reason if you send a drawing to the client, you must send the shape and font files also.

Self-Evaluation Test

Answer the following questions and then compare them to those given at the end of this chapter:

1. The basic objects used in shape files are __________ and __________.

2. Shapes are easy to insert and take less disk space than __________. However, there are certain disadvantages to using shapes. For example, you cannot __________ a shape.

3. The shape number could be any number between 1 and __________ in a particular file, and these numbers cannot be repeated within the same file.

4. The shape file should not contain two __________ with the same name.

5. A hexadecimal number is designated by a leading __________.

6. The maximum number of data bytes is __________ bytes per shape.

7. To define a vector, you need its magnitude and __________.

8. To load the shape file, use the __________ command.

9. Generating shapes with direction vectors has some limitations. For example, you cannot draw an arc or a line that is not along the __________ vectors. These limitations can be overcome by using the __________, which add a lot of flexibility and give a better control over the shapes you want to create.

10. Code 001 activates the __________ mode, and code __________ deactivates the draw mode.

11. The byte that follows the division code divides the __________ vectors.

12. Code 004 is used if you want to multiply the vectors by a certain number. It can also be used to __________ the effect of code 003.

Review Questions

Answer the following questions:

1. Scale factors are __________ within the shape.

2. The number of saves (code 005) must be equal to the number of __________ code __________.

3. The maximum number of saves and restores you can use in a particular shape definition is __________.

4. You can define a subshape like a subroutine in a program. To reference the subshape, use code __________.

5. Vectors can be drawn in the 16 predefined directions only, and the length of the vector cannot exceed __________ units.

6. A nonstandard fractional arc can be generated by using code 00B or __________.

7. Code __________ can be used to define an arc by specifying the displacement of the endpoint of an arc and the bulge factor.

8. The bulge factor can range from -127 to __________.

9. Code 00E or __________ is used when the same text font description is to be used in both horizontal and vertical orientation.

10. The text files are regular __________ files with some additional information about the text font description and the line feed.

11. The shape names in all text font files should be lowercase so that the computer does not have to save the names in its __________.

12. The line feed is used to space the lines so that the characters do not __________.

13. The shape number in the shape definition of the text font corresponds to the __________ code for that character.

14. To recall a subshape, the subshape has to be __________ within the same shape file.

15. Each 45-degrees angle segment is called an __________ .

16. AutoCAD has reserved the ASCII code no. 10 to define the __________ .

17. Shapes cannot be ____________ after insertion.

18. The shape name is the name of the shape in __________ letter.

19. During the definition of the shape, the first line is the __________ line and the second line is the ___________.

Exercises

Exercise 1 *General*

Write a shape file for the uppercase letter M, as shown in Figure 7-21.

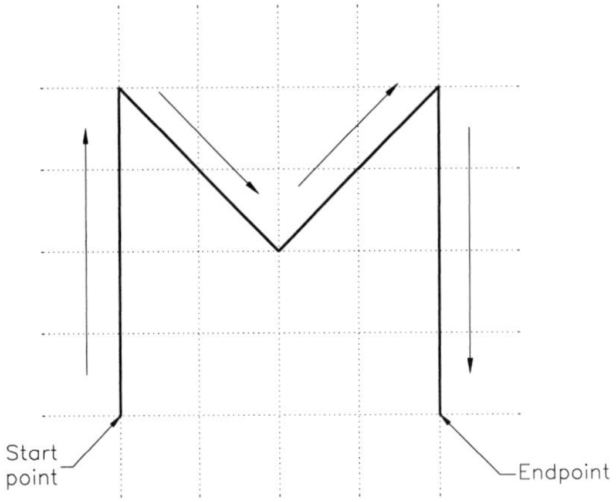

Figure 7-21 *Uppercase letter M*

Exercise 2 *General*

Write a shape file for generating the tapered gib-head key, as shown in Figure 7-22.

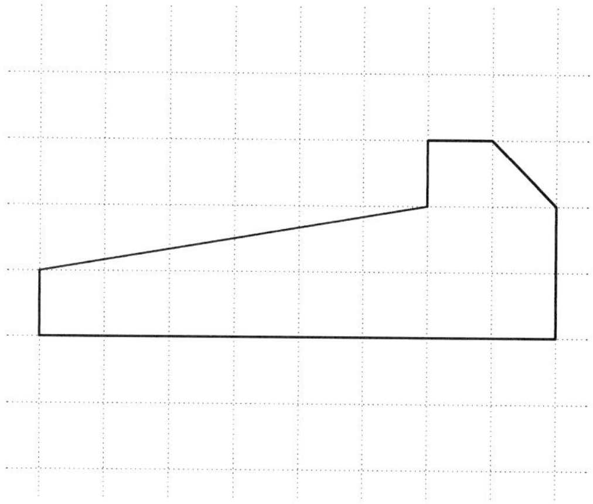

Figure 7-22 Tapered gib-head key

Exercise 3 *General*

Write a text font shape file for the uppercase letter G, as shown in Figure 7-23.

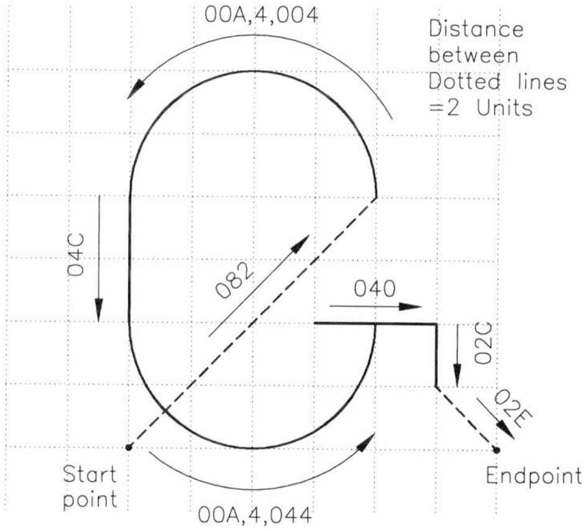

Figure 7-23 Uppercase letter G

Exercise 4 *General*

Write a text font shape file for the uppercase letter W, as shown in Figure 7-24. The font file should be able to generate horizontal and vertical text.

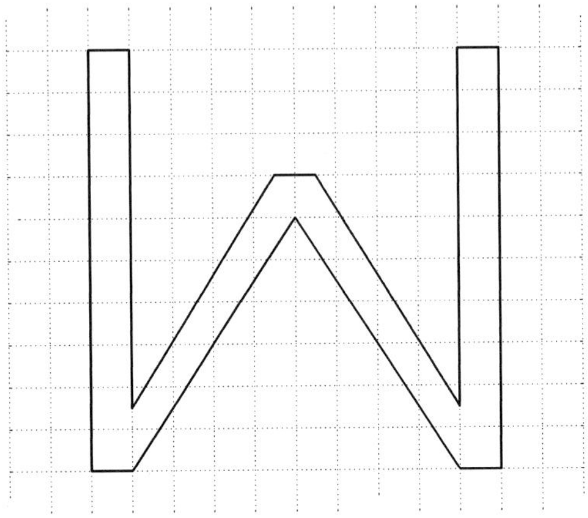

Figure 7-24 *Uppercase letter W*

Answers to Self-Evaluation Test

1. line and arc, **2.** blocks, edit, **3.** 255, **4.** shapes, **5.** zero, **6.** 2000, **7.** direction, **8.** **LOAD**, **9.** direction, special codes, **10.** draw, 002, **11.** whole, **12.** nullify

Chapter 8

Working with AutoLISP

Learning Objectives

After completing this chapter, you will be able to:

- *Perform mathematical operations using AutoLISP.*
- *Use trigonometrical functions in AutoLISP.*
- *Understand the basic AutoLISP functions and their applications.*
- *Load and run AutoLISP programs.*
- *Use the* ***Load/Unload Applications*** *dialog box.*
- *Use flowcharts to analyze problems.*
- *Test a condition using conditional functions.*

ABOUT AutoLISP

Developed by Autodesk, Inc., **AutoLISP** is an implementation of the **LISP** programming language (LISP is an acronym for **LISt Processing**.) The first reference to LISP was made by John McCarthy in the April 1960 issue of *The Communications of the ACM*.

Except for **FORTRAN** and **COBOL**, most of the languages developed in the early 1960s have become obsolete. But LISP has survived and has become a leading programming language for artificial intelligence (AI). Some of the dialects of the LISP programming language are Common LISP, BYSCO LISP, ExperLISP, GCLISP, IQLISP, LISP/80, LISP/88, MuLISP, TLCLISP, UO-LISP, Waltz LISP, and XLISP. XLISP is a public domain LISP interpreter. The LISP dialect that resembles AutoLISP is Common LISP. The AutoLISP interpreter is embedded within the AutoCAD software package. However, AutoCAD LT and AutoCAD Versions 2.17 and lower lack the AutoLISP interpreter; therefore, you can use the AutoLISP programming language only with AutoCAD Release 2.18 and up.

AutoCAD contains most of the commands used to generate a drawing. However, some commands are not provided in AutoCAD. For example, AutoCAD has no command to make global changes in the drawing text objects. With AutoLISP you can write a program in the AutoLISP programming language that will make global or selective changes in the drawing text objects. You can use AutoLISP to write any program or embed it in the menu and thus customize your system to make it more efficient.

The AutoLISP programming language has been used by hundreds of third-party software developers to write software packages for various applications. For example, the author of this text has developed a software package, **SMLayout**, that generates flat layouts of various geometrical shapes like transitions, intersection of pipes and cylinders, elbows, cones, and tank heads. There is a huge demand for AutoLISP programmers as consultants for developing application software and custom menus.

This chapter assumes that you are familiar with AutoCAD commands and AutoCAD system variables. However, you need not be an AutoCAD or programming expert to begin learning AutoLISP. This chapter also assumes that you have no prior programming knowledge. If you are familiar with any other programming language, learning AutoLISP may be easy. A thorough discussion of various functions and a step-by-step explanation of the examples should make it fun to learn. This chapter discusses the most frequently used AutoLISP functions and their application in writing a program. For the functions that are not discussed in this chapter, refer to the *AutoLISP Developer's Guide* from Autodesk. AutoLISP does not require any special hardware. If your system runs AutoCAD, it will also run AutoLISP. To write AutoLISP programs you can use any text editor.

MATHEMATICAL OPERATIONS

A mathematical function constitutes an important feature of any programming language. Most of the mathematical functions commonly used in programming and mathematical calculations are available in AutoLISP. You can use AutoLISP to add, subtract, multiply, and divide numbers. You can also use it to find the sine, cosine, and arctangent of angles expressed in radians. There is a host of other calculations you can do with AutoLISP. This section discusses the most frequently used mathematical functions supported by the AutoLISP programming language.

Addition

Format **(+ num1 num2 num3 - - -)**

This function (+) calculates the sum of all the numbers to the right of the plus (+) sign (num1 + num2 + num3 + . . .). The numbers can be integers or real. If the numbers are integers, then the sum is an integer. If the numbers are real, the sum is real. However, if some numbers are real and some are integers, the sum is real. In the following display, all numbers in the first two examples are integers, so the result is an integer. In the third example, one number is a real number (50.0), so the sum is a real number.

Examples:

Command: (+ 2 5)	returns 7
Command: (+ 2 30 4 50)	returns 86
Command: (+ 2 30 4 50.0)	returns 86.0

Subtraction

Format **(- num1 num2 num3 - - -)**

This function (-) subtracts the second number from the first number (num1 - num2). If there are more than two numbers, the second and subsequent numbers are added and the sum is subtracted from the first number [num1 - (num2 + num3 + . . .)]. In the first of the following examples, 14 is subtracted from 28 and returns 14. Since both numbers are integers, the result is an integer. In the third example, 20 and 10.0 are added, and the sum of these two numbers (30.0) is subtracted from 50, returning a real number, 20.0.

Examples:

Command: (- 28 14)	returns 14
Command: (- 25 7 11)	returns 7
Command: (- 50 20 10.0)	returns 20.0
Command: (- 20 30)	returns -10
Command: (- 20.0 30.0)	returns -10.0

Multiplication

Format **(* num1 num2 num3 - - -)**

This function (*) calculates the product of the numbers to the right of the asterisk (num1 x num2 x num3 x . . .). If the numbers are integers, the product of these numbers is an integer. If one of the numbers is a real number, the product is a real number.

Examples:

Command: (* 2 5)	returns 10

Command: (* 2 5 3)	returns 30
Command: (* 2 5 3 2.0)	returns 60.0
Command: (* 2 -5.5)	returns -11.0
Command: (* 2.0 -5.5 -2)	returns 22.0

Division

Format **(/ num1 num2 num3 - - -)**

This function (/) divides the first number by the second number (num1/num2). If there are more than two numbers, the first number is divided by the product of the second and subsequent numbers [num1 / (num2 x num3 x . . .)]. In the fourth of the following examples, 200 is divided by the product of 5.0 and 4 [200 / (5.0 * 4)].

Examples:

Command: (/ 30)	returns 30
Command: (/ 3 2)	returns 1
Command: (/ 3.0 2)	returns 1.5
Command: (/ 200.0 5.0 4)	returns 10.0
Command: (/ 200 -5)	returns -40
Command: (/ -200 -5.0)	returns 40.0

INCREMENTED, DECREMENTED, AND ABSOLUTE NUMBERS

Incremented Number

Format **(1+ number)**

This function (**1+**) adds 1 (integer) to **number** and returns a number that is incremented by 1. In the second example below, 1 is added to -10.5 and returns -9.5.

Examples:

(1+ 20)	returns 21
(1+ -10.5)	returns -9.5

Decremented Number

Format **(1- number)**

This function (**1-**) subtracts 1 (integer) from the **number** and returns a number that is decremented by 1. In the second example below, 1 is subtracted from -10.5 and returns -11.5

Examples:

(1- 10)	returns 9
(1- -10.5)	returns -11.5

Absolute Number

Format **(abs num)**

The **abs** function returns the absolute value of a number. The number may be an integer number or a real number. In the second example below, the function returns 20 because the absolute value of -20 is 20.

Examples:

(abs 20)	returns 20
(abs -20)	returns 20
(abs -20.5)	returns 20.5

TRIGONOMETRIC FUNCTIONS

sin

Format **(sin angle)**

The **sin** function calculates the sine of an angle, where the angle is expressed in radians. In the second example, the **sin** function calculates the sine of pi (180-degrees) and returns 1.22465e-016.

Examples:

Command: (sin 0)	returns 0.0
Command: (sin pi)	returns 1.22465e-016
Command: (sin 1.0472)	returns 0.866027

cos

Format **(cos angle)**

The **cos** function calculates the cosine of an angle, where the angle is expressed in radians. In the third of the following examples, the **cos** function calculates the cosine of pi (180-degree) and returns -1.0.

Examples:

Command: (cos 0)	returns 1.0
Command: (cos 0.0)	returns 1.0
Command: (cos pi)	returns -1.0
Command: (cos 1.0)	returns 0.540302

atan

Format **(atan num1)**

The **atan** function calculates the arctangent of **num1**, and the calculated angle is expressed in radians. In the second example, the **atan** function calculates the arctangent of 1.0 and returns 0.785398 (radians).

Examples:

Command: (atan 0.5)	returns 0.463648
Command: (atan 1.0)	returns 0.785398
Command: (atan -1.0)	returns -0.785398

You can also specify a second number in the atan function:

Format **(atan num1 num2)**

If the second number is specified, the function returns the arctangent of (num1/num2) in radians. In the first example, the first number (0.5) is divided by the second number (1.0) and the **atan** function calculates the arctangent of the dividend (0.5/1.0 = 0.5).

Examples:

Command: (atan 0.5 1.0)	returns 0.463648 radians
Command: (atan 2.0 3.0)	returns 0.588003 radians
Command: (atan 2.0 -3.0)	returns 2.55359 radians
Command: (atan -2.0 3.00)	returns -0.588003 radians
Command: (atan -2.0 -3.0)	returns -2.55359 radians
Command: (atan 1.0 0.0)	returns 1.5708 radians
Command: (atan -0.5 0.0)	returns -1.5708 radians

angtos

Format **(angtos angle [mode [precision]])**

The **angtos** function returns the angle expressed in radians in a string format. The format of the string is controlled by the **mode** and **precision** settings.

Examples:

(angtos 0.588003 0 4)	returns "33.6901"
(angtos 2.55359 0 4)	returns "146.3099"
(angtos 1.5708 0 4)	returns "90.0002"
(angtos -1.5708 0 2)	returns "270.00"

Note

*In **(angtos angle [mode [precision]])***

***angle** is angle in radians*

mode** is the angtos mode that corresponds to the AutoCAD system variable **AUNITS

The following modes are available in AutoCAD:

ANGTOS MODE	*EDITING FORMAT*
0	*Decimal degrees*
1	*Degrees/minutes/seconds*
2	*Grads*
3	*Radians*
4	*Surveyor's units*

***Precision** is an integer that controls the number of decimal places. Precision corresponds to the AutoCAD system variable **AUPREC**. The minimum value of **precision** is zero and the maximum is eight.*

In the first example mentioned above, angle is 0.588003 radians, mode is 0 (angle in degrees), and precision is 4 (four places after decimal). The function will return 33.6901.

RELATIONAL STATEMENTS

Programs generally involve features that test a particular condition. If the condition is true, the program performs certain functions, and if the condition is not true, then the program performs other functions. For example, the relational statement (if (< x 5)) tests true if the value of the variable x is less than 5. This type of test condition is frequently used in programming. The following section discusses various relational statements used in AutoLISP programming.

Equal to

Format **(= atom1 atom2 - - - -)**

This function (=) checks whether the two atoms are equal. If they are equal, the condition is true and the function will return **T**. If the specified atoms are not equal, the condition is false and the function will return **nil**.

Examples:

(= 5 5)	returns T
(= 5 4.9)	returns nil
(= 5.5 5.5 5.5)	returns T
(= "yes" "yes")	returns T
(= "yes" "yes" "no")	returns nil

Not equal to

Format **(/= atom1 atom2 - - - -)**

This function (/=) checks whether the two atoms are not equal. If they are not equal, the condition is true and the function will return **T**. If the specified atoms are equal, the condition is false and the function will return **nil**.

Examples:

(/= 50 4)	returns T
(/= 50 50)	returns nil
(/= 50 -50)	returns T
(/= "yes" "no")	returns T

Less than

Format **(< atom1 atom2 - - - -)**

This function (**<**) checks whether the first atom **(atom1)** is less than the second atom **(atom2)**. If it is true, then the function will return **T**. If it is not, the function will return **nil**.

Examples:

(< 3 5)	returns T
(< 5 3 4 2)	returns nil
(< "x" "y")	returns T

Less than or equal to

Format **(<= atom1 atom2 - - - -)**

This function (**<=**) checks whether the first atom **(atom1)** is less than or equal to the second atom **(atom2)**. If it is, the function will return **T**. If it is not, the function will return **nil**.

Examples:

(<= 10 15)	returns T
(<= "c" "b")	returns nil
(<= -2.0 0)	returns T

Greater than

Format **(> atom1 atom2 - - - -)**

This function (**>**) checks whether the first atom **(atom1)** is greater than the second atom **(atom2)**. If it is, the function will return **T**. If it is not, then the function will return **nil**. In the first example below, 15 is greater than 10. Therefore, this relational function is true and the function will return **T**. In the second example, 10 is greater than 9, but this number is not greater than the second 9; therefore, this function will return **nil**.

Examples:

(> 15 10)	returns T
(> 10 9 9)	returns nil
(> "c" "b")	returns T

Greater than or equal to

Format **(>= atom1 atom2 - - - -)**

This function (**>=**) checks whether the first atom **(atom1)** is greater than or equal to the second atom **(atom2)**. If it is, the function returns **T**; otherwise, it will return **nil**. In the first example below, 78 is greater than 50, even though 78 is not equal to 50; therefore, it will return **T**.

Examples:

(>= 78 50)	returns T
(>= "x" "y")	returns nil
(>= "78" "80")	returns nil

defun, setq, getpoint, AND Command FUNCTIONS

defun

The **defun** function is used to define a function in an AutoLISP program. The format of the **defun** function is:

(defun name [argument])

Where **Name** ------------ Name of the function
Argument------- Argument list

Examples:

(defun ADNUM ( )
Defines a function **ADNUM** with no arguments or local symbols. This means all b variables used in the program are global variables. A global variable does not lose its value after the program ends.

(defun ADNUM (a b c)
Defines a function **ADNUM** that has three arguments: **a**, **b**, and **c**. The variables **a**, **b**, and **c** receive their value from outside the program.

(defun ADNUM (/ a b)
Defines a function **ADNUM** that has two local variables: **a** and **b**. A local variable is one that retains its value during program execution and can be used within that program only.

(defun C:ADNUM ( )
With **C:** in front of the function name, the function can be executed by entering the name of the function at the AutoCAD Command prompt. If **C**: is not used, the function name has to be enclosed in parentheses.

Note

AutoLISP contains some in-built functions. Do not use any of those names for function or variable names. The following is a list of some of the names reserved for AutoLISP in-built functions. (Refer to the AutoLISP Developer's Guide for a complete list of AutoLISP in-built functions.)

abs	ads	alloc
and	angle	angtos
append	apply	atom
ascii	assoc	atan
atof	atoi	distance
equal	fix	float
if	length	list
load	member	nil
not	nth	null
open	or	pi
read	repeat	reverse
set	type	while

setq

The **setq** function is used to assign a value to a variable. The format of the **setq** function is:

(setq Name Value [Name Value].......)
Where **Name** ------------ Name of variable
Value ------------- Value assigned to variable

The value assigned to a variable can be any expression (numeric, string, or alphanumeric).

Command: **(setq X 12)**
Command: **(setq X 6.5)**
Command: **(setq X 8.5 Y 12)**

In this last expression, the number 8.5 is assigned to the variable **X**, and the number 12 is assigned to the variable **Y**.

Command: **(setq answer "YES")**

In this expression the string value "YES" is assigned to the variable **answer**.

The **setq** function can also be used in conjunction with other expressions to assign a value to a variable. In the following examples, the **setq** function has been used to assign values to different variables.

```
(setq pt1 (getpoint "Enter start point: "))
(setq ang1 (getangle "Enter included angle:"))
(setq answer (getstring "Enter YES or NO: "))
```

Note

*AutoLISP uses some built-in function names and symbols. Do not assign values to any of those functions. The following functions are valid ones, but never be used because the **pi** and **angle** functions are reserved functions and these functions will be redefined.*

(setq pi 3.0)
(setq angle (. . .))

getpoint

The **getpoint** function pauses to enable you to enter the *X*, *Y* coordinates or *X*, *Y*, *Z* coordinates of a point. The coordinates of the point can be entered from the keyboard or by using the screen cursor. The format of the **getpoint** function is:

(getpoint [point] [prompt])
Where **Point** ------------- Enter a point, or select a point
Prompt ---------- Prompt to be displayed on screen

Example:
(setq pt1 (getpoint))
(setq pt1 (getpoint "Enter starting point"))

Note

You cannot enter the name of another AutoLISP routine in response to the getpoint function.

A 2D or a 3D point is always defined with respect to the current user coordinate system (UCS).

Command

The **Command** function is used to execute standard AutoCAD commands from within an AutoLISP program. The AutoCAD command name and the command options have to be enclosed in double quotation marks. The format of the **Command** function is:

(Command "commandname")
Where **Command** ----------------- AutoLISP function
commandname ---------- AutoCAD command

Example:
(Command "line" pt1 pt2 "")
Where **"Line"** ----------- AutoCAD LINE Command
Pt1 ---------------- First point
Pt2 ---------------- Second point
"" ----------------- for Return (two double quotes with no space between)

Chapter 8

Note

Prior to AutoCAD Release 12, the ***Command*** *function could not be used to execute the AutoCAD PLOT command. For example: (Command "plot" . . .) was not a valid statement. In AutoCAD Release 13 and later, you can use plot with the Command function (Command "plot" . . .).*

The ***Command*** *function can also be used to enter data with the* ***TEXT*** *command. For example, enter (command "text""4.0,4"""""AutoCAD Text") at the Command prompt. You will notice that the text is automatically written.*

You cannot use the input functions of AutoLISP with the ***Command*** *function. The input functions are* ***getpoint***, ***getangle***, ***getstring***, *and* ***getint***. *For example, (Command "getpoint" . . .) or (Command "getangle" . . .) are not valid functions. If the program contains such a function, an error message is displayed when the program is loaded. However, you can enclose them in parentheses (Command "text" (getpoint) (getdist) (getangle) "hello, sailer")*

Example 1

Write a program that will prompt you to select three points of a triangle and then draw lines through those points to generate the triangle, as shown in Figure 8-1.

Step 1: Understanding the LISP program

Most programs consist of essentially three parts: input, output, and process. Process includes what is involved in generating the desired output from the given input (Figure 8-2). Before writing a program, you must identify these three parts. In this example, the input to the program is the coordinates of the three points. The desired output is a triangle. The process needed to generate a triangle is to draw three lines from P1 to P2, P2 to P3, and P3 to P1. Identifying these three sections makes the programming process less confusing.

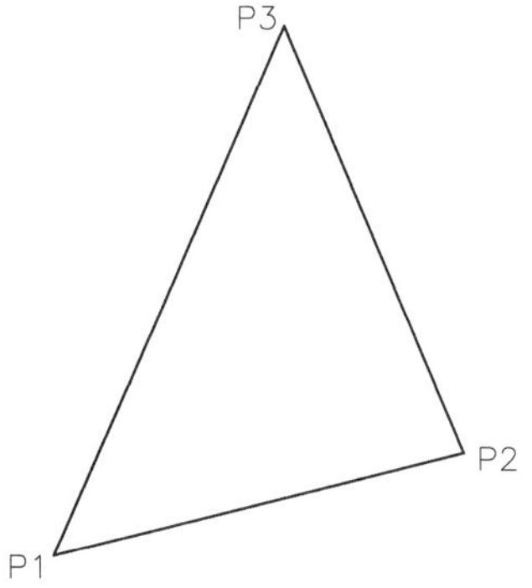

Figure 8-1 *Triangle with points P1, P2, P3*

The process section of the program is vital to the success of the program. Sometimes it is simple, but sometimes it involves complicated calculations. If the program involves many calculations, divide the program into sections (and perhaps subsections) that are laid out in a logical and systematic order. Also, remember that programs need to be edited from time to time. This is the reason, it is wise to document the programs as clearly and unambiguously as possible so that other programmers can

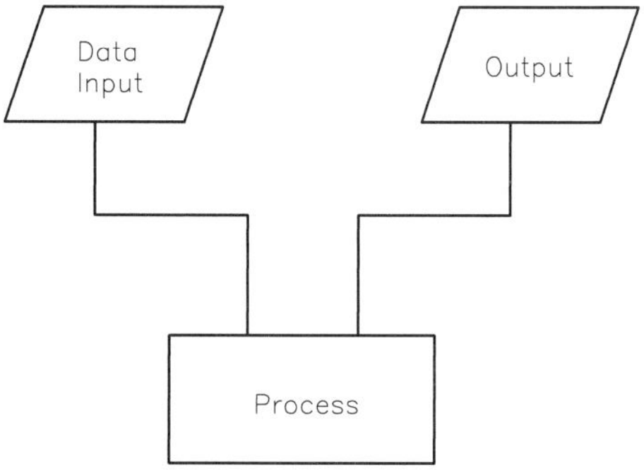

Figure 8-2 *Three elements of a program*

understand what the program is doing at different stages of its execution. Give sketches, and identify points where possible.

Input	**Output**
Location of point P1	
Location of point P2	Triangle P1, P2, P3
Location of point P3	

Process
Line from P1 to P2
Line from P2 to P3
Line from P3 to P1

Step 2: Writing the LISP program

Use any text editor to write the following LISP program. The following file is a listing of the AutoLISP program for Example 1. **The line numbers at the right are not a part of the program; they are shown here for reference only.**

```
1  ;This program will prompt you to enter three points
2  ;of a triangle from the keyboard, or select three points
3  ;by using the screen cursor. P1, P2, P3 are triangle corners.
4
5  (defun c:TRIANG1()
6     (setq P1 (getpoint "\n Enter first point of Triangle: "))
7     (setq P2 (getpoint "\n Enter second point of Triangle: "))
8     (setq P3 (getpoint "\n Enter third point of Triangle: "))
9     (Command "LINE" P1 P2 P3 "C")
10    )
```


Explanation

Lines 1-3

The first three lines are comment lines describing the function of the program. These lines are important because they make it easier to edit a program. Comments should be used when needed. All comment lines must start with a semicolon (;). These lines are ignored when the program is loaded.

Line 4

This is a blank line that separates the comment section from the program. Blank lines can be used to separate different modules of a program. This makes it easier to identify different sections that constitute a program. The blank lines have no effect on the program.

Line 5

(defun c:TRIANG1( )

In this line, **defun** is an AutoLISP function that defines the function **TRIANG1**. **TRIANG1** is the name of the function. The c: in front of the function name **TRIANG1** enables it to be executed like an AutoCAD command. If the c: is missing, the **TRIANG1** command can be executed only by enclosing it in parentheses **(TRIANG1)**. The **TRIANG1** function has three global variables (P1, P2, and P3). When you first write an AutoLISP program, it is a good practice to keep the variables global, because after you load and run the program, you can check the values of these variables by entering an exclamation point (!) followed by the variable name at the Command prompt (Command: !P1). Once the program is tested and it works, you should make the variable local, (defun c:TRIANG1(/ P1 P2 P3)).

Line 6
(setq P1 (getpoint "\n Enter first point of Triangle: "))
In this line, the **getpoint** function pauses for you to enter the first point of the triangle. The prompt, **Enter first point of Triangle,** is displayed in the prompt area of the screen. You can enter the coordinates of this point at the keyboard or select a point by using the screen cursor. The **setq** function then assigns these coordinates to the variable **P1**. **\n** is used for the carriage return so that the statement that follows \n is printed on the next line ("n" stands for "newline".)

Lines 7 and 8
(setq P2 (getpoint "\n Enter second point of Triangle: "))
(setq P3 (getpoint "\n Enter third point of Triangle: "))
These two lines prompt you to enter the second and third corners of the triangle. These coordinates are then assigned to the variables **P2** and **P3. \n** causes a carriage return so that the input prompts are displayed on the next line.

Line 9
(Command "LINE" P1 P2 P3 "C")
In this line, the **Command** function is used to enter the **LINE** command and then draw a line from P1 to P2 and from P2 to P3. "C" (for "close" option) joins the last point, P3, with the first point, P1. All AutoCAD commands and options, when used in an AutoLISP program, have to be enclosed in double quotation marks. The variables P1, P2, P3 are separated by a blank space.

Line 10
This line consists of a closing parenthesis that completes the definition of the function, **TRIANG1**. This parenthesis could have been combined with the previous line. It is good practice to keep it on a separate line so that the end of a definition can be easily identified. In this program there is only one function defined, so it is easy to locate the end of a definition. But in some programs a number of definitions or modules within the same program might need to be clearly identified. The parentheses and blank lines help to identify the start and end of a definition or a section in the program.

Save the file with the extension *.lsp*. There are generally two names associated with an AutoLISP program: the program file name and the function name. For example, *triang1.lsp* is the name of the file, not a function name. All AutoLISP file names have the extension *.lsp*. An AutoLISP file can have one or several functions defined within the same file. For example, TRIANG1 in Example 1 is the name of a function. To execute a function, the AutoLISP program file that defines that function must be loaded. Use the following procedure to load an AutoLISP file when you are in the drawing editor.

LOADING AN AutoLISP PROGRAM

Loading an AutoLISP program can be achieved either through the dialog box or by entering a command at the Command prompt. Choose **Manage > Applications > Load Application** from the **Ribbon** to display the **Load/Unload Applications** dialog box (Figure 8-3).

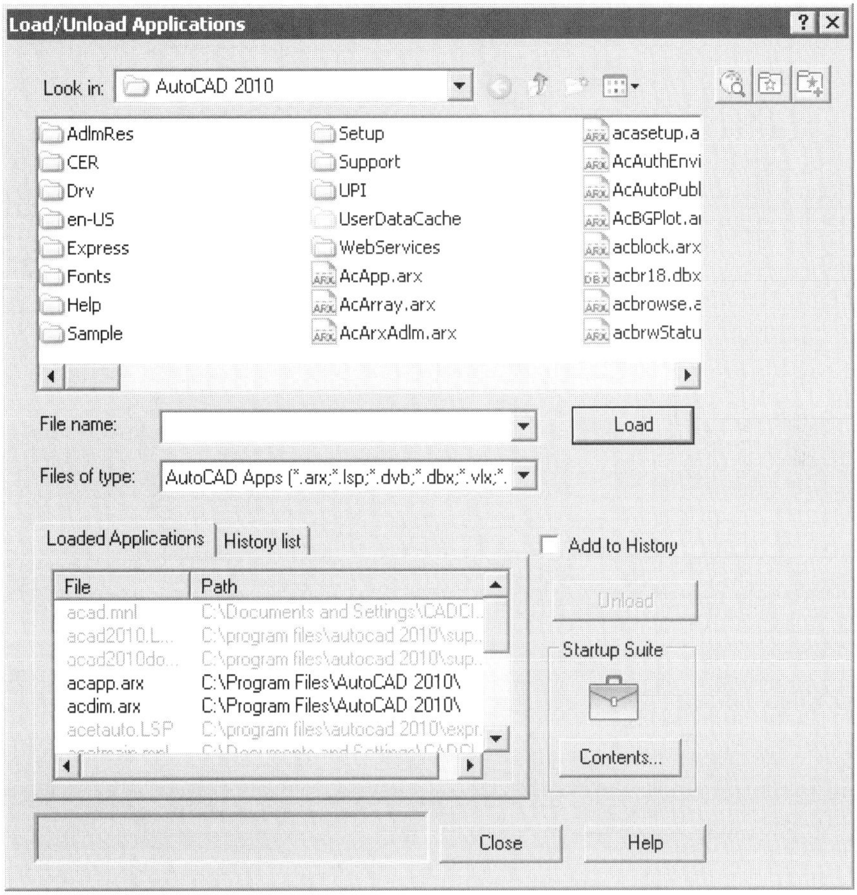

Figure 8-3 *Loading AutoLISP files using the* ***Load/Unload Applications*** *dialog box*

This dialog box can be used to load LSP, VLX, FAS, VBA, DBX, and ObjectARX applications. VBA, DBX, and ObjectARX files are loaded immediately when you select a file. LSP, VLX, and FAS files are queued and loaded when you close the **Load/Unload Applications** dialog box. The top portion of the dialog box lists the files in the selected directory. The file type can be changed by entering the file type (*.lsp) in the **Files of type:** edit box or by selecting the file type in the pop-down list. A file can be loaded by selecting the file and then choosing the **Load** button or by double-clicking on the file. The following is the description of other features of the **Load/Unload Applications** dialog box:

Load

The **Load** button can be used to load and reload the selected files. The files can be selected from the file list box, **Loaded Applications** tab, or **History list** tab. The Object ARX files cannot be reloaded. You must first unload the ObjectARX file and then reload it.

Loaded Applications Tab

When you choose the **Loaded Applications** tab, AutoCAD displays the applications that are currently loaded. You can add files to this list by dragging the file names from the file list box and then dropping them in the **Loaded Applications** list.

History list Tab

When you choose the **History list** tab, AutoCAD displays the list of files that have been previously loaded with **Add to History** check box selected. If the **Add to History** check box is not selected and you drag and drop the files in the History list, the files are loaded but not added to the **History list**. The users can directly reload a file by selecting it from the history list in this tab.

Add to History

When **Add to History** is selected, it adds the files to the **History list** when you drag and drop the files in the **History list**.

Unload

The **Unload** button appears when you choose the **Loaded Applications** tab. To unload an application, select the file name in the **Loaded Applications** list and then choose the **Unload** button. LISP files and the ObjectARX files that are not registered for unloading cannot be unloaded.

Remove

The **Remove** button appears when you choose the **History list** tab. To remove a file from the History List, select the file in the History List and then choose the **Remove** button.

Startup Suite Area

The files in the **Startup Suite** are automatically loaded each time you start AutoCAD. When you choose **Contents** button in this area, AutoCAD displays the **Startup Suite** dialog box that contains a list of files. You can add files to the list by choosing the **Add** button. You can also drag the files from the file list box and drop them in the **Startup Suite**. To add files from the **History list** tab, right-click on the file.

The AutoLISP program can also be loaded using the **LOAD** command in the following format.

Command: **(load "[path]file name")**

Where **Command** ------ AutoCAD command prompt
load ------------- Loads an AutoLISP program file
file name ------- Path and name of AutoLISP program file

The AutoLISP file name and the optional path name must be enclosed in double quotes. The **load** and **file name** must be enclosed in parentheses. If the parentheses are missing, AutoCAD will try to load a shape or a text font file, not an AutoLISP file. The space between **load** and **file name** is not required. If AutoCAD is successful in loading the file, it will display the name of the function in the Command prompt area of the screen.

To run the program, type the name of the function at the AutoCAD Command prompt, and press ENTER (Command: **TRIANG1**). If the function name does not contain **C:** in the program, you can run the program by enclosing the function name in parentheses:

Command: **TRIANG1** or Command: **(TRIANG1)**

Note

Use a forward slash when defining the path for loading an AutoLISP program. For example, if the AutoLISP file TRIANG is in the LISP subdirectory on the C drive, use the following command to load the file. You can also use a double backslash (\\) in place of the forward slash.

Command: **(load "c:/lisp/triang")** or Command: **(load "c:\\lisp\\triang")**

Tip

You can also load an application by using the standard Windows drag and drop technique. To load a LISP program, select the file in the Windows Explorer and then drag and drop it in the graphics window of AutoCAD. The selected program will be automatically loaded.

Exercise 1

General

Write an AutoLISP program that will draw a line between two points (Figure 8-4). The program must prompt the user to enter the *X* and *Y* coordinates of the points.

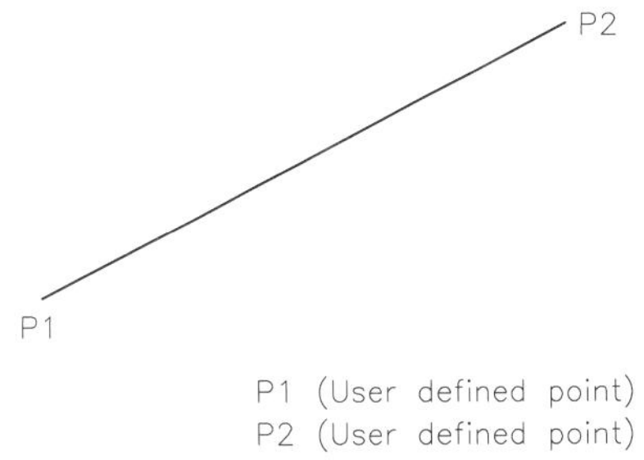

Figure 8-4 *Drawing a line from point P1 to P2*

getcorner, getdist, AND setvar FUNCTIONS

getcorner

The **getcorner** function pauses for you to enter the coordinates of a point. The coordinates of the point can be entered or the screen crosshairs can be used to specify its location. This function requires a base point and displays a rectangle with respect to the base point as you move the screen crosshairs around the screen. The format of the **getcorner** function is given next.

(getcorner point [prompt])

Where **point** ------------------------ Base point
prompt --------------------- Prompt displayed on screen

Examples:

(getcorner pt1)
(setq pt2 (getcorner pt1))
(setq pt2 (getcorner pt1 "Enter second point: "))

Note

The base point and the point that you select in response to the getcorner function are located with respect to the current UCS.

If the point you select is a 3D point with X, Y, and Z coordinates, the Z coordinate is ignored. The point assumes current elevation as its Z coordinate.

getdist

The **getdist** function pauses for you to enter distance, and it then returns the distance as a real number. The format of the **getdist** function is:

(getdist [point] [prompt])

Where **Point** ------------- First point for distance
Prompt ---------- Any prompt that needs to be displayed on screen

Examples:

(getdist)
(setq dist (getdist))
(setq dist (getdist pt1))
(setq dist (getdist "Enter distance"))
(setq dist (getdist pt1 "Enter second point for distance"))

The distance can be entered by selecting two points on the screen. For example, if the assignment is **(setq dist (getdist))**, you can enter a number or select two points. If the assignment is **(setq dist (getdist pt1))**, where the first point (pt1) is already defined, you need to select the second point only. The getdist function will always return the distance as a real number. For example, if the current setting is architectural and the distance is entered in architectural units, the getdist function will return the distance as a real number.

setvar

The assigns a value to an AutoCAD system variable. The name of the system variable must be enclosed in double quotes. The format of the **setvar** function is:

(setvar "variable-name" value)

Where **Variable-name** AutoCAD system variable
Value ------------- Value to be assigned to the system variable

Examples:

(setvar "cmdecho" 0)
(setvar "dimscale" 1.5)
(setvar "ltscale" 0.5)
(setvar "dimcen" -0.25)

Example 2

Write an AutoLISP program that will prompt you to specify the four corner points of a rectangle and then draw lines through those points and the two diagonals of the rectangle. The program should next draw a circle, by first prompting you to select the intersection point of the diagonals as the center of the circle and then asking you to specify the diameter of the circle, as shown in Figure 8-5.

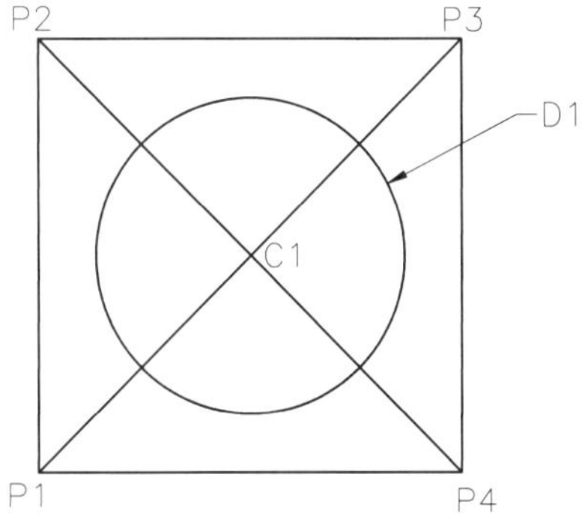

Figure 8-5 *Rectangle with circle centered at the intersection point of the diagonals*

Step 2: Writing the LISP program

Use any text editor to write the following LISP program. The following file is a listing of the AutoLISP program for Example 2. **The line numbers at the right are not a part of the program; they are shown here for reference only.**

```
;This program will prompt you to enter four endpoints of a rectangle,
;the intersection point of the diagonals, and the diameter of the circle.
;Specify the points through keyboard or select the points.
;by using the screen cursor.

(defun c:RECTCIRC()
        (setq P1 (getpoint "\n Enter first point of rectangle:"))
        (setq P2 (getpoint "\n Enter second point of rectangle:"))
        (setq P3 (getpoint "\n Enter third point of rectangle:"))
```

```
10 (setq P4 (getpoint "\n Enter fourth point of rectangle:"))
11 (Command "LINE" P1 P2 P3 P4 P1 P3 "")
12 (Command "LINE" P2 P4 "")
13 (setq C1 (getpoint "\n select the intersection point of the two diagonal:"))
14 (setq D1 (getdist "\n Enter diameter of circle: "))
15 (Command "CIRCLE" C1 "d" D1)
16 )
```

Explanation

Lines 1-4

The first four lines are the comment lines that describe the function of the program. These lines are important because they make it easier to edit a program. Comments should be used when needed. All comment lines must start with a semicolon (;). These lines are ignored when the program is loaded.

Line 5

This is a blank line that separates the comment section from the program. Blank lines can be used to separate different modules of a program. This makes it easier to identify different sections that constitute a program. The blank lines have no effect on the program.

Line 6

defun c:RECTCIRC( )

In this line, **defun** is an AutoLISP function that defines the function **RECTCIRC**. **RECTCIRC** is the name of the function. The c: in front of the function name **RECTCIRC** enables it to be executed like an AutoCAD command. If the c: is missing, the **RECTCIRC** command can be executed only by enclosing it in parentheses (**RECTCIRC**). The **RECTCIRC** function has six global variables (P1, P2, P3, P4, C1, D1). When you first write an AutoLISP program, it is a good practice to keep the variables global, because after you load and run the program, you can check the values of these variables by entering an exclamation point (!) followed by the variable name at the Command prompt (Command: !P1). Once the program is tested and it works, you should make the variable local, (defun c:RECTCIRC(/ P1 P2 P3)).

Line 7

(setq P1 (getpoint "\n Enter first point of rectangle:"))

In this line, the **getpoint** function pauses for you to enter the first point of the rectangle. The prompt, **Enter first point of rectangle**, is displayed in the prompt area of the screen. You can enter the coordinates of this point at the keyboard or select a point by using the screen cursor. The **setq** function then assigns these coordinates to the variable **P1**. **\n** is used for the carriage return so that the statement that follows \n is printed on the next line ("n" stands for "newline".)

Lines 8-10

(setq P2 (getpoint "\n Enter second point of rectangle:"))
(setq P3 (getpoint "\n Enter third point of rectangle:"))
(setq P4 (getpoint "\n Enter fourth point of rectangle:"))

These three lines prompt you to enter the second, third, and fourth corners of the rectangle. These coordinates are then assigned to the variables **P2**, **P3**, and **P4**. **\n** causes a carriage return so that the input prompts are displayed on the next line.

Line 11
(Command "LINE" P1 P2 P3 P4 P1 P3 "")
In this line, the **Command** function is used to enter the **LINE** command and then draw a line from P1 to P2, from P2 to P3, from P3 to P4, from P4 to P1, and from P1 to P3. "" (for ENTER key) ends the **LINE** command. All AutoCAD commands and options, when used in an AutoLISP program, have to be enclosed in double quotation marks. All variables are separated by a blank space.

Line 12
(Command "LINE" P2 P4 "")
In this line, the **LINE** command is invoked and a line is drawn between the points P2 and P4. The "" will end the **LINE** command.

Line 13
(setq C1 (getpoint "\n select the intersection point of the two diagonals:"))
In this line, the **getpoint** function pauses for you to select the intersection point of the diagonals of the rectangle. The prompt, **select the intersection point of the two diagonals:,** is displayed in the prompt area of the screen. You can enter the coordinates of this point through the keyboard or select a point by using the screen cursor. The **setq** function then assigns these coordinates to the variable **C1**. **\n** is used for the carriage return so that the statement that follows \n is printed in the next line ("n" stands for "newline".)

Line 14
(setq D1 (getdist "\n Enter diameter of circle: "))
The **getdist** function pauses for you to enter the diameter of the circle and then the **setq** function assigns that value to variable D1.

Line 15
(Command "CIRCLE" C1 "d" D1)
The **Command** function is used to invoke the **CIRCLE** command and then draw a circle whose center is **C1** and diameter equal to the global variable **D1**.

Line 16
)
This line consists of a closing parenthesis that completes the definition of the function, **RECTCIRC**. This parenthesis could have been combined with the previous line. It is good practice to keep it on a separate line so that the end of a definition can be easily identified. In this program only one function has been defined, so it is easy to locate the end of the definition. But in some programs a number of definitions or modules within the same program might need to be clearly identified. The parentheses and blank lines help to identify the start and end of a definition or a section in the program.

Step 2: Loading the LISP program

Save the file with the extension *.lsp*. To execute a function, The AutoLISP program file that defines the function must be loaded. Use the procedure given next to load the saved AutoLISP file when you are in the drawing editor.

1. Choose **Manage > Applications > Load Application** from the menu bar to display the **Load/Unload Application** dialog box.

2. Browse the *.lsp* file that you saved in this example and choose the **Load** button from the **Load/Unload Application** dialog box. Next, choose the **Close** button.

Step 3: Running the LISP program

To run the program, type the name of the function at the AutoCAD Command prompt, and press ENTER (Command: **RECTCIRC**). If the function name does not contain C: in the program, you can run the program by enclosing the function name in parentheses:

Command: **RECTCIRC** or Command: (**RECTCIRC**)

Example 3

Write an AutoLISP program that will generate a chamfer between two given lines by entering the chamfer angle and the chamfer distance. To generate a chamfer, AutoCAD uses the values assigned to the system variables **CHAMFERA** and **CHAMFERB**. When you select the **CHAMFER** command, the first and second chamfer distances are automatically assigned to the system variables **CHAMFERA** and **CHAMFERB**. The **CHAMFER** command then uses these assigned values to generate a chamfer. However, in most engineering drawings, the preferred way to generate the chamfer is to enter the chamfer length and the chamfer angle, as shown in Figure 8-6.

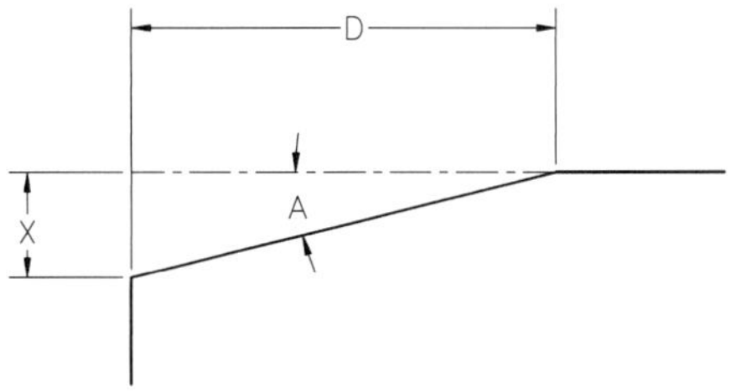

Figure 8-6 *Chamfer with angle A and distance D*

Input	**Output**
First chamfer distance (D)	Chamfer between any two
Chamfer angle (A)	selected lines

Step 1: Understanding the program algorithm

Process

1. Calculate second chamfer distance.
2. Assign these values to the system variables **CHAMFERA** and **CHAMFERB**.
3. Use **CHAMFER** command. to generate chamfer,

Calculations

x/d = tan a

x = d * (tan a)

= d * [(sin a) / (cos a)]

Step 2: Writing the LISP program

Use any text editor to write down the LISP program. The following file is a listing of the program for Example 3. **The line numbers at the right are not a part of the program; they are shown here for reference only.**

```
1  ;This program generates a chamfer by entering
2  ;the chamfer angle and the chamfer distance
3  ;
4  (defun c:chamf (/ d a)
5  (setvar "cmdecho" 0)
6     (graphscr)
7     (setq d (getdist "\n Enter chamfer distance: "))
8     (setq a (getangle "\n Enter chamfer angle: "))
9     (setvar "chamfera" d)
10     (setvar "chamferb" (* d (/ (sin a) (cos a))))
11     (command "chamfer")
12    (setvar "cmdecho" 1)
13    (princ)
14    )
```


Explanation

Line 7

(setq d (getdist "\n Enter chamfer distance: "))

The **getdist** function pauses for you to enter the chamfer distance, and then the **setq** function assigns that value to variable d.

Line 8

(setq a (getangle "\n Enter chamfer angle: "))

The **getangle** pauses for you to enter the chamfer angle, then the **setq** function assigns that value to variable a.

Line 9
(setvar "chamfera" d)
The **setvar** function assigns the value of variable d to AutoCAD system variable **chamfera**.

Line 10
(setvar "chamferb" (* d (/ (sin a) (cos a))))
The **setvar** function assigns the value obtained from the expression **(* d (/ (sin a) (cos a)))** to the AutoCAD system variable **CHAMFERB**.

Line 11
(command "chamfer")
The **Command** function uses the **CHAMFER** command to generate a chamfer.

Step 3: Loading the LISP program
Save the file with the extension *.lsp* and then follow the procedure described in Example 1 to load the lisp file.

Exercise 2 *General*

Write an AutoLISP program that will generate the drawing shown in Figure 8-7. The program should prompt the user to enter points P1 and P2 and diameters D1 and D2.

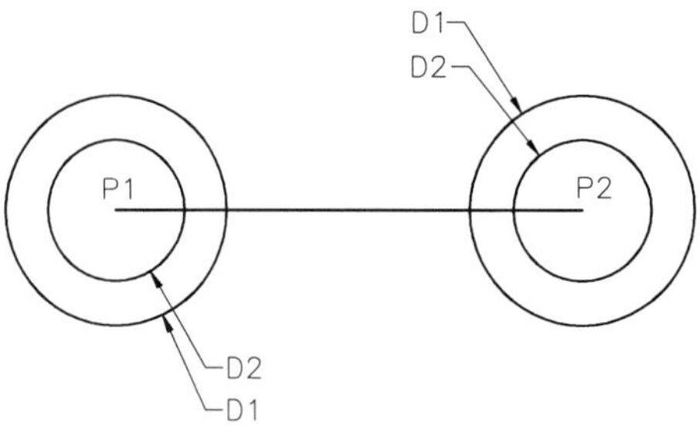

Figure 8-7 *Concentric circles with connecting line*

list FUNCTION

In AutoLISP, the **list** function is used to define a 2D or 3D point. The list function can also be designated by using the single quote character ('), if the expression does not contain any variables or undefined items.

Examples:

(Setq x (list 2.5 3 56))	returns (2.5 3 56)
(Setq x '(2.5 3 56))	returns (2.5 3 56)

Tip

A list when included with integers or real numbers can be enclosed within a single parentheses. A list when represented in the form of a single quote can be given outside the parentheses.

car, cdr, AND cadr FUNCTIONS

car

The **car** function returns the first element of a list. If the list does not contain any elements, the function will return **nil**. Remember that before using this function, you need to list the elements of the function using the **list** function. The format of the **car** function is:

(car list)

Where **car** ---------------- Returns the first element
list ---------------- List of elements

Examples:

(car '(2.5 3 56))	returns 2.5
(car '(x y z))	returns X
(car '((15 20) 56))	returns (15 20)
(car '( ))	returns nil
(car (list 2 3.0 4))	returns 2
(car '(A B C))	returns A

The single quotation mark signifies a list.

cdr

The **cdr** function returns a list with the first element removed from the list. The format of the **cdr** function is:

(cdr list)

Where **cdr** ---------------- Returns a list with the first element removed
list ---------------- List of elements

Examples:

(cdr '(2.5 3 56)	returns (3 56)
(cdr '(x y z))	returns (y z)
(cdr '((15 20) 56))	returns (56)
(cdr '( ))	returns nil

cadr

The **cadr** function performs two operations, **cdr** and **car**, to return the second element of the list. The **cdr** function removes the first element, and the **car** function returns the first element of the new list. The format of the **cadr** function is:

(cadr list)

Where **cadr** -------------- Performs two operations (car (cdr ‘(x y z))
list ---------------- List of elements

Examples:

(cadr ‘(2 3))	returns 3
(cadr ‘(2 3 56))	returns 3
(cadr ‘(x y z))	returns y
(cadr ‘((15 20) 56 24))	returns 56

In these examples, **cadr** performs two functions:

(cadr ‘(x y z)) = (car (cdr ‘(x y z)))
= (car (y z)) returns Y

Note

In addition to the functions ***car***, ***cdr***, *and* ***cadr***, *several other functions can be used to extract different elements of a list. The following is a list of some of the functions where the function f consists of a list ‘((x y) z w)).*

(setq f ‘((x y) z w))

(caar f)=(car (car f))	*returns x*
(cdar f)=(cdr (car f))	*returns (y)*
(cadar)=(car (cdr (car f)))	*returns y*
(cddr f)=(cdr (cdr f))	*returns (w)*
(caddr f)=(car (cdr (cdr f)))	*returns w*
(last f)	*returns w*

graphscr, textscr, princ, AND terpri FUNCTIONS

graphscr

The **graphscr** function switches from the text window to the graphics window, provided the system has only one screen. If the system has two screens, this function is ignored.

textscr

The **textscr** function switches from the graphics window to the text window, provided the system has only one screen. If the system has two screens, this function is ignored.

princ

The **princ** function prints (or displays) the value of the variable. If the variable is enclosed in double quotes, the function prints (or displays) the expression that is enclosed in the quotes. The format of the **princ** function is:

(princ [variable or expression])

Examples:

(princ)	prints a blank on screen
(princ a)	prints the value of the variable 'a' on screen
(princ "Welcome")	prints 'Welcome' on screen

terpri

The **terpri** function prints a new line on the screen, just as **\n** does. This function is used to print the line that follows the **terpri** function.

Examples:

```
(setq p1 (getpoint "Enter first point: "))(terpri)
(setq p2 (getpoint "Enter second point: "))
```

The first line (Enter first point:) will be displayed in the screen's Command prompt area. The **terpri** function causes a carriage return; therefore, the second line (Enter second point:) will be displayed on a new line, just below the first line. If the **terpri** function is missing, the two lines will be displayed on the same line (Enter first point: Enter second point:).

Example 4

Write a program that will prompt you to enter two opposite corners of a rectangle and then draw the rectangle on the screen, as shown in Figure 8-8.

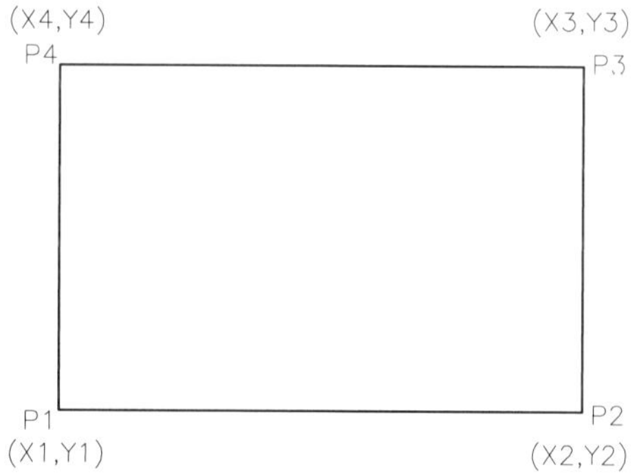

Figure 8-8 *Rectangle with points P1 P2 P3 P4*

Step 1: Understanding the program algorithm

<u>Input</u>	<u>Output</u>
Coordinates of point P1	Rectangle
Coordinates of point P3	

<u>**Process**</u>

1. Calculate the coordinates of the points P2 and P4.
2. Draw the following lines.
   Line from P1 to P2
   Line from P2 to P3
   Line from P3 to P4
   Line from P4 to P1

The *X* and *Y* coordinates of points P2 and P4 can be calculated using the **car** and **cadr** functions. The **car** function extracts the *X* coordinate of a given list, and the **cadr** function extracts the *Y* coordinate.

X coordinate of point p2
x2 = x3
x2 = car (x3 y3)
x2 = car p3

Y coordinate of point p2
y2 = y1
y2 = cadr (x1 y1)
y2 = cadr p1

X coordinate of point p4
x4 = x1
x4 = car(x1 y1)
x4 = car p1

Y coordinate of point p4
y4 = y3
y4 = cadr (x3 y3)
y4 = cadr p3

Therefore, points p2 and p4 are:
p2 = (list (car p3) (cadr p1))
p4 = (list (car p1) (cadr p3))

Step 2: Writing the LISP program

Use any text editor to write the LISP program. The following file is a listing of the program for Example 4. **The line numbers at the right are for reference only; they are not a part of the program.**

```
;This program will draw a rectangle. User will
;be prompted to enter the two opposite corners
;
```

```
4  (defun c:RECT1(/ p1 p2 p3 p4)
5      (graphscr)
6      (setvar "cmdecho" 0)
7      (prompt "RECT1  command  draws  a rectangle")(terpri)
8     (setq p1 (getpoint "Enter first corner"))(terpri)
9     (setq p3 (getpoint "Enter opposite corner"))(terpri)
10     (setq p2 (list (car p3) (cadr p1)))
11     (setq p4 (list (car p1) (cadr p3)))
12   (command "pline" p1 p2 p3 p4 "c")
13  (setvar "cmdecho" 1)
14  (princ)
15  )
```

Explanation

Lines 1-3

The first three lines are comment lines that describe the function of the program. All comment lines that start with a semicolon are ignored when the program is loaded.

Line 4

(defun c:RECT1(/ p1 p2 p3 p4)

The **defun** function defines the function **RECT1**.

Line 5

(graphscr)

This function switches the text screen to the graphics screen, if the current screen happens to be a text screen. Otherwise, this function has no effect on the display screen.

Line 6

(setvar "cmdecho" 0)

The **setvar** function assigns the value 0 to the **CMDECHO** system variable, which turns the echo off. When **CMDECHO** is off, Command prompts are not displayed in the Command prompt area of the screen.

Line 7

(prompt "RECT1 command draws a rectangle")(terpri)

The **prompt** function will display the information in double quotes ("RECT1 command draws a rectangle"). The function **terpri** causes a carriage return so that the next text is printed in a separate line.

Line 8

(setq p1 (getpoint "Enter first corner"))(terpri)

The **getpoint** function pauses for you to enter a point (the first corner of the rectangle), and the **setq** function assigns that value to variable p1.

Line 9

(setq p3 (getpoint "Enter opposite corner"))(terpri)

The **getpoint** function pauses for you to enter a point (the opposite corner of the rectangle), and the **setq** function assigns that value to variable p3.

Line 10
(setq p2 (list (car p3) (cadr p1)))
The **cadr** function extracts the *Y* coordinate of point p1, and the car function extracts the *X* coordinate of point p3. These two values form a list and the **setq** function assigns that value to variable p2.

Line 11
(setq p4 (list (car p1) (cadr p3)))
The **cadr** function extracts the *Y* coordinate of point p3, and the **car** function extracts the *X* coordinate of point p1. These two values form a list and the **setq** function assigns that value to variable p4.

Line 12
(command "line" p1 p2 p3 p4 "c")
The command function uses the AutoCAD **LINE** command to draw lines between points p1, p2, p3, and p4. The c (close) joins the last point, p4, with the first point, p1.

Line 13
(setvar "cmdecho" 1)
The **setvar** function assigns a value of 1 to the AutoCAD system variable **CMDECHO**, which turns the echo on.

Line 14
(princ)
The princ function prints a blank on the screen. If this line is missing, AutoCAD will print the value of the last expression. This value does not affect the program in any way. However, it might be confusing at times. The **princ** function is used to prevent display of the last expression in the command prompt area.

Line 15
The closing parenthesis completes the definition of the function **RECT1** and ends the program.

Note

In this program, the rectangle is generated after you define the two corners of the rectangle. The rectangle is not dragged as you move the screen crosshairs to enter the second corner. However, the rectangle can be dragged by using the ***getcorner*** *function, as shown in the following program listing:*

```
;This program will draw a rectangle with the
;drag mode on and using getcorner function
;
```

```
(defun c:RECT2(/ p1 p2 p3 p4)
   (graphscr)
   (setvar "cmdecho" 0)
   (prompt "RECT2 command draws a rectangle")(terpri)
   (setq p1 (getpoint "Enter first corner"))(terpri)
   (setq p3 (getcorner p1 "Enter opposite corner" ))(terpri)
   (setq p2 (list (car p3) (cadr p1)))
   (setq p4 (list (car p1) (cadr p3)))
(command "line" p1 p2 p3 p4 "c")
(setvar "cmdecho" 1)
(princ)
)
```

Step 3: Loading the LISP program

Save the file with the extension **.lsp* and then load the file using the **APPLOAD** command, as described in Example 1.

getangle AND getorient FUNCTIONS

getangle

The **getangle** function pauses for you to enter the angle. Then it returns the value of that angle in radians. The format of the **getangle** function is:

(getangle [point] [prompt])

Where **point** ------------- First point of the angle
prompt ---------- Any prompt that needs to be displayed on screen

Examples:

```
(getangle)
(setq ang (getangle))
(setq ang (getangle pt1))————————    pt1 is a predefined point
(setq ang (getangle "Enter taper angle"))
(setq ang (getangle pt1 "Enter second point of angle"))
```

The angle you enter is affected by the angle setting. The angle settings can be changed using the **UNITS** command or by changing the value of the **ANGBASE** and **ANGDIR** system variables. Following are the default settings for measuring an angle:

The angle is measured with respect to the positive *X* axis (3 o'clock position). The value of this setting is saved in the **ANGBASE** system variable.

The angle is positive if it is measured in the counterclockwise direction and negative if it is measured in the clockwise direction. The value of this setting is saved in the **ANGDIR** system variable.

If the angle has a default setting [Figure 8-9(a)], the **getangle** function will return 2.35619 radians for an angle of 135.

Example:
(setq ang (getangle "Enter angle")) returns 2.35619 for an angle of 135-degree

Figure 8-9(b) shows the new settings of the angle, where the *Y* axis is 0-degree and the angles measured clockwise are positive. The **getangle** function will return 3.92699 for an angle of 135-degree. The **getangle** function calculates the angle in the counterclockwise direction, ignoring the direction set in the **ANGDIR** system variable, with respect to the angle base as set in the system variable **ANGBASE** [Figure 8-10(a) and Figure 8-10(b)].

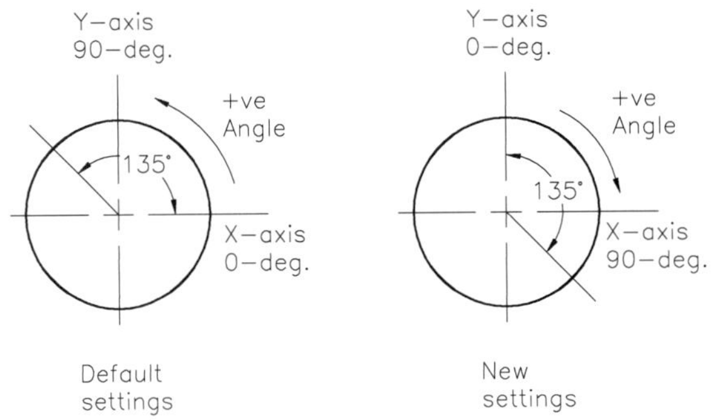

Figure 8-9(a) *Figure 8-9(b)*

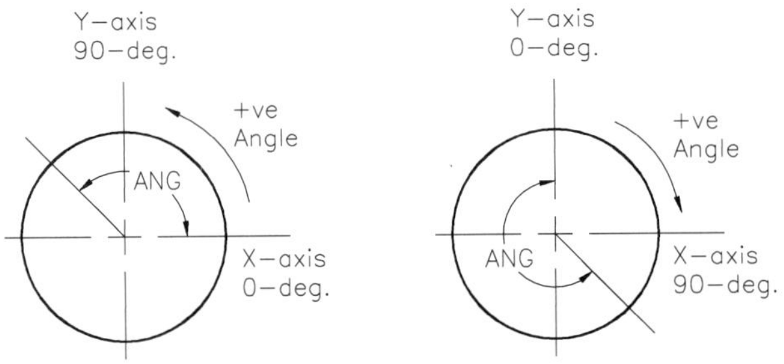

Figure 8-10(a) *Figure 8-10(b)*

Example:
(setq ang (getangle "Enter angle")) returns 3.92699

getorient

The **getorient** function pauses for you to enter the **angle** and then it returns the value of that angle in radians. The format of the **getorient** function is:

(getorient [point] [prompt])

Where **point** ------------- First point of the angle
prompt ---------- Any prompt that needs to be displayed on the screen

Examples:

(getorient)
(setq ang (getorient))
(setq ang (getorient pt1))
(setq ang (getorient "Enter taper angle"))
(setq ang (getorient pt1 "Enter second point of angle"))

The **getorient** function is just like the **getangle** function. Both return the value of the angle in radians. However, the **getorient** function always measures the angle with a positive *X* axis (3 o'clock position) and in a counterclockwise direction. It ignores the **ANGBASE** and **ANGDIR** settings. If the settings have not been changed, as shown in Figure 8-11(a) (default settings for **ANGDIR** and **ANGBASE**), for an angle of 135-degree the **getorient** function will return 2.35619 radians. If the settings are changed, as shown in Figure 8-11(b), for an angle of 135-degree the **getorient** function will return 5.49778 radians. Although the settings have been changed where the angle is measured with the positive *Y* axis and in a clockwise direction, the **getorient** function ignores the new settings and measures the angle from positive *X* axis and in a counterclockwise direction.

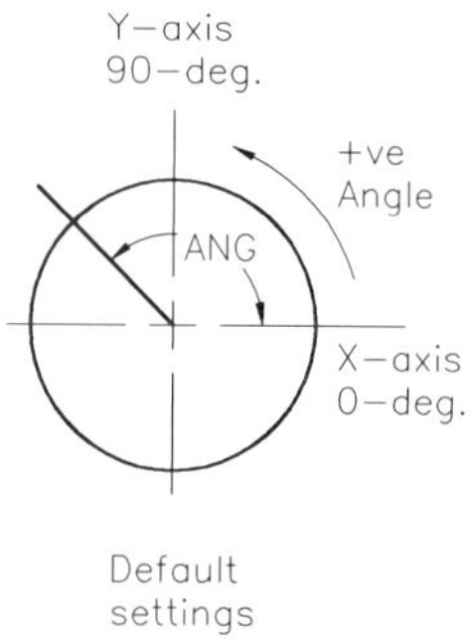

Figure 8-11(a)

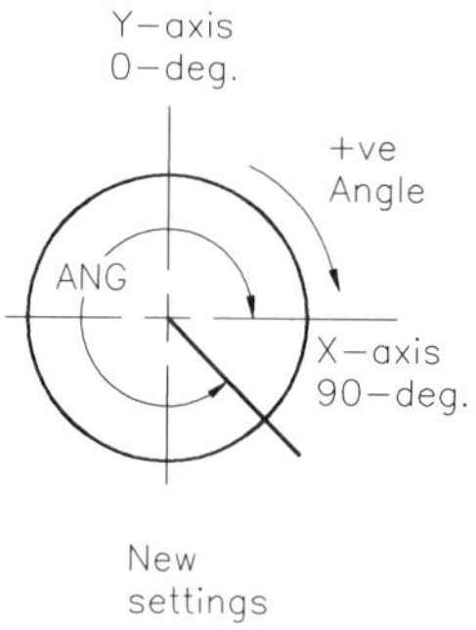

Figure 8-11(b)

Chapter 8

Note

For the **getangle** *and* **getorient** *functions, you can enter the angle by typing the angle or by selecting two points on the screen. If the assignment is* (**setq ang** (**getorient pt1**)), *where the first point pt1 is already defined, you will be prompted to enter the second point. You can enter this point by selecting a point on the screen or by entering the coordinates of the second point.*

180-degree is equal to pi (3.14159) radians. To calculate an angle in radians, use the following relation:

Angle in radians = (pi x angle)/180

getint, getreal, getstring, AND getvar FUNCTIONS

getint

The **getint** function pauses for you to enter an integer. The function always returns an integer. If you enter a real number, it will inform you that it requires an integer value. The format of the **getint** function is:

(getint [prompt])
Where **prompt** ---------- Optional prompt that you want to display on screen

Examples:

```
(getint)
(setq numx (getint))
(setq numx (getint "Enter number of rows: "))
(setq numx (getint "\n Enter number of rows: "))
```

getreal

The **getreal** function pauses for you to enter a real number and it returns a real number. If you happen to enter an integer, it returns a real number. The format of the **getreal** function is:

(getreal [prompt])
Where **prompt** ---------- Optional prompt that is displayed on screen

Examples:

```
(getreal)
(setq realnumx (getreal))
(setq realnumx (getreal "Enter num1: "))
(setq realnumx (getreal "\n Enter num2: "))
```

getstring

The **getstring** function pauses for you to enter a string value, and it always returns a string, even if the string you enter contains numbers only. The format of the **getstring** function is:

(getstring [cr] [prompt])

Where **cr** ------------ May be T or Nil. The default value, if not specified, is Nil. If the value is T, the input string can have blanks. Also note that the string must be terminated by ENTER.

prompt ---------- Optional prompt that is displayed on screen

Examples:

(getstring)
(setq answer (getstring))
(setq answer (getstring "Enter Y for yes, N for no:))
(setq answer (getstring "\n Enter Y for yes, N for no:))

Note

The maximum length of the string is 256 characters. If the string exceeds 256 characters, the exceeding characters are ignored.

getvar

The **getvar** function lets you retrieve the value of an AutoCAD system variable. The format of the **getvar** function is:

(getvar "variable")

Where **variable** --------- AutoCAD system variable name

Examples:

(getvar)	
(getvar "dimcen")	returns 0.09
(getvar "ltscale")	returns 1.0
(getvar "limmax")	returns 12.00,9.00
(getvar "limmin")	returns 0.00,0.00

Note

The system variable name should always be enclosed in double quotes.

You can retrieve only one variable value in one assignment. To retrieve the values of several system variables, use a separate assignment for each variable.

polar AND sqrt FUNCTIONS

Polar

The **polar** function defines a point at a given angle and distance from the given point (Figure 8-12). The angle is expressed in radians, measured positive in the counterclockwise direction (assuming default settings for **ANGBASE** and **ANGDIR**). The format of the **polar** function is:

(polar point angle distance)

Where **point** ------------- Reference point
angle ------------- Angle the point makes with the referenced point
distance --------- Distance of the point from the referenced point

Examples:

(polar pt1 ang dis)
(setq pt2 (polar pt1 ang dis))
(setq pt2 (polar '(2.0 3.25) ang dis))

sqrt

The **sqrt** function calculates the square root of a number, and the value this function returns is always a real number, Figure 8-13. The format of the **sqrt** function is:

(sqrt number)

Where **number** --------- Number you want to find the square root of
-------------------- (real or integer)

Examples:

(sqrt 144)	returns 12.0
(sqrt 144.0)	returns 12.0
(setq x (sqrt 57.25))	returns 7.566373
(setq x (sqrt (* 25 36.5)))	returns 30.207615
(setq x (sqrt (/ 7.5 (cos 0.75))))	returns 3.2016035

(setq hyp (sqrt (+ (* base base) (* ht ht))))

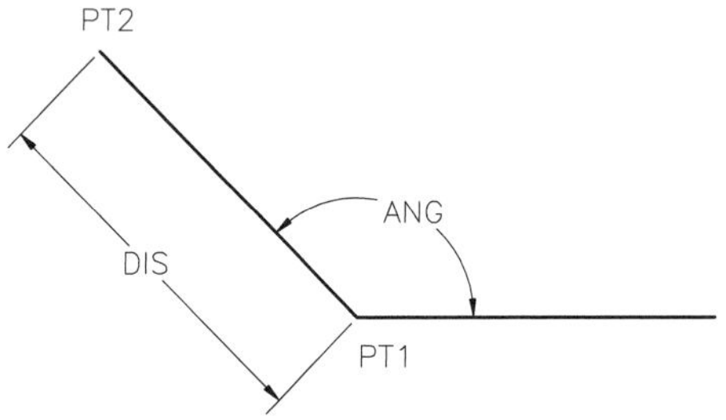

Figure 8-12 *Using the* ***polar*** *function to define a point*

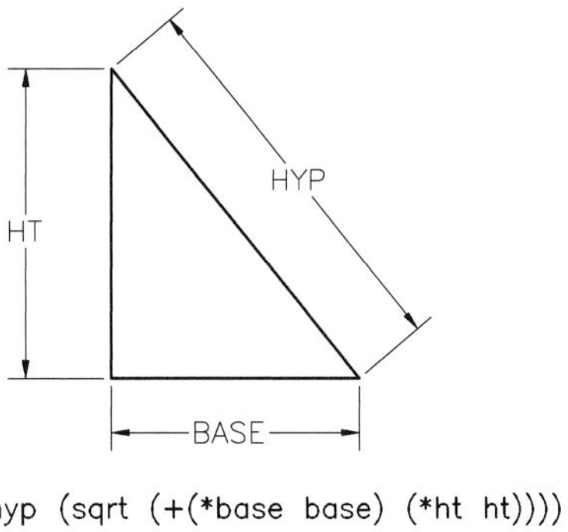

Figure 8-13 *Application of the* ***sqrt*** *function*

Example 5

Write an AutoLISP program that will draw a triangle inside a circle (Figure 8-14). The triangle should be a right-angled triangle. The program should prompt you to enter the radius and the center point of the circle.

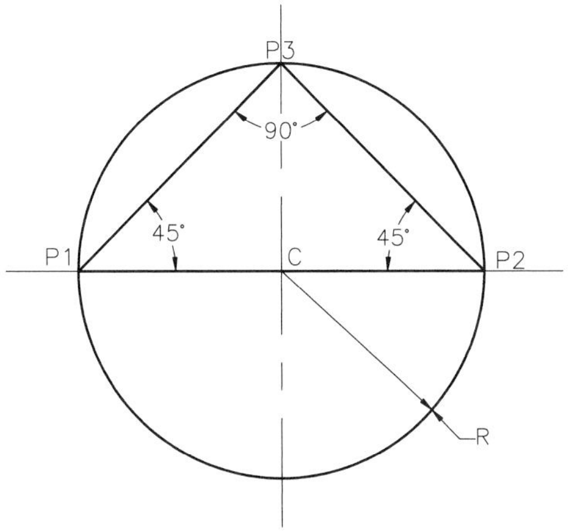

Figure 8-14 *Right angled triangle inside a circle*

Step 1: Turning on OSNAP

Right-click on the **OSNAP** button of the status bar in the AutoCAD window and choose the **Settings** option from the shortcut menu. Select the **Quadrant** check box in the **Drafting Settings** dialog box and choose the **OK** button. Next, turn on the OSNAP from the status bar.

Step 2: Writing the LISP program

Use any text editor to write the LISP program. The following file is the listing of the AutoLISP program for Example 5.

```
;This program will draw a right-angled triangle
; inside a circle.
;
(defun dtr (a)
        (* a (/ pi 180.0))
)
(defun c:TRGCIR()
(setvar "cmdecho" 0)
(graphscr)
        (setq R(getdist "\n Enter circle radius: "))
        (setq C(getpoint "\n Enter center of circle: "))
        (setq P1(polar C (dtr 210) R))
        (setq P2(polar C (dtr 330) R))
        (setq P3(polar C (dtr 90) R))
```

```
(command "CIRCLE" c r)
(command "LINE" P1 P2 P3 "C")
(setvar "cmdecho" 1)
(princ)
)
```

Step 3: Loading the LISP program

Save the file in the *.lsp* format and then load it using the **APPLOAD** command, as described in Example 1.

Example 6

Write an AutoLISP program that will draw an equilateral triangle outside a circle (Figure 8-15). The sides of the triangle are tangent to the circle. The program should prompt you to enter the radius and the center point of the circle.

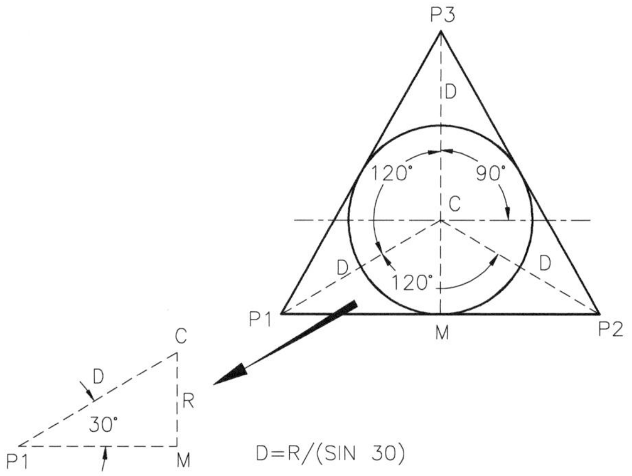

Figure 8-15 *Equilateral triangle outside a circle*

Step 1: Writing the LISP program

Use any text editor to write the LISP program. The following file is the listing of the AutoLISP program for Example 6.

```
;This program will draw a triangle outside
;the circle with the lines tangent to the circle
;
(defun dtr (a)
   (* a (/ pi 180.0))
)
(defun c:trgcir()
(setvar "cmdecho" 0)
(graphscr)
   (setq r(getdist "\n Enter circle radius: "))
```

```
    (setq c(getpoint "\n Enter center of circle: "))
    (setq d(/ r (sin(dtr 30))))
    (setq p1(polar c (dtr 210) d))
    (setq p2(polar c (dtr 330) d))
    (setq p3(polar c (dtr 90) d))
  (command "circle" c r)
  (command "line" p1 p2 p3 "c")
  (setvar "cmdecho" 1)
  (princ)
  )
```

Step 2: Loading the LISP program

Save the LISP file and then load it using the **APPLOAD** command as described in Example 1.

Exercise 3 *General*

Write an AutoLISP program that will draw an isosceles triangle P1,P2,P3. The base of the triangle (P1,P2) makes an angle B with the positive *X* axis (Figure 8-16). The program should prompt you to enter the starting point, P1, length L1, and angles A and B.

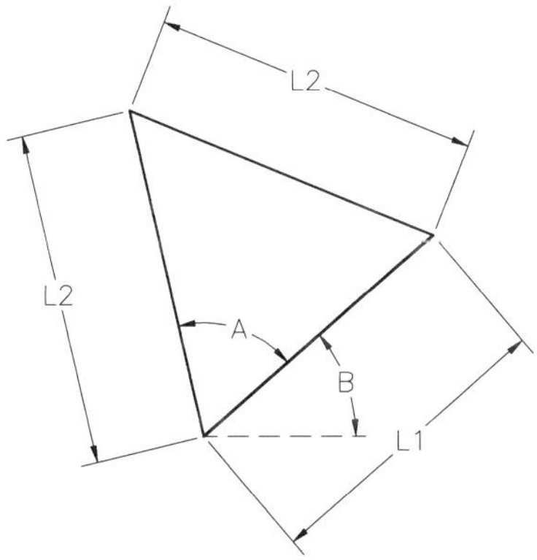

Figure 8-16 *Isosceles triangle at an angle*

Exercise 4 *General*

Write a program that will draw a slot with centerlines. The program should prompt you to enter slot length, slot width, and the left center point of the slot (Figure 8-17).

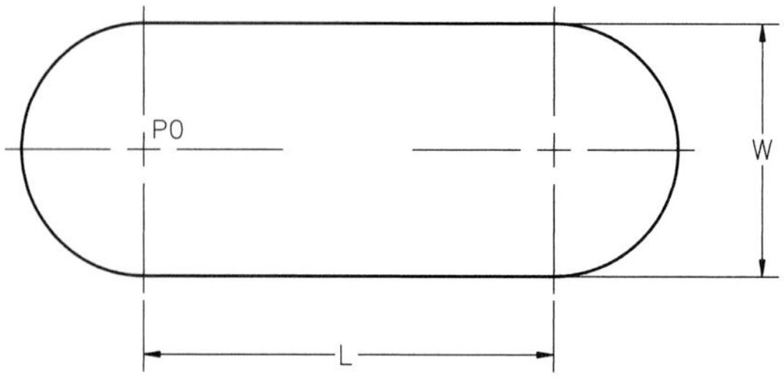

Figure 8-17 *Slot of length L and width W*

itoa, rtos, strcase, AND prompt FUNCTIONS

itoa

The **itoa** function changes an integer into a string and returns the integer as a string. The format of the **itoa** function is:

(itoa number)

Where **Number** --------- The integer number that you want to convert into a string

Examples:

(itoa 89) returns "89"
(itoa -356) returns "-356"

(setq intnum 7)
(itoa intnum) returns "7"

(setq intnum 345)
(setq intstrg (itoa intnum)) returns "345"

rtos

The **rtos** function changes a real number into a string and the function returns the real number as a string. The format of the **rtos** function is:

(rtos realnum)

Where **Realnum** -------- The real number that you want to convert into a string

Examples:

(rtos 50.6) returns "50.6"
(rtos -30.0) returns "-30.0"
(setq realstrg (rtos 5.25)) returns "5.25"

(setq realnum 75.25)
(setq realstrg (rtos realnum)) returns "75.25"

The **rtos** function can also include mode and precision. If mode and precision are not provided, then the current AutoCAD settings are used. The format of the **rtos** function with mode and precision is:

(rtos realnum [mode [precision]])

Where **realnum** --------- Real number
mode ------------- Unit mode, like decimal, scientific
precision ------- Number of decimal places or denominator of fractional units

strcase

The **strcase** function converts the characters of a string into uppercase or lowercase. The format of the **strcase** function is:

(strcase string [true])

Where **String** ------------ String that needs to be converted to uppercase or
-------------------- lowercase
True -------------- If it is not nil, all characters are converted to lowercase

The **true** is optional. If it is missing or if the value of **true** is nil, the string is converted to uppercase. If the value of true is not nil, the string is converted to lowercase.

Examples:

(strcase "Welcome Home") returns "WELCOME HOME"

(setq t 0)
(strcase "Welcome Home" t) returns "welcome home"
(strcase "Welcome Home" a) returns "WELCOME HOME"

(setq answer (strcase (getstring "Enter Yes or No: ")))

prompt

The **prompt** function is used to display a message on the screen in the Command prompt area. The contents of the message must be enclosed in double quotes. The format of the **prompt** function is given next.

(prompt message)

Where **message** --------- Message that you want to display on the screen

Examples:

(prompt "Enter circle diameter: ")
(setq d (getdist (prompt "Enter circle diameter: ")))

Note

On a two-screen system, the prompt function displays the message on both screens.

Example 7

Write a program that will draw two circles of radii r1 and r2, representing two pulleys that are separated by a distance d. The line joining the centers of the two circles makes an angle **a** with the *X* axis, as shown in Figure 8-18.

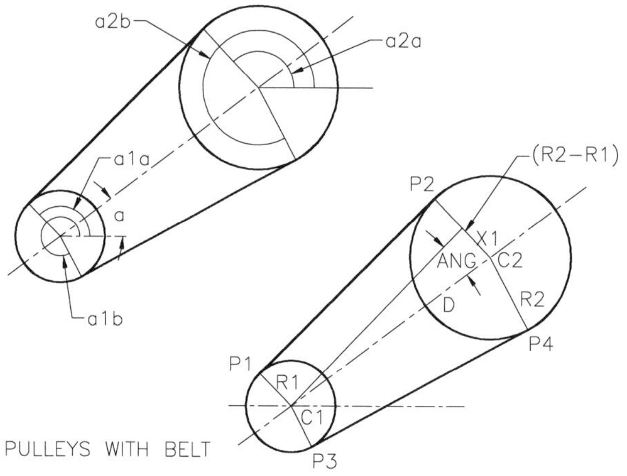

Figure 8-18 *Two circles with tangent lines*

Step 1: Understanding the program algorithm

Input
Radius of small circle - r1
Radius of large circle - r2
Distance between circles - d
Angle of center line - a
Center of small circle - c1

Output
Small circle of radius - r1
Large circle of radius - r2
Lines tangent to circles

Process
1. Calculate distance x1, x2.
2. Calculate angle ang.
3. Locate point c2 with respect to point c1.

4. Locate points p1, p2, p3, p4.
5. Draw small circle with radius r1 and center c1.
6. Draw large circle with radius r2 and center c2.
7. Draw lines p1 to p2 and p3 to p4.

Calculations

x1 = r2 - r1
x2 = SQRT [d^2 - (r2 - r1)^2]
tan ang = x1 / x2
ang = atan (x1 / x2)
a1a = 90 + a + ang
a1b = 270 + a - ang
a2a = 90 + a + ang
a2b = 270 + a - ang

Step 2: Writing the LISP program

The following file is a listing of the AutoLISP program for Example 7. **The line numbers at the right are not a part of the program; they are shown here for reference only.** Use any text editor to write down the following LISP file.

```
;This program draws a tangent (belt) over two
;pulleys that are separated by a given distance.
;
;This function changes degrees into radians
(defun dtr (a)
  (* a (/ pi 180.0))
  )
;End of dtr function
;The belt function draws lines that are tangent to circles
(defun c:belt()
   (setvar "cmdecho" 0)
   (graphscr)
   (setq r1(getdist "\n Enter radius of small pulley: "))
   (setq r2(getdist "\n Enter radius of larger pulley: "))
   (setq d(getdist "\n Enter distance between pulleys: "))
   (setq a(getangle "\n Enter angle of pulleys: "))
   (setq c1(getpoint "\n Enter center of small pulley: "))
   (setq x1 (- r2 r1))
   (setq x2 (sqrt (- (* d d) (* (- r2 r1) (- r2 r1)))))
   (setq ang (atan (/ x1 x2)))
   (setq c2 (polar c1 a d))
   (setq p1 (polar c1 (+ ang a (dtr 90)) r1))
   (setq p3 (polar c1 (- (+ a (dtr 270)) ang) r1))
```

```
24 (setq p2 (polar c2 (+ ang a (dtr 90)) r2))
25 (setq p4 (polar c2 (- (+ a (dtr 270)) ang) r2))
26 ;
27 ;The following lines draw circles and lines
28 (command "circle" c1 p3)
29 (command "circle" c2 p2)
30 (command "line" p1 p2 "")
31 (command "line" p3 p4 "")
32 (setvar "cmdecho" 1)
33 (princ))
```

Explanation

Line 5
(defun dtr (a)
In this line, the **defun** function defines a function, **dtr (a)**, that converts degrees into radians.

Line 6
(* a (/ pi 180.0))
(/ pi 180) divides the value of **pi** by 180, and the product is then multiplied by angle **a** (180-degree is equal to **pi** radians).

Line 10
(defun c:belt()
In this line, the function **defun** defines a function, c:belt, that generates two circles with tangent lines.

Line 18
(setq x1 (- r2 r1))
In this line, the function **setq** assigns a value of r2 - r1 to variable x1.

Line 19
(setq x2 (sqrt (- (* d d) (* (- r2 r1) (- r2 r1)))))
In this line, **(- r2 r1)** subtracts the value of r1 from r2 and **(* (- r2 r1) (- r2 r1))** calculates the square of (- r2 r1). **(sqrt (- (* d d) (* (- r2 r1) (- r2 r1))))** calculates the square root of the difference, and **setq x2** assigns the product of this expression to variable x2.

Line 20
(setq ang (atan (/ x1 x2)))
In this line, **(atan (/ x1 x2))** calculates the arctangent of the product of **(/ x1 x2)**. The function **setq ang** assigns the value of the angle in radians to variable **ang**.

Line 21
(setq c2 (polar c1 a d))
In this line, **(polar c1 a d)** uses the **polar** function to locate point c2 with respect to c1 at a distance d and making an angle a with the positive *X* axis.

Line 22
(setq p1 (polar c1 (+ ang a (dtr 90)) r1))
In this line, **(polar c1 (+ ang a (dtr 90)) r1))** locates point p1 with respect to c1 at a distance r1 and making an angle **(+ ang a (dtr 90))** with the positive *X* axis.

Line 28
(command "circle" c1 p3)
In this line, the **Command** function uses the **CIRCLE** command to draw a circle with center c1 and a radius defined by the point p3.

Line 30
(command "line" p1 p2 "")
In this line, the **Command** function uses the **LINE** command to draw a line from p1 to p2. The pair of double quotes ("") at the end introduces a Return, which terminates the **LINE** command.

Step 3: Loading the LISP program

Save the program and then load the program as described in Example 1.

Note
Remember that you need to turn off the ***Object Snap*** *mode before running the program to get the desired figure.*

Exercise 5 — *General*

Write an AutoLISP program that will draw two lines tangent to two circles, as shown in Figure 8-19. The program should prompt you to enter the circle diameters and the center distance between the circles.

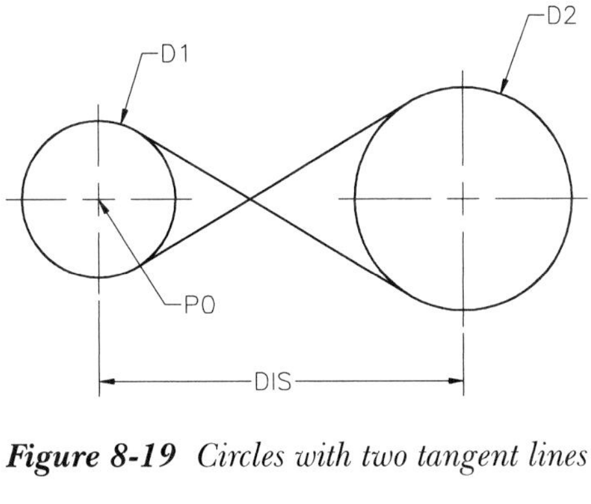

Figure 8-19 *Circles with two tangent lines*

Note
You will have to turn off the object snaps to get the desired figure.

FLOWCHARTS

A flowchart is a graphical representation of an algorithm. It is used to analyze a problem systematically. It gives a better understanding of the problem, especially if the problem involves some conditional statements. It consists of standard symbols that represent a certain function in the program. For example, a rectangle is used to represent a process that takes place when the program is executed. The blocks are connected by lines indicating the sequence of operations. Figure 8-20 gives the standard symbols that can be used in a flowchart.

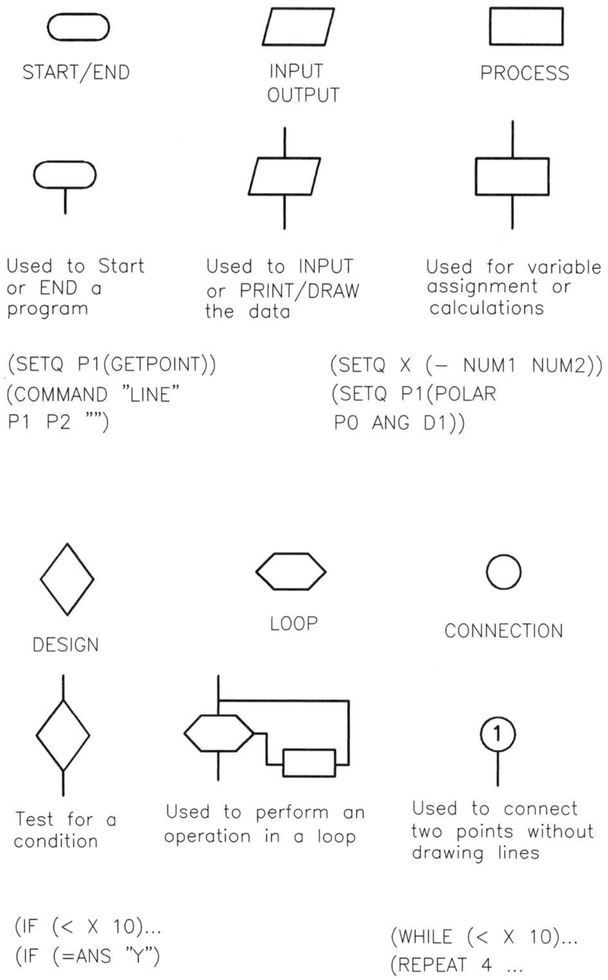

***Figure 8-20** Flowchart symbols*

CONDITIONAL FUNCTIONS

The relational functions discussed earlier in the chapter establish a relationship between two atoms. For example, (< x y) describes a test condition for an operation. To use such functions in a meaningful way a conditional function is required. For example, (if (< x y) (setq z (- y x)) (setq z (- x y))) describes the action to be taken when the condition is true (T) and when it is false (nil). If the condition is true, then z = y - x. If the condition is not true, then z = x - y. Therefore, conditional functions are very important for any programming language, including AutoLISP.

if

The **if** function (Figure 8-21) evaluates the first expression (then) if the specified condition returns "true," it evaluates the second expression (else) if the specified condition returns "nil." The format of the **if** function is:

(if condition then [else])

Where **condition** ------- Specified conditional statement
then -------------- Expression evaluated if the condition returns T
else --------------- Expression evaluated if the condition returns nil

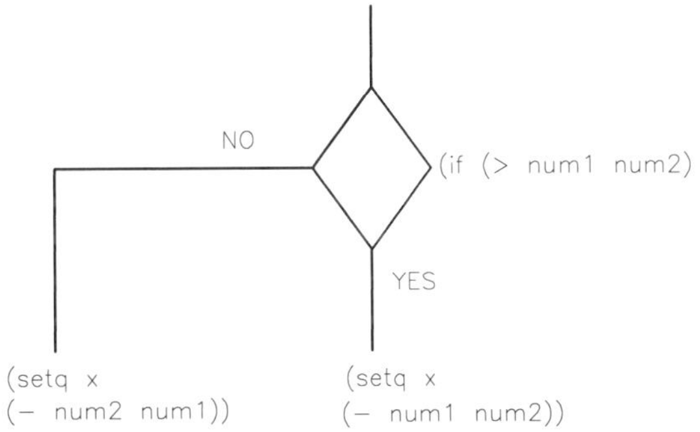

Figure 8-21 *The* ***if*** *function*

Examples

(if (= 7 7) ("true")) returns "true"
(if (= 5 7) ("true") ("false")) returns "false"
(setq ans "yes")
(if (= ans "yes") ("Yes") ("No")) returns "Yes"
(setq num1 8)
(setq num2 10)
(if (> num1 num2)
  (setq x (- num1 num2))
  (setq x (- num2 num1))
) returns 2

Example 8

Write an AutoLISP program that will subtract a smaller number from a larger number. The program should also prompt you to enter two numbers.

Step 1: Understanding program algorithm and flowchart

Input
Number (num1)
Number (num2)

Output
x = num1 - num2
or
x = num2 - num1

Process
If num1 > num2 then x = num1 - num2
If num1 < num2 then x = num2 - num1

The flowchart in Figure 8-22 describes the process involved in writing the program using standard flowchart symbols.

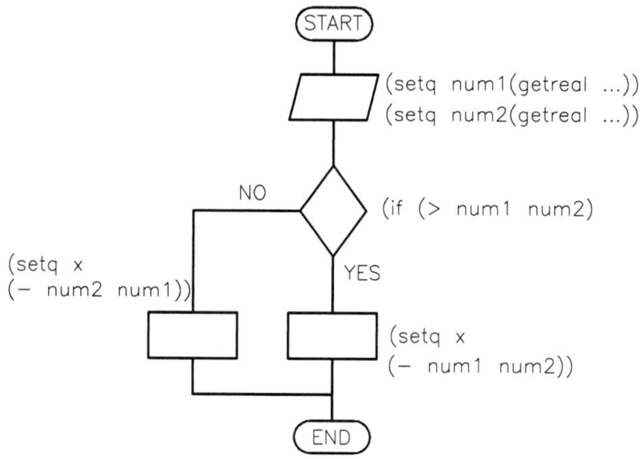

Figure 8-22 Flowchart for Example 8

Step 2: Writing the LISP program

Use any text editor to write the LISP file. The following file is a listing of the program for Example 8. **The line numbers at the right are not a part of the program; they are shown here for reference only.**

```
;This program subtracts smaller number
;from larger number
;
(defun c:subnum( )
```

```
5  (setvar "cmdecho" 0)
6  (setq num1 (getreal "\n Enter first number: "))
7  (setq num2 (getreal "\n Enter second number: "))
8  (if (> num1 num2)
9    (setq x (- num1 num2))
10    (setq x (- num2 num1))
11    )
12   (princ x)
13 (setvar "cmdecho" 1)
14 (princ)
15 )
```

Explanation

Line 8

(if (> num1 num2)

In this line, the **if** function evaluates the test expression **(> num1 num2)**. If the condition is true, it returns **T**. If the condition is not true, it returns nil.

Line 9

(setq x (- num1 num2))

This expression is evaluated if the test expression **(if (> num1 num2)** returns T. The value of the variable num2 is subtracted from num1, and the resulting value is assigned to variable x.

Line 10

(setq x (- num2 num1))

This expression is evaluated if the test expression **(if (> num1 num2)** returns nil. The value of variable num1 is subtracted from num2, and the resulting value is assigned to variable x.

Line 11

)

The closing parenthesis completes the definition of the **if** function.

Step 3: Loading the LISP program

Save the file and then load the file using the **APPLOAD** command, as described in Example 1.

Example 9

Write an AutoLISP program that will enable you to multiply or divide two numbers (Figure 8-23). The program should prompt you to enter the choice of multiplication or division. The program should also display an appropriate message if you do not enter the right choice.

Step 1: Making the flowchart

A flowchart can be designed as shown in Figure 8-23.

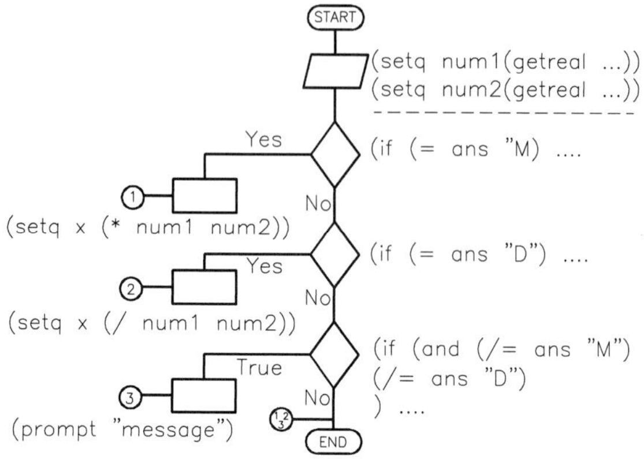

Figure 8-23 *Flow diagram for Example 9*

Step 2: Writing the lisp file

The following file is a listing of the AutoLISP program for Example 9:

```
;This program multiplies or divides two given numbers
(defun c:mdnum()
    (setvar "cmdecho" 0)
    (setq num1 (getreal "\n Enter first number: "))
    (setq num2 (getreal "\n Enter second number: "))
    (prompt "Do you want to multiply or divide. Enter M or D: ")
    (setq ans (strcase (getstring)))
    (if (= ans "M")
      (setq x (* num1 num2))
    )
    (if (= ans "D")
      (setq x (/ num1 num2))
    )
    (if (and (/= ans "D")(/= ans "M"))
      (prompt "Sorry! Wrong entry, Try again")
      (princ x)
    )
(setvar "cmdecho" 1)
(princ)
)
```

Step 3: Loading the lisp file

Save the file and then load the file using the **APPLOAD** command, as described in Example 1.

progn

The **progn** function can be used with the **if** function to evaluate several expressions. The format of the **progn** function is:

(progn expression expression . . .)

The **if** function evaluates only one expression if the test condition returns "true." The **progn** function can be used in conjunction with the **if** function to evaluate several expressions. Here is an example to demonstrate the use of the **progn** function with the **if** function.

Example:

```
(defun c:IFPRGN()
        (setq p1(getint "Enter the integer"))
        (If (>= p1 5)
        (progn (command "line" "2,2" "3,3" "")
                (command "rec" "3,3" "6,6") )

        (progn (command "circle" "3,3" 1)
                (command "line" "3,3" "5,5" ""))
                )
        )       ; End of program
```

while

The **while** function evaluates a test condition (Figure 8-24). If the condition is true (expression does not return nil), the operations that follow the while statement are repeated until the test expression returns nil. The format of the **while** function is:

(while testexpression operations)

Where **Testexpression** ----------- Expression that tests a condition
Operations ---------------- Operations to be performed until the test expression returns nil

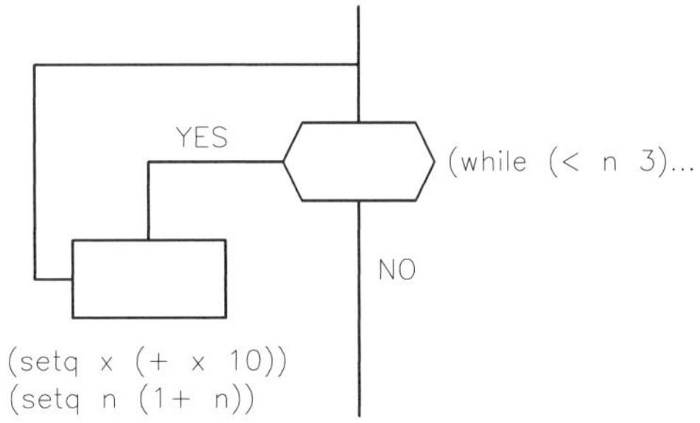

***Figure 8-24 while** function*

Example:

```
(while (= ans "yes")
  (setq x (+ x 1))
  (setq ans (getstring "Enter yes or no: "))
)
(while (< n 3)
      (setq x (+ x 10))
      (setq n (1+ n))
)
```

Example 10

Write an AutoLISP program that will find the nth power of a given number. The power is an integer. The program should prompt you to enter the number and the nth power (Figure 8-25).

Step 1: Understanding the program algorithm and flowchart

Input
Number x
nth power n

Output
product x^n

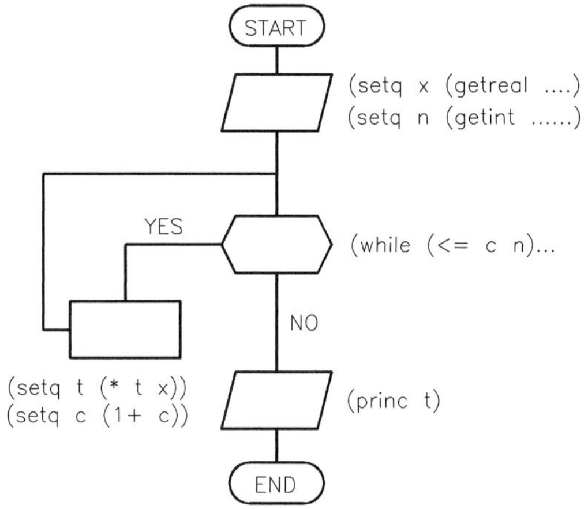

Figure 8-25 Flowchart for Example 10

Process

1. Set the value of t = 1 and c = 1.
2. Multiply t * x and assign that value to the variable t.
3. Repeat the process until the counter c is less than or equal to n.

Step 2: Writing the lisp program

The following file is a listing of the AutoLISP program for Example 10.

```
;This program calculates the nth
;power of a given number
(defun c:npower()
    (setq x(getreal "\n Enter a number: "))
    (setq n(getint "\n Enter Nth power-integer number: "))
    (setq t 1) (setq c 1)
    (while (<= c n)
      (setq t (* t x))
      (setq c (1+ c))
    )
    (princ t)
(setvar "cmdecho" 1)
(princ)
)
```

Step 3: Loading the LISP file

Save the file and then load the file with the help of the **APPLOAD** command.

Example 11

Write an AutoLISP program that will generate the holes of a bolt circle (Figure 8-26). The program should prompt you to enter the center point of the bolt circle, the bolt circle diameter, the bolt circle hole diameter, the number of holes, and the start angle of the bolt circles.

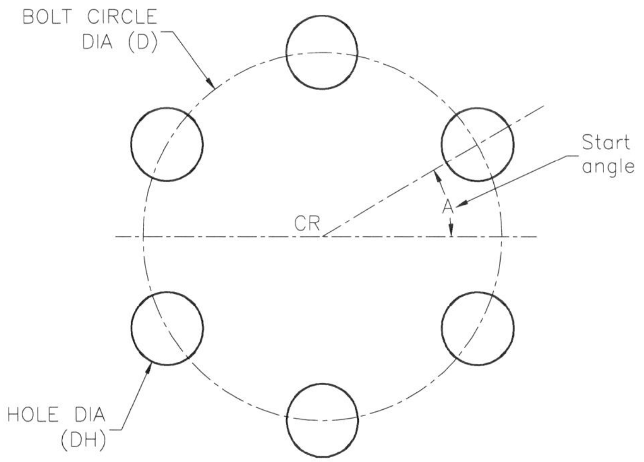

Figure 8-26 *Bolt circle with six holes*

Step 1: Writing the LISP program

```
;This program generates the bolt circles
;
(defun c:bc1( )
(graphscr)
(setvar "cmdecho" 0)
   (setq cr(getpoint "\n Enter center of Bolt-Circle: "))
   (setq d(getdist "\n Dia of Bolt-Circle: "))
   (setq n(getint "\n Number of holes in Bolt-Circle: "))
   (setq a(getangle "\n Enter start angle: "))
   (setq dh(getdist "\n Enter diameter of hole: "))
   (setq inc(/ (* 2 pi) n))
   (setq ang 0)
   (setq r (/ dh 2))
(while (< ang (* 2 pi))
   (setq p1 (polar cr (+ a inc) (/ d 2)))
   (command "circle" p1 r)
   (setq a (+ a inc))
   (setq ang (+ ang inc))
   )
(setvar "cmdecho" 1)
(princ)
)
```

Step 2: Loading the LISP file

Save the file and then load it using the **APPLOAD** command.

repeat

The **repeat** function evaluates the expressions **n** number of times as specified in the **repeat** function (Figure 8-27). The variable **n** must be an integer. The format of the **repeat** function is:

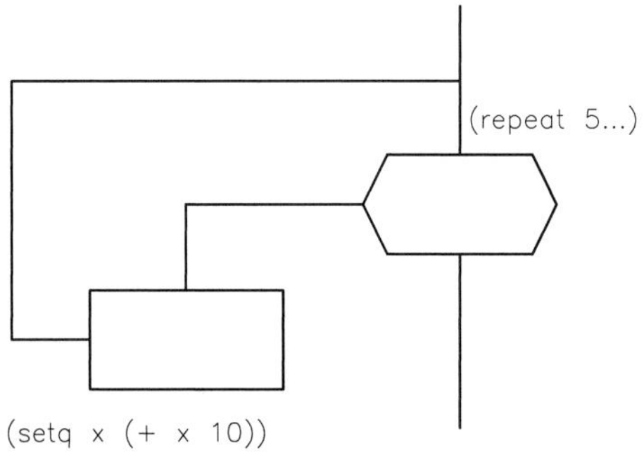

***Figure 8-27 repeat** function*

repeat n

Where ------------------------------ n is an integer that defines the number of times the expressions are to be evaluated

Example:

```
(repeat 5
 (setq x (+ x 10))
 )
```

Note

*AutoCAD allows you to load the specified AutoLISP programs automatically, each time you start AutoCAD. For example, if you are working on a project and you have loaded an AutoLISP program, the program will autoload if you start another drawing. You can enable this feature by adding the name of the file to the **Startup Suite** of the **Load/Unload Application** dialog box. For details, see the **Load/Unload Application** discussed earlier in this chapter.*

Example 12

Write an AutoLISP program that will generate a given number of concentric circles. The program should prompt you to enter the center point of the circles, the start radius, and the radius increment (Figure 8-28).

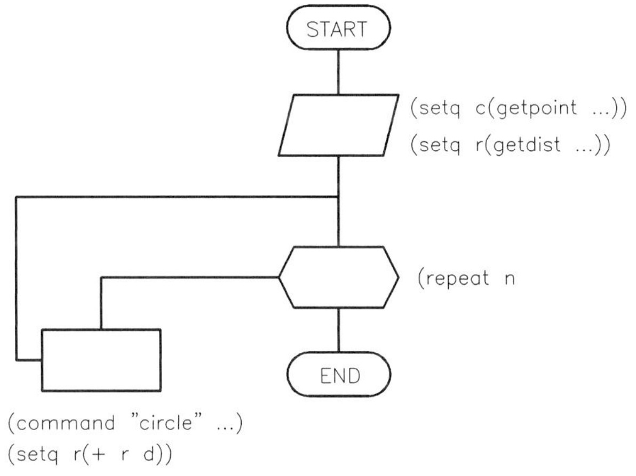

Figure 8-28 *Flowchart for Example 12*

The following file is a listing of the AutoLISP program for Example 12.

```
;This program uses the repeat function to draw
;a given number of concentric circles.
(defun c:concir( )
```

```
(graphscr)
(setvar "cmdecho" 0)
(setq c (getpoint "\n Enter center point of circles: "))
(setq n (getint "\n Enter number of circles: "))
(setq r (getdist "\n Enter radius of first circle: "))
(setq d (getdist "\n Enter radius increment: "))
(repeat n
  (command "circle" c r)
  (setq r (+ r d))
  )
  (setvar "cmdecho" 1)
  (princ)
  )
```

Self-Evaluation Test

Answer the following questions and then compare them to those given at the end of this chapter:

1. LISP is a leading programming language for __________ .

2. The AutoLISP interpreter is embedded within the __________ software package.

3. The function __________ checks whether the two atoms are equal or not.

4. AutoLISP contains some __________ functions.

5. If some built-in functions are used with **setq** then there is every possibility that reserved functions will be __________ .

6. When using the **LOAD** function, the AutoLISP file name and the optional path name must be enclosed in __________ .

7. The command __________ is used to load a LISP file.

8. The **strcase** function converts the characters of a string into __________ .

9. The **Progn** function can be used in conjunction with the __________ function.

10. The **setq** function is used to assign a value to a __________ .

Review Questions

Answer the following questions:

1. Fill in the blanks:

Command: (+ 2 30 5 50)	returns	____________	
Command: (+ 2 30 4 55.0)	returns	____________	
(- 20 40)	returns	____________	
(- 30.0 40.0)	returns	____________	
(* 72 5 3 2.0)	returns	____________	
(* 7 -5.5)	returns	____________	
(/ 299 -5)	returns	____________	
(/ -200 -9.0)	returns	____________	
(1- 99)	returns	____________	
(1- -18.5)	returns	____________	
(abs -90)	returns	____________	
(abs -27.5)	returns	____________	
(sin pi)	returns	____________	
(sin 1.5)	returns	____________	
(cos pi)	returns	____________	
(cos 1.2)	returns	____________	
(atan 1.1 0.0)	returns	____________	radians
(atan -0.4 0.0)	returns	____________	radians
(angtos 1.5708 0 5)	returns	____________	
(angtos -1.5708 0 3)	returns	____________	
(< “x” “y”)	returns	____________	
(>= 80 90 79)	returns	____________	

2. The __________ function pauses to enable you to enter the *X*, *Y* coordinates or *X*, *Y*, *Z* coordinates of a point.

3. The __________ function is used to execute standard AutoCAD commands from within an AutoLISP program.

4. In an AutoLISP expression, the AutoCAD command name and the command options have to be enclosed in double quotation marks. (T/F).

5. The **getdist** function pauses for you to enter a __________ and it then returns the distance as a real number.

6. The __________ function assigns a value to an AutoCAD system variable. The name of the system variable must be enclosed in __________ .

7. The **cadr** function performs two operations, __________ and __________, to return the second element of the list.

8. The __________ function prints a new line on the screen just as \n.

9. The __________ function pauses for you to enter the angle, then it returns the value of that angle in radians.

10. The __________ function always measures the angle with a positive *X* axis and in a counterclockwise direction.

11. The __________ function pauses for you to enter an integer. The function always returns an integer, even if the number that you enter is a real number.

12. The __________ function lets you retrieve the value of an AutoCAD system variable.

13. The __________ function defines a point at a given angle and distance from the given point.

14. The __________ function calculates the square root of a number and the value this function returns is always a real number.

15. The __________ function changes a real number into a string and the function returns the real number as a string.

16. The **if** function evaluates the test expression **(> num1 num2)**. If the condition is true, it returns __________; if the condition is not true, it returns__________.

17. The __________ function can be used with the **if** function to evaluate several expressions.

18. The **while** function evaluates the test condition. If the condition is true (expression does not return nil) the operations that follow the while statement are __________ until the test expression returns __________.

19. The **repeat** function evaluates the expressions n number of times as specified in the **repeat** function. The variable n must be a real number. (T/F).

Exercises

Exercise 6 *General*

Write an AutoLISP program that will draw three concentric circles with center C1 and diameters D1, D2, D3 (Figure 8-29). The program should prompt you to enter the coordinates of center point C1 and the circle diameters D1, D2, D3.

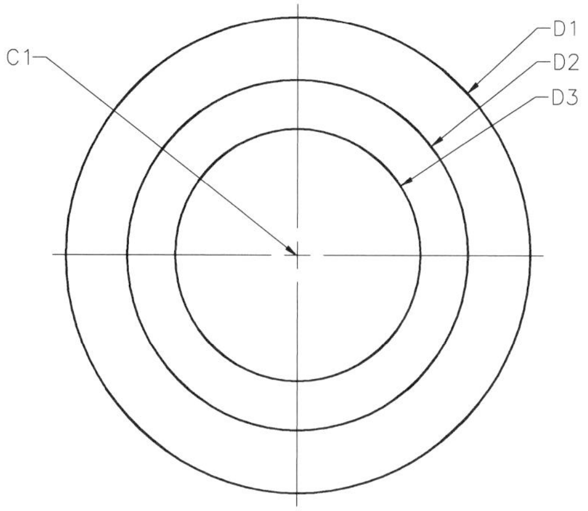

***Figure 8-29** Three concentric circles with diameters D1, D2, D3*

Exercise 7 *General*

Write an AutoLISP program that will draw a line from point P1 to point P2 (Figure 8-30). Line P1, P2 makes an angle A with the positive *X* axis. Distance between the points P1 and P2 is L. The diameter of the circles is D1 (D1 = L/4).

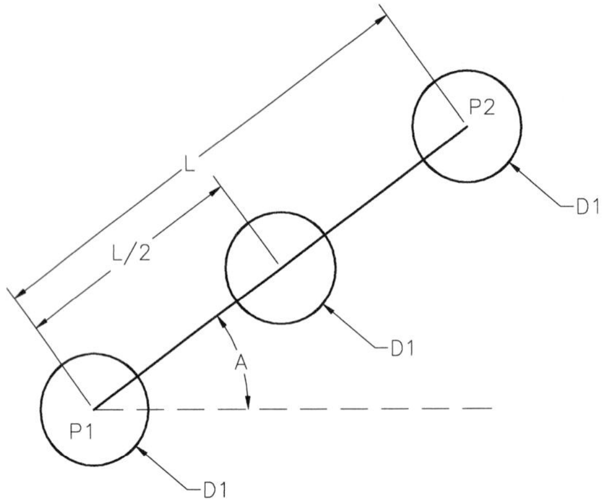

Figure 8-30 *Circles and line making an angle A with X-axis*

Exercise 8 *General*

Write an AutoLISP program that will draw an isosceles triangle P1, P2, P3 (Figure 8-31). The program should prompt you to enter the starting point P1, length L1, and the included angle A.

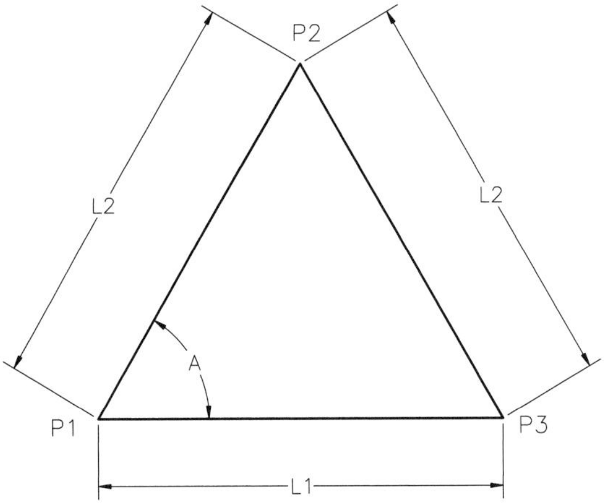

Figure 8-31 *Isosceles triangle*

Exercise 9 *General*

Write a program that will draw a slot with center lines. The program should prompt you to enter slot length, slot width, and the layer name for centerlines (Figure 8-32).

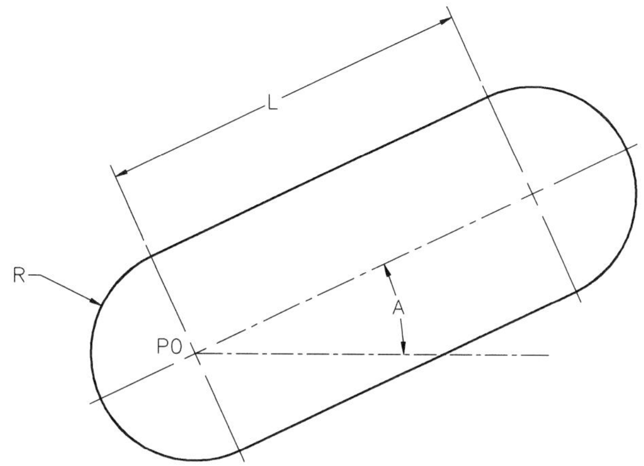

Figure 8-32 *Slot of length L and radius R*

Exercise 10 *General*

Write an AutoLISP program that will draw a line and then generate a given number of lines (N), parallel to the first line (Figure 8-33).

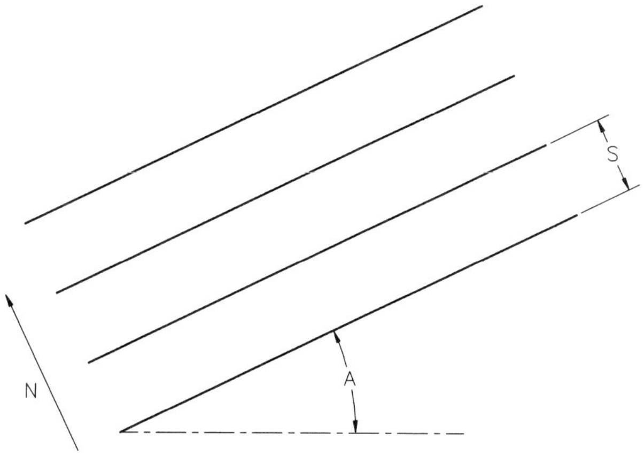

Figure 8-33 *N number of lines offset at a distance S*

Exercise 11 *General*

Write an AutoLISP program that will generate the holes, as shown in Figure 8-34. The program should prompt the user to enter the center point of the hole; diameters of the holes at the center and at the circumference; the angle betwwen the holes; start angle of the hole from the X axis; and the hole circle diameter.

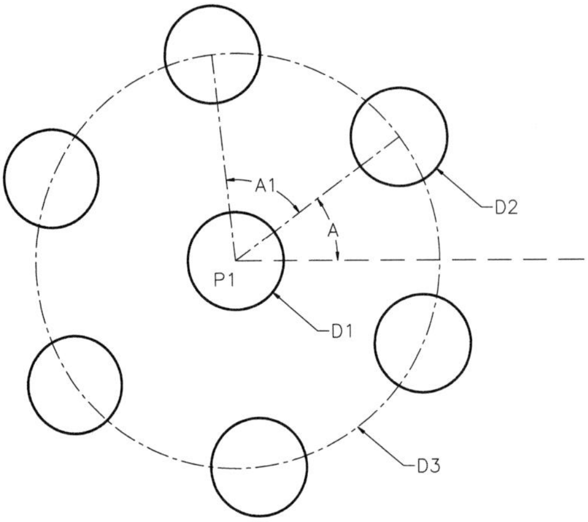

Figure 8-34 *Holes at the center and the circumference*

Answers to Self-Evaluation Test

1. Artificial Intelligence, **2.** AutoCAD, **3.** /=, **4.** built-in, **5.** Redefined, **6.** double quotes, **7. APPLOAD**, **8.** Uppercase or Lowercase, **9. if**, **10.** variable

Chapter 9

Working with Visual LISP

Learning Objectives

After completing this chapter, you will be able to:

- *Start Visual LISP in AutoCAD.*
- *Use the Visual LISP Text Editor.*
- *Load and run Visual LISP programs.*
- *Load existing AutoLISP files.*
- *Use the Visual LISP Console.*
- *Use the Visual LISP formatter and change formatting options.*
- *Debug a Visual LISP program and trace the variables.*

Visual LISP

Visual LISP is another step forward to extend the customizing power of AutoCAD. Since AutoCAD Release 2.0 in the mid 1980s, AutoCAD users have used AutoLISP to write programs for their applications; architecture, mechanical, electrical, air conditioning, civil, sheet metal, and hundreds of other applications. AutoLISP has some limitations. For example, when you use a text editor to write a program, it is difficult to locate and check parentheses, AutoLISP functions, and variables. Debugging is equally problematic, because it is hard to find out what the program is doing and what is causing an error in the program. Generally, programmers will add some statements in the program to check the values of variables at different stages of the program. Once the program is successful, the statements are removed or changed into comments. Balancing parentheses and formatting the code are some other problems with traditional AutoLISP programming.

Visual LISP has been designed to make programming easier and more efficient. It has a powerful text editor and a formatter. The text editor allows color coding of parentheses, function names, variables and other components of the program. The formatter formats the code in a easily readable format. It also has a watch facility that allows you to watch the values of variables and expressions. Visual LISP has an interactive and intelligent console that has made programming easier. It also provides context-sensitive help for AutoLISP functions and apropos features for symbol name search. The debugging tools have made it easier to debug the program and check the source code. These and other features have made Visual LISP the preferred tool for writing LISP programs. Visual LISP works with AutoCAD and it contains its own set of windows. AutoCAD must be running when you are working with Visual LISP.

Overview of Visual LISP

As you proceed in this chapter, you will learn about some features unique to Visual LISP that are not available in AutoLISP. You will learn how to start Visual LISP, use the Visual LISP text editor to write a LISP Program, and then how to load and run the programs.

Starting Visual LISP

Ribbon:	Manage > Applications > Visual LISP Editor
Menu Bar:	Tools > AutoLISP > Visual LISP Editor
Command:	VLIDE or VLISP

1. Start AutoCAD.

2. Choose **Manage > Applications > Visual LISP Editor** from the **Ribbon**. You can also start Visual LISP by entering the **VLIDE** command at the Command prompt. AutoCAD displays the Visual LISP window, Figure 9-1. If it is the first time in the current drawing session that you are loading Visual LISP, the **Trace** window may appear, displaying information about the current release of Visual LISP and errors that might be encountered when loading Visual LISP. The Visual LISP window has four areas: menu, toolbars, console window, trace window, and status bar, see Figure 9-1. The following is a short description of these areas.

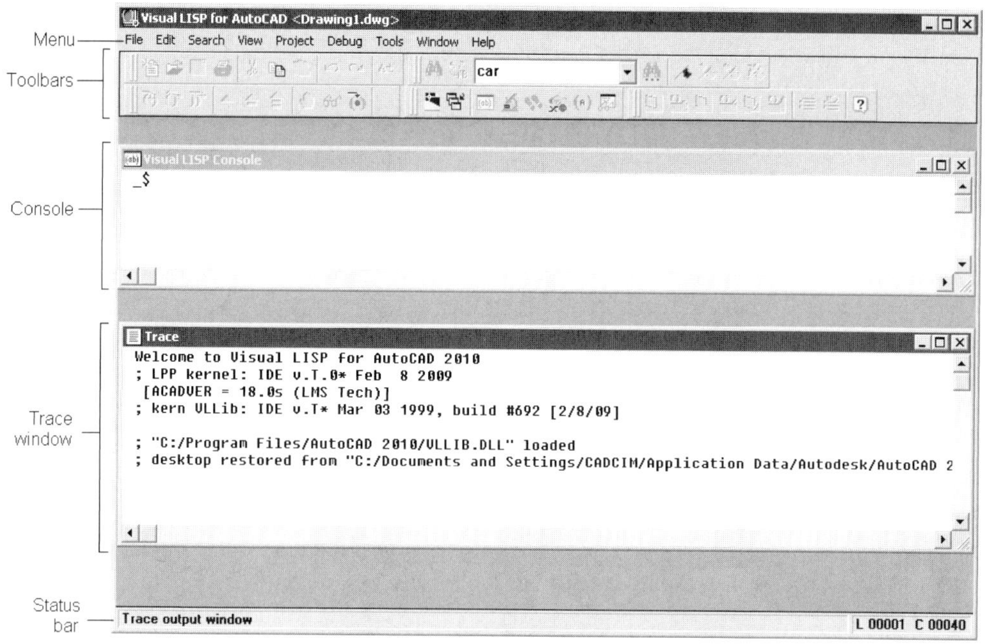

Figure 9-1 *The* ***Visual LISP*** *window*

Menu. The menu bar is at the top of the Visual LISP window and lists different menu items. If you select a menu, AutoCAD displays the items in that menu. Also, the function of the menu item is displayed in the status bar that is located at the bottom of the Visual LISP window.

Toolbars. The toolbars provide a quick and easy way to invoke Visual LISP commands. Visual LISP has 5 toolbars; **Standard**, **Search**, **Tools**, **Debug**, and **View**. Depending on the window that is active, the display of the toolbars changes. If you move the mouse pointer over a tool and position it there for a few seconds, a tool tip is displayed that indicates the function of that tool. A more detailed description of the tool appears in the status bar.

Console window. The **Visual LISP Console** window is contained within the Visual LISP window. It can be scrolled and positioned anywhere in the Visual LISP window. The **Console** window can be used to enter AutoLISP or Visual LISP commands. For example, if you want to add two numbers, enter (+ 2 9.5) next to the dollar ($) sign and then press the ENTER key. Visual LISP will return the result and display it in the same window.

Status bar. When you select a menu item or a tool in the toolbar, Visual LISP displays the related information in the status bar that is located near the bottom of the Visual LISP window.

Using the Visual LISP Text Editor

1. Choose the **New File** option from the **File** menu. The text editor window is displayed, see Figure 9-2. The default file name is **Untitled-0** and is displayed at the top of the window.

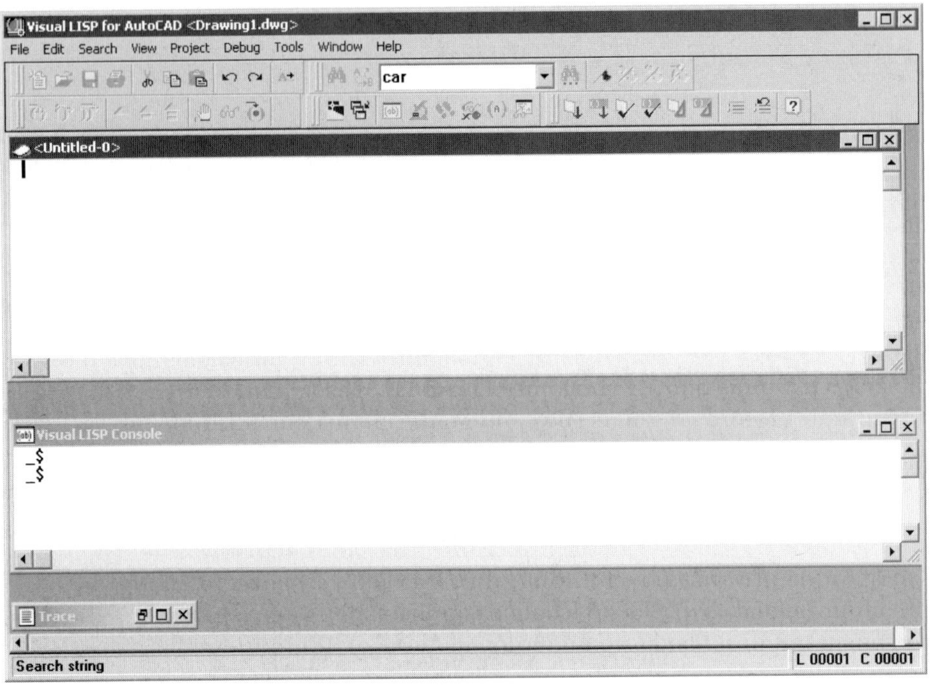

Figure 9-2 *The* ***Visual LISP*** *text editor*

2. To activate the editor, click anywhere in the **Visual LISP** text editor.

3. Type the following program and notice the difference between the **Visual LISP** text editor and the text editors you have been using to write AutoLISP programs (like Notepad).

```
;;;This program will draw a triangle
(defun tr1 ()
(setq p1(list 2.0 2.0))
(setq p2(list 6.0 3.0))
(setq p3(list 4.0 7.0))
(command "line" p1 p2 p3 "c")
)
```

4. Choose the **File > Save** from the menu bar. In the **Save As** dialog box, enter the name of the file, **Triang1.lsp**. After you save the file, the file name is displayed at the top of the editor window, Figure 9-3.

Loading and Running Programs

1. Make sure the **Visual LISP** text editor window is active. If not, click a point anywhere in the window to make it active.

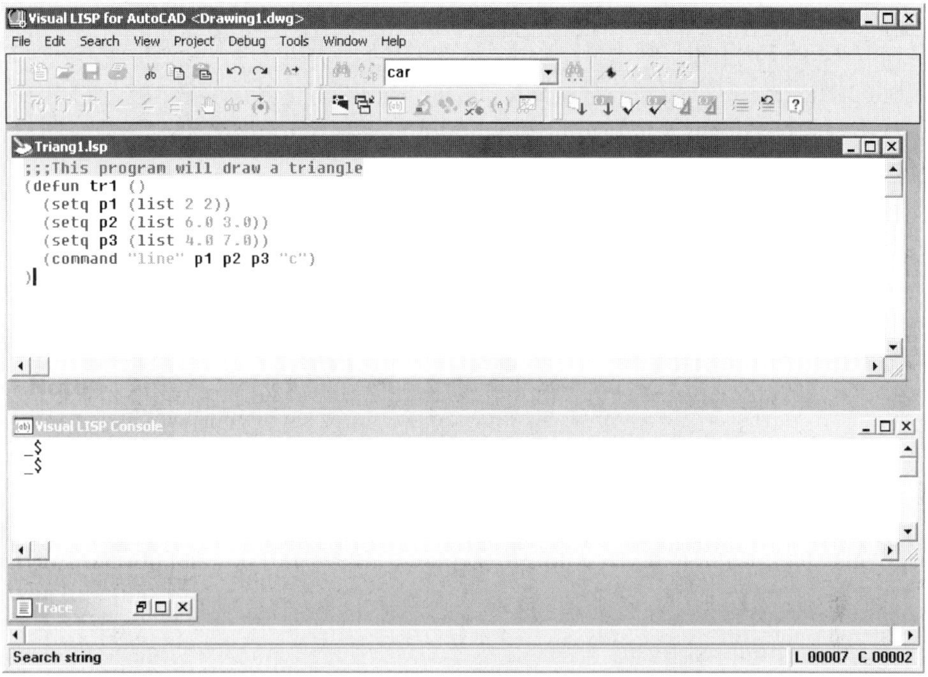

Figure 9-3 *Writing program code in the* ***Visual LISP*** *text editor*

2. Choose **Tools > Load Text in Editor** from the menu bar of the **Visual LISP** editor. You can also load the program by choosing **Load active edit window** in the **Tools** toolbar. Visual LISP displays a message in the **Console** window indicating that the program has been loaded. If there was a problem in loading the program, an error message will appear.

3. To run the program, enter the function name **(tr1)** at the console prompt (_$ sign). The function name must be enclosed in parentheses. The program will draw a triangle in AutoCAD. To view the program output, switch to AutoCAD by choosing **Activate AutoCAD** in the **View** toolbar. You can also run the program from AutoCAD. Switch to AutoCAD and enter the name of the function **TR1** at the Command prompt. AutoCAD will run the program and draw a triangle on the screen, see Figure 9-4.

Exercise 1 *General*

Write a LISP program that will draw an equilateral triangle. The program should prompt the user to select the starting point (bottom left corner) and the length of one of the sides of the triangle, Figure 9-5.

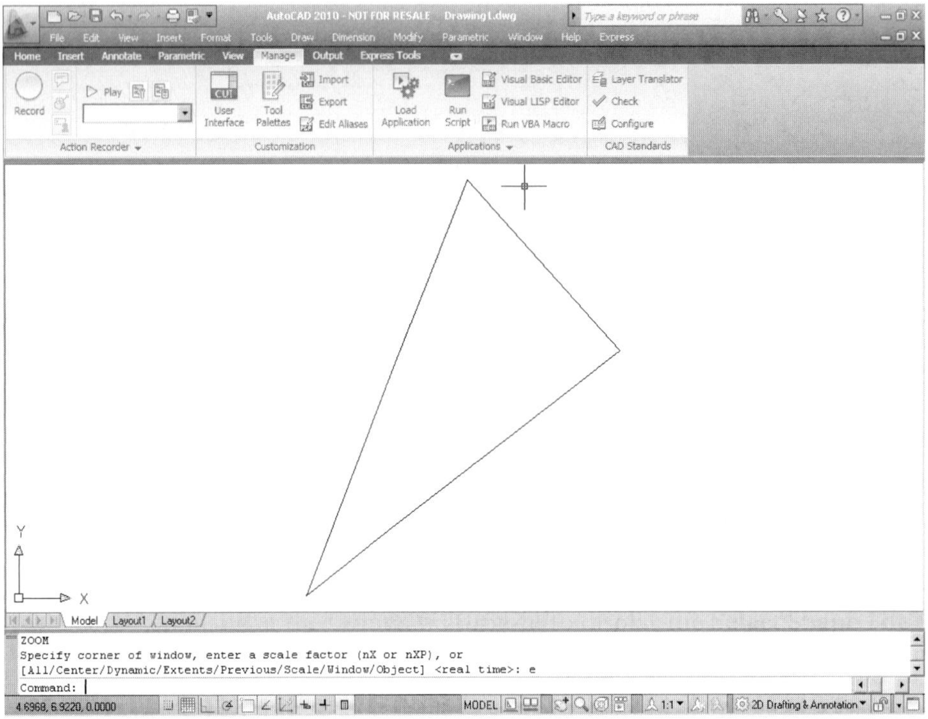

Figure 9-4 Program output in AutoCAD

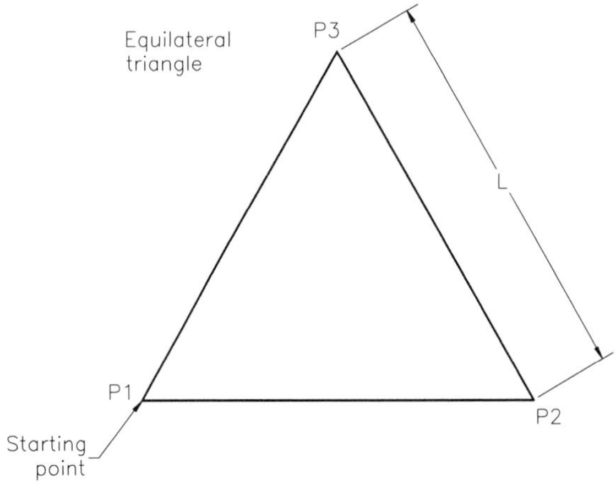

Figure 9-5 Equilateral triangle to be drawn for Exercise 1

Loading Existing AutoLISP Files

You can also load the AutoLISP files in Visual LISP and then edit, debug, load, and run the programs from Visual LISP.

1. Start AutoCAD and choose **Manage > Applications > Visual LISP Editor** from the **Ribbon**. You can also start Visual LISP by entering the **VLIDE** command at the Command prompt.

2. In the **Visual LISP** text editor, choose **Open File** from the **File** menu. The **Open file to edit/view** dialog box is displayed on the screen.

3. Select the AutoLISP program that you want to load and then choose **Open**. The program is loaded in the **Visual LISP** text editor.

4. To format the program in the edit window, choose the **Format edit window** tool from the **Tools** toolbar. The program automatically gets formatted.

5 To load the program, choose **Load active edit window** in the **Tools** toolbar or choose **Load Text in Editor** from the **Tools** menu.

6. To run the program, enter the function name at the console prompt ($) in the Visual LISP **Console** window and then press ENTER. If it gives nil, this means it has executed the drawing on the graphical screen.

Visual LISP Console

Most of the programming in Visual LISP is done in the **Visual LISP** text editor. However, you can also enter the code in the Visual LISP console and see the results immediately. For example, if you enter **(sqrt 37.2)** at the console prompt (after the _$ sign) and press the ENTER key, Visual LISP will evaluate the statement and return the value 6.09918. Similarly, enter the code **(setq x 99.3)**, **(+ 38 23.44)**, **(- 23.786 35)**, **(abs -37.5)**, **(sin 0.333)** and see the results. Each statement must be entered after the $ sign, as shown in Figure 9-6.

Features of the Visual LISP Console

1. In the Visual LISP Console, you can enter more than one statement on a line and then press ENTER. Visual LISP will evaluate all statements and then return the value. Enter the following statement after the $ sign.

   _$(setq x 37.5) (setq y (/ x 2))

   Visual LISP will display the value of X and Y in the console, refer to Figure 9-7.

2. If you want to find the value of a variable in the code being developed, enter the variable name after the $ sign and Visual LISP will return the value and display it in the console. For example, if you want to find the value of X, enter X after the $ sign.

3. Visual LISP allows you to write an AutoLISP expression on more than one line by pressing CTRL+ENTER at the end of the line. For example, if you want to write the following two expressions on two lines, enter the first line and then press CTRL+ENTER. You will notice that Visual LISP does not display the $ sign in the next line, indicating continuation of the statement on the previous line. Now, enter the next line and press ENTER, see Figure 9-7.

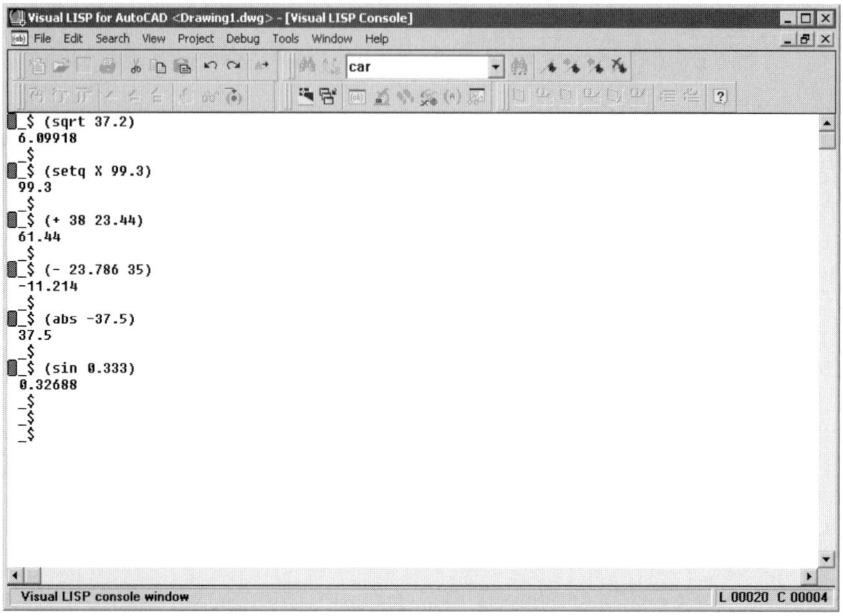

Figure 9-6 *Entering the LISP code in Visual LISP Console*

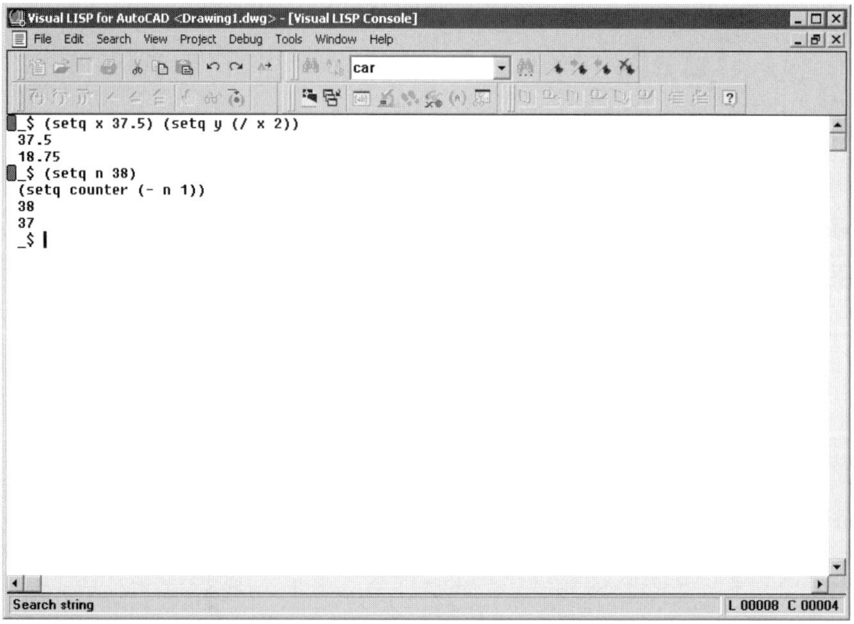

Figure 9-7 *Entering the LISP code in Visual LISP Console*

_$ (setq n 38) Press the CTRL+ENTER keys
(setq counter (- n 1))

4. To retrieve the text that has been entered in the previous statement, press the TAB key while at the Console prompt (_$). Each time you press TAB, the previously entered text is displayed. You can continue pressing TAB to cycle through the text that has been entered in the Visual LISP console. If the same statement has been entered several times, pressing the TAB key will display the text only once and then jump to the statement before that. Pressing the SHIFT+TAB key displays the text in the opposite direction.

5. Pressing ESC clears the text following the console prompt (_$). However, if you press SHIFT+ESC, the text following the console prompt is not evaluated and stays as it is. The cursor jumps to the next Console prompt. For example, if you enter the expression **(setq x 15)** and then press ESC, the text is cleared. However, if you press SHIFT+ESC, the expression (setq x 15) is not evaluated and the cursor jumps to the next console prompt without clearing the text.

6. The TAB key also lets you do an associative search for the expression that starts with the specified text string. For example, if you want to search the expression that started with (sin, enter it at the console prompt (_$) and then press ENTER. Visual LISP will search the text in the **Console** window and return the statement that starts with (sin. If it cannot find the text, nothing happens, except maybe a beep.

7. If you right-click anywhere in the Visual LISP Console or press SHIFT+F10, Visual LISP displays the context menu, Figure 9-8. You can use the context menu to perform several tasks like **Cut**, **Copy**, **Paste**, **Clear Console Window**, **Find**, **Inspect**, **Add Watch**, **Apropos Window**, **Symbol Service**, **Undo**, **Redo**, **AutoCAD Mode**, and **Toggle Console Log**.

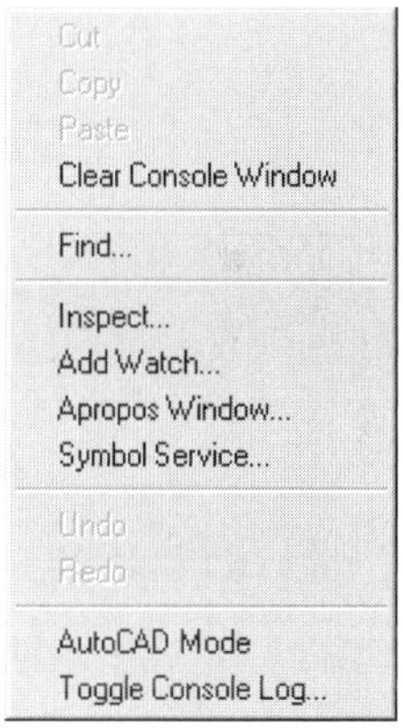

Figure 9-8 *Context menu displayed when you right-click*

8. One of the important features of Visual LISP is the facility of context-sensitive help available for the Visual LISP functions. If you need help with any Visual LISP function, enter or select the function and then choose the **Help** button in the **Tools** toolbar. Visual LISP will display the help for the selected function, Figure 9-9.

9. Visual LISP also allows you to keep a record of the Visual LISP Console activities by saving them on the diskette as a log files (*.log*). To create a log file, choose **Toggle Console Log** in the **File** menu or choose **Toggle Console Log** from the contextual menu. The Visual LISP **Console** window must be active to create a log file.

10. As you enter the text, Visual LISP automatically assigns a color to it based on the nature of the text string. For example, if the text is a built-in function or a protected symbol, the color assigned is blue. Similarly, text strings are magenta, integers are green, parentheses are red, real numbers are teal, and comments are magenta with shaded background.

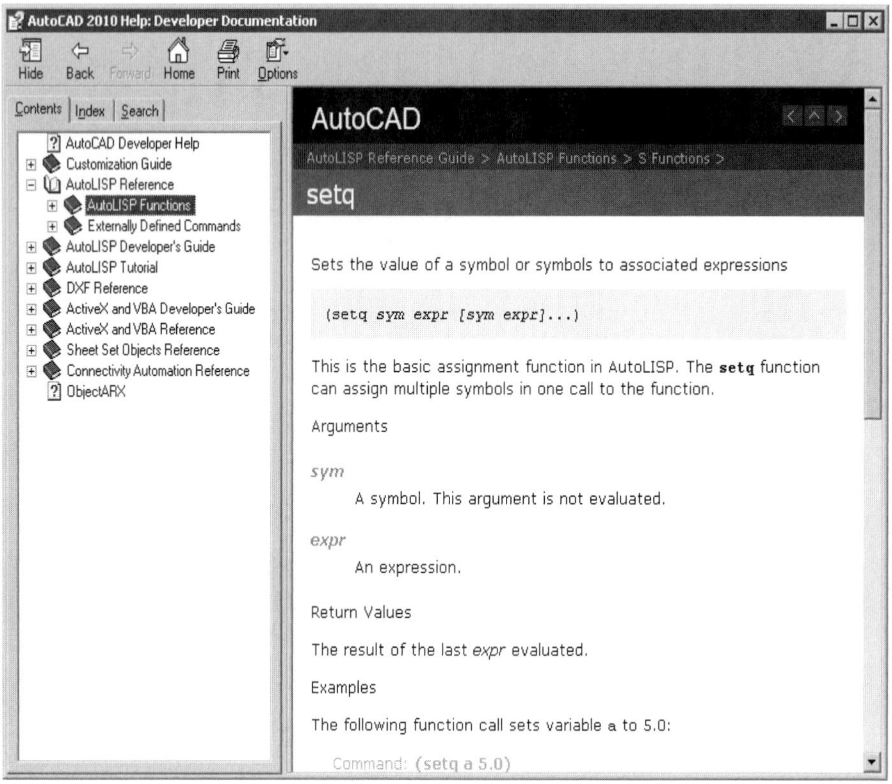

Figure 9-9 *Context-sensitive help screen*

11. Although native AutoLISP and Visual LISP are separate programming environments, you can transfer any text entered in the Visual LISP console to AutoCAD Command line. To accomplish this, enter the code at the console prompt and then choose **AutoCAD Mode** in the context menu or from the **Tools** menu. The console prompt is replaced with the Command prompt. Press the **Tab** key to display the text that was entered at the console prompt. Press ENTER to switch to the AutoCAD screen. The text is displayed at AutoCAD's Command prompt.

Exercise 2 *General*

Enter the following statements in the Visual LISP console at the console prompt (_$):

```
(+ 2 30 4 38.50)
(- 20 39 32)
(* 2.0 -7.6 31.25)
(/ -230 -7.63 2.15)
(sin 1.0472)
```

```
(atan -1.0)
(< 3 10)
(<= -2.0 0)
(setq variable_a 27.5)
(setq variable_b (getreal "Enter a value:"))
```

Using the Visual LISP Text Editor

You can use the **Visual LISP** text editor to enter the programming code. It has several features that are not available in other text editors. For example, when you enter text, it is automatically assigned a color based on the nature of the text string. If the text is in parenthesis, it is assigned red color. If the text is a Visual LISP function, it is assigned blue color. The **Visual LISP** text editor also allows you to execute an AutoLISP function without leaving the text editor, and also checks for balanced parentheses. These features make it an ideal tool for writing Visual LISP programs.

Color Coding

The **Visual LISP** text editor provides color coding for files that are recognized by Visual LISP. Some of the files that it recognizes are LISP, DCL, SQL, and C language source files. When you enter text in the text editor, Visual LISP automatically determines the color and assigns it to the text, Figure 9-10.

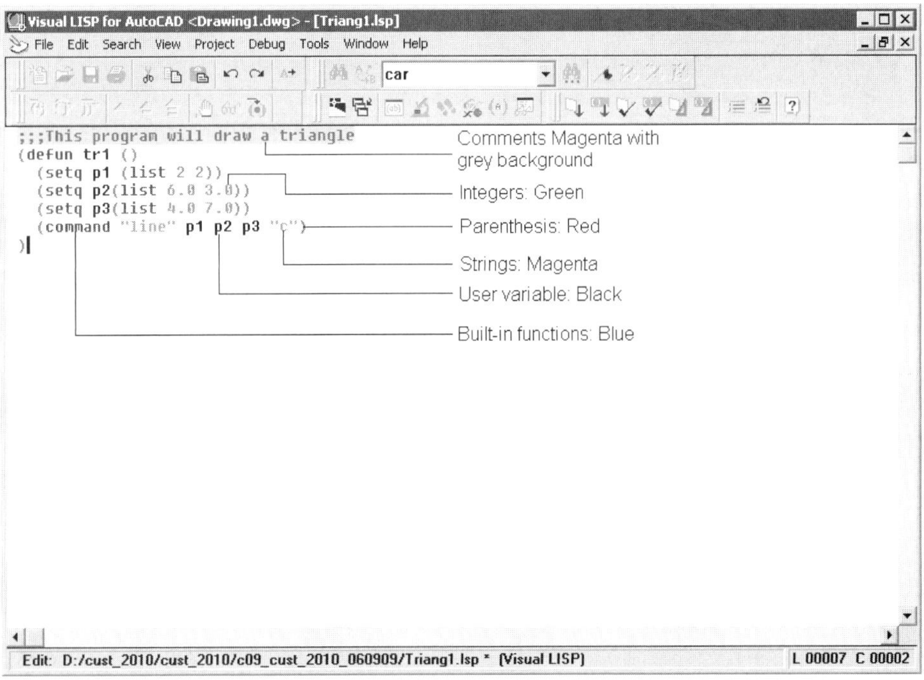

Figure 9-10 *Color coding for LISP code*

The following table shows the default color scheme.

Visual LISP Text Element	Color
Parentheses	Red
Built-in functions	Blue
Protected symbols	Blue
Comments	Magenta with gray background
Strings	Magenta
Real numbers	Teal
Integers	Green
Unrecognized items	Black (like variables)

Words in the Visual LISP Text Editor

In most of the text editors, the words are separated by space. In the **Visual LISP** text editor, the word has a different meaning. A word is a set of characters that are separated by one or more of the following special characters:

Character Description	Visual LISP Special Characters
Space	
Tab	
Single quote	'
Left parenthesis	(
Right parenthesis	)
Double quotes	"
Semicolon	;
Newline	\n

For example, if you enter the text (Command"Line"P1 P2 P3"c"), Visual LISP treats Command and Line as separate words because they are separated by double quotes. In a regular text editor (Command"Line" would have been treated as a single word. Similarly, in **(setq p1,** setq and p1 are separate words because they are separated by a space. If there was no space, Visual LISP will treat **(setqp1** as one word.

***Figure 9-11** Context menu displayed upon right-clicking*

Context-Sensitive Help

One of the important features of Visual LISP is the facility of context-sensitive help available for the Visual LISP functions. If you need help with any Visual LISP function, enter or select the function in the text editor and then choose the **Help** button in the **Tools** toolbar. Visual LISP will display the help for the selected function.

Context Menu

If you right-click anywhere in the **Visual LISP** text editor or press SHIFT+F10, Visual LISP displays the context menu, see Figure 9-11. You can use the context menu to perform several tasks like **Cut**, **Copy**, **Paste**, **Find**, **Go to Last Edited**, **Toggle Breakpoint**, **Inspect**, **Add Watch**, **Apropos Window**, **Symbol Service**, **Undo**, and **Redo**. The table given next contains a short description of these functions.

Functions in the context menu	Description of the function
Cut	Moves the text to the clipboard
Copy	Copies the text to the clipboard
Paste	Pastes the contents of the clipboard at the cursor position
Go to Last Edited	Retrives the last edited state of the text editor
Find	Finds the specified text
Inspect	Opens the **Inspect** dialog box
Add Watch	Opens the **Add Watch** dialog box
Apropos Window	Opens **Apropos options** dialog box
Symbol Service	Opens **Symbol Service** dialog box
Undo	Reverses the last operation
Redo	Reverses the previous Undo
AutoCAD Mode	Transfers all the text to the AutoCAD command prompt for execution
Toggle Console Log	Creates a log file from the outputs of the context window

Visual LISP Formatter

You can use the **Visual LISP Formatter** to format the text entered in the text editor. When you format the text, the formatter automatically arranges the AutoLISP expressions in a way that makes it easy to read and understand the code. It also makes it easier to debug the program when the program does not perform the intended function.

Running the Formatter

With the formatter you can format all the text in the **Visual LISP** text editor or the selected text. To format the text, the text editor window must be active. If it is not, click a point anywhere in the window to make it active. To format all the text, choose the **Format edit window** button in the **Tools** toolbar or choose **Format code in Editor** from the **Tools** menu. To format part of the text in the text editor, select the text and then select the **Format selection** tool in the **Tools** toolbar or choose **Format code in Selection** from the **Tools** menu. Figure 9-12 shows the **Tools** menu and **Tools** toolbar.

In Visual LISP, you can have five types of comments and depending on the comment in the program code, Visual LISP will format and position the comment text according to its type. Following is the list of various comment types available in Visual LISP:

Comment formatting	Description
; Single semicolon	The single semicolon is indented as specified in the Single-semicolon comment indentation of AutoLISP Format option.

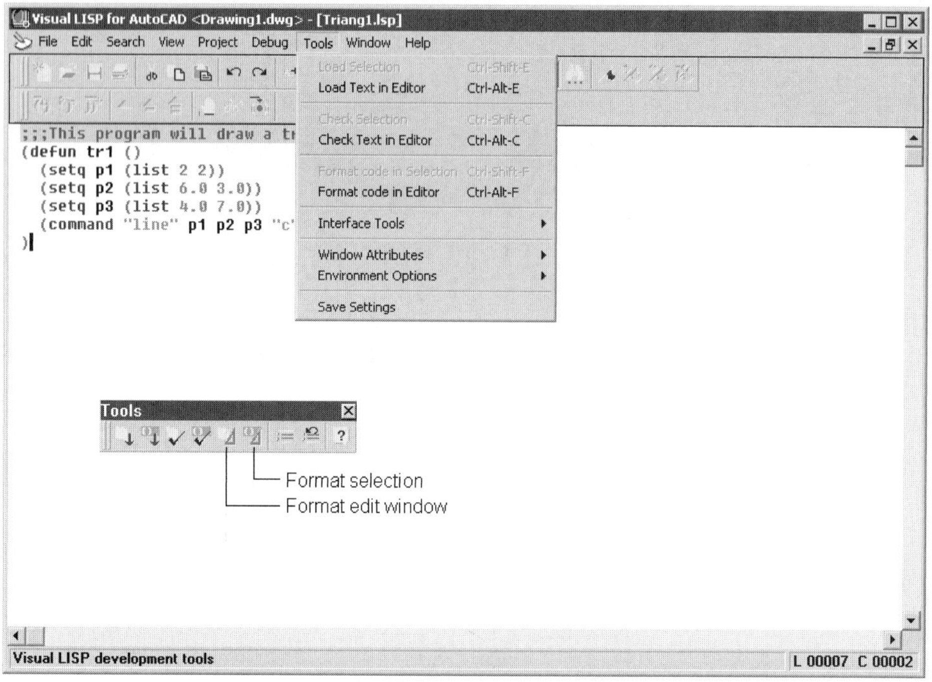

Figure 9-12 *Using the* ***Format selection*** *or* ***Format edit window*** *tools*

;; Current column comment	The comment is indented at the same position as the previous line of the program code.
;;; Heading comment	The comment appears without any indentation.
;| Inline comment|;	The Inline comments appear as any other expression.
;_ Function closing comment	The function closing comment appears just after the previous function, provided Insert form-closing comment is on in the AutoLISP Format options.

Two text editor windows are shown in Figure 9-13. The window at the top has the code without formatting. The text window at the bottom has been formatted. If you compare the two, you will see the difference formatting makes in the text display.

Changing Formatting Options

You can change the formatting options through the **Format options** dialog box (Figure 9-14) that can be invoked by choosing the **Tools > Environment Options > Visual LISP Format Options** from the menu bar. In this dialog box, if you choose the **More options** button, this dialog box expands, as shown in Figure 9-15 and lists more formatting options.

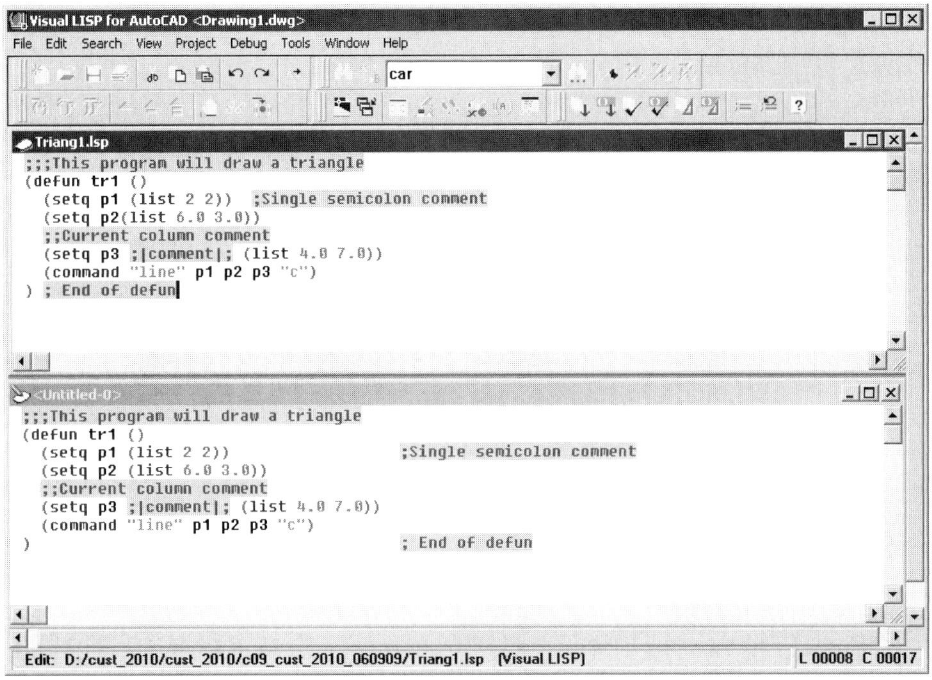

Figure 9-13 *Top window shows code without formatting and the bottom window shows the same code after formatting the programming code*

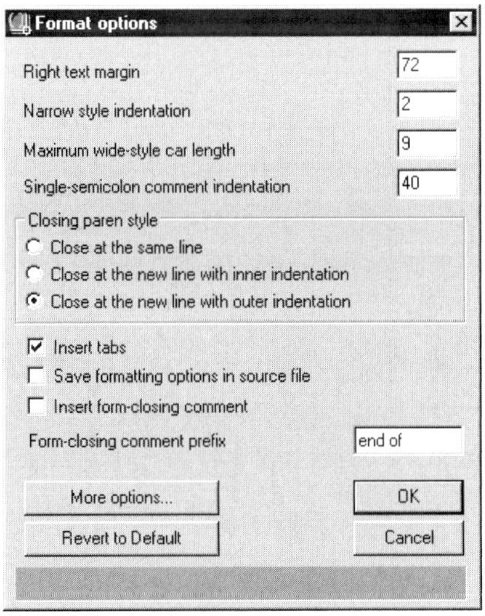

Figure 9-14 *The* ***Format options*** *dialog box*

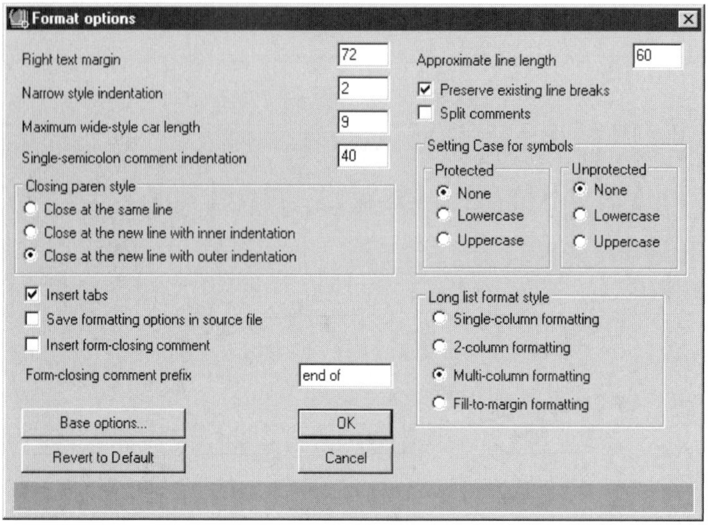

Figure 9-15 *The* ***Format options*** *dialog box with more options*

DEBUGGING THE PROGRAM

When you write a program, it is rare that the program will run the first time. Even if it runs, it may not perform the desired function. The error could be a syntax error, missing parentheses, a misspelled function name, improper use of a function, or simply a typographical error. It is time consuming and sometimes difficult to locate the source of the problem. Visual LISP comes with several debugging tools that make it easier to locate the problem and debug the code. Make sure the **Debug** toolbar is displayed on the screen. If it is not, in the **Visual LISP** window, invoke the **View** menu and then choose **Toolbars** to display the **Toolbars** dialog box. Select the **Debug** check box, choose the **Apply** button, and then choose the **Close** button to exit the dialog box. The **Debug** toolbar will appear on the screen. Figure 9-16 shows the toolbar with tool tips.

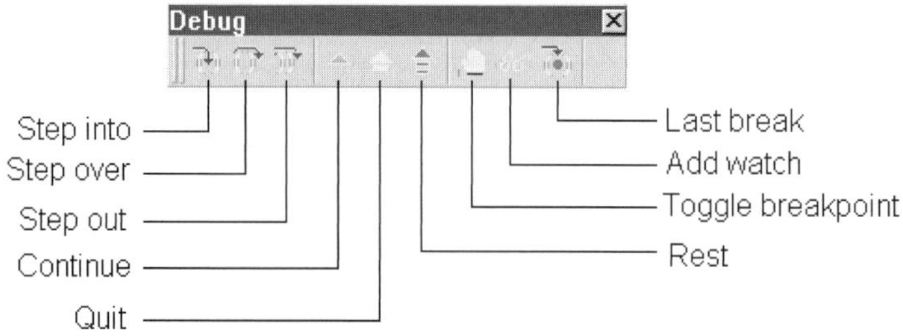

Figure 9-16 *Tools in the* ***Debug*** *toolbar*

1. To try some of the debugging tools, enter the following program in the Visual LISP Text Editor and then save the file as *triang2.lsp*.

```
;;; This program will draw a triangle and an arc
(defun tr2 ()
  (setq p1 (getpoint "\nEnter first point p1:"))
  (setq p2 (getpoint "\nEnter second point p2:"))
  (setq p3 (getpoint "\nEnter third point p3:"))
  (command "arc" p1 p2 p3)
  (command "line" p1 p2 p3 "c")
)
```

2. To format the code, choose the **Format edit window** tool in the **Tools** toolbar. You can also format the code by selecting **Format code in Editor** in the **Tools** menu.

3. Load the program by choosing the **Load active edit window** tool in the **Tools** toolbar. You can also load the program from the **Tools** menu by selecting **Load Text in Editor**.

4. In the **Visual LISP Text Editor**, position the cursor at the end of the following line and then choose the **Toggle breakpoint** tool in the **Debug** toolbar. You can also select it from the **Debug** menu. Visual LISP inserts a breakpoint at the cursor location.

```
(setq p2 (getpoint "\Enter second point p2:"))|
```

5. In the **Visual LISP Console**, at the console prompt, enter the name of the function and then press ENTER to run the program.

   _$ **(tr2)**

   The AutoCAD window will appear and the first two prompts in the program (Enter first point p1: and Enter second point p2:) will be displayed in the Command prompt area. Specify two points and the **Visual LISP** window will appear. Notice, the line that has the breakpoint at the end is highlighted, as shown in Figure 9-17.

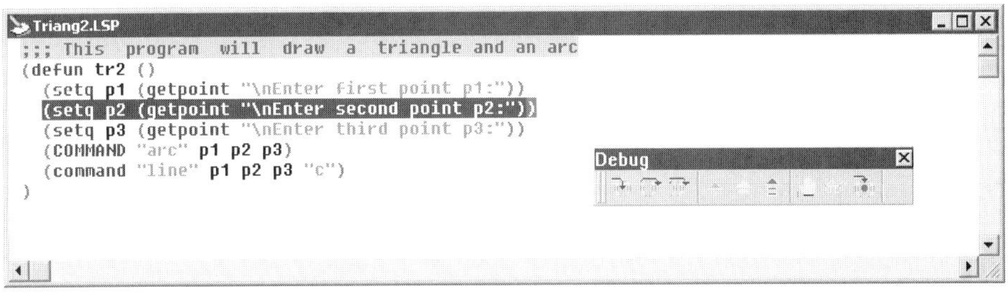

Figure 9-17 *Using **Step into** button to start execution until it encounters the next expression*

6. Choose the **Step into** button in the **Debug** toolbar or choose **Step Into** from the **Debug** menu. You can also press F8 to invoke the **Step Into** command. The program starts execution until it encounters the next expression. The next expression that is contained within the inside parentheses is highlighted, see Figure 9-18. Notice the position of the cursor. It is just before the expression **(setq p3 (getpoint "\nEnter third point p3:"))**.

Chapter 9

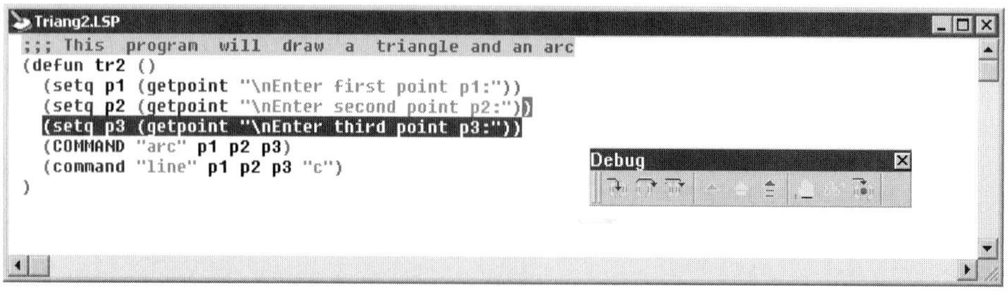

***Figure 9-18** Using the **Step into** button again, the cursor jumps to the start point of the next expression*

7. Choose the **Step into** button again. Cursor moves toward the right and highlights the expression **(getpoint "\nEnter third point p3:")**, see Figure 9-19.

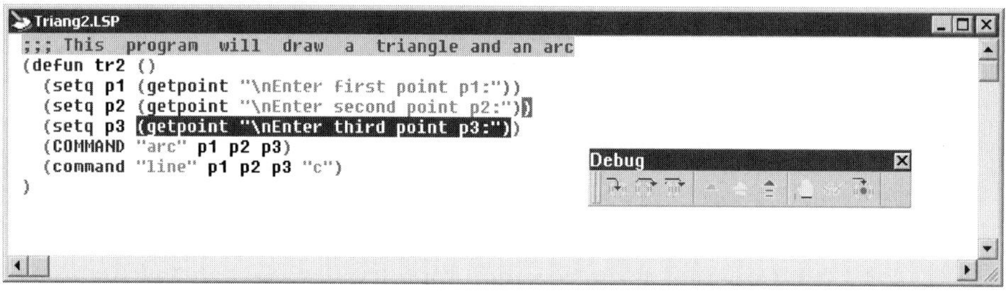

***Figure 9-19** Using the **Step into** button again, the vertical bar is displayed to the right of **Step Indicator** symbol*

8. Choose the **Step Into** button again. AutoCAD window appears and the third prompt in the program (Enter third point p3:) is displayed in the Command prompt area. Specify the third point, and the **Visual LISP** window returns. Notice the position of the cursor. It is just after the expression **(getpoint "\nEnter third point p3:")**.

9. Choose the **Step Into** button again to step through the program. The entire line is highlighted and the cursor moves to the end of the expression **(getpoint "Enter third point p3:")**.

10. Choose the **Step Into** button again. The next line of the program is highlighted and the cursor is at the start of the statement.

11. To step over the highlighted statement, choose the **Step over** button in the **Debug** toolbar, or select **Step Over** in the **Debug** menu. You can also press SHIFT+F8 to invoke this command.

12. Choose the **Step over** button until you go through all the statements and the entire program is highlighted. You will notice that the program is executed in the AutoCAD window.

GENERAL RECOMMENDATIONS FOR WRITING THE LISP FILE

1. Built-in functions and arguments should be spaced.
2. Arguments should be spaced.
3. Every function must start with an open parenthesis.
4. Every function must end with a closed parenthesis.

Example 1

Write a LISP program to draw the I-section, as shown in Figure 9-20. The starting point is p1. The point P1, length, width, and the thickness of flange and web are user-defined values that are to be entered at the Command prompt.

Step 1: Understanding the program algorithm

Before writing the program, it is very important to understand what is given, what is the desired output, and how we can get that output from the given information. Analyze the program as follows:

Input

P1 Starting point of the I-section
L Length of the I-section
W Width of the I-section
T1 Thickness of the flange
T2 Thickness of the web

Output

I section, as shown in Figure 9-20

Process

Define all points like p2, p3, p4, p5, p6, p7, p8, p9, p10, p11 and p12
Draw lines between points

Step 2: Writing the Visual LISP program

Choose **Manage > Applications > Visual LISP Editor** from the **Ribbon** or enter **VLIDE** at the Command prompt to display the **Visual LISP** window. Select **New File** from the **File** menu to display the **Visual LISP** Text Editor window. Write the following program in the **Visual LISP** Text Editor.

```
(defun isec ()
(setq  p1  (getpoint "\n Enter the starting point of the I-section: ")
       l  (getdist "\n Enter the length of the I-section: ")
       w  (getdist "\n Enter the width of the I-section: ")
       t1  (getdist "\n Enter the thickness of the flange: ")
```

Chapter 9

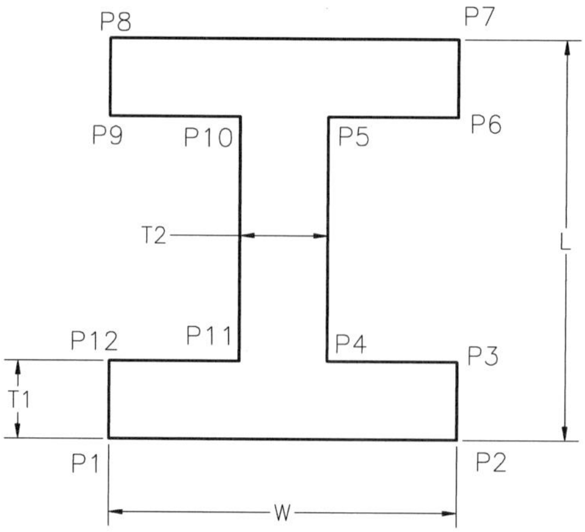

***Figure 9-20** I-section for Example 1*

```
        t2 (getdist "\n Enter the thickness of the web: ")
        p2 (list (+ (car p1) w) (cadr p1))
        p3 (list (car p2) (+ (cadr p2) t1))
        p4 (list (- (car p3) (/ (- w t2) 2)) (cadr p3))
        p5 (list (car p4) (+ (cadr p4) (- l (* 2 t1))))
        p6 (list (car p3) (cadr p5))
        p7 (list (car p6) (+ (cadr p6) t1))
        p8 (list (car p1) (+ (cadr p1) l))
        p9 (list (car p8) (- (cadr p8) t1))
        p10 (list (- (car p5) t2) (cadr p5))
        p11 (list (- (car p4) t2) (cadr p4))
        p12 (list (car p1) (cadr p11))
  )
  (command "PLINE" p1 p2 p3 p4 p5 p6 p7 p8 p9 p10 p11 p12 p1 "")
)
```

Step 3: Loading and running the Visual LISP program

Save the file as *isec.lsp* using the **Save** or **Save As** option in the **File** menu. To load the program, choose **Load Text in Editor** in the **Tools** menu. You can also load the program by choosing **Load active edit window** in the **Tools** toolbar. Visual LISP displays a message in the **Console** window indicating that program has been loaded. If the **Console** window does not appear on the screen, select the **Tile Horizontally** option from the **Window** toolbar to display it on the screen. It also gives the error message, if there is any error in the program.

The step-by-step debugging can be done as described under the heading *Debugging the Program* The mismatch in the parentheses can be corrected by choosing the **Format edit window** in the **Tools** toolbar. The nature of the error and its meaning are given at the end of this chapter under the heading *Error Codes and Messages*. To run the program, enter the function name **(isec)** at the console prompt adjacent to the $ sign. The function name must be enclosed in parentheses. When you enter the function name and press enter, the AutoCAD screen is displayed prompting you to enter the information about the I-section. After you enter the information, the I-section will be drawn on the screen.

Tip

*1. This program can also be entered using any text editor like Notepad. The program can be loaded by entering the **APPLOAD** command at the Command prompt, as described in the AutoLISP chapter.*

*2. The above example can also be solved by defining only half of the I-section and then mirroring the I-section with the **MIRROR** command to get the complete I-section. You can also draw only one-fourth of the I-section and then use the **MIRROR** command to get the remaining section.*

Step 4: Writing the Visual LISP program using the POLAR function

The points can also be calculated by using the **POLAR** function and then draw the lines to create the I-section. In this program d1= distance between P2 and P3, d2 = distance between P3 and P4, and d3 = distance between P4 and P5. Notice the **dtr** function used to convert the degrees into radians. The following file is the listing of the program using **POLAR** function.

```
(defun dtr (a)
  (* a (/ pi 180.0))
)
(defun isec ()
  (setq  p1 (getpoint "\n Enter the starting point of the I-section: ")
         l (getdist "\n Enter the length of the I-section: ")
         w  (getdist "\n Enter the width of the I-section: ")
         t1 (getdist "\n Enter the thickness of the flange: ")
         t2 (getdist "\n Enter the thickness of the web: ")
  )
  (setq  d1 t1
         d2 (- (/ w 2.0) (/ t2 2.0))
         d3 (- l (* 2.0 t1))
  )
  (setq  p2  (polar p1 (dtr 0) w)
         p3  (polar p2 (dtr 90) d1)
         p4  (polar p3 (dtr 180) d2)
         p5  (polar p4 (dtr 90) d3)
         p6  (polar p5 (dtr 0) d2)
         p7  (polar p6 (dtr 90) d1)
         p8  (polar p7 (dtr 180) W)
         p9  (polar p8 (dtr 270) d1)
         p10 (polar p9 (dtr 0) d2)
         p11 (polar p10 (dtr 270) d3)
         p12 (polar p11 (dtr 180) d2)
```

```
  )
  (command "PLINE" p1 p2 p3 p4 p5 p6 p7 p8 p9 p10 p11 p12 p1 "")
)
```

Example 2

Write a Visual LISP program that will extrude the I-section of Example 1 and create a 3D shape of the section, Figure 9-21. The additional input to the program is vpoint (v), extrusion height (h), and extrusion angle (a).

Step 1: Understanding the program algorithm

In addition to the inputs of Example 1, add the Vpoint, Extrusion, and Angle of Extrusion as additional inputs in the program. Also add the commands corresponding to vpoint, extrusion, zoom, and hide. Vpoint, extrusion height, and angle of extrusion have been represented as **v**, **h**, and **a** respectively. Angle has been entered using the **getreal** function to accept the angle in decimal degrees. Do not use the **getangle** function because this function returns the angle in radians. The extrusion angle in the **EXTRUDE** command must be in degrees.

Step 2: Writing the Visual LISP program

Choose **Manage > Applications > Visual LISP Editor** from the **Ribbon** or enter **VLIDE** at the Command prompt to display the **Visual LISP** window. Select **New File** from the **File** menu to display the **Visual LISP Text Editor** window. Now write the program as shown below in the **Visual LISP Text Editor** or open the LISP file of Example 1 and add the new lines to this program.

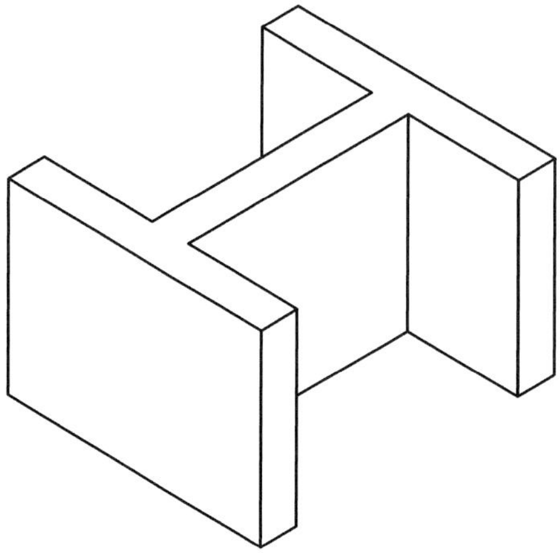

Figure 9-21 *3D I-section for Example 2*

```
(defun dtr (a)
  (* a (/ pi 180.0)))
(defun isec3d ()
```

```
(setq p1 (getpoint "\n Enter the starting point of the I-section: ")
      l  (getdist "\n Enter the length of the I-section: ")
      w  (getdist "\n Enter the width of the I-section: ")
      t1 (getdist "\n Enter the thickness of the flange: ")
      t2 (getdist "\n Enter the thickness of the web: ")
      h  (getdist "\n Enter Extrusion Height: ")
      v  (getpoint "\n Enter the Viewpoint: ")
      a  (getreal "\n Enter Angle of Extrusion: ")
)
(setq d1 t1
      d2 (- (/ w 2.0) (/ t2 2.0))
      d3 (- l (* 2.0 t1))
)
(setq p2  (polar p1 (dtr 0) w)
      p3  (polar p2 (dtr 90) d1)
      p4  (polar p3 (dtr 180) d2)
      p5  (polar p4 (dtr 90) d3)
      p6  (polar p5 (dtr 0) d2)
      p7  (polar p6 (dtr 90) d1)
      p8  (polar p7 (dtr 180) W)
      p9  (polar p8 (dtr 270) d1)
      p10 (polar p9 (dtr 0) d2)
      p11 (polar p10 (dtr 270) d3)
      p12 (polar p11 (dtr 180) d2)
)
(command "PLINE" p1 p2 p3 p4 p5 p6 p7 p8 p9 p10 p11 p12 p1 "")
(command "VPOINT" v)
(command "EXTRUDE" "All" "" h a)
(command "ZOOM" "All")
(command "HIDE")
)
```

Step 3: Loading and running the Visual LISP program

Save the file as *isec3d.lsp* using the **Save** or **Save As** option in the **File** menu. To load the program, select **Load Text in Editor** in the **Tools** menu. You can also load the program by choosing **Load active edit window** in the **Tools** toolbar. **Visual LISP** displays a message in the **Console** window indicating that program has been loaded. To run the program enter the function name **(isec3d)** at the console prompt adjacent to $ sign. The function name must be enclosed in parentheses. The program will prompt the user to enter the starting point followed by length, width, thicknesses, vpoint, extrusion height and angle at the Command prompt. After entering all the inputs, a 3D I-section will be displayed on the AutoCAD screen.

Tip

Step by step debugging can be done by choosing the ***Toggle breakpoint*** *tool in the* ***Debug*** *toolbar as described under the heading* ***Debugging the Program*** *(Page 9-16).*

Example 3

Write a Visual LISP program to draw a C-section, as shown in the Figure 9-22. The program should prompt the user to enter the starting point, length, width, and thickness.

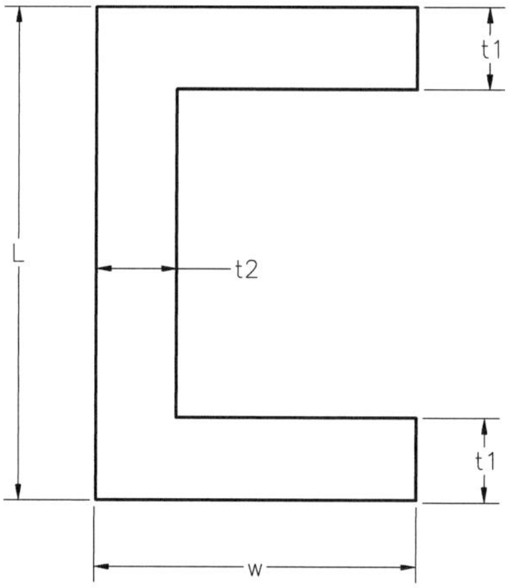

Figure 9-22 *C-section*

Step 1: Writing the Visual LISP program

Select **Manage > Applications > Visual LISP Editor** from the **Ribbon** or enter **VLIDE** at the Command prompt to display the **Visual LISP** window. Select **New File** from the **File** menu to display the **Visual LISP Text Editor** window. Now write the program in the **Visual LISP Text Editor**. You can also use any text editor like Notepad to write the LISP program. The following file is the listing of the program that will draw a c-section.

```
;;This program draws a C-Section
(defun csec ()
  (setq p1 (getpoint "\n Enter the point:"))
  (setq L (getdist "\n Enter the length:"))
  (setq W (getdist "\n Enter the width:"))
  (setq t1 (getdist "\n Enter the thickness of the flange:"))
  (setq t2 (getdist "\n Enter the thickness of the web:"))
  (setq p2 (list (+ (car p1) w) (cadr p1)))
  (setq p3 (list (car p2) (+ (cadr p2) t1)))
  (setq p4 (list (- (car p3) (- w t2)) (cadr p3)))
  (setq p5 (list (car p4) (+ (cadr p4) (- L (* 2 t1)))))
  (setq p6 (list (+ (car p5) (- w t2)) (cadr p5)))
  (setq p7 (list (car p6) (+ (cadr p6) t1)))
```

```
  (setq p8 (list (car p1) (cadr p7)))
  (command "pline" p1 p2 p3 p4 p5 p6 p7 p8 "c")
)
```

Step 2: Loading and Running the Visual LISP program

Save the file as *csec.lsp* using the **Save** or **Save As** option in the **File** menu. To load the program, choose **Load Text in Editor** in the **Tools** menu. You can also load the program by choosing **Load active edit window** in the **Tools** toolbar. Visual LISP displays a message in the **Console** window indicating that program has been loaded. To run the program, enter the function name **(csec)** at the console prompt adjacent to the $ sign. The function name must be enclosed in parentheses.

Example 4

Write a visual LISP program by using the **POLAR** function that will draw and extrude the C-section of Example 3 and mirror the C-section to create a 3D shape, as shown in Figure 9-23. The additional input of the program is vpoint (v) and extrusion height (h).

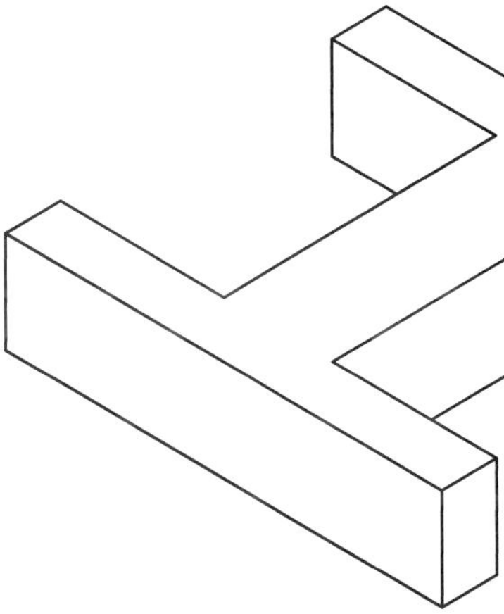

Figure 9-23 *Resulting model for Example 4*

Step 1: Writing the Visual LiSP program

Choose **Manage > Applications > Visual LISP Editor** from the **Ribbon** or enter **VLIDE** at the Command prompt to display the **Visual LISP** window. Select **New File** from the **File** menu to display the **Visual LISP** Text Editor window. Now write the program, as shown below in the **Visual LISP** Text Editor.

```
(defun dtr(a)
(* a (/ pi 180.0))
)
```

```
(defun csec3d ()
        (setq p1 (getpoint "\n Enter the starting point of the C-section: ")
            l  (getdist "\n Enter the length of the C-section: " )
            w  (getdist "\n Enter the width of the C-section: ")
            t1 (getdist "\n Enter the thickness of the flange: ")
            t2 (getdist "\n Enter the thickness of the web: ")
            h  (getdist "\n Enter the height of the C-section: ")
            v  (getpoint "\n Enter the Viewpoint: ")
        )
        (setq d1 t1
            d2 (- l (* t1 2.0))
            d3 (- w t2)
    )
        (setq p2 (polar p1 (dtr 90) l)
             p3 (polar p2 (dtr 0) w)
             p4 (polar p3 (dtr 270) d1)
             p5 (polar p4 (dtr 180) d3)
             p6 (polar p5 (dtr 270) d2)
            p7 (polar p6 (dtr 0) d3)
            p8 (polar p7 (dtr 270) d1)
        )
        (command "PLINE" p1 p2 p3 p4 p5 p6 p7 p8 p1 "")
        (command "VPOINT" v)
        (command "EXTRUDE" "ALL" "" h "")
        (command "MIRROR" "ALL" "" p1 p2 "N" "")
        (command "UNION" "ALL" "")
        (command "ZOOM" "All")
        (command "HIDE")
)
```

Step 3: Loading and running the Visual LISP program

Save the file as *csec3d.lsp* using the **Save** or **Save As** option in the **File** menu. To load the program, choose **Tools > Load Text in Editor**. You can also load the program by choosing **Load active edit window** in the **Tools** toolbar. **Visual LISP** displays a message in the **Console** window indicating that program has been loaded. To run the program, enter the function name **(csec3d)** at the console prompt adjacent to $ sign. The function name must be enclosed in parentheses. The program will prompt the user to enter the starting point followed by length, width, thicknesses, vpoint, extrusion height and angle at the Command prompt. After entering all the inputs, a 3D I-section will be displayed on the AutoCAD screen.

Exercise 3 *General*

Write a Visual LISP program to draw a T-section, as shown in Figure 9-24. The program should prompt the user to specify the starting point, length (L), width (w), and thickness (t1 and t2) of the T-section.

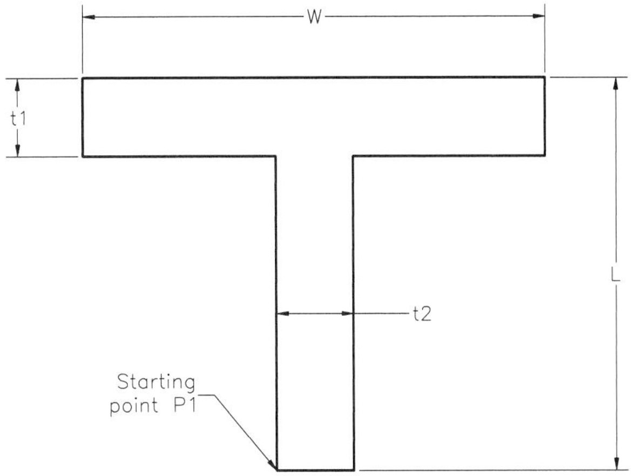

Figure 9-24 *T-section for Exercise 3*

Exercise 4

General

Write a Visual LISP program to draw a 3D T-section with the same specifications as Exercise 3. In addition to the inputs in Exercise 3, the program should also prompt the user to specify vpoint, extrusion height, and angle of extrusion.

Exercise 5

General

Write a Visual LISP program to draw an L-section, as shown in Figure 9-25. The program should prompt the user to specify starting point, length, width, and thicknesses of the L-section.

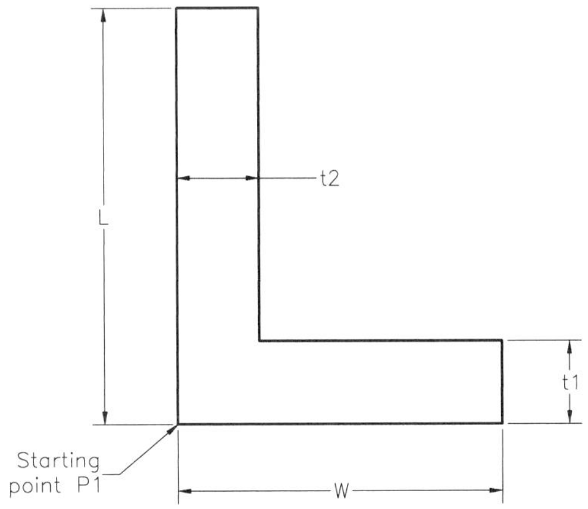

Figure 9-25 *L-section for Exercise 5*

Example 5

Write a Visual LISP program that will generate a flat layout drawing of a transition and then dimension the layout. The transition and the layout without dimensions are shown in Figure 9-26.

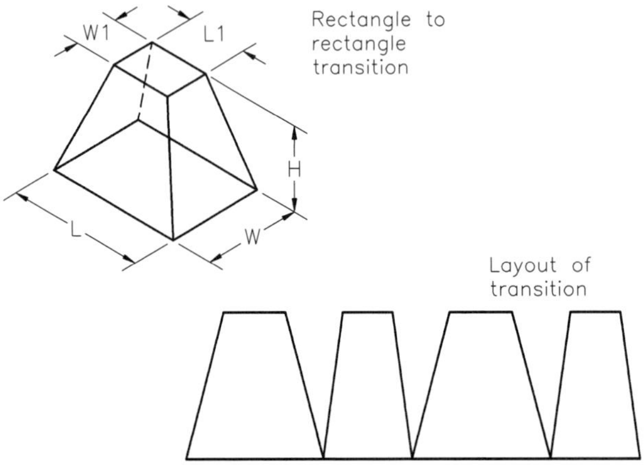

Figure 9-26 *Flat layout of a transition*

Step 1: Writing the Visual LISP program

Choose **Manage > Applications > Visual LISP Editor** from the **Ribbon** or enter **VLIDE** at the Command prompt to display the **Visual LISP** window. Select **New File** from the **File** menu to display the **Visual LISP** text editor. Now write the program in the **Visual LISP** text editor. You can also use any text editor like Notepad to write the LISP program. The following file is the listing of the **LISP** program for Example 5. The LISP programs do not need to be in lowercase letters. They can be uppercase or a combination of uppercase and lowercase.

```
;This program generates flat layout of
;a rectangle to rectangle transition
;
(defun TRANA ()
(graphscr)
(setvar "cmdecho" 0)
(setq L (getdist "\n Enter length of bottom rectangle: "))
(setq W (getdist "\n Enter width of bottom rectangle: "))
(setq H (getdist "\n Enter height of transition: "))
(setq L1 (getdist "\n Enter length of top rectangle: "))
(setq W1 (getdist "\n Enter width of top rectangle: "))

(setq x1 (/ (- w w1) 2))
(setq y1 (/ (- l l1) 2))
(setq d1 (sqrt (+ (* h h) (* x1 x1))))
```

```
(setq d2 (sqrt (+ (* d1 d1) (* y1 y1))))
(setq p1 (sqrt (- (* d2 d2) (* y1 y1))))
(setq p2 (sqrt (- (* d2 d2) (* x1 x1))))

(setq t1 (+ l1 y1))
(setq t2 (+ l w))
(setq t3 (+ l x1 w1))
(setq t4 (+ l x1))
(setq pt1 (list 0 0))
(setq pt2 (list y1 p1))
(setq pt3 (list t1 p1))
(setq pt4 (list l 0))
(setq pt5 (list t4 p2))
(setq pt6 (list t3 p2))
(setq pt7 (list t2 0))
(command "layer" "make" "ccto" "c" "1" "ccto" "")
(command "line" pt1 pt2 pt3 pt4 pt5 pt6 pt7 "c")

(setq sf (/ (+ l w) 12))
(setvar "dimscale" sf)
(setq c1 (list 0 (- 0 (* 0.75 sf))))
(setq c7 (list (- 0 (* 0.75 sf)) 0))
(setq c8 (list (- l (* 0.75 sf)) 0))

(command "layer" "make" "cctd" "c" "2" "cctd" "")
(command "dim" "hor" pt1 pt2 c1 "" "base" pt3 "" "base" pt4 "" "exit")
(command "dim" "hor" pt4 pt5 c1 "" "base" pt6 "" "base" pt7 "" "exit")
(command "dim" "vert" pt1 pt2 pt2 "" "exit")
(command "dim" "vert" pt4 pt5 pt5 "" "exit")
(command "dim" "aligned" pt1 pt2 c7 "" "exit")
(command "dim" "aligned" pt4 pt5 c8 "" "exit")
(setvar "cmdecho" 1)
(princ))
```

Step 2: Loading and Running the Visual LISP program

Save the file as *trana.lsp* using the **Save** or **Save As** option in the **File** menu. To load the program, choose **Load Text in Editor** in the **Tools** menu. You can also load the program by choosing **Load active edit window** in the **Tools** toolbar. Visual LISP displays a message in the **Console** window indicating that the program has been loaded. To run the program enter the function name **(trana)** at the console prompt adjacent to the $ sign. The function name must be enclosed in parentheses.

Example 6

Write a Visual LISP program that can generate a flat layout of a cone, as shown in Figure 9-27. The program should also dimension the layout.

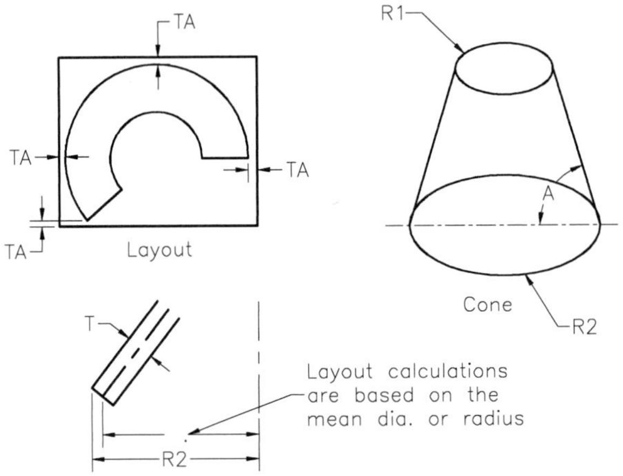

Figure 9-27 *Flat layout of a cone*

Step 1: Writing the Visual LISP program

```
;The following file is a listing of the Visual LISP program for Example 6:
;This program generates layout of a cone
;
;DTR function changes degrees to radians
(defun DTR (a)
 (* PI (/ A 180.0))
 )
;RTD Function changes radians to degrees
(defun rtd (a)
(* a (/ 180.0 pi))
)

(defun tan (a)
(/ (sin a) (cos a))
)
(defun cone-1p ()
(graphscr)
(setvar "cmdecho" 0)
(setq r2 (getdist "\n Enter outer radius at larger end: "))
(setq r1 (getdist "\n Enter inner radius at smaller end: "))
(setq t1 (getdist "\n Enter sheet thickness:-"))
(setq a (getangle "\n Enter cone angle:-"))
;this part of the program calculates various parameters
```

```
;needed in calculating the strip layout
(setq x0 0)
(setq y0 0)
(setq sf (/ r2 3))
(setvar "dimscale" sf)
(setq ar a)
(setq tx (/ (* t1 (sin ar)) 2))
(setq rx2 (- r2 tx))
(setq rx1 (+ r1 tx))
(setq w (* (* 2 pi) (cos ar)))
(setq rl1 (/ rx1 (cos ar)))
(setq rl2 (/ rx2 (cos ar)))

;this part of the program calculates the x-coordinate
;of the points
  (setq x1 (+ x0 rl1)
      x3 (+ x0 rl2)
      x2 (- x0 (* rl1 (cos (- pi w))))
      x4 (- x0 (* rl2 (cos (- pi w))))
  )

;this part of the program calculates the y-coordinate
;of the points
  (setq y1 y0
      y3 y0
      y2 (+ y0 (* rl1 (sin (- pi w))))
      y4 (+ y0 (* rl2 (sin (- pi w))))
      )

  (setq p0 (list x0 y0)
      p1 (list x1 y1)
      p2 (list x2 y2)
      p3 (list x3 y3)
      p4 (list x4 y4)
      )
  (command "layer" "make" "ccto" "c" "1" "ccto" "")
  (command "arc" p1 "c" p0 p2)
  (command "arc" p3 "c" p0 p4)
  (command "line" p1 p3 "")
  (command "line" p2 p4 "")

  (setq f1 (/ r2 24))
  (setq f2 (/ r2 2))
  (setq d1 (list (+ x3 f2) y3))
  (setq d2 (list x0 (- y0 f2)))
```

```
(command "layer" "make" "cctd" "c" "2" "cctd" "")
(setvar "dimtih" 0)
(command "dim" "hor" p0 p1 d2 "" "baseline" p3 "" "baseline" p2 "" "baseline" p4 ""
"exit")
(command "dim" "vert" p0 p2 d1 "" "baseline" p4 "" "exit")
(setvar "dimscale" 1)
(setvar "cmdecho" 1)
(princ)
)
```

Step 2: Loading and running the Visual LISP program

Save the file as *cone-1p.lsp* using the **Save** or **Save As** option in the **File** menu. To load the program, choose **Load Text in Editor** in the **Tools** menu. You can also load the program by choosing **Load active edit window** in the **Tools** toolbar. Visual LISP displays a message in the **Console** window indicating that program has been loaded. To run the program enter the function name **(cone-1p)** at the console prompt adjacent to the $ sign. The function name must be enclosed in parentheses.

TRACING VARIABLES

Sometimes it becomes necessary to trace the values of the variables used in the program. For example, let us assume that the program crashes or does not perform as desired. Now you want to locate where the problem occurred. One of the ways you can do this is by tracing the value of the variables used in the program. The following steps illustrate this procedure:

1. Open the file containing the program code in the **Visual LISP** Text Editor. You can open the file by selecting **Open File** from the **File** menu.

2. Load the program by choosing **Load Text in Editor** from the **Tools** menu or by choosing **Load active edit window** tool in the **Tools** toolbar.

3. Run the program by entering the name of the function (tr2) at the console prompt (_$) in the **Visual LISP Console** window. The program will perform the operations as defined in the program. In this program, it will draw a triangle and an arc between the user-specified points. Now, to find out the coordinates of the points p1, p2, and p3, you can use the **Watch window** tool in the **View** toolbar.

4. Highlight the variable **p1** anywhere in the **Visual LISP** window after running the program and choose the **Watch window** button in the **View** toolbar. You can also choose **Watch Window** in the **View** menu. The **Watch** window appears, which displays the *X*, *Y*, and *Z* coordinates of the p1 variable in the list box. This window also provides four buttons on top of the list box.

5. Position the cursor near (in front, middle, or end) the variable p2 and then choose the **Add Watch** button in the **Watch** window; the **Add Watch** dialog box is displayed. Choose the **OK** button from this dialog box. The coordinate values of the selected variable are

listed in the **Watch** window, see Figure 9-28. Similarly, you can trace the value of other variables in the program.

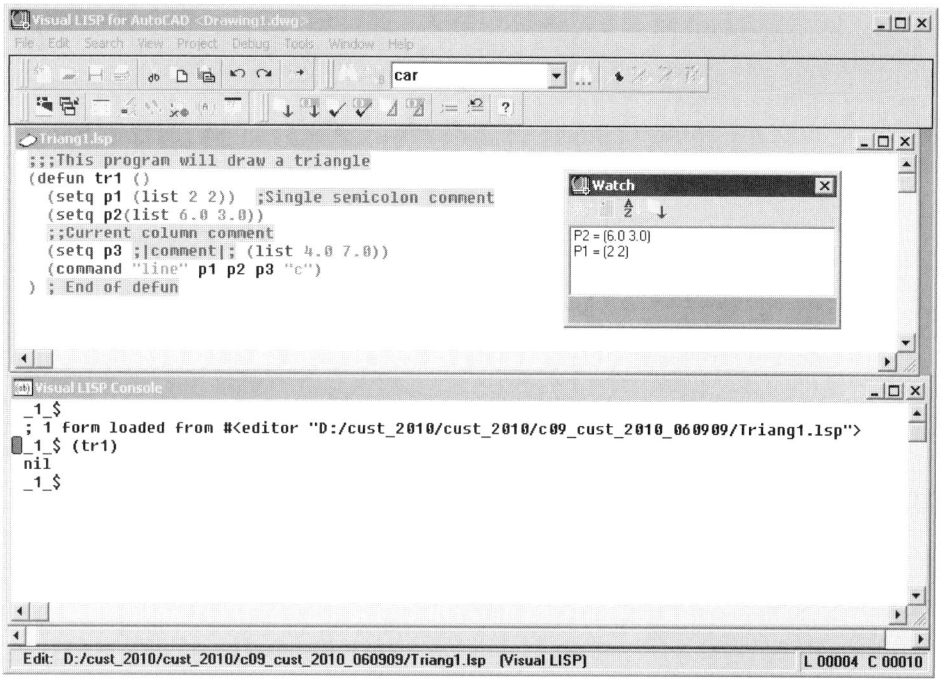

*Figure 9-28 Using **Add Watch** to trace variables in the program*

VISUAL LISP ERROR CODES AND MESSAGES

This section describes error messages that you may encounter at the console prompt while running the Visual LISP program and their corresponding meaning. The errors are arranged depending upon the frequency with which they occur during running the Visual LISP program. The common errors encountered have been listed first.

Malformed list
Premature ending of the list. Common cause is the mismatch between the number of opening and closing parentheses.

Malformed string
Premature ending of the string being read from a file.

Null function
A function evaluated with nil definition.

Too few arguments
The built-in function given less arguments.

String too long
The system variable SETVAR has been given too long string.

Too many arguments
The built-in function has been given too many arguments.

Invalid argument
The type of argument is improper.

Invalid argument list
The function has been given an invalid argument list.

Incorrect number of arguments
It is expected to provide one argument within the quote function, but some other arguments have also been given.

Incorrect number of arguments to a function
Mismatch between the number of formal arguments specified and the number of arguments to the user-defined function.

Function cancelled
The user entered CTRL+C or ESC in response to an input prompt.

Extra right paren
One or more extra right parentheses have been encountered.

Exceeded maximum string length
A string given to a function is greater than 132 characters.

Divide by zero
Division by zero is invalid.

Divide overflow
Division by a very small value has resulted in a invalid quotient.

Console break
The user entered CTRL+C while a function was in progress.

Bad argument type
A function was passed an incorrect type of argument.

Bad association list
The list of the assoc function does not consist of key value lists.

Visual LISP stack overflow
The program has exceeded the Visual LISP stack storage space. This can be due to very large function arguments list.

Bad ENTMOD list
The argument passed to entmod is not a proper entity data list.

Bad conversion code
An invalid space identifier was passed to the trans function.

Bad entmod list value
One of the sublists in the association list passed to entmod contains an improper value.

Bad function
The first element in the list is not a valid function name.

Bad list
An improper list was given to a function.

Bad ssget list
The argument passed to (ssget"x") is not a proper entity data.

Bad ssget list value
One of the sublists in the association list passed to (ssget"x") contains an improper value.

Bad ssget mode string
Caused when ssget was passed an invalid string in the mode argument.

Base point is required
The getcorner function was called without the required base point argument.

Can't evaluate expression
A decimal point was improperly placed, or some other expression was poorly formed.

Can't open file for input- LOAD failed
The file name in the load function could not be found, or the user does not have read access to the file.

Input aborted
An error or premature end of file condition has been detected, causing termination of the file input.

Invalid character
An expression contains an improper character.

Invalid dotted pair
A real number begins with a decimal point; you must use a leading zero in such a case.

Misplaced dot
A real number begins with a decimal point. You must use a leading zero in such a case.

Self-Evaluation Test

Answer the following questions and then compare them to those given at the end of this chapter:

1. Visual LISP has a powerful __________ and a __________ .

2. The text editor allows the __________ of parentheses, function names, variables and other components.

3. The formatter __________ the code in an easily readable format.

4. Values can be traced with the help of __________ tool in the view toolbar.

5. Step-by-step debugging can be done with the __________ tool in the **Debug** toolbar.

6. What are the different ways to start Visual LISP?
   1. __________ 2. __________

7. Name the four areas of the **Visual LISP** window.
   1. __________ 2. __________
   3. __________ 4. __________

8. Name the toolbars available in Visual LISP
   1. __________ 2. __________
   3. __________ 4. __________
   5. __________

Review Questions

Answer the following questions:

1. What are the different ways of loading a Visual LISP program?
   1. __________ 2. __________

2. How can you run a Visual LISP program?
   ____

3. Can you open an AutoLISP file in the **Visual LISP** text editor?
   ____

4. Explain the function of the **Visual LISP Console**.
   ____
   ____

5. Can you enter several LISP statements in the **Visual LISP Console**? If yes, give an example.
   ____

6. In the Visual LISP Console, how can you enter a LISP statement in more than one line?
   ____

7. What is a context menu and how can you display it?
   ____
   ____

8. In the **Visual LISP** text editor, what are the default colors assigned to the following?

   Parentheses ________________ LISP functions ______________________
   Comments __________________ Integers ____________________________

9. What is the function of the **Visual LISP Formatter** and how can you format the text?
   ____
   ____

10. How can you get the heading comment and function closing comment?
    ____

11. How can you change the formatting options?
    ____

12. How can you debug a Visual LISP program?

____

____

13. How can you trace the value of variables in the Visual LISP program?

____

____

Exercises

Exercise 6 *General*

Write a Visual LISP program that will draw the staircase shown in Figure 9-29. When you run the program, the user should be prompted to enter the rise and run of the staircase and the number of steps. The user should also be prompted to enter the length of the landing area.

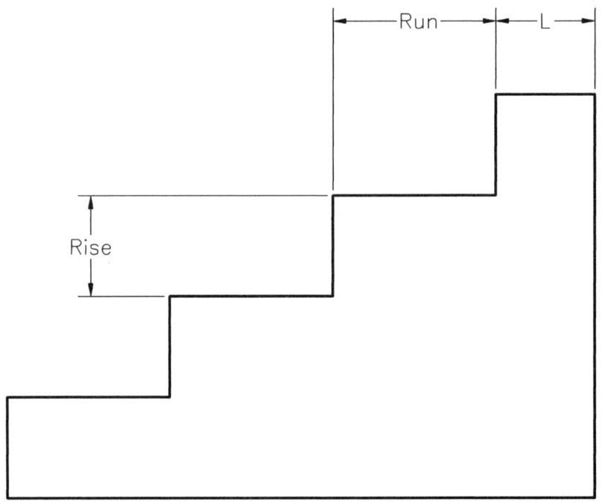

Figure 9-29 *Drawing for Exercise 6*

Exercise 7 *General*

Write a Visual LISP program that will draw the picture frame of Figure 9-30. The program should prompt the user to enter the length, width, and the thickness of the frame.

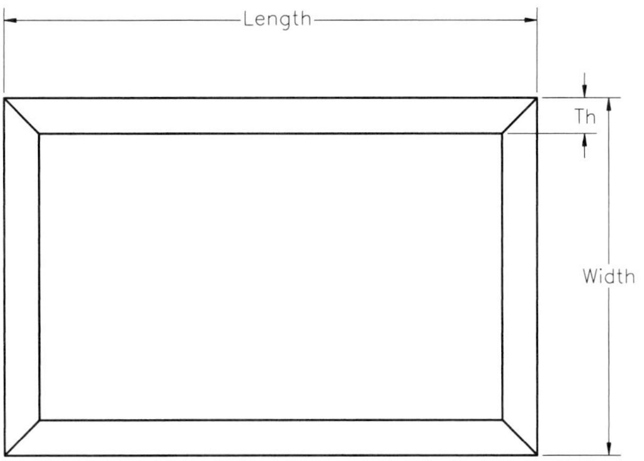

Figure 9-30 *Drawing for Exercise 7*

Exercise 8 *General*

Write a Visual LISP program that will draw a three-dimensional table, as shown in Figure 9-31. The program should prompt the user to enter the length, width, height, thickness of tabletop, and the size of the legs. The program should also use the **VPOINT** command to change the view direction to display the 3D view of the table.

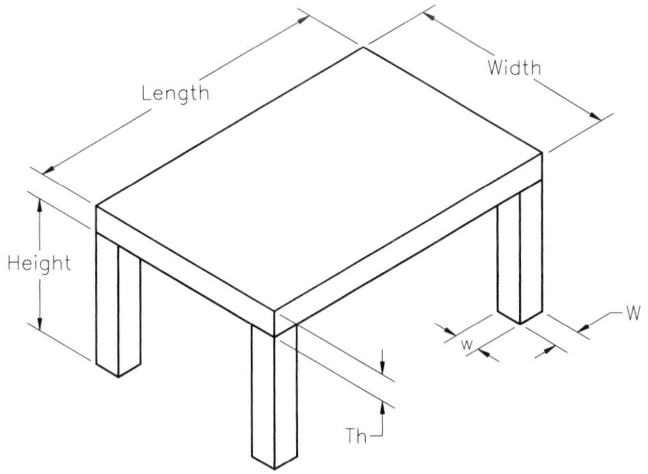

Figure 9-31 *Drawing for Exercise 8*

Exercise 9 *General*

Write a Visual LISP program that will draw the figure, as shown in Figure 9-32 with center lines and dimensions. Assume, L5=D1, L3=1.5*D1, L2=10*D1, L1= L2-D1, L4=L3+D1.

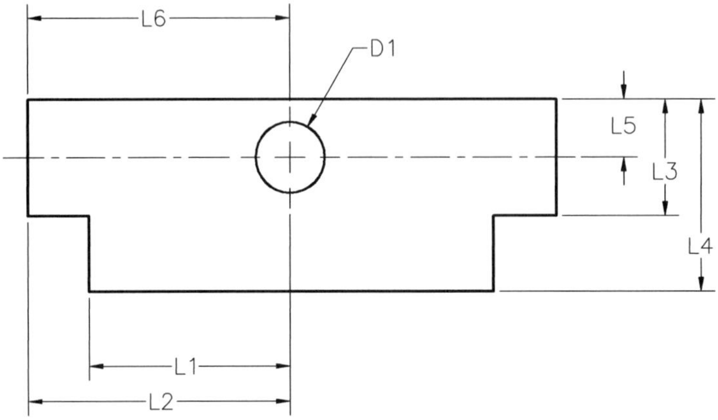

***Figure 9-32** Drawing for Exercise 9*

Exercise 10 *General*

Write a Visual LISP program that will draw a hub with the key slot, as shown in Figure 9-33. The program should prompt the user to enter the values for P0 (Center of hub/shaft, D1 (Diameter of shaft), D2 (Outer diameter of hub), W (Width of key), and T (Height of key). The program should also draw the centerlines in Center layer (Green color) and draw dimension T and W in Dim layer (Magenta color).

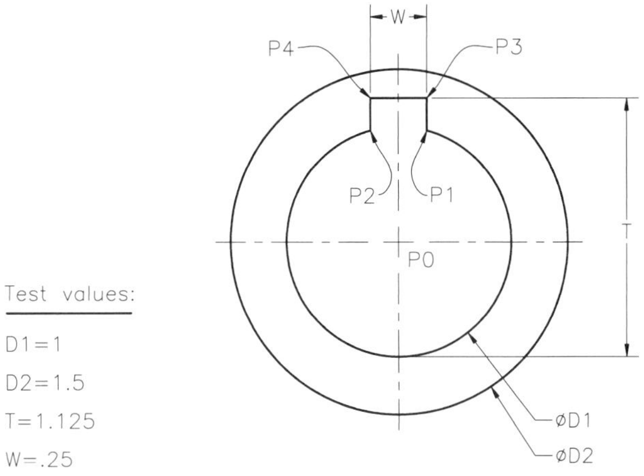

Figure 9-33 *Drawing for Exercise 10*

Exercise 11 *General*

Write a Visual LISP program that will draw an equilateral triangle inside the circle (Figure 9-34). The program should prompt you to enter the radius and the center point of the circle.

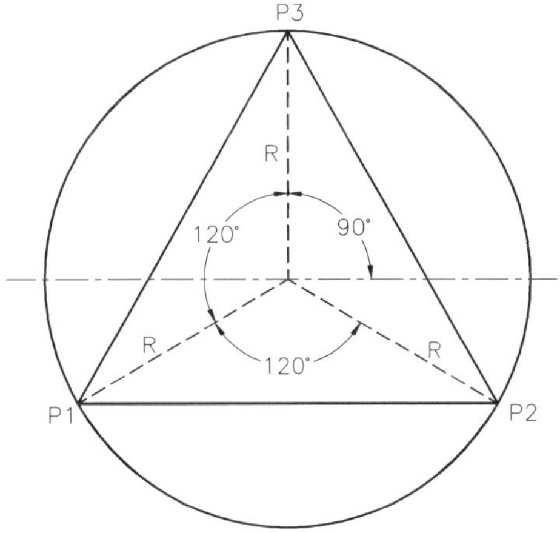

Figure 9-34 *Equilateral triangle inside a circle*

Exercise 12 *General*

Write a Visual LISP program that will draw two lines tangent to two circles, as shown in Figure 9-35. The program should prompt you to enter the circle diameters and the center distance between the circles.

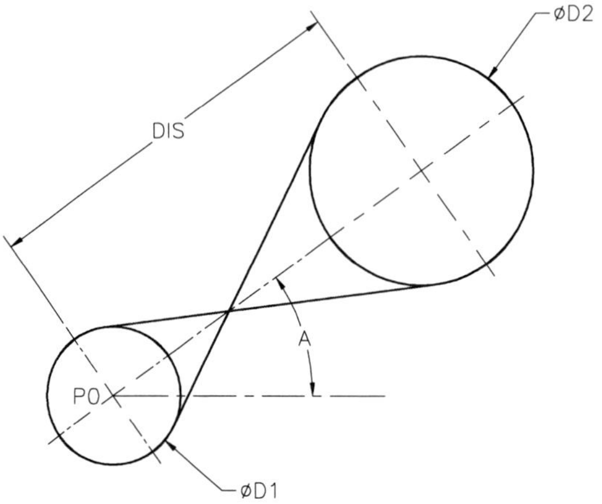

Figure 9-35 *Circles at angle A, with tangent lines*

Answers to Self-Evaluation Test

1. Text Editor, **Formatter**, **2.** color coding, **3.** formats, **4. Watch Window**, **5. Toggle Breakpoint**, **6. Manage > Applications > Visual LISP Editor** from the **Ribbon**, **VLIDE**, **7.** menu, toolbars, console window, and status bar, **8. Standard**, **Search**, **Tools**, **Debug**, **View**

Chapter 10

Visual LISP: Editing the Drawing Database

Learning Objectives

After completing this chapter, you will be able to:

- *Edit the drawing database using Visual LISP.*
- *Use the* ***ssget****,* ***sslength****,* ***ssname****,* ***ssadd****,* ***ssdel****,* ***entget****,* ***assoc****,* ***cons****,* ***subst****,* ***entnext****,* ***entlast*** *and* ***entmod*** *functions.*
- *Retrieve information from the drawing database.*
- *Edit the database and substitute the values back into the drawing database.*

EDITING THE DRAWING DATABASE

In addition to writing programs to create new commands, you can use the Visual LISP programming language to edit the drawing database. This is a powerful tool to make changes in a drawing. For example, you can write a program that will delete all text objects in the drawing or change the layer and color of all circles just by entering one command. Once you understand how AutoCAD stores the information of the drawing objects and how it can be retrieved and edited, you can manipulate the database any way you want.

This chapter discusses some of the commands that are frequently used to edit the drawing database. For other commands, not discussed in this section, please refer to the online reference manual, Visual LISP Programmers Reference, published by Autodesk.

ssget

The **ssget** function enables you to select any number of objects in a drawing. The object selection modes (window, crossing, previous, last, etc.) and the points that define the corners of the window can be included in the **ssget** assignment. The format of the **ssget** function is:

(ssget [selection-mode] [point1 point2])

Where **selection-mode** ---------- Object selection mode (w, c, l, p, etc.)
point 1 --------------------- First point of window (optional)
point 2 --------------------- Second point of window (optional)

Examples

(ssget)	For general object selection
(ssget "L")	For selecting last object
(ssget "p")	For selecting previous selection set
(ssget "w" (list 0 0) (list 12.0 9.0))	Object selection using window object selection mode, where the window is defined by points 0,0 and 12.0,9.0
(ssget "c" pt1 pt2)	Object selection using crossing object selection mode, where the window is defined by predefined points pt1 and pt2.

Example 1

Write a Visual LISP program that will erase all objects within the drawing limits, limmax, and limmin. Use the **ssget** function to select the objects. Assume that the objects have already been drawn within the limits.

Step 1: Writing the Visual LISP program

Choose **Manage > Applications > Visual LISP Editor** from the **Ribbon** or enter **VLIDE** at the Command prompt to display the **Visual LISP** window. Choose **New File** from the **File** menu to display the Visual LISP text editor. Now write the program in the **Visual LISP** Text Editor, as shown next.

The following file is a listing of the **Visual LISP** program for Example 1. **The line numbers are not a part of the program; they are shown here for reference only.**

```
1  ;This program will delete all objects
2  ;that are within the drawing limits
3  ;
4  (defun delall ()
5     (setvar "cmdecho" 0)
6     (setq pt1 (getvar "limmin"))
7     (setq pt2 (getvar "limmax"))
8     (setq ss1 (ssget "c" pt1 pt2))
9     (command "erase" ss1 "")
10    (command "redraw")
11    (setvar "cmdecho" 1)
12    (princ)
13 )
```

Explanation

Lines 1-3

The first three lines are comment lines that describe the function of the program. Notice that all comment lines start with a semicolon (;).

Line 4

(defun c:delall()

In this line, the **defun** function defines function the **delall**.

Line 6

(setq pt1 (getvar "limmin"))

The **getvar** function secures the value of the lower left corner of the drawing limits (limmin;) and the **setq** function assigns that value to the variable **pt1**.

Line 7

(setq pt2 (getvar "limmax"))

The **getvar** function secures the value of the upper right corner of the drawing limits (limmax) and the **setq** function assigns that value to the variable **pt2**.

Line 8

(setq ss1 (ssget "c" pt1 pt2))

The **ssget** function uses the "crossing" objection selection mode to select the objects that are within or touching the window defined by points pt1 and pt2. The **setq** function then assigns this object selection set to the variable **ss1**.

Line 9

(command "erase" ss1 "")

The **command** function uses AutoCAD's **ERASE** command to erase the predefined object selection set **ss1**.

Line 10
(command "redraw")
In this line the **command** function uses AutoCAD's **REDRAW** command to redraw the screen and get rid of the blip marks left after erasing the objects.

Step 2: Loading the Visual LISP program
Save the file as *delall.lsp* using the **Save** or **Save As** option in the **File** menu. To load the program, select **Load Text in Editor** in the **Tools** menu. You can also load the program by choosing the **Load active edit window** button in the **Tools** toolbar. Visual LISP displays a message in the **Console** window indicating that the program has been loaded. It also gives an error message, if it detects any error in the program. The nature of the errors and their meanings are given at the end of Chapter 9: *Working with Visual Lisp*. Mismatch in the parentheses can be corrected by choosing the **Format edit window** in the **Tools** toolbar.

To run the program, enter the function name (**delall**) at the console prompt $ sign. The function name must be enclosed in parentheses. After writing the function name, press ENTER and then choose **Activate AutoCAD** from the **Window** menu to display the AutoCAD screen.

You can see the objects within the limits already erased.

ssget "X"

The **ssget "X"** function enables you to select specified types of objects in the entire drawing database, even if the layers are frozen or turned off. The format of the **ssget "X"** function is:

(ssget "X" specified-criteria)

Where **X** ---------------------------- Filter mode of the **ssget** function
specified-criteria -------- List of the specified criteria for selecting the objects

Examples:

(ssget "X" (list (cons 0 "TEXT")))	Returns a selection set that consists of all TEXT objects in the drawing
(ssget "X" (list (cons 7 "ROMANC")))	Returns a selection set that consists of all TEXT objects in the drawing with the text style name ROMANC
(ssget "X" (list (cons 0 "LINE")))	returns a selection set that consists of all LINE objects in the drawing
(ssget "X" (list (cons 8 "OBJECT")))	returns a selection set that consists of all objects in the OBJECT layer.

The **ssget "X"** function can contain more than one selection criterion. This option can be used to select a specific set of objects in a drawing. For example, if you want to select all the line objects in the OBJECT layer, there are two selection criteria. The first is that the object has to be a line, the second is that the line object has to be in the OBJECT layer.

(ssget "X" (list (cons 0 "LINE")(cons 8 "OBJECT")))

Group Codes for ssget "X"

The following table is a list of AutoCAD group codes that can be used with the function **ssget "X"**:

Group Code	Code Function
0	Object type
2	Block name for block reference
3	Dimension object DIMSTYLE name
6	Linetype name
7	Text style name
8	Layer name
38	Elevation
39	Thickness
62	Color number
66	Attributes
210	3D extrusion direction

Example 2

Write a Visual LISP program that will erase all text objects in a drawing on a specified layer. Use the **filter** option of the **ssget** function (ssget "X") to select text objects in the specified layer. Assume that some objects have already been drawn on any specified layer.

Step 1: Writing the Visual LISP program

Choose **Manage > Applications > Visual LISP Editor** from the **Ribbon** or enter **VLIDE** at the Command prompt to display the **Visual LISP** window. Choose **New File** from the **File** menu to display the **Visual LISP** Text Editor. Now, write the program as shown below.

The following file is a listing of the Visual LISP program for Example 2:

```
;This program will delete all text
;in the user-specified layer
;
(defun deltext ()
   (setvar "cmdecho" 0)
   (setq layer (getstring "\n Enter layer name: "))
   (setq ss1 (ssget "x" (list (cons 8 layer) (cons 0 "text"))))
   (command "erase" ss1 "")
   (command "redraw")
```

Chapter 10

```
(setvar "cmdecho" 1)
(princ))
```

Step 2: Loading the Visual LISP program

Save the file as *deltext.lsp* using the **Save** or **Save As** option in the **File** menu. To load the program, choose **Load Text in Editor** in the **Tools** menu. You can also load the program by choosing the **Load active edit window** button in the **Tools** toolbar. Visual LISP displays a message in the **Console** window indicating that the program has been loaded. It also gives an error message, if it detects any error in the program. To run the program enter the function name (**deltext**) at the console prompt $ sign. The function name must be enclosed in parentheses. After entering the function name, press ENTER and then choose **Activate AutoCAD** from the **Window** menu to display AutoCAD screen. You can see that the text has been erased.

sslength

The **sslength** function determines the number of objects in a selection set and returns an integer corresponding to the number of objects found. The format of the **sslength** function is given next.

(sslength selection-set)

Where **selection-set** --- Name of the selection set

Examples:

(setq ss1 (ssget))
(setq num (sslength ss1)) — returns the number of objects in the predefined selection set ss1

(setq ss2 (ssget "l"))
(setq num (sslength ss2)) — returns the number of objects (1) in selection set ss2, where selection set ss2 has been defined as the last object in the drawing

ssname

The **ssname** function returns the name of the object, from a predefined selection set, as referenced by the index that designates the object number. The format of the **ssname** function is given next.

(ssname <u>selection-set</u> index)

Where **selection-set** --- A predefined selection set
index ------------ Index designates the object number in a selection set

Examples:

(setq ss1 (ssget))
(setq index 0)

(setq entname (ssname ss1 index)) returns the name of the first object contained in the predefined selection set ss1

Note
*If the index is 0, the **ssname** function returns the name of the first object in the selection set. Similarly, if the index is 1, it returns the name of the second object and so on.*

ssadd

The **ssadd** function adds an object (entity) to a selection set, or creates a new selection set. The syntax for the **ssadd** function is:

(ssadd entity-name selection-set)

When **ssadd** is used without arguments, a new selection set is created with no member. If only entity name arguments are used, a selection set is created with one member.

Examples:

(setq all(ssget))
Select objects: All
Selection set=30

A selection set has been created with the name *All*, which contains specified objects, say two circles.

Now we want to add a new circle to the existing selection set.

(setq b(entsel))
Select objects: Use any object selection method to select the circle.
(<entity name: 1ffo5a0>(10.977 4.99 0))
(setq c(car b))
entity name: 1ffo5a0
(setq d(ssadd c all))
(Command "erase" all "")

A new circle is added to the existing selection set.

ssdel

The **ssdel** function enables the user to delete any object or entity from any selection set. It functions like **ssadd**. The syntax for the **ssdel** function is:

(ssdel entity-name selection-set)

Examples:

Consider a case in which you have created three circles and have converted them into a selection set. Now, consider that you want to delete one of the circles from this selection set.

```
(setq all(ssget))
Select objects: All
(setq b(entsel))
Select objects: Use any object selection method to select the circle to be deleted.
(<entity name : 1ffo5a0>(2.455 5.99 0))
(setq c(car b))
Entity name : 1ffo5a0
(setq d(ssdel c all))
selection set=36
(Command "erase" all "")
```

Only two circles will be erased because one circle was removed from the selection set.

entget

The **entget** function retrieves the object list from the object name. The name of the object can be obtained by using the function **ssname**. The format of the **entget** function is given next.

(entget <u>object-name</u>)

Where **object-name**---- Name of the object obtained by ssname

Examples:

(setq ss1 (ssget))
(setq index 0)
(setq entname (ssname ss1 index))
(setq entlist (entget entname)) returns the list of the first object from the variable, entname, and assigns the list to the variable, **entlist**

assoc

The **assoc** function searches for a specified code in the object list and returns the element that contains that code. The format of the **assoc** function is given next.

(assoc code <u>object-list</u>)

Where **code** -------------- AutoCAD's object code
object-list ------- List of an object

Examples:

(setq ss1 (ssget))
(setq index 0)
(setq entname (ssname ss1 index))
(setq entlist (entget entname))

(setq entasso (assoc 0 entlist)) returns the element associated with AutoCAD's object code 0, from the list defined by the variable **entlist**

cons

The **cons** function constructs a new list from the given elements or lists. The format of the **cons** function is given next.

(cons <u>first-element</u> <u>second-element</u>)
Where **first-element**-------------- First element or list
second-element--------------------- Second element or list

Examples:

(cons 'x 'y)	returns (X . Y)
(cons '(x y) 'z)	returns ((X Y) . Z)
(cons '(x y z) '(0.5 5.0))	returns ((X Y Z) 0.5 5.0)

subst

The **subst** function substitutes the new item in place of old items. The old items can be a single item or multiple items, provided they are in the same list. The format of the **subst** function is given next.

(subst <u>new-item</u> <u>old-item</u> <u>list</u>)
Where **new-item** -------- New item that will replace old items
old-item --------- Old items that are to be replaced
list ---------------- Object list name or list

Examples:

(setq entlist '(x y x))
(setq newlist (subst 'z 'x entlist) returns (z y z); the subst function replaces x in the object list (entlist) by z

entmod

The **entmod** function updates the drawing by writing the modified list back to the drawing database. The format of the **entmod** function is given next.

(entmod <u>object-list</u>)
Where **object-list** ------- Name of the modified object list

Example 3

Write a Visual LISP program that will enable you to change the height of a text object. The program should prompt you to enter the new height of the text.

Step 1: Understanding the program algorithm

<u>Input</u>
New text height
Text object

<u>Output</u>
Text with new text height

<u>Process</u>

1. Select the text object; obtain the name of the object using the function **ssname**.
2. Extract the list of the object using the function **entget**.
3. Separate the element associated with AutoCAD object code 0 from the list using the function **assoc**.
4. Construct a new element where the height of text is changed to a new height using the function **cons**.
5. Substitute the new element back into the original list using the **subst** function.
6. Update the drawing database using the **entmod** function.

Step 2: Writing the Visual LISP program

Choose **Manage > Applications > Visual LISP Editor** from the **Ribbon** or enter **VLIDE** at the Command prompt to display the **Visual LISP** window. Choose **New File** from the **File** menu to display the **Visual LISP** Text Editor. Now, write the program in the **Visual LISP** Text Editor. The following file is a listing of the Visual LISP program for Example 3.

```
;This program changes the height of the
;selected text, only one text at a time.
;
(defun chgtext1 ()
(setvar "cmdecho" 0)
(setq newht (getreal "\n Enter new text height: "))
(setq ss1 (ssget))
(setq name (ssname ss1 0))
(setq ent (entget name))
(setq oldlist (assoc 40 ent))
(setq conlist (cons (car oldlist) newht))
(setq newlist (subst conlist oldlist ent))
(entmod newlist)
(setvar "cmdecho" 1)
(princ)
)
```

Step 3: Loading the Visual LISP program

Save the file as *chgtext1.lsp* using the **Save** or **Save As** option in the **File** menu. To load the program, select **Load Text in Editor** in the **Tools** menu. You can also load the program by choosing **Load active edit window** button in the **Tools** toolbar. Visual LISP displays a message in the **Console** window indicating that the program has been loaded. It also gives an error message, if it detects any error in the program. To run the program, enter the function name (**chgtext1**) at the console prompt $ sign. The function name must be enclosed in parentheses. After writing the function name, press ENTER. You will automatically get into the drawing editor and you will be prompted to enter the text height and select an object. Select the text object and press ENTER. You will notice that the text height has changed.

RETRIEVING AND EDITING THE DATABASE

To change the objects in a drawing, you need to understand the structure of the drawing database and how it can be manipulated. Once you understand this concept, it is easy to edit the drawing database and the drawing. The following step-by-step explanation describes the process involved in changing the height of a selected text object in a drawing. Assume that the text that needs to be edited is "CHANGE TEXT" and that this text is already drawn on the screen. The height of the text is 0.3 units. Before going through the following steps, load the Visual LISP program from Example 3, and run it so that the variables are assigned a value.

Step 1

Select the text using the function **ssget** or **ssget "X"** and assign it to variable ss1. AutoCAD creates a selection set that could have one or more objects. In line 7 **(setq ss1 (ssget))** of the program for Example 3, the selection set is assigned to variable ss1. Use the following command to check the variable ss1.

Command: **!ss1**
<Selection set: 2>

Step 2

There could be several objects in a selection set and these objects need to be separated, one at a time, before any change is made to an object. This is made possible using the function **ssname** which extracts the name of an object. The index number used in the function **ssname** determines the object whose name is being extracted. For example, if the index is 0 the **ssname** function will extract the name of the first object, if the index is 1 the **ssname** function will extract the name of second object, and so on. In line 8 **(setq name (ssname ss1 0))** of the program, the **ssname** extracts the name of the first object and assigns it to the variable name. Use the following command to check the variable name.

Command: **!name**
<Object name: 60000018>

Step 3

Extract the object list using the function **entget**. In line 9 **(setq ent (entget name))** of the program, the value of the list has been assigned to the variable **ent**. Use the following command to check the value of the variable ent.

Command: **!ent**
((-1.<Object name: 600000018> (0 . "TEXT") (8 . "0") (10 4.91227 5.36301 0.0) **(40 . 0.3)** (1 . "CHANGE TEXT") (50 .0.0) (41 . 1.0) (51 .0.0) (7 . "standard") (71 .0)) (72 . 1) (11 6.51227 5.36302 0.0) (210 0.0 0.0 1.0))

This list contains all the information about the selected text object (CHANGE TEXT), but you are only interested in changing the height of the text. Therefore, you need to identify the element that contains the information about the text height (40 . 0.3) and separate that from the list.

Step 4
Use the function **assoc** to separate the element that is associated with code 40 (text height). The statement in line 10 **(setq oldlist (assoc 40 ent))** of the program uses the assoc function to separate the value and assign it to the variable **oldlist**. Use the following command to check the value of this variable.

Command: **!oldlist**
(40 . 0.3)

Step 5
The **(40 . 0.3)** element consists of the code for text (40), and the text height (0.3). To change the height of the old text, the text height value (0.3) needs to be replaced by the new value. This is accomplished by constructing a new list as described in line 11 **(setq conlist (cons (car oldlist) newht))** of the program. This line also assigns the new element to variable **conlist**. For example, if the value assigned to variable newht is 0.5, the new element will be (40 . 0.5). Use the following command to check the value of conlist.

Command: **!conlist**
(40 . 0.5)

Step 6
After constructing the new element, use the **subst** function to substitute the new element back into the original list, **ent**. This is accomplished by line 12 **(setq newlist (subst conlist oldlist ent))** of the program. Use the following command to check the value of the variable **newlist**.

Command: **!newlist**
((-1.<Object name: 600000018> (0 . "TEXT") (8 . "0") (10 4.91227 5.36301 0.0) **(40 . 0.5)** (1 . "CHANGE TEXT") (50 .0.0) (41 . 1.0) (51 .0.0) (7 . "standard") (71 .0)) (72 . 1) (11 6.51227 5.36302 0.0) (210 0.0 0.0 1.0))

Step 7
The last step is to update the drawing database. Do this by using the function **entmod**, as shown in line 13 **(entmod newlist)** of the program.

SOME MORE FUNCTIONS TO RETRIEVE ENTITY DATA

entnext

The **entnext** function allows the programmer to extract main entity and sub-entity names from the database. The **entnext** function returns the name of the next non-deleted entity or first drawn entity. (**entnext**(**entnext**)) returns the next to the next non-deleted entity or next drawn entity. The format of the **entnext** function is given next.

(entnext <entity name>)

entlast

This function returns the name of the last non-deleted entity from the database. The format of the **entlast** function is given next.

(entlast)

entsel

The **entsel** function permits the user to select one entity. The format of the **entsel** function is given next.

(entsel [prompt])

Prompt is optional.

Example 4

Write a Visual LISP program that will enable you to change the height of all text objects in a drawing. The program should prompt you to enter the new text height.

Step 1: Making the flowchart

The following flowchart, Figure 10-1 gives the procedure to write the Visual LISP program (Figure 10-1).

Step 2: Writing the Visual LISP program

Choose **Manage > Applications > Visual LISP Editor** from the **Ribbon** or enter **VLIDE** at the Command prompt to display the **Visual LISP** window. Choose **New File** from the **File** menu to display the **Visual LISP** Text Editor. Now write the program as shown below in the **Visual LISP** Text Editor. **The line numbers are not a part of the program; they are shown here for reference only.**

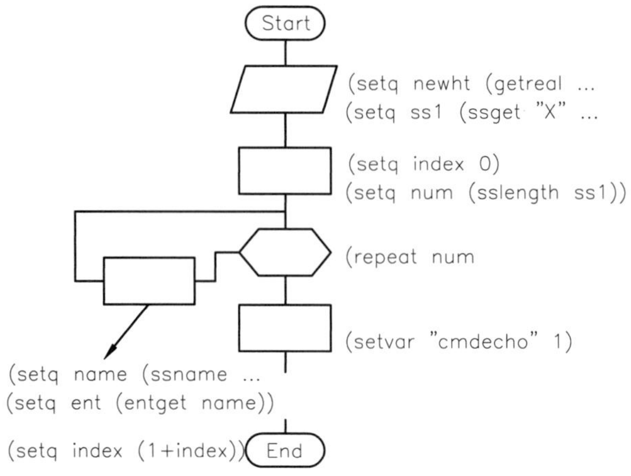

***Figure 10-1** Flowchart for Example 4*

```
1 ;This program changes the height of
2 ;all text objects in a drawing.
3 ;
4 (defun chgtext2 ()
5   (setvar "cmdecho" 0)
6   (setq newht (getreal "\n Enter new text height: "))
7   (setq ss1 (ssget "x" (list (cons 0 "text"))))
8   (setq index 0)
9   (setq num (sslength ss1))
10  (repeat num
11    (setq name (ssname ss1 index))
12    (setq ent (entget name))
13    (setq oldlist (assoc 40 ent))
14    (setq conlist (cons (car oldlist) newht))
15    (setq newlist (subst conlist oldlist ent))
16    (entmod newlist)
17    (setq index (1+ index))
18  )
19  (setvar "cmdecho" 1)
20  (princ)
21 )
```

Explanation

Line 7

(setq ss1 (ssget "x" (list (cons 0 "text"))))

The **ssget "X"** function filters the text objects from the drawing database. The **setq** function assigns that selected set of text objects to variable **ss1**.

Line 8
(setq index 0)
The **setq** function sets the value of the **INDEX** variable to 0. This variable is used later to select different objects.

Line 9
(setq num (sslength ss1))
The function **sslength** determines the number of objects in the selection set **ss1**, and the **setq** function assigns that number to the **num** variable.

Line 10
(repeat num
The **repeat** function will repeat the processes defined within the repeat function **num** number of times.

Step 3: Loading the Visual LISP program
Save the file as *chgtext2.lsp* using the **Save** or **Save As** option in the **File** menu. To load the program, choose **Load Text in Editor** in the **Tools** menu. You can also load the program by choosing the **Load active edit window** button in the **Tools** toolbar. To run the program, enter the function name **(chgtext2)** at the console prompt $ sign. The function name must be enclosed in parentheses. After entering the function name, press ENTER. You will automatically get into the drawing editor and you will be prompted to enter the text height. You will notice that the text height of all text objects has changed.

Example 5

Write a Visual LISP program that will enable you to change the text height of the selected text objects in the drawing.

Step 1: Making the flowchart
The flowchart can be designed, as shown in the Figure 10-2.

Step 2: Writing the Visual LISP program
Choose **Manage > Applications > Visual LISP Editor** from the **Ribbon** or enter **VLIDE** at the Command prompt to display the **Visual LISP** window. Choose **New File** from the **File** menu to display the **Visual LISP** Text Editor. Now write the program in the **Visual LISP** Text Editor, as shown below.

```
(defun chgtext3 ()
(setvar “cmdecho” 0)
(setq newht (getreal “\n Enter new text height: “))
(setq ss1 (ssget))
(setq index 0)
(setq num (sslength ss1))
```

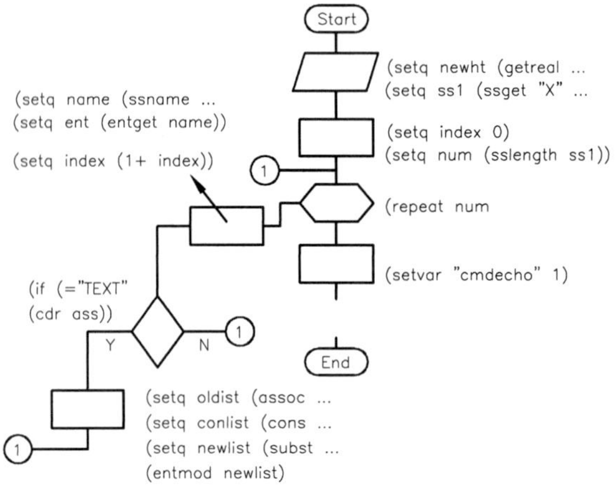

Figure 10-2 *Flowchart for Example 5*

```
(repeat num
  (setq name (ssname ss1 index))
  (setq ent (entget name))
  (setq ass (assoc 0 ent))
  (setq index (1+ index))
  (If (= "TEXT" (cdr ass))
      (progn
      (setq oldlist (assoc 40 ent))
      (setq conlist (cons (car oldlist) newht))
      (setq newlist (subst conlist oldlist ent))
      (entmod newlist)
      )
    )
  )
(setvar "cmdecho" 1)
(princ)
)
```

Step 3: Loading the Visual LISP program

Save the file as *chgtext3.lsp* using the **Save** or **Save As** option in the **File** menu. To load the program, choose **Load Text in Editor** in the **Tools** menu. You can also load the program by choosing the **Load active edit window** button in the **Tools** toolbar. To run the program, enter the function name (**chgtext3**) at the console prompt $ sign. After entering the function name, press ENTER. You will automatically get into the drawing editor where you will be prompted to enter the text height and then select objects. Give the text height and then select the object. The height of the selected text will change.

Example 6

Write a Visual LISP program to change the center point of a circle by manipulating database of the circle. The prompt should ask the user to enter a new center point of the circle.

Step 2: Writing the Visual LISP program

Choose **Manage > Applications > Visual LISP Editor** from the **Ribbon** or enter **VLIDE** at the Command prompt to display the **Visual LISP** window. Choose **New File** from the **File** menu to display the **Visual LISP** Text Editor. Now write the program as shown below in the **Visual LISP** Text Editor.

```
(Defun radcir ()
        (setq p1(entsel "select the circle")
        p2(car p1)
        p3(entget p2)
        p4(assoc 10 p3)
        p5(getpoint "\n enter the new center point of the circle")
        p6(cons 10 p5)
        p3(subst p6 p4 p3)
        );setq
(entmod p3)
);defun
```

Step 4: Loading the Visual LISP program

Save the file as *radcir.lsp* using the **Save** or **Save As** option in the **File** menu. To load the program, choose **Load Text in Editor** in the **Tools** menu. You can also load the program by choosing the **Load active edit window** button in the **Tools** toolbar. To run the program, enter the function name **(radcir)** at the **Console** prompt $ sign. After writing the function name, press ENTER. You will automatically get into the drawing editor where you will be prompted to select the circle and then specify the new center point. Select the circle and specify the new center point and its center point will change.

Self- Evaluation Test

Answer the following questions and then compare them to those given at the end of this chapter:

1. In addition to writing programs to create new commands, you can use the Visual LISP programming language to edit the drawing database. (T/F)

2. The __________ function enables you to select any number of objects in a drawing.

3. The __________ function enables you to select specified types of objects in the entire drawing database, even if the layers are frozen or turned off.

4. The __________ function determines the number of objects in a selection set and returns an integer corresponding to the number of objects found.

5. The __________ function returns the name of the object, from a predefined selection set, as referenced by the index that designates the object number.

Review Questions

Answer the following questions:

1. The __________ function gives the name of the first drawn entity.

2. The __________ function gives the name of the last drawn entity.

3. The __________ function retrieves the object list from the object name.

4. The __________ function searches for a specified code in the object list and returns the element that contains that code.

5. The __________ function constructs a new list from the given elements or lists.

6. The __________ function substitutes the new item in place of old items.

Exercises

Exercise 1 *General*

Write a Visual LISP program that will enable you to change the layer of the selected objects in a drawing. The program should prompt you to enter the new layer name.

Exercise 2 *General*

Write a Visual LISP program that will change the text style name of the selected text objects in a drawing. The program should prompt you to enter the new text style.

Exercise 3 *General*

Write a Visual LISP program that will change the layer of the selected objects in a drawing to a new layer. You should be able to enter the new layer by selecting an object in that layer.

Answers to Self-Evaluation Test

1. T, **2. ssget**, **3. ssget "X"**, **4. sslength**, **5. ssname**

Chapter 11

Creating Programmable Dialog Boxes Using the Dialog Control Language

Learning Objectives

After completing this chapter, you will be able to:

- *Write programs using dialog control language.*
- *Use predefined attributes.*
- *Load a dialog control language (DCL) file.*
- *Display new dialog boxes.*
- *Use standard button subassemblies.*
- *Use AutoLISP functions to control dialog boxes.*
- *Manage dialog boxes with AutoLISP.*
- *Use tiles, buttons, and attributes in DCL programs.*

DIALOG CONTROL LANGUAGE

Dialog control language **(DCL)** files are ASCII files that contain the descriptions of dialog boxes. A DCL file can contain the description of single or multiple dialog boxes. There is no limit to the number of dialog box descriptions that can be defined in a DCL file. The suffix of a DCL file is *.dcl* (such as *ddosnap.dcl*).

This chapter assumes that you are familiar with AutoCAD commands, AutoCAD system variables, and AutoLISP programming. You need not be a programming expert to learn to write programs for dialog boxes in DCL or to control the dialog boxes through AutoLISP programming. However, knowledge of any programming language should help you to understand and learn DCL. This chapter introduces you to the basic concepts of developing a dialog box, frequently used attributes, and tiles. A thorough discussion of DCL functions and a step-by-step explanation of examples should make it easy for you to learn DCL. For those functions not discussed in this chapter, you can refer to the *AutoCAD 2010 Customization Guide*. To write programs in DCL, you need no special software or hardware. If AutoCAD is installed on your computer, you can write DCL files. To write DCL files, you can use any text editor.

DIALOG BOX

A dialog control language (DCL) file contains the description of how the dialog boxes will appear on the screen. These boxes can contain buttons, text, lists, edit boxes, rows, columns, sliders, and images. A sample dialog box is shown in Figure 11-1.

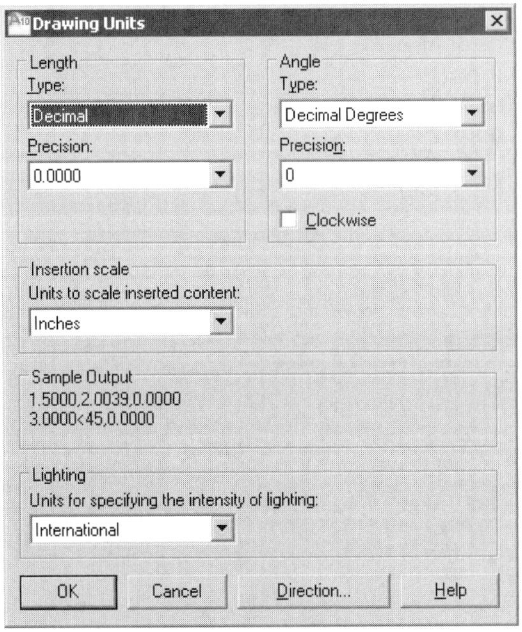

Figure 11-1 *The* ***Drawing Units*** *dialog box*

You do not need to specify the size and layout of a dialog box or its component parts. Sizing is done automatically when the dialog box is loaded on the screen. By itself, the dialog box cannot perform the functions it is designed for. The functions of a dialog box are controlled by a program written in the AutoLISP programming language or with the AutoCAD development system (ADS), or ARX. For example, if you load a dialog box and select the Cancel button, it will not perform the cancel operation. The instructions associated with a button or any part of the dialog box are handled through the functions provided in AutoLISP, ADS, or ARX. Therefore, AutoLISP or ADS are needed to control the dialog boxes, and you should have, in addition to an understanding of DCL, a good knowledge of AutoLISP or ADS to develop new dialog boxes or edit existing ones.

Dialog boxes are not dependent on the platform; therefore, they can run on any system that supports AutoCAD. However, depending on the graphical user interface (GUI) of the platform, the appearance of the dialog boxes might change from one system to another. The functions defined in the dialog box will still work without making any changes in the dialog box or the application program (AutoLISP or ADS) that uses these dialog boxes.

DIALOG BOX COMPONENTS

The two major components of a dialog box are the tiles and the box itself. The tiles can be arranged in rows and columns in any desired configuration. They can also be enclosed in boxes or borders to form subassemblies, giving them a tree structure, Figure 11-2(a).

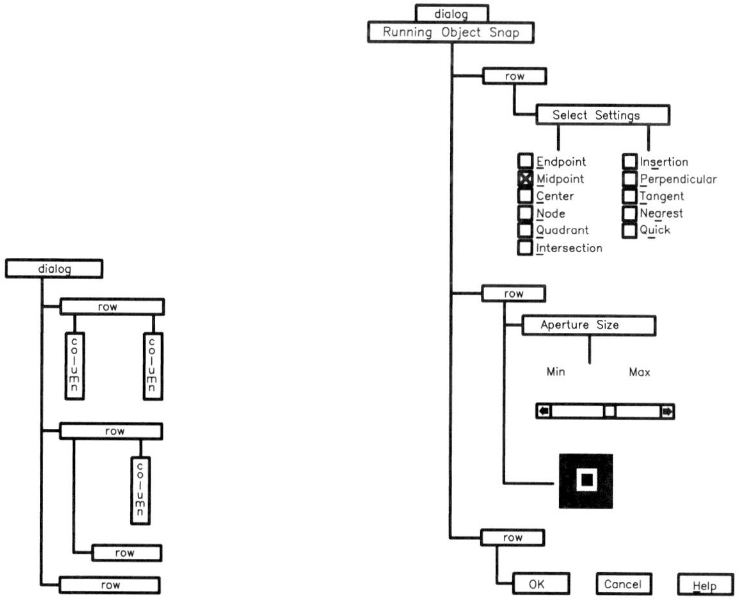

Figure 11-2(a) *Tree structure of a dialog box*

The basic tiles, such as buttons, lists, edit boxes, and images, are predefined by the programmable dialog box (PDB) facility of AutoCAD. These buttons are described in the file *base.dcl*. The layout and function of a tile is determined by the attribute assigned to it. For example, the height attribute controls the height of the tile. Similarly, the label attribute specifies the text that is associated with the tile. Some of the components of a dialog box are shown in Figure 11-2(b). Following this figure is a list of predefined tiles and their format in DCL.

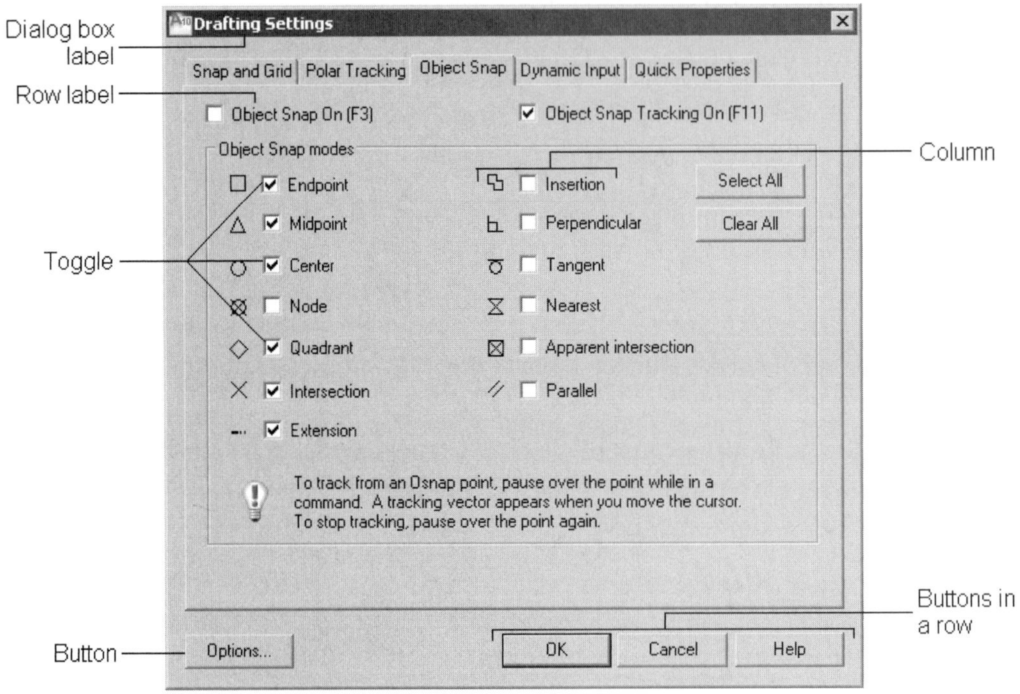

Figure 11-2(b) *Components of a dialog box*

Predefined Tile	**DCL Format**
Button	button
Edit box	edit_box
Image button	image_button
List box	list_box
Pop-up list	popup_list
Radio button	radio_button
Slider	slider
Toggle	toggle
Column	column
Boxed column	boxed_column
Row	row
Boxed row	boxed_row
Radio column	radio_column
Boxed radio column	boxed_radio_column
Radio row	radio_row

Boxed radio row	boxed_radio_row
Image	image
Text	text
Spacer	spacer

BUTTON AND TEXT TILES

Button Tile

Format in DCL: **button**

The button tile consists of a rectangular box that resembles a push button. The button's label appears inside the button. For example, in the **OK** button of a dialog box, the label OK appears inside the button. If you choose the **OK** button in a dialog box, it performs the functions defined in the dialog box and clears the dialog box from the screen. Similarly, if you select the **Cancel** button, it cancels the dialog box without taking any action.

Note

A dialog box should contain at least one ***OK*** *button or a button that is equivalent to it. This allows you to exit the dialog box when you are done using it.*

Text Tile

Format in DCL: **text**

The text tile is used to display information or a title in a dialog box. It has limited application in the dialog boxes because most of the tiles have their own label attributes for tiling. However, if you need to display any text string in the dialog box, you can do so with the text tile. The text tile is used extensively in AutoCAD alert boxes to display warnings or error messages.

Note

An alert box must contain an ***OK*** *button or a* ***Cancel*** *button to end the dialog box.*

TILE ATTRIBUTES

The appearance, size, and function performed by a dialog box tile depend on the attributes that have been assigned to the tile. For example, if a button has been assigned the **fixed_width** attribute, the width of the box surrounding the button will not stretch through the entire length of the dialog box. Similarly, the **height** attribute determines the height of the tile, and the key attribute assigns a name to the tile that is then used by the application program. The tile attribute consists of two parts: the name of the attribute and the value assigned to the attribute. For example, in the expression **fixed_width = true**, **fixed_width** is the name of the attribute and **true** is the value assigned to the attribute. Attribute names are like variable names in programming, and the values assigned to these variables must be of a specific type.

Types of Attribute Values

Integer. Unlike integer values in programming (1, 15, 22), the numeric values assigned to attributes can be in both integers and real numbers.

Examples:
width = 15
height = 10

Real Number. The real values assigned to attributes should always be fractional real numbers with a leading digit.

Examples:
aspect_ratio = 0.75

Quoted String. The string values assigned to an attribute consist of a text string that is enclosed in double quotes.

Examples:
key = "accept"
label = "OK"

Reserved Word

Dialog control language uses some reserved words as identifiers. These identifiers are alphanumeric characters that start with a letter.

Examples:
is_default = true
fixed_width = true

Note

Reserved names are case-sensitive. For example, "is_default = true" is not the same as "is_default = True". The reserved word is "true," not "True" or "TRUE".

Like reserved words, attribute names are also case-sensitive. For example, the attribute is "key", not "Key" or "KEY".

PREDEFINED ATTRIBUTES

To facilitate writing programs in dialog control language, AutoCAD has provided some predefined attributes that are defined in the programmable dialog box (PDB) package that comes with the AutoCAD software. Some of these attributes can be used with any tile and some only with a particular type of tile. The values assigned to various attributes in a DCL file are used by the application program to handle the tiles or the dialog box. Therefore you must use the correct attributes and assign an appropriate value to these attributes. The following is a list of some of the frequently used predefined attributes defined in the PDB facility of AutoCAD.

action
alignment
allow_accept
key
label
layout

aspect_ratio
color
edit_limit
edit_width
fixed_height
fixed_width
height
is_cancel
is_default
list
max_value
min_value
mnemonic
multiple_select
small_increment
tabs
value
width

key, label, AND is_default ATTRIBUTES

key Attribute

Format in DCL
key

Examples:
key = "accept"
key = "XLimit"

The **key** attribute assigns a name to a tile. The name must be enclosed in double quotes. This name can then be used by the application program to handle the tile. A dialog box can have any number of key values, but within a particular dialog box the values used for the key attributes must be unique. In the example key = "accept", a string value "accept" is assigned to the **key** attribute. If there is another key attribute in the dialog box, you must assign it a different value.

label Attribute

Format in DCL
label

Examples:
label = "OK"
label = "Hello DCL Users"

Sometimes it is necessary to display a label in a dialog box. The **label** attribute can be used in a boxed column, boxed radio column, boxed radio row, boxed row, button tile, dialog box, or edit box. Some of the frequently used label attributes are described next.

Use of the label Attribute in a Dialog Box. When the **label** attribute is used in a dialog box, it is displayed in the top border or the title bar of the dialog box. The label must be a string enclosed in double quotes. Use of the label in a dialog box is optional; in case of default, no title is displayed in the dialog box.

Example:

```
welcome : dialog {
    label = "Sample Dialog Box";
```

In this example, the label "Sample Dialog Box" is displayed at the top of the dialog box.

Use of the label Attribute in a Boxed Column. When the label attribute is used in a boxed column, the label is displayed within a box in the upper left corner of the column; the box consists of a single line at the top of the column. The label must be a quoted string, and the default is a set of a quoted string (" "). If the default is used, only the box is displayed, without any label.

Example:

```
: text {
   label = "Welcome to the world of DCL";
}
```

In this example, the label "Welcome to the world of DCL" is displayed within a box in the upper left corner of the column.

Use of the label Attribute in a Button. When the label attribute is used in a button, the label is displayed inside the button. The label must be a quoted string and has no default.

Example:

```
: button {
  key = "accept";
  label = "OK";
}
```

In this example, the label "OK" is displayed inside and in the center of the button tile.

is_default Attribute

Format in DCL
is_default

Example:
is_default = true

The **is_default** attribute is used for the button of a dialog box. In the example, the value assigned to the **is_default** attribute is true. Therefore, this button will be automatically selected when you press ENTER. For example, if you load a dialog box on the screen, one way to exit and accept the values of the dialog box is to choose the **OK** button. You can accomplish the same thing by pressing ENTER. This action is made possible by assigning the value, **true**, to the **is_default** attribute. In a dialog box, the default button can be recognized by the thick border drawn around the text string.

Note

In a dialog box, only one button can be assigned the true value for the ***is_default*** *attribute.*

fixed_width AND alignment ATTRIBUTES

fixed_width Attribute

Format in DCL
fixed_width

Example:
fixed_width = true

This attribute controls the width of the tile. If the value of this attribute is set to true, the width of the tile does not extend across the complete width of the dialog box. The width of the tile is automatically adjusted to the length of the text string that is displayed in the tile.

alignment Attribute

Format in DCL	**Example:**
alignment	alignment = centered
	alignment = right

The value assigned to the **alignment** attribute determines the horizontal or vertical position of the tile in a row or column. For a row, the values that can be assigned to this attribute are left, right, and centered. The default value of the alignment attribute is left; this forces the tile to be displayed left-justified. For a column, the possible values of the **alignment** attribute are top, bottom, and centered. The default value is centered.

Example 1

Using dialog control language (DCL), write a program for the dialog box shown in Figure 11-3. The dialog box has two text labels and an **OK** button to end the dialog box.

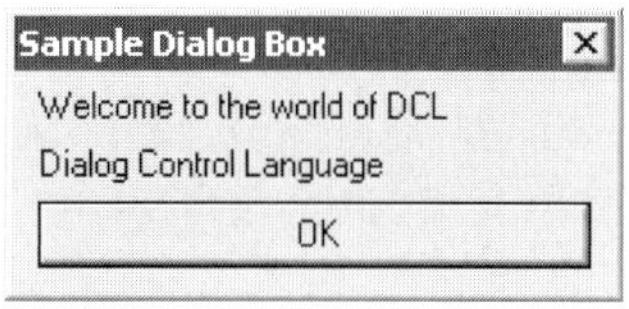

Figure 11-3 *Dialog box for Example 1*

The following file is a listing of the DCL file for the dialog box of Example 1. The name of this DCL file is *dclwel1.dcl*. **The line numbers on the right are not a part of the file; they are for reference only.**

```
1   welcome1 : dialog {
2           label = "Sample Dialog Box";
3           : text {
4           label = "Welcome to the world of DCL";
5           }
6           : text {
7             label = "Dialog Control Language";
8           }
9           : button {
10            key = "accept";
11            label = "OK";
12           is_default = true;
13          }
14  }
```

Explanation

Line 1

welcome1 : dialog {

In this line, **welcome1** is the name of the dialog box and the definition of the dialog box is contained within the braces. The open brace in this line starts the definition of this dialog box.

Line 2
label = "Sample Dialog Box";
In this line, **label** is the label attribute, and "**Sample Dialog Box**" is the string value assigned to the label attribute. This string will be displayed in the title bar of the dialog box. The label description must be enclosed in quotes. If this line is missing, no title is displayed in the title bar of the dialog box.

Lines 3-5
: text {
label = "Welcome to the world of DCL";
}
These three lines define a text tile with the label description. In the first line, **text** refers to the text tile. The line that follows it, **label = "Welcome to the world of DCL";** defines a label for this tile that will be displayed left-justified in the dialog box. The closing brace completes the definition of this tile.

Lines 9-13
: button {
key = "accept";
label = "OK";
is_default = true;
}
These five lines define the attributes of the button tile. In the first line, **button** refers to the button tile. In the second line, **key = "accept"**; specifies an ASCII name, **"accept"**, that will be used by the application program to refer to this tile. The next line, **is_default = true;**, specifies that this button is the default button. It is automatically selected if you press ENTER.

The closing brace in line 14 completes the definition of the dialog box.

LOADING A DCL FILE

As with an AutoLISP file, you can load a DCL file from the AutoCAD drawing editor. But note that the directory in which the DCL file is saved should be mentioned in the AutoCAD support file search path. If the directory is not mentioned in the support file path, the file name with the complete path in which it is saved needs to be provided to load the dialog box. A DCL file can contain the definition of one or several dialog boxes. There is no limit to the number of dialog boxes you can define in a DCL file. The format of the command for loading a dialog box is given next.

(load_dialog filename)
Where **load_dialog** ---- Load command for loading a dialog file
filename -------- Name of the DCL file, with or without the file extension (.dcl)

Example:
(load_dialog "dclwell.dcl") or (load_dialog "dclwell")

In this example, **dclwell** is the name of the DCL file and **.dcl** is the file extension for DCL files.

Note

*When you use the **load_dialog** and **new_dialog** functions to load and display a DCL program, the AutoCAD screen will freeze. To avoid this, it is recommended to use the AutoLISP program to load and display a DCL program. See "Using AutoLISP Function to Load a DCL File" at the end of this section). To preview the DCL dialog box, open the *.dcl in the **Visual LISP** editor. Next, choose **Tools > Interface Tools > Preview DCL in Editor**. Choose **OK** in the **Enter the dialog name** dialog box; the dialog box will be displayed in the AutoCAD environment.*

The file name can be with or without the DCL file extension (**welcome1** or **welcome1.dcl**). The load function returns an integer value that is used as a handle in the **new_dialog** function and the **unload_dialog** function. This integer will be referred to as **dcl_id** in subsequent sections. You do not need to use the name dcl_id for this integer; it could be any name (dclid, or just id).

DISPLAYING A NEW DIALOG BOX

The load_dialog function loads the DCL file, but it does not display it on the screen. The format of the command used to display a new dialog box is:

(new_dialog dlgname dcl_id)

Where **new_dialog** ----- Command to display a dialog box
dlgname -------- Name of the dialog box
dcl_id ------------ Integer returned in the load_dialog function

Example:
(new_dialog "welcome1" 1)

In this example, assume that dcl_id is 1, an integer returned by the load_dialog function. You can use the (load_dialog) and (new_dialog) commands to load the DCL file in Example 1. In the following command sequence, assume that the integer (dcl_id) returned by the (load_dialog) function is 3. Figure 11-4 shows the dialog box after entering the following two command lines.

```
Command: (load_dialog "dclwell.dcl")
3
Command: (new_dialog "welcome1" 3)
```

Note that loading a dialog box using this method freezes the AutoCAD environment. To avoid this, it is recommended that you preview the dialog box using the **Visual LISP** editor. To preview the DCL dialog box, open the program in the **Visual LISP** editor. Next, choose

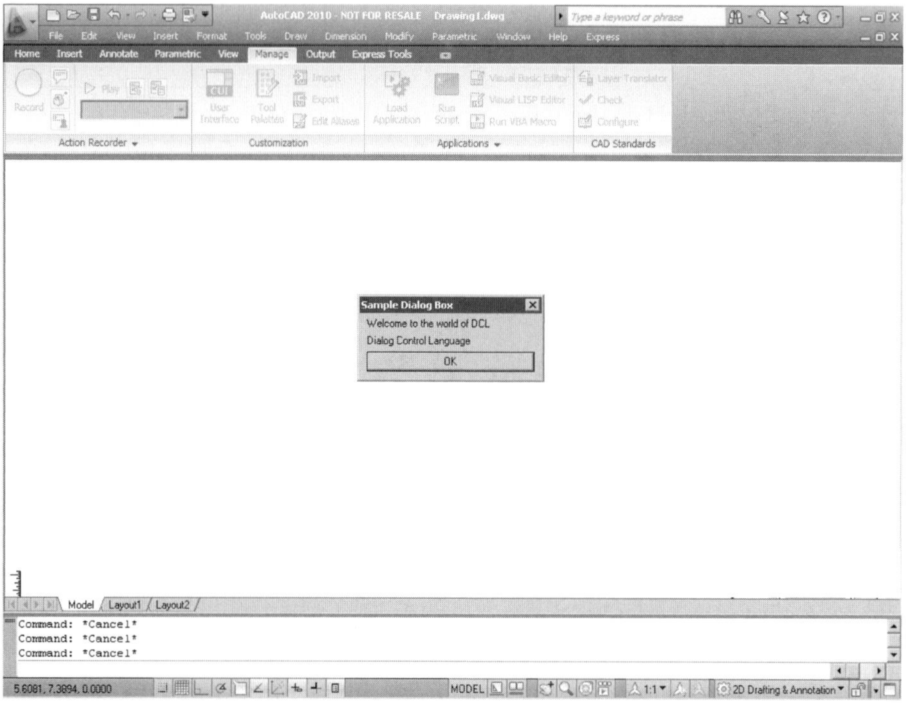

Figure 11-4 *Dialog box as displayed on the screen*

Tools > Interface Tools > Preview DCL in Editor; the dialog box will be displayed in the AutoCAD environment.

In this example, notice that the **OK** button stretches across the complete width of the dialog box. This button would look better if its width were limited to the width of the string "OK". This can be accomplished by using the **fixed_width = true;** attribute in the definition of this button. You can also use the **alignment = centered** attribute to display the **OK** button label center-justified. The following file is a listing of the DCL file where the OK label for the **OK** button is center-justified and the width of the box does not stretch across the width of the dialog box (Figure 11-5).

```
welcome2 : dialog {
      label = "Sample Dialog Box";
     : text {
        label = "Welcome to the world of DCL";
     }
     : text {
     label = "Hello - DCL";
     }
     : button {
     key = "accept";
```

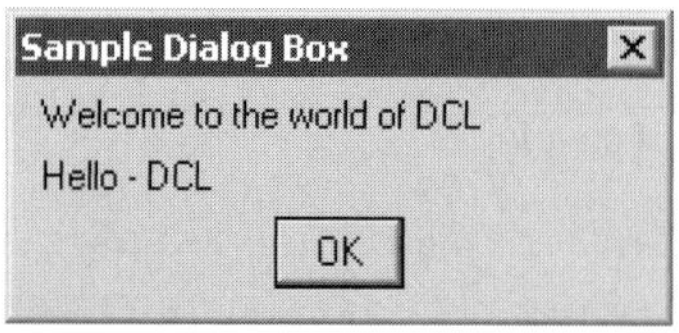

Figure 11-5 *Dialog box with fixed width for the* ***OK*** *button*

```
      label = "OK";
        is_default = true;
      fixed_width = true;                  // (Controls width)
      alignment = centered;                // (Controls justification)
      }
}
```

Using the AutoLISP Function to Load a DCL File

You can use the following AutoLISP program to load and display the dialog box without freezing the screen.

```
(defun c:load_dcl( / dcl_id )
  (setq dcl_id (load_dialog "dclwel2.dcl"))      (Loads the DCL file)
  (new_dialog "welcome2" dcl_id)                 (Initializes the dialog box)
  (start_dialog)                                 (Displays the dialog box)
  (princ)
  )
```

Where -------------------- **load_dcl** is the name of the AutoLISP function.
-------------------- **dclwel2** is the name of the DCL file that you want to load.
-------------------- **welcome2** is the name of the dialog as defined in the DCL file.

USE OF STANDARD BUTTON SUBASSEMBLIES

Some standard button subassemblies are predefined in the *base.dcl* file. You can use these standard buttons in your DCL file to maintain consistency among various dialog boxes. One such predefined button is **ok_cancel**, which displays the **OK** and **Cancel** buttons in the dialog box (Figure 11-6). The following file is the listing of the DCL file of Example 1, using the **ok_cancel** predefined standard button subassembly:

```
welcome3 : dialog {
    label = "Sample Dialog Box";
    : text {
      label = "Welcome to the world of DCL";
    }
    : text {
        label = "Dialog Control Language";
       alignment = left;
    }
    ok_cancel;
}
```

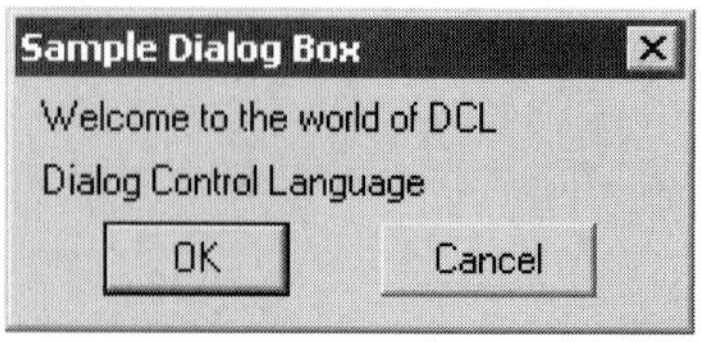

***Figure 11-6** Dialog box with the **OK** and **Cancel** buttons*

The following is a list of the standard button subassemblies that are predefined in the **base.dcl** file. These buttons are also referred to as dialog exit buttons because they are used to exit a dialog box.

OK Button
Format in DCL: **ok_only**

OK and Cancel
Format in DCL: **ok_cancel**

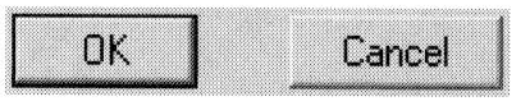

OK, Cancel, and Help Buttons
Format in DCL: **ok_cancel_help**

AutoLISP FUNCTIONS

load_dialog

The AutoLISP function **load_dialog** is used to **load a DCL file** that is specified in the load_dialog function. In the following examples, the name of the file that AutoCAD loads is "dclwel1". The extension (*.dcl*) of the file is optional. When the DCL file is loaded successfully, AutoCAD returns an integer that identifies the dialog box.

Format in DCL
(load_dialog filename)

Examples:
(load_dialog "dclwel1.dcl")
(load_dialog "dclwel1")

unload_dialog

The AutoLISP function **unload_dialog** is used to **unload a DCL file** that is specified in the unload_dialog function. The file is specified by the variable (dcl_id) that identifies a DCL file.

Format in DCL
(unload_dialog dcl_id)

Example:
(unload_dialog dcl_id)

new_dialog

The AutoLISP function **new_dialog** is used to initialize a dialog box and then display it on the screen. In the following file, **"welcome1"** is the name of the **dialog box**. (Note: welcome1 is not the name of the DCL file.) The variable **dcl_id** contains an integer value that is returned when the DCL file is loaded.

Format in DCL	**Example:**
(new _dialog "dialogname" dcl_id)	(new_dialog "welcome1" dcl_id)

start_dialog

The AutoLISP function **start_dialog** is used in an AutoLISP program to accept user input from the dialog box. For example, if you choose **OK** from the dialog box, the **start_dialog** function retrieves the value of that tile and uses it to take action and to end the dialog box.

Format in DCL
(start_dialog)

done_dialog

The AutoLISP function **done_dialog** is used to terminate the display of the dialog box from the screen. This function must be defined within the action expression, as shown in the following example.

Format in DCL	**Example:**
(done_dialog)	(action_tile "accept" "(done_dialog)")

action_tile

The AutoLISP function **action_tile** is used to associate an action expression with a tile in the dialog box. In the following example, the **action_tile** function associates the tile "accept" with the action expression (**done_dialog**) that terminates the dialog box. The "accept" is the name of the tile assigned to the **OK** button in the DCL file.

Format in DCL	**Example:**
(action_tile tile-name action-expression)	(action_tile"accept" "(done_dialog)")

MANAGING DIALOG BOXES WITH AutoLISP

When you load the DCL file of Example 1 and choose the **OK** button, it does not perform the desired function (exit from the dialog box). This is because a dialog box cannot, by itself, execute the AutoCAD commands or the functions assigned to a tile. An application program is required to handle a dialog box. These application programs can be written in AutoLISP or ADS. The functions defined in AutoLISP and ADS can be used to load a DCL file, display the dialog box on the screen, prompt user input, set values in tiles, perform action associated with user input, and execute AutoCAD commands. Example 2 describes the use of an AutoLISP program to handle a dialog box.

Example 2

Write an AutoLISP program that will handle the dialog box and perform the functions shown in the dialog box of Example 1.

The following file is a listing of the DCL file of Example 1. This DCL file defines the dialog box **welcome1**, which contains only one action tile: **"OK"**. If you choose the **OK** button, you must be able to exit the dialog box. As just mentioned, the dialog box will not perform by itself the function assigned to the **OK** button unless you write an application that will execute the functions defined in the dialog box.

```
welcome1   : dialog {
      label = "Sample Dialog Box";
      : text {
         label = "Welcome to the world of DCL";
      }
      : text {
      label = "Dialog Control Language";
      }
       : button {
           key = "accept";
           label = "OK";
           is_default = true;
      }
}
```

The following file is a listing of the AutoLISP program that loads the DCL file **(dclwel1)** of Example 1, displays the dialog box **welcome1** on, and defines the action for the **OK** button. **The line numbers on the right are not a part of the program; they are for reference only.**

```
1  (defun C:welcome1 ( / dcl_id)
2  (setq dcl_id (load_dialog "dclwel1.dcl"))
3  (new_dialog "welcome1" dcl_id)
4  (action_tile
5     "accept"
6     "(done_dialog)")
7  (start_dialog)
8  (unload_dialog dcl_id)
9  (princ)
10 )
```

Explanation

Line 1

(defun C:welcome1 (/ dcl_id)

In this line, **defun** is an AutoLISP function that defines the function welcome. With the C: in front of the function name, the **welcome1** function can be executed like an AutoCAD command. The **welcome** function has one local variable, **dcl_id**.

Line 2
(setq dcl_id (load_dialog "dclwell.dcl"))
In this line, **(load_dialog "dclwell.dcl")** loads the DCL file **dclwell.dcl** and returns a positive integer. The **setq** function assigns this integer value to the local variable **dcl_id**.

Line 3
(new_dialog "welcome1" dcl_id)
In this line, the AutoLISP function **new_dialog** loads the dialog box **welcome1** that is defined in the DCL file (line 1 of DCL file). The variable **dcl_id** is an integer that identifies the DCL file.

Note
The dialog name (welcome1) in the AutoLISP program must be the same as the dialog name in the DCL file (welcome1).

Lines 4-6
(action_tile
"accept"
"(done_dialog)")
In the DCL file of Example 1, the **OK** button has been assigned an ASCII name, **accept** (key = "accept"). The first two lines associate the key (the **OK** button) with the action expression. The action_tile initializes the association between the **OK** button and the action expression (done_dialog). If you select the **OK** button from the dialog box, the AutoLISP program reads that value and performs the function defined in the statement (done_dialog). The **done_dialog** function ends the dialog box.

Line 7
(start_dialog)
The **start_dialog** function enables the AutoLISP program to accept your input from the dialog box.

Line 8
(unload_dialog dcl_id)
This statement unloads the DCL file identified by the integer value of dcl_id.

Lines 9 and 10
(princ)
)
The **princ** function displays a blank on the screen. You use this to prevent display of the last expression in the command prompt area of the screen. If the **princ** function is not used, AutoCAD will display the value of the last expression. The closing parenthesis in the last line completes the definition of the welcome function.

Note that the program given above is a LISP program. Therefore, to load this program in AutoCAD, first you need to invoke the **Load/Unload Application** dialog box and then use the command mentioned in the LISP program to execute it.

ROW AND BOXED ROW TILES

Row Tile

Format in DCL: **row**

In a DCL file, several tiles can be grouped together to form a composite row or a composite column that is treated as a single tile. A row tile consists of several tiles grouped together in a horizontal row.

Boxed Row Tile

Format in DCL: **boxed_row**

In a boxed row, the tiles are grouped together in a row and a border is drawn around them, forming a box shape. If the boxed row has a label, it will be displayed left-justified at the top of the box, above the border line. If no label attribute is defined, only the box is displayed around the tile. On some systems, depending on the graphical user interface (GUI), the label may be displayed inside the border.

COLUMN, BOXED COLUMN, AND TOGGLE TILES

Column Tile

Format in DCL: **column**

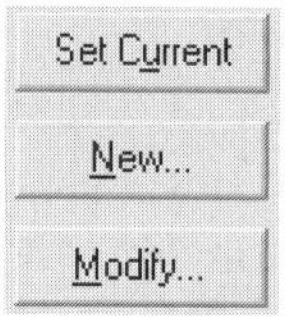

In a column tile, the tiles are grouped together in a vertical column to form a composite tile.

Boxed Column Tile

Format in DCL: **boxed_column**

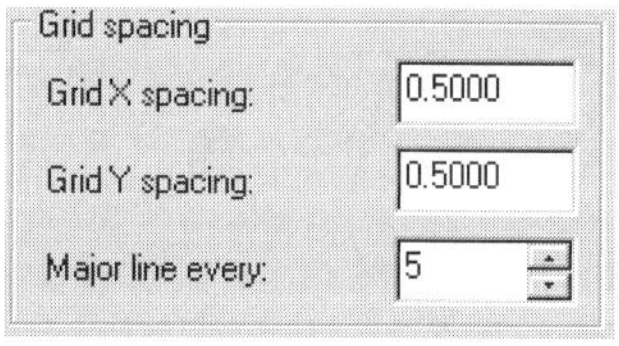

In a boxed column, the tiles are grouped together in a column and a border is drawn around the tiles. If the boxed column has a label, it will be displayed left-justified at the top of the box, above the border line. If there is no label attribute defined, only the box is displayed around the tile. On some systems, depending on the graphical user interface (GUI), the label may be displayed inside the border.

Toggle Tile

Format in DCL: **toggle**

A toggle tile in a dialog box displays a small box on the screen with an optional label on the right of the box. Although the label is optional, you should label the toggle box so that users know what function is assigned to the toggle box. A toggle box has two states: on and off. When the function is turned on, a check mark is displayed in the box. Similarly, when a function is turned off, no check mark is displayed. The on or off state of the toggle box represents a Boolean value. When the toggle box is on, the Boolean value is 1; when the toggle box is off, the Boolean value is 0.

MNEMONIC ATTRIBUTE

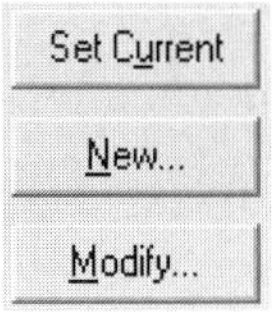

Format in DCL
mnemonic

Example:
mnemonic = "U"

A dialog box can have several tiles with labels. One of the ways you can select a tile is by using the arrow to highlight the tile and then pressing the **accept** key. This is possible if you have a pointing device like a digitizer or a mouse. If you do not have a pointing device, it may not be possible to select a tile. However, you can select a tile by using the mnemonic key assigned to the tile. For example, the mnemonic character for the **New** tile is **N**. If you press the **N** key at the keyboard, it will highlight the **New** tile. The mnemonic character is designated by underlining one of the characters in the label. The underlining is done automatically once you define the mnemonic attribute for the tile. You can select only one character **in the label** as a mnemonic character and you must use different mnemonic characters for different tiles. If a dialog box has two tiles with the same mnemonic character, only one tile will be selected when you press the mnemonic key. The mnemonic characters are not case-sensitive; therefore, they can be uppercase (**N**) or lowercase (**n**). However, the character in the label you select as a mnemonic character should be capitalized for easy identification.

Note
Some graphical user interfaces (GUIs) do not support mnemonic attributes. On such systems, you cannot select a tile by pressing the mnemonic key. However, you can still use pointing devices to select the tile.

Example 3 *General*

Write a DCL program for the object snap dialog box shown in Figure 11-7(a). The object snap tiles are arranged in two columns in a boxed row.

Before writing a program, especially a DCL program for a dialog box, you should determine the organization of the dialog box. It is given in Example 3. But when you develop a dialog box yourself, you must be careful when organizing the tiles in the dialog box. The structure of the DCL program depends on the desired output. In Example 3, the desired output is shown in Figure 11-7(a). This dialog box has two rows and two columns. The first row has two columns and the second row has no columns, as shown in Figure 11-7(b).

The following file is a listing of the DCL file for Example 3. **The line numbers are not a part of the file; they are shown for reference only.**

```
osnapsh : dialog {
      label = "Running Object Snaps";
   : boxed_row {
      label = "Select Object Snaps";
```

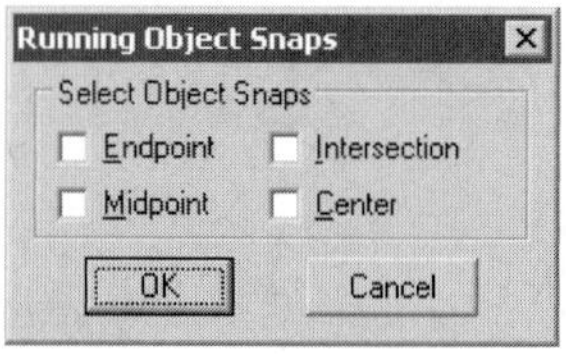

Figure 11-7(a) *Dialog box for object snaps*

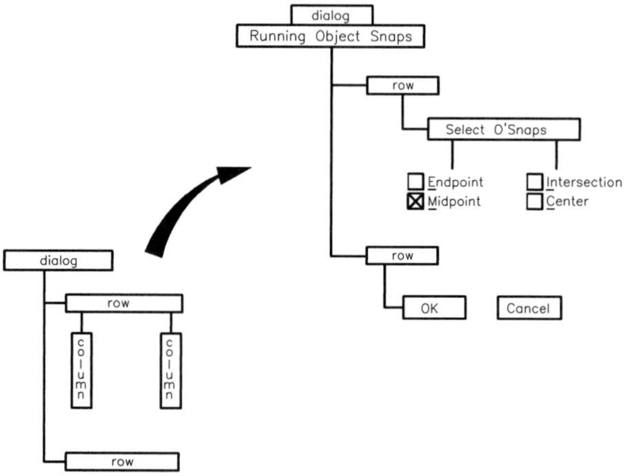

Figure 11-7(b) *Dialog box for object snaps*

```
: column {
      : toggle {
       label = "Endpoint";
       key = "Endpoint";
       mnemonic = "E";
       fixed_width = true;
             }
      : toggle {
       label = "Midpoint";
       key = "Midpoint";
       mnemonic = "M";
       fixed_width = true;
             }
      }
: column {
      : toggle {
       label = "Intersection";
       key = "Intersection";
       mnemonic = "I";
       fixed_width = true;
             }
```

```
26          : toggle {
27           label = "Center";
28           key = "Center";
29           mnemonic = "C";
30           fixed_width = true;
31             }
32          }
33      }
34        ok_cancel;
35    }
```

Explanation

Lines 3 and 4

: boxed_row {
label = "Select Object Snaps";

The **boxed_row** is a predefined cluster tile that draws a border around the tiles. The second line, **label = "Select Object Snaps";**, will display the label **(Select Object Snaps)** left-justified at the top of the box. The **label** is a predefined DCL attribute.

Lines 5-7

: column {
: toggle {
label = "Endpoint";

The **column** is a predefined tile that will arrange the tiles within it into a vertical column. The **toggle** is another predefined tile that displays a small box in the dialog box with an optional text on the right side of the box. The **label** attribute will display the label **(Endpoint)** to the right of the toggle box.

Lines 8-11

key = "Endpoint";
mnemonic = "E";
fixed_width = true;
}

The key attribute assigns a name **(Endpoint)** to the tile. This name is then used by the application program to handle the tile. The second line, **mnemonic = "E";**, defines the keyboard mnemonic for **Endpoint**. This causes the character **E** of **Endpoint** to be displayed underlined in the dialog box. The attribute **fixed_width** controls the width of the tile. If the value of this attribute is true, the tile does not stretch across the width of the dialog box. The closing brace (}) on the next line completes the definition of the toggle tile.

Lines 31-35

```
            }
        }
    }
  ok_cancel;
}
```

The closing brace on line 31 completes the definition of the toggle tile, and the closing brace on line 32 completes the definition of the column of line 19. The closing brace on line 33 completes the definition of the **boxed_row** of line 3. The predefined tile, **ok_cancel**, displays the **ok** and **cancel** tiles in the dialog box. The closing brace on the last line completes the definition of the dialog box.

After writing the program, use the following commands to load the DCL file and display the dialog box on the screen. The name of the file is assumed to be *osnapsh.dcl*. If AutoCAD is successful in loading the file, it will return an integer. In the following example, the integer AutoCAD returns is assumed to be 1. The screen display after loading the dialog box is shown in Figure 11-8. When you load the program, AutoCAD will freeze up. Use the **Windows Task Manager** to kill the AutoCAD program, see the note on page 11-11.

Command: (load_dialog "osnapsh.dcl")
1
Command: (new_dialog "osnapsh" 1)

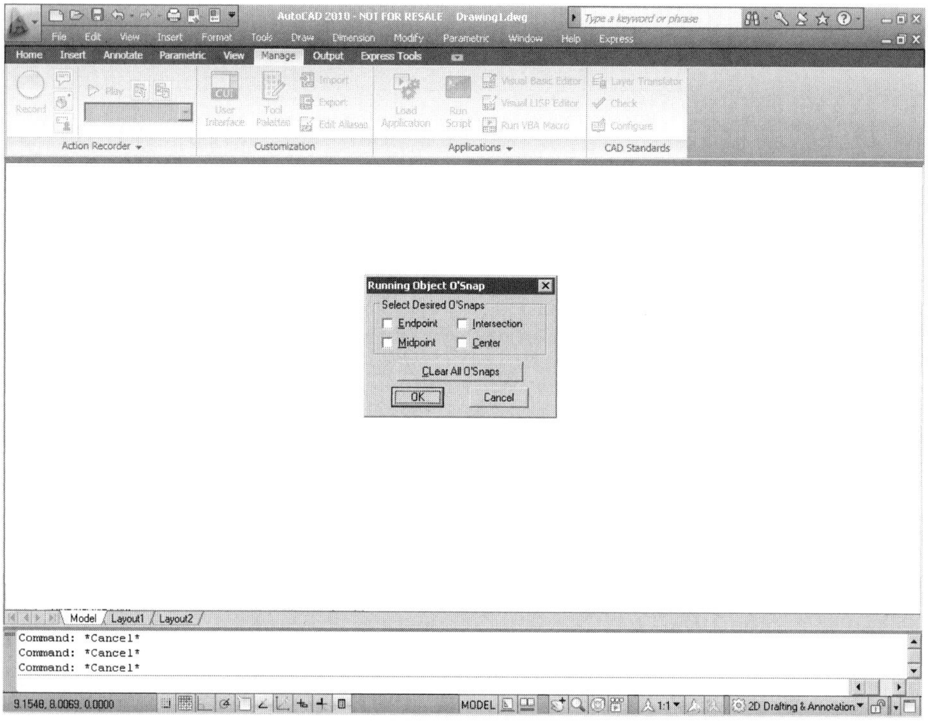

Figure 11-8 *Screen display with dialog box for Example 3*

AutoLISP FUNCTIONS

logand and logior

These AutoLISP functions, **logand** and **logior**, are used to obtain the result of **logical bitwise AND** and **logical bitwise inclusive OR** of a list of numbers.

Examples of logand function

(logand 2 7)	will return	2
(logand 8 15)	will return	8
(logand 6 15 7)	will return	6
(logand 6 15 1)	will return	0
(logand 1 4)	will return	0
(logand 1 5)	will return	1

Examples of logior function

(logior 2 7)	will return	7
(logior 8 15)	will return	15
(logior 6 15 7)	will return	15
(logior 6 15 1)	will return	15
(logior 1 4)	will return	5
(logior 1 5)	will return	5

One application of bit-codes is found in the object snap modes. The following is a list of bit-codes assigned to different object snap modes:

Snap Mode	Bit Code	Snap Mode	Bit Code
None	0	Intersection	32
Endpoint	1	Insertion	64
Midpoint	2	Perpendicular	128
Center	4	Tangent	256
Node	8	Nearest	512
Quadrant	16	Quick	1024
Apparent Intersection	2048	Extension	4096
Parallel	8192		

The system variable **OSMODE** can be used to specify a snap mode. For example, if the value assigned to **OSMODE** is 4, the Object Snap mode is center. The bit-codes can be combined to produce multiple object snap modes. For example, you can get endpoint, midpoint, and center object snaps by adding the bit-codes of endpoint, midpoint, and center (1 + 2 + 4 = 7) and assigning that value (7) to the **OSMODE** system variable. In AutoLISP, the existing snap modes information can be extracted by using the **logand** function. For example, if **OSMODE** is set to 7, the **logand** function can be used to check different bit-codes that represent object snaps.

(logand 1 7)	will return	1	(Endpoint)
(logand 2 7)	will return	2	(Midpoint)
(logand 4 7)	will return	4	(Center)

atof and rtos Functions

The **atof** function converts a string into a real number and the rtos function converts a number into a string.

Examples:
(atof "3.5") will return 3.5
(rtos 8 15) will return "8.1500"

get_tile and set_tile Functions

The **get_tile** function retrieves the current value of a dialog box tile and the **set_tile** function sets the value of a dialog box tile.

Examples:
(get_tile "midpoint")
(set_tile "xsnap" value)

Example 4

Write an AutoLISP program that will handle the dialog box and perform the functions as described in the dialog box of Example 3.

The following file is a listing of the AutoLISP program for Example 4. **The line numbers on the right are not a part of the program; they are only for your reference.**

```
;;Lisp program for setting Osnaps
;;Dialog file name is osnapsh.dcl
(defun c:osnapsh ( / dcl_id)
(setq dcl_id (load_dialog "osnapsh.dcl"))
(new_dialog "osnapsh" dcl_id)
;;Get the existing value of object snaps and
;;then write the values to the dialog box
(setq osmode (getvar "osmode"))
(if (= 1 (logand 1 osmode))
   (set_tile "Endpoint" "1")
   )
(if (= 2 (logand 2 osmode))
   (set_tile "Midpoint" "1")
   )
(if (= 32 (logand 32 osmode))
   (set_tile "Intersection" "1")
   )
(if (= 4 (logand 4 osmode))
   (set_tile "Center" "1")
   )
```

```
22 ;;Read the values as set in the dialog box and
23 ;;assign those values to AutoCAD variable osmode
24 (defun setvars ()
25 (setq osmode 0)
26 (if (= "1" (get_tile "Endpoint"))
27    (setq osmode (logior osmode 1))
28    )
29 (if (= "1" (get_tile "Midpoint"))
30    (setq osmode (logior osmode 2))
31    )
32 (if (= "1" (get_tile "Intersection"))
33    (setq osmode (logior osmode 32))
34    )
35 (if (= "1" (get_tile "Center"))
36    (setq osmode (logior osmode 4))
37    )
38   (setvar "osmode" osmode)
39   )
40
41 (action_tile "accept" "(setvars) (done_dialog)")
42 (start_dialog)
43 (princ)
44 )
```

Explanation

Lines 1 and 2

;;Lisp program for setting Osnaps
;;Dialog file name is osnapsh.dcl

These lines are comment lines, and all comment lines start with a semicolon. AutoCAD ignores the lines that start with a semicolon.

Lines 3-5

(defun c:osnapsh (/ dcl_id)
(setq dcl_id (load_dialog "osnapsh.dcl"))
(new_dialog "osnapsh" dcl_id)

In line 3, **defun** is an AutoLISP function that defines the **osnapsh** function. Because of the c: in front of the function name, the **osnapsh** function can be executed like an AutoCAD command. The **osnapsh** function has one local variable, **dcl_id**. In line 4, the **(load_dialog "osnapsh.dcl")** loads the DCL file osnapsh.dcl and returns a positive integer. The **setq** function assigns this integer to the local variable, **dcl_id**. In line 5, the AutoLISP **new_dialog** function loads the **osnapsh** dialog box that is defined in the DCL file (line 1 of the DCL file). The variable **dcl_id** has an integer value that identifies the DCL file.

Line 8

(setq osmode (getvar "osmode"))

In this line, **getvar "osmode"** has the value of the AutoCAD system variable **osmode**, and the

setq function sets the **osmode** variable equal to that value. Note that the first **osmode** is just a variable, whereas the second osmode, in quotes **("osmode")**, is the system variable.

Lines 9-11
(if (= 1 (logand 1 osmode))
(set_tile "Endpoint" "1")
)
In line 9, the **(logand 1 osmode)** will return 1 if the bit-code of endpoint (1) is a part of the **OSMODE** value. For example, if the value assigned to **OSMODE** is 7 (1 + 2 + 4 = 7), (logand 1 7) will return 1. If **OSMODE** is 6 (2 + 4 = 6), then (logand 1 6) will return 0. The AutoLISP **if** function checks whether the value returned by **(logand 1 osmode)** is 1. If the function returns **T** (true), the instructions described in the second line are carried out, see Figure 11-9. Line 10 sets the value of the **Endpoint** tile to 1; that displays a check mark in the toggle box. The closing parenthesis in line 11 completes the definition of the **if** function. If the expression **(if (= 1 (logand 1 osmode))** returns **nil**, the program skips to line 12 of the program.

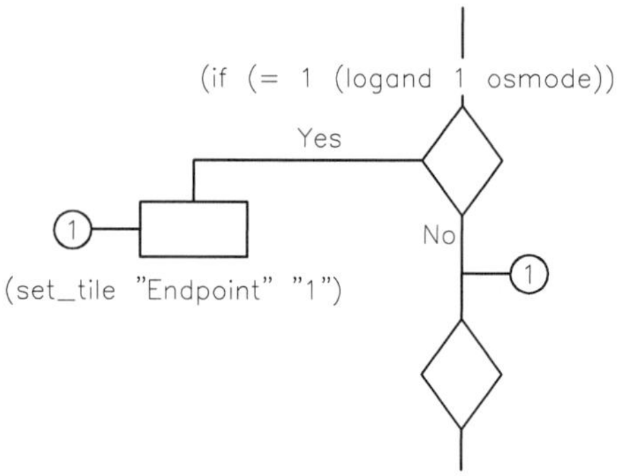

***Figure 11-9** Flowchart explaining lines 9 through 11*

Lines 24 and 25
(defun setvars ( )
(setq osmode 0)
Line 24 defines a **setvars** function, and line 25 sets the value of the **osmode** variable to zero.

Lines 26-28
(if (= "1" (get_tile "Endpoint"))
(setq osmode (logior osmode 1))
)
In line 26, the **(get_tile "Endpoint")** obtains the value of the toggle tile named **Endpoint**. If the Endpoint toggle tile is on, the value it returns is 1; if the toggle tile is off, the value it returns is 0. The **gen_tile** function returns the value as string (0 or 1). In line 27, the **setq** function sets the value of **osmode** to the value returned by **(logior osmode 1)**. For example, if the initial value of osmode is 0, then **(logior osmode 1)** will return 1. Similarly, if the initial value of **osmode** is 1, then **(logior osmode 2)** will return 3.

Lines 41 and 42
(action_tile "accept" "(setvars) (done_dialog)")
(start_dialog)
Line 42, **(start_dialog)**, starts the dialog box. In the dialog file the name assigned to the **OK** button is **"accept"**. When you select the **OK** button in the dialog box, the program executes the **setvars** function that updates the value of the **OSMODE** system variable and sets the

selected object snaps and then **(done_dialog)** will terminate the dialog box. If you choose the **Cancel** button, the dialog box will terminate without altering the value of **OSMODE**.

PREDEFINED RADIO BUTTON, RADIO COLUMN, BOXED RADIO COLUMN, AND RADIO ROW TILES

Predefined Radio Button Tile

Format in DCL: **radio_button**

The **radio button** is a predefined active tile. Radio buttons can be arranged in a row or in a column. The unique characteristic of radio buttons is that only one button can be selected at a time. For example, if there are buttons for scientific, decimal, and engineering units, only one can be selected. If you select the decimal button, the other two buttons will be turned off automatically. Because of this special characteristic, radio buttons must be used only in a radio row or in a radio column. The label for the radio button is optional; in most systems the label appears to the right of the button.

Predefined Radio Column Tile

Format in DCL: **radio_column**

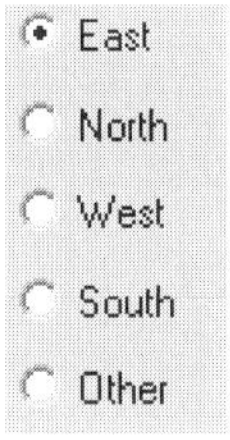

The **radio column** is an active predefined tile where the radio button tiles are arranged in a column and only one button can be selected at a time. When the radio buttons are arranged in a column, the buttons are next to each other vertically and are easy to select. Therefore, you should arrange radio buttons in a column to make it easy to select a radio button.

Predefined Boxed Radio Column Tile

Format in DCL: **boxed_radio_column**

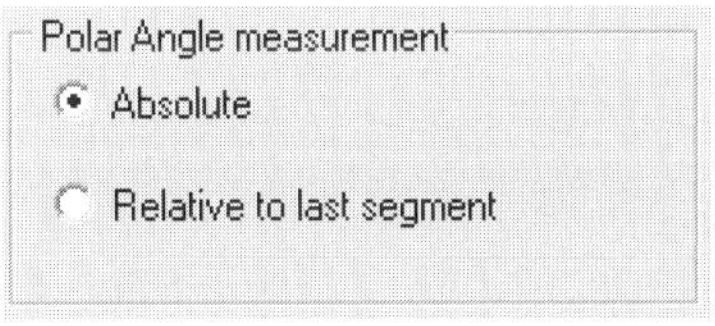

The **boxed radio column** is an active predefined tile where a border is drawn around the radio column.

Predefined Radio Row Tile

Format in DCL: **radio_row**

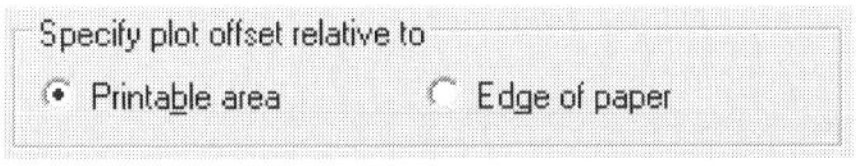

The **radio row** consists of radio button tiles arranged in a row. Only one button can be selected at a time. The radio row can become quite long if there are several radio button tiles and labels. On selecting a radio button, the cursor travel will increase because the buttons are not immediately next to each other. Therefore, you should avoid using radio button tiles in a row, especially if there are more than two.

Example 5

Write a DCL program for a dialog box that will enable you to select different units and unit precision, as shown in Figure 11-10. Also, write an AutoLISP program that will handle the dialog box.

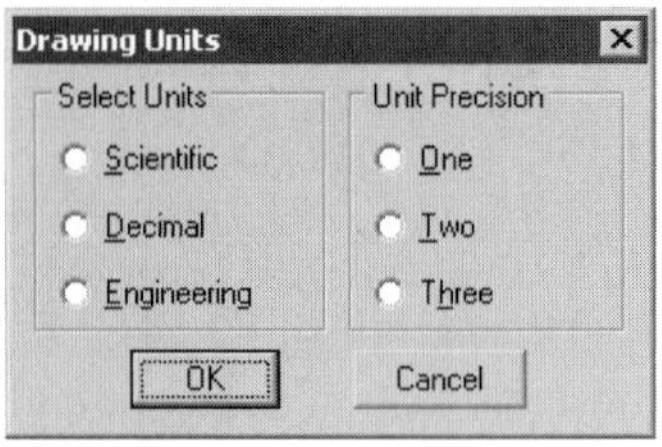

Figure 11-10 *Dialog box for Example 5*

The following file is a listing of the DCL program for the dialog box shown in Figure 11-10. The name of the dialog box is *dwgunits*. **The line numbers on the right are not a part of the file; they are shown here only for reference.**

```
dwgunits : dialog {
 label = "Drawing Units";
 : row {
  : boxed_column {
   label = "Select Units";
   : radio_column {
    : radio_button {
     key = "scientific";
     label = "Scientific";
     mnemonic = "S";
     }
    : radio_button {
     key = "decimal";
     label = "Decimal";
     mnemonic = "D";
     }
    : radio_button {
     key = "engineering";
     label = "Engineering";
     mnemonic = "E";
     }
    }
  }
  : boxed_column {
   label = "Unit Precision";
   : radio_column {
    : radio_button {
```

```
28          key = "one";
29          label = "One";
30          mnemonic = "O";
31          }
32        : radio_button {
33          key = "two";
34          label = "Two";
35          mnemonic = "T";
36          }
37        : radio_button {
38          key = "three";
39          label = "Three";
40          mnemonic = "h";
41          }
42        }
43      }
44    }
45    ok_cancel;
46  }
```

Explanation

Lines 4 and 5

: boxed_column {
  label = "Select Units";

The **boxed_column** is an active predefined tile that draws a border around the column. Line 5, **label = "Select Units";**, displays the label **(Select Units)** at the top of the column. The **label** is a predefined DCL attribute.

Lines 6-8

: radio_column {
  : radio_button {
    key = "scientific";

The **radio_column** is a predefined active tile that will arrange the tiles within it in a vertical column and draw a border around the column. The **radio_button** is another active predefined tile that displays a radio button in the dialog box, with an optional text to the right of the button. The key attribute assigns a name (**scientific**) to the tile. This name is then used by the application program to handle the tile.

Lines 9-11

    label = "Scientific";
    mnemonic = "S";
  }

The label attribute will display the label (**Scientific**) on the right of the toggle box. The second line, **mnemonic = "S";**, defines the keyboard mnemonic for **Scientific**. This causes the letter S of **Scientific** to be displayed underlined in the dialog box. The closing brace (}) on the next line completes the definition of the toggle tile.

Lines 41-46

```
                }
            }
        }
    }
  ok_cancel;
}
```

The closing brace on line 41 completes the definition of the radio button and the closing brace on the second line completes the definition of the radio column. The closing brace on the next line completes the definition of the **boxed_column**, and the closing brace on the next line completes the definition of the row on line 3 of the DCL file. The predefined tile, **ok_cancel**, displays the **OK** and **Cancel** tiles in the dialog box. The last closing brace completes the definition of the dialog box.

Use the Visual LISP editor to preview the dialog box. This dialog box is shown in Figure 11-11.

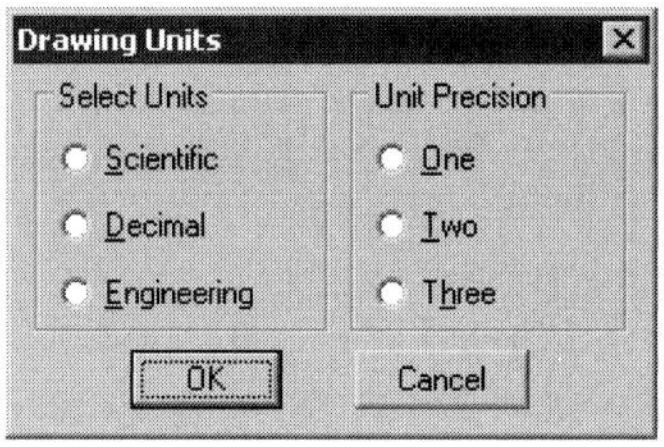

Figure 11-11 *Dialog box for Example 5*

The following file is a listing of the AutoLISP program that loads, displays, and handles the dialog box for Example 5. **The line numbers are not a part of the file; they are shown here for reference only.**

```
;;Lisp program dwgunits.lsp for setting units
;;and precision. Dialog file name dwgunits.dcl
;
(defun c:dwgunits ( / dcl_id)
(setq dcl_id (load_dialog "dwgunits.dcl"))
(new_dialog "dwgunits" dcl_id)
;
;;Get the existing values of lunits and luprec
;;and turn the corresponding radio_button on
;
(setq lunits (getvar "lunits"))
(if (= 1 lunits)
   (set_tile "scientific" "1")
```

```
14     )
15   (if (= 2 lunits)
16     (set_tile "decimal" "1")
17     )
18   (if (= 3 lunits)
19     (set_tile "engineering" "1")
20     )
21   ;
22   (setq luprec (getvar "luprec"))
23   (if (= 1 luprec)
24     (set_tile "one" "1")
25     )
26   (if (= 2 luprec)
27     (set_tile "two" "1")
28     )
29   (if (= 3 luprec)
30     (set_tile "three" "1")
31     )
32
33   ;;Read the value of the radio_buttons and
34   ;;assign it to AutoCAD lunit and luprec variables
35   ;
36   (action_tile "scientific" "(setq lunits 1)")
37   (action_tile "decimal" "(setq lunits 2)")
38   (action_tile "engineering" "(setq lunits 3)")
39   ;
40   (action_tile "one" "(setq luprec 1)")
41   (action_tile "two" "(setq luprec 2)")
42   (action_tile "three" "(setq luprec 3)")
43   (action_tile "accept" "(done_dialog)")
44   ;
45   (start_dialog)
46   (setvar "lunits" lunits)
47   (setvar "luprec" luprec)
48   (princ)
49   )
```

Explanation

Lines 1-3

;;Lisp program dwgunits.lsp for setting units
;;and precision. Dialog file name dwgunits.dcl
;

The first three lines of this program are comment lines, and all comment lines start with a semicolon. AutoCAD ignores them.

Lines 4-6
(defun c:dwgunits (/ dcl_id)
(setq dcl_id (load_dialog "dwgunits.dcl"))
(new_dialog "dwgunits" dcl_id)
In line 4, **defun** is an AutoLISP function that defines the **dwgunits** function, which has one local variable, **dcl_id**. The **c:** in front of the function name, **dwgunits**, makes the **dwgunits** function act like an AutoCAD command. In the next line, **(load_dialog "dwgunits.dcl")** loads the DCL file *dwgunits.dcl* and returns a positive integer. The **setq** function assigns this integer to the local variable, **dcl_id**. In the next line, the AutoLISP **new_dialog** function displays the **dwgunits** dialog box that is defined in the DCL file (line 1 of DCL file). The **dcl_id** variable is an integer that identifies the DCL file.

Line 11
(setq lunits (getvar "lunits"))
In this line, **(getvar "lunits")** has the value of the AutoCAD system variable, **lunits**, and the **setq** function sets the **lunits** variable equal to that value. The first lunits is a variable, whereas the second **lunits**, in quotes (**"lunits"**), is a system variable.

Lines 12-14
(if (= 1 lunits)
   (set_tile "scientific" "1")
)
The **if** function (AutoLISP function) checks whether the value of the variable **lunits** is 1. If the function returns **T** (true), the instructions described in the next line are carried out. Line 13 sets the value of the tile named **"scientific"** equal to 1, which turns the corresponding radio button on, see Figure 11-12. The closing parenthesis in line 14 completes the **if** function. If the **if** function, **(if (= 1 lunits)**, returns **nil**, the program skips to line number 14.

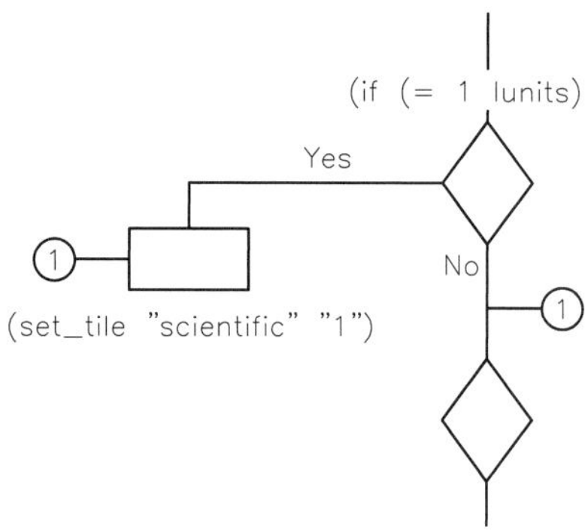

Figure 11-12 *Flowchart explaining lines 12 through 14*

Line 22
(setq luprec (getvar "luprec"))
In this line, **(getvar "luprec")** has the value of the AutoCAD system variable **luprec**, and the **setq** function sets the luprec variable equal to that value. The first **luprec** is a variable, whereas the second luprec, in quotes (**"luprec"**), is a system variable.

Lines 23-25
(if (= 1 luprec)
(set_tile "one" "1")
)
In line 23, the **if** function checks whether the value of the **luprec** variable is 1. If the function returns **T** (true), the instructions described in the next line are carried out. This line sets the value of the tile named **"one"** to 1; that turns the corresponding radio button on. The closing parenthesis in the next line completes the definition of the **if** function. If the **if** function returns **nil**, the program skips to line 25, see Figure 11-13.

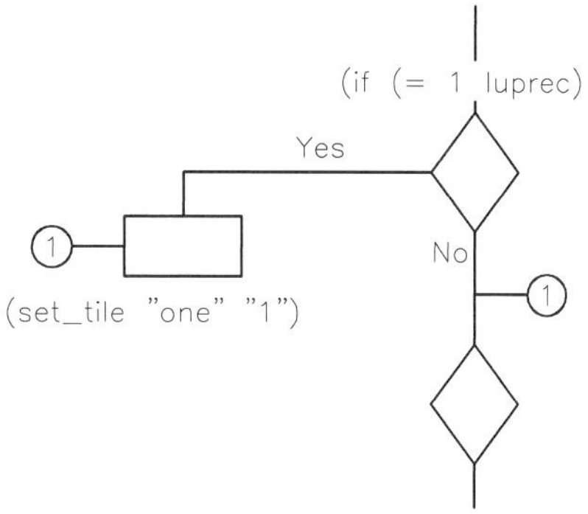

***Figure 11-13** Flowchart explaining lines 23 through 25*

Line 36
(action_tile "scientific" "(setq lunits 1)")
If the radio button tile named **"scientific"** is turned on, the **setq** function sets the value of the AutoCAD system variable **lunits** to 1. The **lunits** system variable controls the drawing units. The following is a list of the integer values that can be assigned to the **LUNITS** system variable.

1 Scientific
2 Decimal
3 Engineering
4 Architectural
5 Fractional

Line 40
(action_tile "one" "(setq luprec 1)")
If the radio button tile named **"one"** is turned on, the **setq** function sets the value of the AutoCAD system variable **luprec** to 1. The **lunits** system variable controls the number of decimal places in a decimal number or the denominator of a fractional or architectural unit.

Lines 43-47
(action_tile "accept" "(done_dialog)")
;
(start_dialog)
(setvar "lunits" lunits)
(setvar "luprec" luprec)
In the dialog file, the name assigned to the **OK** button is **"accept"**. When you choose the **OK** button in the dialog box, the program executes the function defined in the **dwgunits** dialog box. The **(start_dialog)** function starts the dialog box. The **(setvar "lunits" lunits)** and **(setvar "luprec" luprec)** set the values of the **lunits** and **luprec** system variables to lunits and luprec, respectively.

EDIT BOX TILE

Format in DCL: **edit_box**

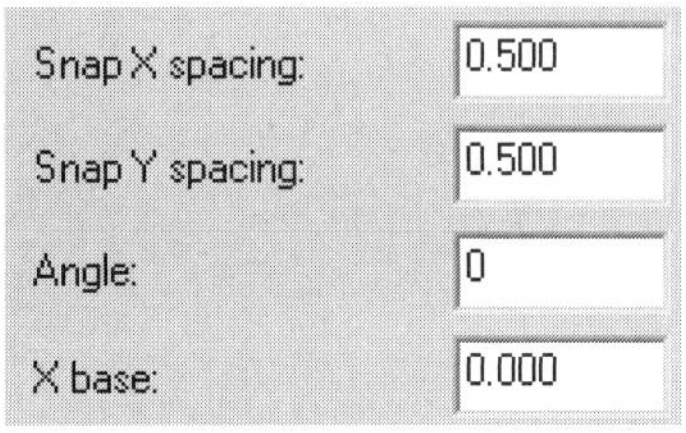

The **edit box** is a predefined active tile that enables you to enter or edit a single line of value. If the text is longer than the length of the edit box, the text will automatically scroll to the right or left horizontally. The label for the edit box is optional, and it is displayed to the left of the edit box.

width AND edit_width ATTRIBUTES

width Attribute

Format in DCL: **width** **Example**: width = 22

The **width** attribute is used to keep the width of the tile to a desired size. The value assigned to the **width** attribute can be a real number or an integer that represents the distance in character width. The value assigned to the **width** attribute defines the minimum width of the tile. For example, in this case, the minimum width of the tile is 22. However, the tile will automatically stretch if more space is available. It will retain the size of 22 only if the **fixed_width** attribute is assigned to the tile. The **width** attribute can be used with any tile.

edit_width Attribute

Format in DCL: **edit_width** **Example**: edit_width = 10

The **edit_width** attribute is used with the predefined edit box tiles, and it determines the size of the edit box in character width units. If the width of the edit box is 0, or if it is not specified and the **fixed_width** attribute is not assigned to the tile, the tile will automatically stretch to fill the available space. When the edit box is stretched, the PDB facility inserts spaces between the edit box and the label so that the box is right justified and the label is left justified.

Example 6 *General*

Write a DCL program for a dialog box shown in Figure 11-14 that will enable you to turn the snap and grid on and off. You should also be able to edit the X and Y values of snap and grid. Also, write an AutoLISP program that will load, display, and handle the dialog box.

The following file is a listing of the DCL file for Example 6.

```
dwgaids : dialog {
  label = “Drawing Aids”;
  : row {
    : boxed_column {
      label = “SNAP”;
      fixed_width = true;
```

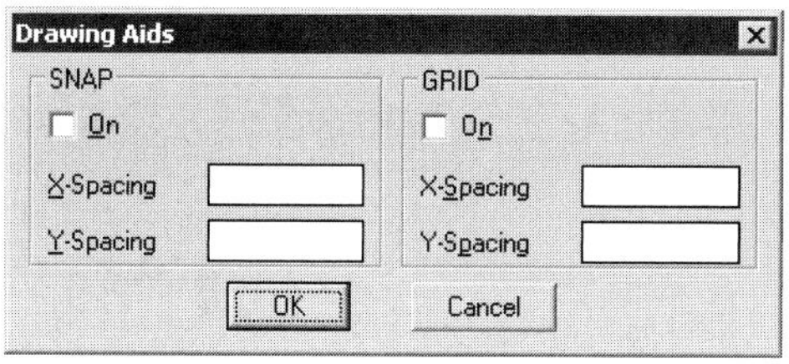

Figure 11-14 *Dialog box for Example 6*

```
    width = 22;
    : toggle {
      label = "On";
      mnemonic = "O";
      key = "snapon";
      }
    : edit_box {
      label = "X-Spacing";
      mnemonic = "X";
      key = "xsnap";
      edit_width = 10;
      }
    : edit_box {
      label = "Y-Spacing";
      mnemonic = "Y";
      key = "ysnap";
      edit_width = 10;
      }
    }
  : boxed_column {
    label = "GRID";
    fixed_width = true;
    width = 22;
    : toggle {
      label = "On";
      mnemonic = "n";
      key = "gridon";
      }
    : edit_box {
      label = "X-Spacing";
      mnemonic = "S";
      key = "xgrid";
      edit_width = 10;
      }
    : edit_box {
```

```
        label = "Y-Spacing";
        mnemonic = "p";
        key = "ygrid";
        edit_width = 10;
        }
      }
    }
    ok_cancel;
  }
```

The following file is a listing of the AutoLISP program for Example 6. When the program is loaded and run, it will load, display, and control the dialog box (Figure 11-15).

```
;;Lisp program for Drawing Aids dialog box
;;Dialog file name is dwgaids.dcl
;
(defun c:dwgaids( / dcl_id snapmode xsnap ysnap
 orgsnapunit gridmode gridsnap xgrid ygrid orggridunit)
(setq dcl_id (load_dialog "dwgaids.dcl"))
(new_dialog "dwgaids" dcl_id)

;;Get the existing value of snapmode and snapunit
;;and write those values to the dialog box
(setq snapmode (getvar "snapmode"))
(if (= 1 snapmode)
   (set_tile "snapon" "1")
   (set_tile "snapon" "0")
   )
(setq orgsnapunit (getvar "snapunit"))
(setq xsnap (car orgsnapunit))
(setq ysnap (cadr orgsnapunit))
(set_tile "xsnap" (rtos xsnap))
(set_tile "ysnap" (rtos ysnap))
;
;;Get the existing value of gridmode and gridunit
;;and write those values to the dialog box
(setq gridmode (getvar "gridmode"))
(if (= 1 gridmode)
   (set_tile "gridon" "1")
   (set_tile "gridon" "0"))
(setq orggridunit (getvar "gridunit"))
(setq xgrid (car orggridunit))
(setq ygrid (cadr orggridunit))
(set_tile "xgrid" (rtos xgrid))
(set_tile "ygrid" (rtos ygrid))
;;Read the values set in the dialog box and
```

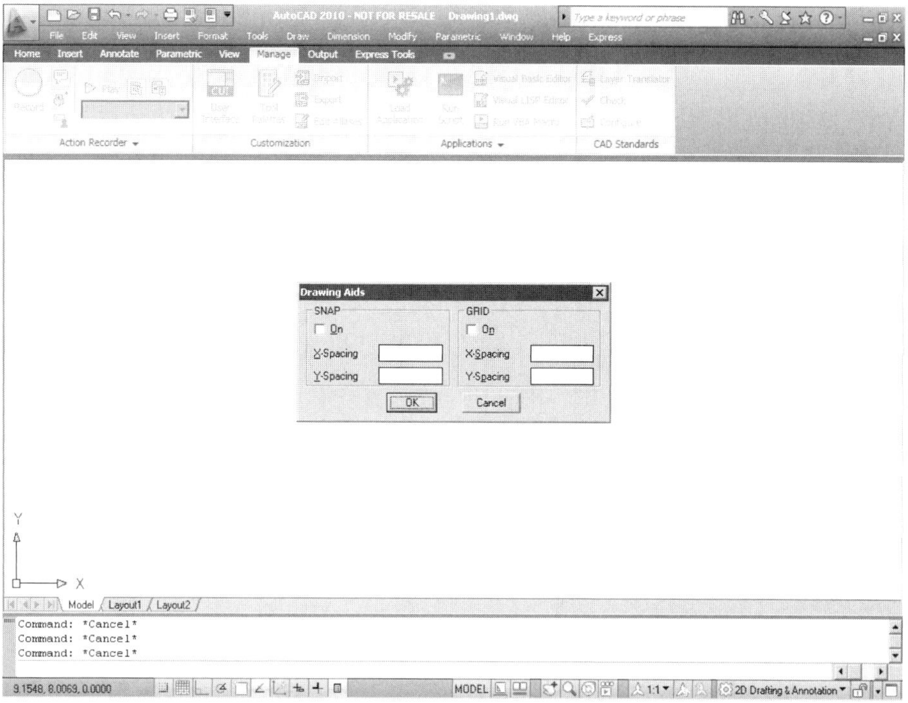

Figure 11-15 *Screen display with the dialog box for Example 6*

```
;;then change the associated AutoCAD variables
(defun setvars ()
(setq xsnap (atof (get_tile “xsnap”)))
(setq ysnap (atof (get_tile “ysnap”)))
(setvar “snapunit” (list xsnap ysnap))
(if (= “1” (get_tile “snapon”))
   (setvar “snapmode” 1)
   (setvar “snapmode” 0))
(setq xgrid (atof (get_tile “xgrid”)))
(setq ygrid (atof (get_tile “ygrid”)))
(setvar “gridunit” (list xgrid ygrid))
(if (= “1” (get_tile “gridon”))
   (progn
     (setvar “gridmode” 0)
     (setvar “gridmode” 1))
   (setvar “gridmode” 0)
   )
(action_tile “accept” “(setvars) (done_dialog)”)
(start_dialog)
(princ)
)
```

SLIDER AND IMAGE TILES

Slider Tile

Format in DCL
slider

A slider tile is an active predefined tile that consists of an indicator, a slider bar (rectangular strip) in which the slider moves, and the direction arrows. The slider tile can be used to obtain a string value. The value is determined by the position of the indicator box in the slider bar. The string value returned by the slider tile can then be used by the application program. For example, you can use the value returned by the aperture slider tile to set the value of the **APERTURE** system variable. The indicator box can be dragged to the left or right by positioning the arrow on the indicator box and then moving the arrow while holding the pick button down. You can also press the direction arrows to move the slider in increments. The slider tiles can be horizontal or vertical. If the slider tile is horizontal, the values increase from left to right; if the slider is vertical, the values increase from bottom to top. The values returned by the slider tile are always integers.

Image Tile

Format in DCL
image

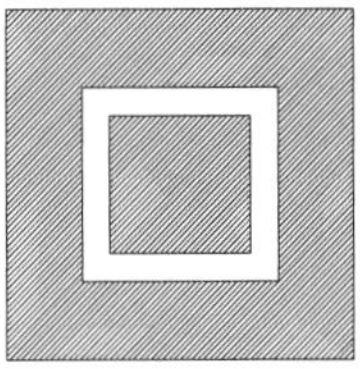

The image tile is a predefined tile that is used to display graphical information. For example, it can be used to display the aperture box, linetypes, icons, and text fonts in dialog boxes. It consists of a rectangular box and the vector graphics that are displayed within the box.

min_value, max_value, small_increment, and big_increment Attributes

min_value and max_value Attributes

Format in DCL	**Examples:**
min_value	min_value = 2
max_value	max_value = 15

The **min_value** attribute and the **max_value** attribute are predefined attributes that specify the minimum and maximum value that the **slider** tile will return. In the previous example, the minimum value is 2 and the maximum value is 15. If you do not assign these attributes to a slider tile, the slider automatically assumes the default values. For the **min_value** attribute the default value is 0, for the **max_value** attribute, the default value is 10,000.

small_increment and big_increment Attributes

Format in DCL	**Examples:**
small_increment	small_increment = 1
big_increment	big_increment = 1

The value assigned to the **small_increment** attribute and the **big_increment** attribute determines the increment value of the slider incremental control. For example, if the increment is 1, the values returned by the slider will be in the increment of 1. If these attributes are not assigned to a slider tile, the slider automatically assumes the default values. The default value of **small_increment** is one-hundredth (1/100) of the slider range and the default value of **big_increment** is one-tenth (1/10) of the slider range.

aspect_ratio and color Attributes

aspect_ratio Attribute

Format in DCL
aspect_ratio

Example:
aspect_ratio = 1

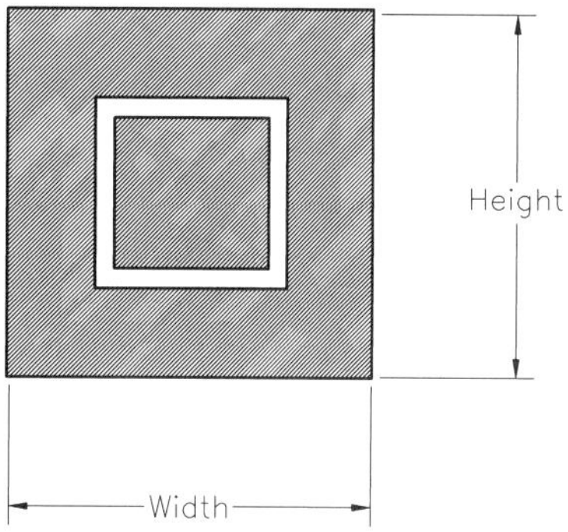

The **aspect_ratio** is a predefined attribute that can be used with an image. The aspect ratio is the ratio of the width of the image box to the height of the image box (width/height). You can control the size of the image box by assigning an integer value to the width attribute and the height attribute. You can also control the size of the image box by assigning an integer value to one of the attributes (width or height) and assigning a real or integer value to the **aspect_ratio** attribute. For example, if the value assigned to the height attribute is five and the value assigned to the **aspect_ratio** is 0.5, the width of the image box will be 2.5 (width = height x **aspect_ratio**). In this case, you do not need to specify the width attribute.

color Attribute

Format in DCL
color

Example:
color = 2

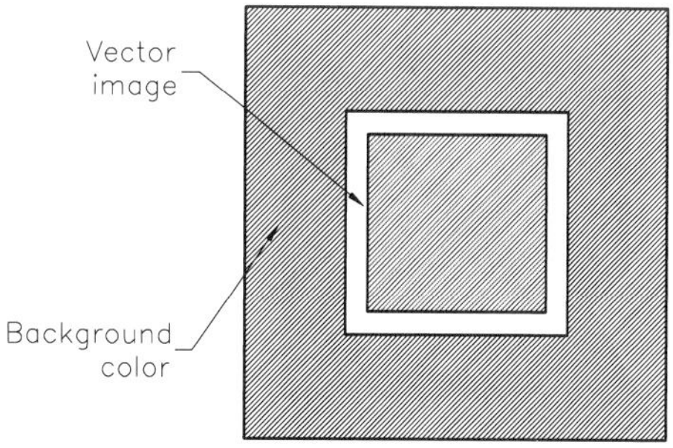

The color is a predefined attribute that can be used with the image box to control the background color. The integer number assigned to the color attribute specifies an AutoCAD color index. In the above example, the background color is yellow (color number 2). The color of the vector image that is drawn inside the image box is determined by the color specified in the **vector_image** attribute.

Example 7 *General*

Write a DCL program for the dialog box in Figure 11-16(a) that will enable you to change the size of the aperture box.

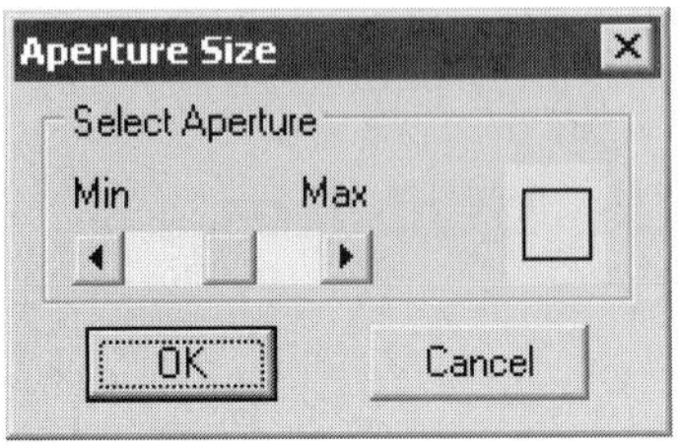

Figure 11-16(a) *Dialog box for Example 7*

The dialog box shown in Figure 11-16(a) has two rows and two columns. The first column has two items, label **(Min .... Max)** and the **slider bar**. The second column has only one item (the image box). The second row has the **OK** and **Cancel** buttons. As shown in Figure 11-16(b), the dialog box also has a dialog label (Aperture Size) and a row label (Select Aperture).

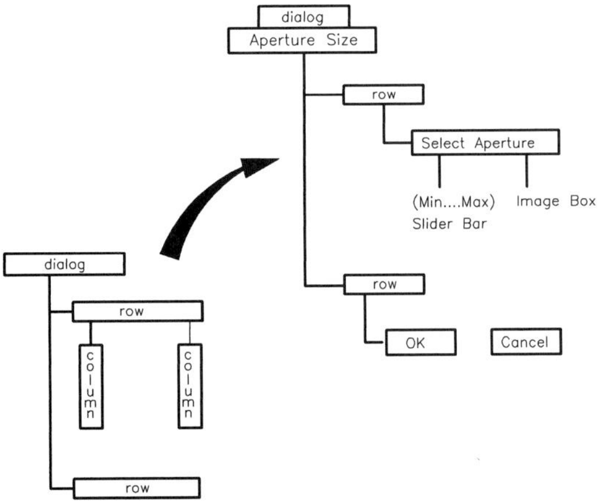

Figure 11-16(b) *Layout of dialog box for Example 7*

The file given next is a listing of the DCL file for Example 7. **The line numberson the right are not a part of the file; they are shown here only for reference.**

```
aprtsize : dialog {                                    1
  label = "Aperture Size";                             2
```

```
3  : boxed_row {
4   label = "Select Aperture";
5  : column {
6     fixed_width = true;
7    : text {
8     label = "Min        Max";
9     alignment = centered;
10     }
11   : slider {
12     key = "aperture_slider";
13     min_value = 2;
14     max_value = 17;
15     width = 15;
16     height = 1;
17     big_increment = 1;
18     fixed_width = true;
19     fixed_height = true;
20     }
21   }
22   : image {
23     key = "aperture_image";
24     aspect_ratio = 1;
25     width = 5;
26     color = 2;
27     }
28   }
29   ok_cancel;
30 }
```

Lines 11-12
: slider {
key = "aperture_slider";
In the first line, **: slider** starts the definition of the slider tile. The **slider** is an active predefined tile that consists of a slider bar, a small indicator box, and the direction arrows at the ends of the slider bar. The second line, **key = "aperture_slider"**, assigns the name, **aperture_slider**, to this slider tile.

Lines 13-14
min_value = 2;
max_value = 17;
The **min_value** is a predefined attribute that specifies the minimum value that the **slider** tile will return. Similarly, the **max_value** attribute specifies the maximum value that the slider tile will return. In the above two lines, the minimum value is 2 and the maximum value is 17.

Lines 15-17
width = 15;
height = 1;
big_increment = 1;
The first line specifies the width of the slider tile and the second line specifies the height of the slider tile. The third line, **big_increment = 1;**, defines the increment of the indicator box in the slider bar.

Lines 22-23
: image {
key = "aperture_image";
The first line, **: image {**, starts the definition of the image box and the second line assigns a name, **aperture_image**, to the image tile. The image is a predefined tile that is used to display graphical information.

Lines 24-26
aspect_ratio = 1;
width = 5;
color = 2;
The first line, **aspect_ratio = 1;**, defines the ratio of the width of the image box to the height of the image box. The second line, **width = 5;**, specifies the width of the image box. Since the aspect ratio is 1, the height of the image box is 5 (width/height = aspect_ratio). The third line, **color = 2;**, assigns the AutoCAD color number 2 (yellow) to the background of the image box.

AutoLISP FUNCTIONS

dimx_tile and dimy_tile

The AutoLISP function, **dimx_tile**, obtains the width dimension (x_aperture) of the specified image box along the X-axis, and the **dimy_tile** function obtains the height dimension (y_aperture) along the Y-axis. In the following examples, "aperture_image" is the name of the image tile that is assigned by using the **key** attribute in the DCL file (key = "aperture_image").

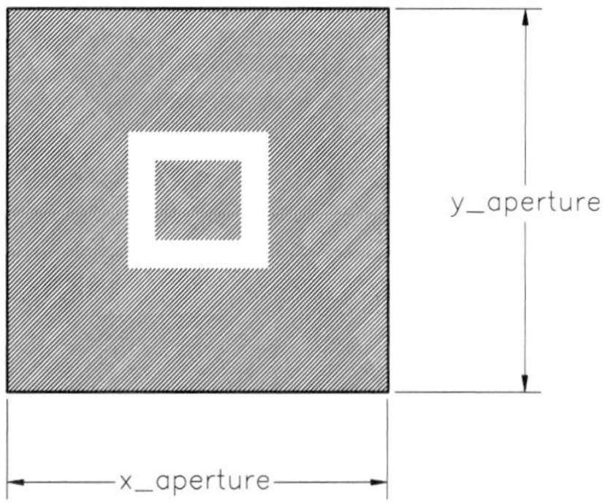

Format	**Examples:**
(dimx_tile tilename)	(dimx_tile "aperture_image")
(dimy_tile tilename)	(dimy_tile "aperture_image")

vector_image

The AutoLISP function **vector_image** draws a vector (line) in the active image box, between the points that are defined in the **vector_image** function. In the following example, AutoCAD will draw a line from point 1,1 to point 3,3 of the image box and the color of this vector will be AutoCAD's color number 1 (red).

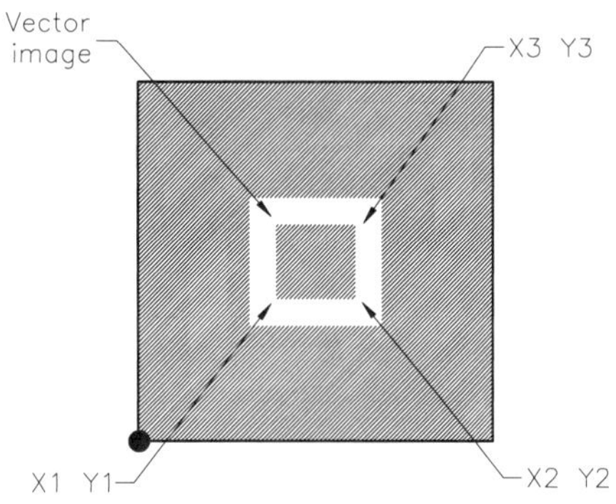

Format
(vector_image x1 y1 x2 y2 color)

Example:
(vector_image 1 1 3 3 1)
Where **1 1** ---------------- X and Y coordinates of first point
3 3 ---------------- X and Y coordinates of second point
1 ------------------ AutoCAD color index

fill_image

The AutoLISP function, **fill_image**, fills the rectangle in the active image box with the color specified in the **fill_image** function. For example, in the following example the active image box will be filled with yellow color (AutoCAD's color number 2). The filled rectangle is determined by the coordinates of the first point (x1 y1) and the second point (x2 y2), the two opposite corners of the rectangle.

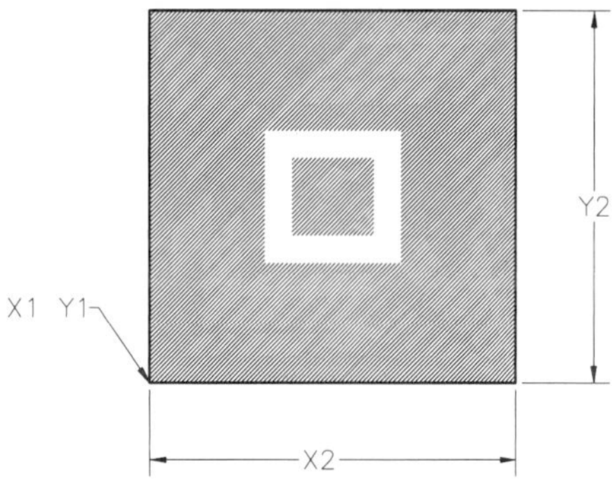

Format
(fill_image x1 y1 x2 y2 color)

Example:
(fill_image 0 0 x_aperture y_aperture 2)

start_image

The AutoLISP function, **start_image**, starts the image whose name is specified in the **start_image** function. In the following example, the name of the image tile is **"aperture_image"** which is assigned by using the **key** attribute in the DCL file (key = "apertue_ige").

Format
(start_image)

Example
(start_image "aperture_image")

end_image

The AutoLISP function **end_image** ends the active image whose name is specified in the **end_image** function. In the following example, the name of the active image tile is **"aperture_image"**.

Format	**Example:**
(end_image)	(end_image "aperture_image")

$value

The **$value** function is an AutoLISP action expression that retrieves the string value from the tile of a dialog box. The tile could be an edit box or a toggle box. In Example 8, the **$value** function retrieves the current value of the active tile and the setq function assigns that value to the variable, **aprt_size**.

Format	**Example:**
$value	(setq aprt_size $value)

Example 8

Write an AutoLISP program that will load, display, and control the dialog box given Example 7 (Figure 11-16(a)). The flowchart and screen display for this example are shown in Figures 11-17 and 11-18.

The following file is a listing of the AutoLISP file for Example 8. **The line numbers are not a part of the file; they are shown here for reference only.**

```
;;APRTSIZE.LSP, AutoLISP program for Aperture
;; Dialog Box. DCL file name — APRTSIZE.DCL
;
(defun c:aprtsize ( )
  (setq dcl_id (load_dialog "aprtsize"))
  (new_dialog "aprtsize" dcl_id)
;
;Obtain value of the system variable "aperture",
;calculate X and Y values of vector image, and draw
;vector image in the image box.
  (setq aprt_size (getvar "aperture"))
  (if (> aprt_size 15)
    (setq aprt_size 15)
  )
  (setq x_aperture (dimx_tile "aperture_image"))
  (setq y_aperture (dimy_tile "aperture_image"))
  (set_tile "aperture_slider" (itoa aprt_size))
  (setq x1 (- (/ x_aperture 2) aprt_size))
  (setq x2 (+ (/ x_aperture 2) aprt_size))
```

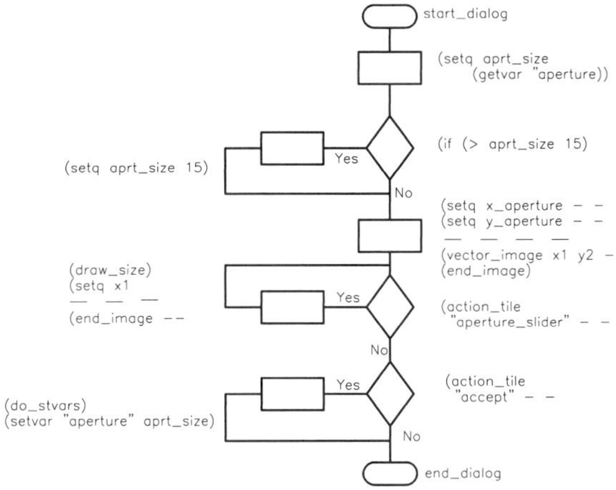

Figure 11-17 *Flowchart for Example 8*

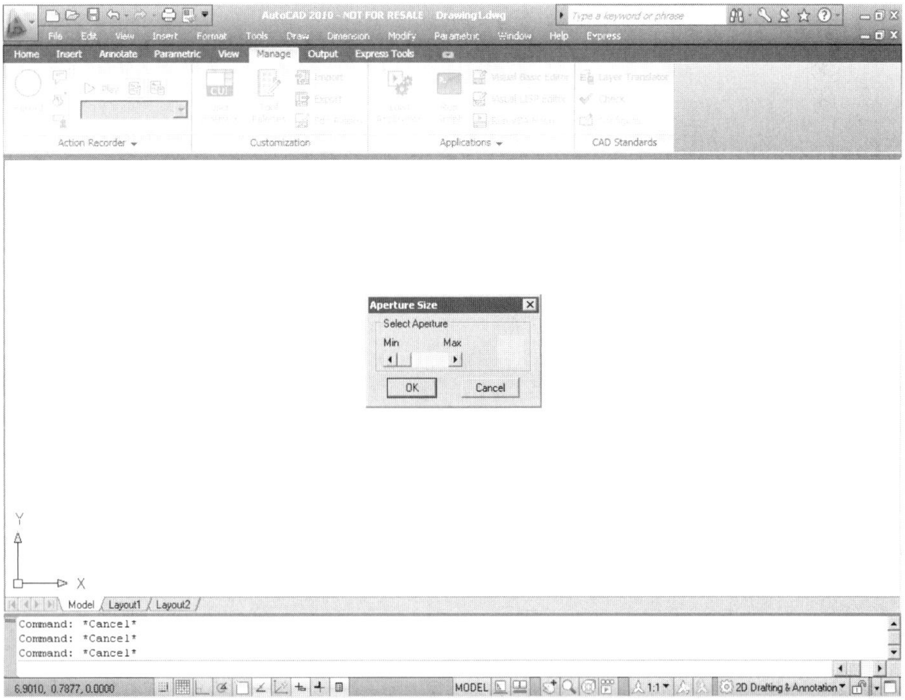

Figure 11-18 *Screen display with the dialog box for Example 8*

```
(setq y1 (- (/ y_aperture 2) aprt_size))
(setq y2 (+ (/ y_aperture 2) aprt_size))
(start_image "aperture_image")
(fill_image 0 0 x_aperture y_aperture 2)
```

```
24   (vector_image x1 y1 x2 y1 1)
25   (vector_image x2 y1 x2 y2 1)
26   (vector_image x2 y2 x1 y2 1)
27   (vector_image x1 y2 x1 y1 1)
28   (end_image)
29   (action_tile "aperture_slider"
30     "(draw_size (setq aprt_size (atoi $value)))")
31   (action_tile "accept" "(do_setvars)(done_dialog)")
32   (start_dialog)
33  (princ)
34  )
35  ;Set aperture variable "aperture" equal to aprt_size
36  (defun do_setvars ( )
37   (setvar "aperture" aprt_size)
38   )
39  ;
40  ;Calculate the X and Y coordinates of vector image
41  ;and draw the aperture image in the image box.
42  (defun draw_size (aprt_size)
43   (setq x1 (- (/ x_aperture 2) aprt_size))
44   (setq x2 (+ (/ x_aperture 2) aprt_size))
45   (setq y1 (- (/ y_aperture 2) aprt_size))
46   (setq y2 (+ (/ y_aperture 2) aprt_size))
47   (start_image "aperture_image")
48   (fill_image 0 0 x_aperture y_aperture -2)
49   (vector_image x1 y1 x2 y1 1)
50   (vector_image x2 y1 x2 y2 1)
51   (vector_image x2 y2 x1 y2 1)
52   (vector_image x1 y2 x1 y1 1)
53   (end_image)
54   )
```

Lines 11-13

(setq aprt_size (getvar "aperture"))
(if (> aprt_size 15)
  (setq aprt_size 15)

In the first line, the AutoLISP function **getvar** obtains the value of the **"aperture"** system variable and the **setq** function assigns that value to the **aprt_size** variable. The second line, **(if (> aprt_size 15)**, checks whether the value of the variable aprt_size is greater than 15. If it is, the third line, **(setq aprt_size 15)**, sets the value of the **aprt_size** variable equal to 15.

The value of the "aperture" system variable must be equal to or less than 15 to display the vector image in the image box because the maximum size of the image box in this example is 15.

Lines 15-16
(setq x_aperture (dimx_tile "aperture_image"))
(setq y_aperture (dimy_tile "aperture_image"))
In the first line, the AutoLISP function, **dimx_tile**, retrieves the X-dimension of the image tile and the setq function assigns that value to the **x_aperture** variable. Similarly, the **dimy_tile** function retrieves the Y-dimension of the image tile.

Line 17
(set_tile "aperture_slider" (itoa aprt_size))
The AutoLISP function **itoa** changes the integer value of the **aprt_size** variable into a string and the **set_tile** function assigns that value to **"aperture_slider"**. The **aperture_slider** is the name of the slider tile in the DCL file (DCL file of Example 7).

Lines 18-21
(setq x1 (- (/ x_aperture 2) aprt_size))
(setq x2 (+ (/ x_aperture 2) aprt_size))
(setq y1 (- (/ y_aperture 2) aprt_size))
(setq y2 (+ (/ y_aperture 2) aprt_size))
The first line calculates the *X* coordinate of the lower left corner of the aperture box. The value is calculated by dividing the **x_aperture** distance by 2 and then subtracting the distance **aprt_size** from it. The **aprt_size** has the same value as that of the system variable, "aperture". Aperture is the distance of the sides of the aperture box from the crosshair lines, measured in pixels.

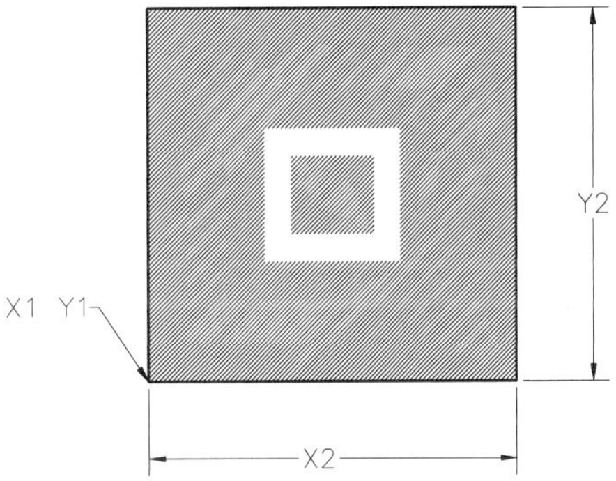

Similarly, the second line calculates the *X* coordinate of the lower-right corner. The next two lines calculate the *Y* coordinates of these two points.

Lines 22-23
(start_image "aperture_image")
(fill_image 0 0 x_aperture y_aperture 2)
The AutoLISP function **start_image** starts the image box and **aperture_image** is the name of the image box as defined in the DCL file. The **fill_image** function fills the image within the specified coordinates with AutoCAD color index 2 (yellow).

Lines 27-28
(vector_image x1 y2 x1 y1 1)
(end_image)
The AutoLISP function **vector_image** draws a vector (line) from the point with coordinates x1 y2 to the point with coordinates x1 y1. The **end_image** function ends the image.

Lines 29-31
(action_tile "aperture_slider"
"(draw_size (setq aprt_size (atoi $value)))")
(action_tile "accept" "(do_setvars)(done_dialog)")
The AutoLISP action expression **$value** retrieves the string value of the active slider tile and the **atoi** function returns the integer value of the string. This value is then used by the **draw_size** function to draw the vector image of the aperture box. The **aperture_slider** is the name of the slider tile defined in the DCL file. The third line executes the **do_setvars** function and ends the dialog box when you choose **OK** from the dialog box. The **draw_size** and **do_setvars** functions are defined in the program.

Lines 36-37
(defun do_setvars ( )
(setvar "aperture" aprt_size)
The defun function defines the **do_setvars** function. The **setvar** function sets the value of the AutoCAD system variable, **"aperture"**, equal to **aprt_size**.

Lines 42-43
(defun draw_size (aprt_size)
(setq x1 (- (/ x_aperture 2) aprt_size))
The first line defines the **draw_size** function with one argument, **aprt_size**, and the second line calculates the value of X-coordinate of the first vector.

Self- Evaluation Test

Answer the following questions and then compare them to those given at the end of this chapter:

1. A dialog control language (DCL) file can contain descriptions of multiple files. (T/F)

2. In a DCL file, you do not need to specify the size of a dialog box. (T/F)

3. Dialog boxes are not dependent on the platform. (T/F)

4. Dialog boxes can contain several OK buttons. (T/F)

5. The numeric values assigned to the attributes in a DCL file can be both integers and real numbers. (T/F)

6. The reserved names in DCL are case-sensitive. (T/F)

7. The label attribute used in a button tile has no default value. (T/F)

8. In a dialog box, only one button can be assigned the true value for the **is_default** attribute. (T/F)

9. The **load_dialog** function displays the dialog box on the screen. (T/F)

10. You cannot select a tile by using the mnemonic key assigned to the tile. (T/F)

Review Questions

Answer the following questions:

1. Mnemonic characters are case-sensitive. (T/F)

2. Bit-codes cannot be combined to produce multiple object snaps. (T/F)

3. The radio button is a predefined attribute. (T/F)

4. You should arrange the radio buttons in a column setting. (T/F)

5. The edit box active tile allows you to enter or edit multiple lines of text. (T/F)

6. The width attribute will make the tile stretch automatically to fill the entire width of the dialog box. (T/F)

7. The slider tile returns a string value. (T/F)

8. The aspect ratio is the ratio of the width of the image box to the height of the image box. (T/F)

9. The color attribute can be used with an image box to control the background color. (T/F)

10. The **vector_image** function can be used to draw vectors in the Drawing Editor. (T/F)

11. The AutoLISP function, **$value**, retrieves the real value from the tile of a dialog box. (T/F)

12. The basic tiles, such as buttons, edit boxes, and images, are predefined by the __________ facility of AutoCAD.

13. The label of a button appears _________ the button.

14. The _________ attribute assigns a name to the tile.

15. The label attribute in a boxed column must be a _________ string.

16. The _________ attribute controls the width of a tile.

17. The format of the command for loading a dialog box is _________.

18. The AutoLISP function _________ is used to initialize a dialog box and then display it on the screen.

19. The AutoLISP function _________ is used to accept user input from the dialog box.

20. The AutoLISP function _________ is used to associate an action expression with a tile in the dialog box.

21. The AutoLISP function _________ can be used to obtain the result of logical bitwise AND.

22. The edit_width attribute determines the size of the edit box in _________ units.

23. The image tile is used to display _________ information.

24. The default value of the max_value attribute is _________.

25. The default value of the big_increment attribute is _________ of the slider range.

26. The AutoLISP function _________ can be used to fill the image box with a color.

Exercises

Exercise 1 *General*

Using dialog control language (DCL), write a program for the dialog box in Figure 11-19. Also, write an AutoLISP program that will load, display, and control the dialog box and perform the functions shown in the dialog box.

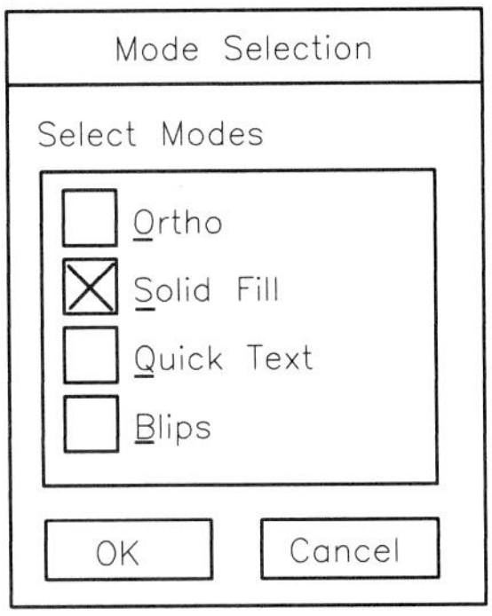

Figure 11-19 *Dialog box for mode selection*

Exercise 2 *General*

Write a DCL program for the **Isometric Snap/Grid** dialog box shown in Figure 11-20. Also, write an AutoLISP program that will load, display, and control the dialog box and perform the functions shown in the dialog box.

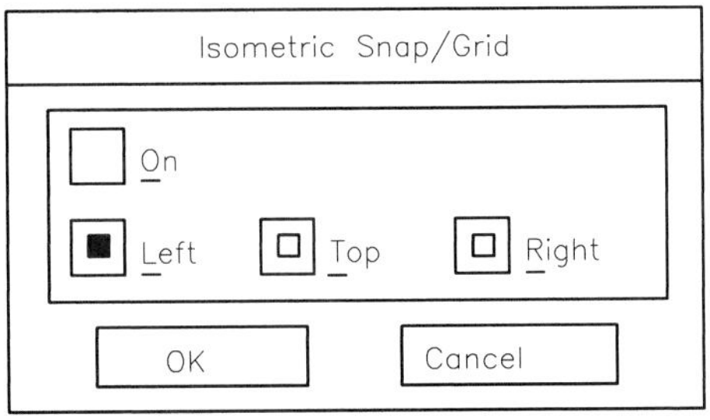

Figure 11-20 *Dialog box for isometric snap/grid*

Exercise 3 *General*

Write a DCL program for the dialog box in Figure 11-21 that will enable you to insert a block. Also, write an AutoLISP program that will load, display, and control the dialog box. The values shown in the edit boxes for insertion point, scale, and rotation are the default values.

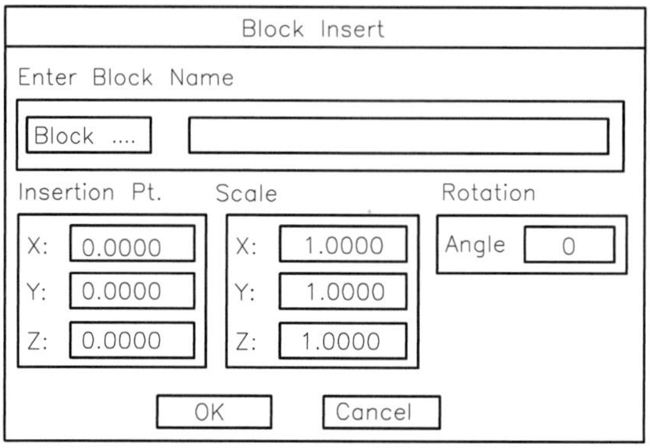

Figure 11-21 *Dialog box for inserting blocks*

Exercise 4 *General*

Write a DCL program for the dialog box in Figure 11-22 that will enable you to select the angle and angle precision. Also, write an AutoLISP program that will load, display, and control the dialog box.

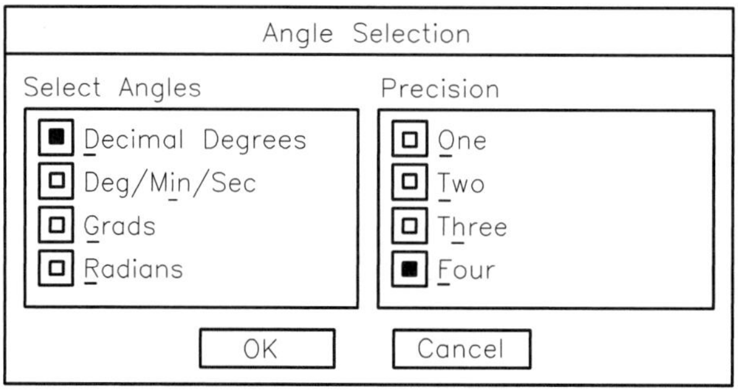

Figure 11-22 *Dialog box for angle selection*

Answers to Self-Evaluation Test

1. T, **2.** T, **3.** T, **4.** T, **5.** T, **6.** T, **7.** T, **8.** T, **9.** F, **10.** F

Chapter 12

DIESEL: A String Expression Language

Learning Objectives

After completing this chapter, you will be able to:

- *Use DIESEL to customize a Status Bar.*
- *Use the **MODEMACRO** system variable.*
- *Write macro expressions using DIESEL.*
- *Use AutoLISP with **MODEMACRO**.*

DIESEL

DIESEL (Direct Interpretively Evaluated String Expression Language) is a string expression language. It can be used to display a user-defined text string (macro expression) in the Status Bar by altering the value of the AutoCAD system variable **MODEMACRO**. The value assigned to **MODEMACRO** must be a string, and the output thus generated will be a string. It is fairly easy to write a macro expression in DIESEL, and it is an important tool for customizing AutoCAD. However, DIESEL is slow, and it is not intended to function like AutoLISP or DCL. You can use AutoLISP to write and assign a value to the **MODEMACRO** variable, or you can write the definition of the **MODEMACRO** expression in the menu files. A detailed explanation of the DIESEL functions and the use of DIESEL in writing a macro expression is discussed later in this chapter.

STATUS BAR

When you are in AutoCAD, a Status Bar is displayed at the bottom of the graphics screen (Figure 12-1). This line contains some useful information and tools that make it easy to change the status of some AutoCAD functions. To change the status, you must click on the buttons. For example, if you want to display grid lines on the screen, click on the **GRID** button. The Status Bar contains the following information:

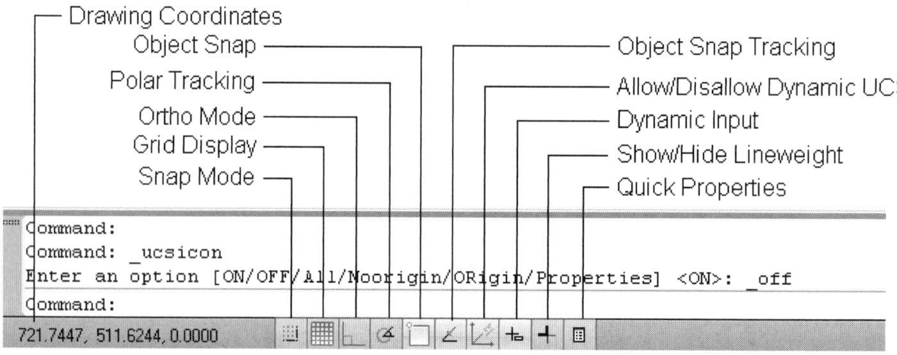

Figure 12-1 Default Status Bar display

Drawing Coordinates. The coordinate information displayed in the left corner of the Status Bar can be static or dynamic. If the coordinate display is static, the coordinate values displayed in the Status Bar change only when you specify a point. However, if the coordinate display is dynamic (default setting), AutoCAD constantly displays the absolute coordinates of the graphics cursor with respect to the UCS origin. AutoCAD can also display the polar coordinates (length<angle) if you are in an AutoCAD command. Some of the buttons available on the status bar are discussed next.

Snap Mode. If **Snap Mode** is on, the cursor snaps to the snap point.

Grid Display. If **Grid Display** is on, grid lines are displayed on the screen.

Ortho Mode. If **Ortho Mode** is on, a line can be drawn in a vertical or horizontal direction.

Polar Tracking. When the **Polar Tracking** is on, the cursor snaps to polar angles as set in the **Polar Tracking** button of the **Drafting Settings** dialog box.

Object Snap. If **Object Snap** is on, you can use the running object snaps. If Object Snap is off, the running object snaps are temporarily disabled. The status of Object Snap (Off or On) does not prevent you from using regular object snaps.

Object Snap Tracking. When **Object Snap Tracking** is on, you can track from a point. The object snap must be set before using this option.

Allow/Disallow Dynamic UCS. When **Allow/Disallow Dynamic UCS** is on, you can ensure that the XY plane of the UCS gets dynamically aligned with the selected face of the model.

Dynamic Input. When **Dynamic Input** is on, you can invoke the commands using pointer inputs and enter values using the pointer inputs.

Show/Hide Lineweight. When **Show/Hide Lineweight** is on, the objects are displayed with the assigned width.

Quick Properties. When **Quick Properties** is on, the **Quick Properties** panel is displayed for the selected object.

MODEMACRO SYSTEM VARIABLE

The AutoCAD system variable **MODEMACRO** can be used to display a new text string in the Status Bar. You can also display the value returned by a macro expression using the DIESEL language, which is discussed in a later section of this chapter. **MODEMACRO** is a system variable and you can assign a value to this variable by entering **MODEMACRO** at the Command prompt or by using the **SETVAR** command. For example, if you want to display **Customizing AutoCAD** in the Status Bar, enter **SETVAR** at the Command: prompt and then press ENTER. AutoCAD will prompt you to enter the name of the system variable. Enter **MODEMACRO** and then press ENTER again. Now you can enter the text you want to display in the Status Bar. After you enter **Customizing AutoCAD** and press ENTER, the Status Bar will display the new text.

Command: **SETVAR**
Enter Variable name or [?]: **MODEMACRO**
Enter new value for MODEMACRO, or . for none<" ">: **Customizing AutoCAD**

You can also enter **MODEMACRO** at the Command prompt and then enter the text that you want to display in the Status Bar.

Command: **MODEMACRO**
Enter new value for MODEMACRO, or . for none<"">: **Customizing AutoCAD**

Once the value of the **MODEMACRO** variable is changed, it retains that value until you enter a new value, start a new drawing, or open an existing drawing file. If you want to display the standard text in the Status Bar, enter a period (.) at the prompt **Enter new value for MODEMACRO, or . for none <"">:**. The value assigned to the **MODEMACRO** system variable is not saved with the drawing, in any configuration file, or anywhere in the system.

Command: **MODEMACRO**
Enter new value for MODEMACRO, or . for none<"">:

CUSTOMIZING THE STATUS BAR

The information contained in the Status Bar can be divided into two parts: toggle functions and coordinate display. The toggle functions part consists of the status of **Snap Mode**, **Grid Display**, **Ortho Mode**, **Polar Tracking**, **Object Snap**, **Object Snap Tracking**, **Allow/Disallow Dynamic UCS**, **Dynamic Input**, **Show/Hide Lineweight**, and **Quick Properties** (Figure 12-1). The coordinate display displays the *X*, *Y*, and *Z* coordinates of the cursor. The Status Bar can be customized to your requirements by assigning a value to the **MODEMACRO** system variable. The value assigned to this variable is displayed left-justified in the status bar at the bottom of the AutoCAD window. The number of characters that can be displayed in the Status Bar depends on the system display and the size of the AutoCAD window. The coordinate display field cannot be changed or edited.

The information displayed in the Status Bar is a valuable resource. Therefore, you must be careful when selecting the information to be displayed in the Status Bar. For example, when working on a project, you may like to display the name of the project in the Status Bar. If you are using several dimensioning styles, you could display the name of the current dimensioning style (DIMSTYLE) in the Status Bar. Similarly, if you have several text files with different fonts, the name of the current text file (TEXTSTYLE) and the text height (TEXTSIZE) can be displayed in the Status Bar. Sometimes, in 3D drawings, if you need to monitor the viewing direction (VIEWDIR), the camera coordinate information can be displayed in the Status Bar. Therefore, the information to be displayed in the Status Bar depends on the drawing requirements. AutoCAD lets you customize this line and have any information displayed in the Status Bar that you think is appropriate for your application.

MACRO EXPRESSIONS USING DIESEL

You can also write a macro expression using DIESEL to assign a value to the **MODEMACRO** system variable. The macro expressions are similar to AutoLISP functions, with some differences. For example, the drawing name can be obtained by using the AutoLISP statement **(getvar dwgname)**. In DIESEL, the same information can be obtained by using the macro expression **$(getvar,dwgname)**. However, unlike the case with AutoLISP, the DIESEL macro expressions return only string values. The format of a macro expression is given next.

$(function-name,argument1,argument2, ...)

Example
$(getvar,dwgname)

Here, **getvar** is the name of the DIESEL string function and **dwgname** is the argument of the function. There must not be any spaces between different elements of a macro expression. For example, spaces between the **$** sign and the open parentheses are not permitted. Similarly, there must not be any spaces between the comma and the argument, **dwgname**. All macro expressions must start with a **$** sign.

The following example illustrates the use of a macro expression using DIESEL to define and then assign a value to the **MODEMACRO** system variable.

Example 1

Using the **MODEMACRO** command, redefine the Status Bar to display the following information in the Status Bar:

Project name (**Cust-Acad**)
Name of the drawing (**DEMO**)
Name of the current layer (**OBJ**)

Note that in this example the project name is **Cust-Acad**, the drawing name is **DEMO**, and the current layer name is **OBJ**.

Before entering the **MODEMACRO** command, you need to determine how to retrieve the required information from the drawing database. For example, here the project name **(Cust-Acad)** is a user-defined name that lets you know the name of the current project. This project name is not saved in the drawing database. The name of the drawing can be obtained using the DIESEL string function **getvar $(getvar,dwgname)**. Similarly, the **getvar** function can also be used to obtain the name of the current layer, **$(getvar,clayer)**. Once you determine how to retrieve the information from the system, you can use the **MODEMACRO** system variable to obtain the new Status Bar. For Example 1, the following DIESEL expression will define the required Status Bar.

Command: **MODEMACRO**
Enter new value for MODEMACRO, or . for none<“”>: **Cust-AcadN:$(GETVAR,dwgname)**
L:$(GETVAR,clayer)

Explanation

Cust-Acad

Cust-Acad is assumed to be the project name you want to display in the Status Bar.

N:$(GETVAR,dwgname)
Here, N: is used as an abbreviation for the drawing name. The **getvar** function retrieves the name of the drawing from the system variable **dwgname** and displays it in the Status Bar, next to N:.

L:$(GETVAR,clayer)
Here L: is used as an abbreviation for the layer name. The **getvar** function retrieves the name of the current layer from the system variable **clayer** and displays it in the Status Bar.

The specified information will be added to the Status Bar and the new Status Bar is shown in Figure 12-2.

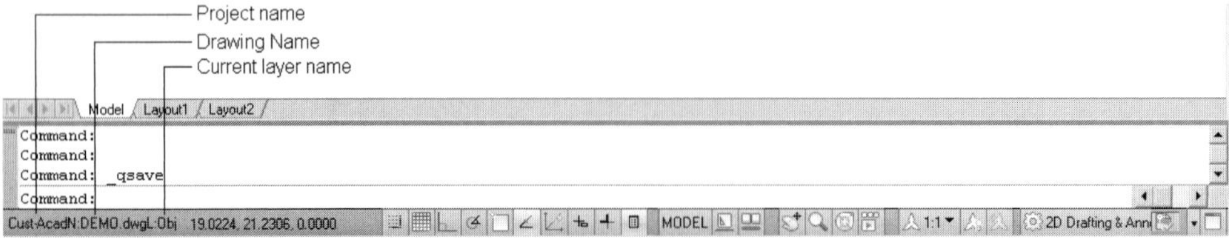

***Figure 12-2** New Status Bar for Example 1*

Example 2

Using the **MODEMACRO** command, redefine the Status Bar to display the following information in it, see Figure 12-3.

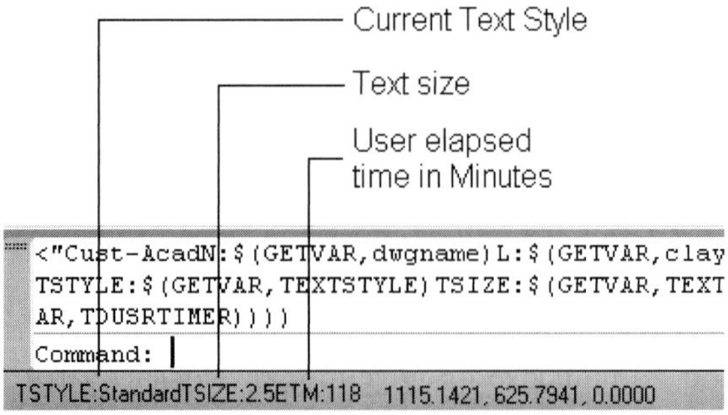

***Figure 12-3** Status Bar for Example 2*

Name of the current textstyle
Size of text
User-elapsed time in minutes

In this example, the abbreviations for text style, text size, and user-elapsed time in minutes are TSTYLE:, TSIZE:, and ETM:, respectively.

Command: **MODEMACRO**
Enter new value for MODEMACRO, or . for none<“”>: **TSTYLE:$(GETVAR,**

TEXTSTYLE)TSIZE:$(GETVAR,TEXTSIZE) ETM:$(FIX,$(*,60,$(*,24,$(GETVAR, TDUSRTIMER))))

Explanation

TSTYLE:$(GETVAR,TEXTSTYLE)
The **getvar** function obtains the name of the current textstyle from the system variable **TEXTSTYLE** and displays it next to **TSTYLE**: in the Status Bar.

TSIZE:$(GETVAR,TEXTSIZE)
The **getvar** function obtains the current size of the text from the system variable **TEXTSIZE** and then displays it next to **TSIZE**: in the Status Bar.

ETM:$(FIX,$(*,60,$(*,24,$(GETVAR,TDUSRTIMER))))
The **getvar** function obtains the user-elapsed time from the system variable **TDUSRTIMER** in the following format:

<Number of days>.<Fraction>

Example:
0.03206400 (time in days)
To change this time into minutes, multiply the value obtained from the system variable **TDUSRTIMER** by 24 to change it into hours, and then multiply the product by 60 to change the time into minutes. To express the minutes value without a decimal, determine the integer value using the DIESEL string function FIX.

Example:
Assume that the value returned by the system variable **TDUSRTIMER** is 0.03206400. This time is in days. Use the following calculations to change the time into minutes, and then express the time as an integer:

0.03206400 days x 24 = 0.769536 hr
0.769536 hr x 60 = 46.17216 min
integer of 46.17216 min = 46 min

USING AutoLISP WITH MODEMACRO

Sometimes the DIESEL expressions can be as long as those shown in Example 1 and Example 2. It takes time to type the DIESEL expression, and there is always a possibility of errors in entering the expression. Also, if you need several different Status Bar displays, you need to type them every time you want a new Status Bar display. This can be time-consuming and sometimes confusing.

To make it convenient to change the Status Bar display, you can use AutoLISP to write a DIESEL expression. It is easier to load an AutoLISP program, and it also eliminates any errors that might be caused by typing a DIESEL expression. The following example illustrates the use of AutoLISP to write a DIESEL expression to assign a new value to the **MODEMACRO** system variable.

Example 3

Using AutoLISP, redefine the value assigned to the **MODEMACRO** system variable to display the following information in the Status Bar:

Name of the current text style
Size of text
User-elapsed time in minutes

In this example the abbreviations for text style, text size, and user-elapsed time in minutes are TSTYLE:, TSIZE:, and ETM:, respectively.

The following file is a listing of the AutoLISP program for Example 3. The name of the file is *etm.lsp*. **The line numbers are not a part of the file; they are shown here for reference only.**

```
1  (defun c:etm ( )
2  (setvar "MODEMACRO"
3  (strcat
4    "TSTYLE:$(getvar,textstyle)"
5    " TSIZE:$(getvar,textsize)"
6    " ETM:$(fix,$(*,60,$(*,24,
7    $(getvar,tdusrtimer))))"
8    )
9   )
10 )
```

Explanation

Line 3
(strcat
The AutoLISP function **strcat** links the string value of lines 4 through 7 and returns a single string that becomes a DIESEL expression for the **MODEMACRO** command.

Line 4
"TSTYLE:$(getvar,textstyle)"
This line is a DIESEL expression in which **getvar**, a DIESEL string function, retrieves the value of the system variable **textstyle** and **$(getvar,textstyle)** is replaced by the name of the textstyle. For example, if the textstyle is STANDARD, the line will return "**TSTYLE:STANDARD**". This is a string because it is enclosed in quotes.

Lines 6 and 7
" ETM:$(fix,$(*,60,$(*,24,
$(getvar,tdusrtimer))))"
These two lines return **ETM:** and the time in minutes as a string. The **fix** is a DIESEL string function that changes a real number to an integer.

To load this AutoLISP file (*etm.lsp*), use the following commands. In this example, the file name and the function name are the same (ETM).

Command: **(load "ETM")**
ETM
Command: **ETM**

DIESEL EXPRESSIONS IN MENUS

You can also define a DIESEL expression in the screen, tablet, pull-down, or button menu. When you select the menu item, it will automatically assign the value to the **MODEMACRO** system variable and then display the new Status Bar. The following example illustrates the use of the DIESEL expression in the screen menu.

Example 4

Write a DIESEL macro for the screen menu that displays the following information in the Status Bar, see Figure 12-4.

Macro-1	**Macro-2**	**Macro-3**
Project name	Pline width	Dimtad
Drawing name	Fillet radius	Dimtix
Current layer	Offset distance	Dimscale

The following file is a listing of the screen menu that contains the definition of three DIESEL macros for Example 4. This menu can be loaded using the **MENU** command and then entering the name of the menu file. If you select the first item, DIESEL1, it will display the new Status Bar.

```
***MENUGROUP=MENU1
***POP1
[*DIESEL*]
[DIESEL1:]^C^CMODEMACRO;$M=Cust-Acad,N:$(GETVAR,DWGNAME)+
,L:$(GETVAR,CLAYER);
[DIESEL2:]^C^CMODEMACRO;$M=PLWID:$(GETVAR,PLINEWID),+
FRAD:$(GETVAR,FILLETRAD),OFFSET:$(GETVAR,OFFSETDIST),+
LTSCALE:$(GETVAR,LTSCALE);
[DIESEL3:]^C^CMODEMACRO;$M=DTAD:$(GETVAR,DIMTAD),+
DTIX:$(GETVAR,DIMTIX),DSCALE:$(GETVAR,DIMSCALE);
```

Note

Write this program in any text editor like Notepad and then save it as a .mnu file. The process of creating the screen menu files and executing it is discussed in detail in Chapter 19.

***Figure 12-4** Status Bar for Example 4*

MACROTRACE SYSTEM VARIABLE

The **MACROTRACE** variable is an AutoCAD system variable that can be used to debug a DIESEL expression. The default value of this variable is 0 (off). It is turned on by assigning a value of 1 (on). When on, the **MACROTRACE** system variable will evaluate all DIESEL expressions and display the results in the command prompt area. For example, if you have defined several DIESEL expressions in a drawing session, all of them will be evaluated at the same time and the messages, if any, will be displayed in the command prompt area.

Example:
Command: **MODEMACRO**
Enter new value for MODEMACRO, or . for none<“”>: **$(getvar,dwgname),$(getvar clayer)**

Note that in this DIESEL expression, a comma is missing between getvar and clayer. If the **MACROTRACE** system variable is on, the following message will be displayed in the Command prompt area.

Eval: $(GETVAR,DWGNAME)
====>UNNAMED

Eval: $(GETVAR CLAYER)
Err: $(GETVAR CLAYER)??

This error message gives you an idea about the location of the error in a DIESEL expression. In the previous example, the first part of the expression successfully returns the name of the drawing (unnamed) and the second part of the expression results in the error message. This confirms that there is an error in the second part of the DIESEL expression. You can further determine the cause of the error by comparing it with the error messages in the following table.

Error Message	**Description**
$?	Syntax error
$?(func,??)	Incorrect argument to function
$(func)??	Unknown function
$(++)	Output string too long

DIESEL STRING FUNCTIONS

Like AutoLISP, you can use DIESEL functions to do some mathematical operations, retrieve the values from the drawing database, and display the values in the Status Bar in a predetermined order. For example, you can add, subtract, multiply, and divide the numbers. You can also obtain the values of some system variables and display them in the Status Bar. The maximum number of parameters that the DIESEL expression can contain is 10. This number includes the name of the function. The following are some frequently used DIESEL string functions.

Addition

Format **$(+,num1,num2,num3 - - -)**

This function **(+)** calculates the sum of the numbers that are to the right of the plus (+) sign. The numbers can be integers or real.

Example:

$(+,2,5)	returns 7
$(+,2,5,50.75)	returns 57.75

Note
You can test the calculations by assigning the DIESEL expression to the ***MODEMACRO*** *system variable.*

Command: **MODEMACRO**
New value for MODEMACRO, or . for none<“”>: **$(+,2,5)**

AutoCAD will return a value of 7, which will be displayed in the Status Bar. It is important to note that the values returned by the DIESEL string expressions are string values.

Subtraction

Format **$(-,num1,num2,num3,- - -)**

This function **(-)** subtracts the second number from the first number. If there are more than two numbers, the second and the subsequent numbers are added and their sum is subtracted from the first number.

Examples:

$(-,28,14)	returns 14
$(-,25,7,11.5)	returns 6.5

Multiplication

Format **$(*,num1,num2,num3,- - -)**

This function **(*)** calculates the product of the numbers that are to the right of the asterisk.

Examples:

$(*,2,5)	returns 10
$(*,2,5,3.0)	returns 30
$(*,2,5,3.25)	returns 32.5

Division

Format **$(/,num1,num2,num3 - - -)**

This function **(/)** divides the first number by the second number. If there are more than two numbers, the first number is divided by the product of the second and subsequent numbers.

Examples:

$(/,3,2)	returns 1.5
$(/,3.0,2)	returns 1.5
$(/,200,5,4)	returns 10
$(/,200.0,5.5)	returns 36.36363636
$(/,200,-5)	returns -40
$(/,-200,-5.0)	returns 40

Relational Statements

Some DIESEL expressions involve features that test a particular condition. If the condition is true, the expression performs a certain function; if the condition is not true, the expression performs another function. The following are the various relational statements used in DIESEL expressions.

Equal to

Format **$(=,num1,num2)**

This function **(=)** checks whether the two numbers are equal. If they are, the condition is true and the function returns **1**. Similarly, if the specified numbers are not equal, the condition is false and the function returns **0**.

Examples:

$(=,5,5)	returns 1
$(=,5,4.9)	returns 0
$(=,5,-5)	returns 0

Not equal to

Format **$(!=,num1,num2)**

This function **(!=)** checks whether the two numbers are **not equal**. If they are not equal, the condition is true and the function returns **1**; otherwise the function returns **0**.

Examples:

$(!=,50,4)	returns 1
$(!=,50,50)	returns 0
$(!=,50,-50)	returns 1

Less than

Format **$(<,num1,num2)**

This function **(<)** checks whether the first number **(num1)** is less than the second number **(num2)**. If it is true, the function returns **1**; otherwise the function returns **0**.

Examples:

$(<,3,5)	returns 1
$(<,5,3)	returns 0
$(<,3.0,5)	returns 1

Less than or equal to

Format **$(<=,num1,num2)**

This function **(<=)** checks whether the first number **(num1)** is less than or equal to the second number **(num2)**. If it is true, the function returns **1**; otherwise the function returns **0**.

Examples:

$(<=,10,15) returns 1
$(<=,19,10) returns 0
$(<=,-2.0,0) returns 1

Greater than

Format **$(>,num1,num2)**

This function **(>)** checks whether the first number **(num1)** is greater than the second number **(num2)**. If it is true, the function returns **1**; otherwise the function returns **0**.

Examples:

$(>,15,10) returns 1
$(>,20,30) returns 0

Greater than or equal to

Format **$(>=,num1,num2)**

This function **(>=)** checks whether the first number **(num1)** is greater than or equal to the second number **(num2)**. If it is true, the function returns **1**; otherwise the function returns **0**.

Examples:

$(>=,78,50) returns 1
$(>=,78,88) returns 0

eq function

Format **$(eq,value1,value2)**

This function **(eq)** checks whether the two string values are equal. If they are, the condition is true and the function returns **1**; otherwise the function returns **0**.

Examples:

$(eq,5,5) returns 1
$(eq,yes,yes) returns 1
$(eq,yes,no) returns 0

angtos

Format **$(angtos,angle[,mode,precision])**

The **angtos** function returns the angle expressed in radians in a string format. The format of the string is controlled by the mode and precision settings.

Examples:

$(angtos,0.588003,0,4)	returns	33.6901
$(angtos,-1.5708,1,2)	returns	270d0'

Note
The following modes are available in AutoCAD.

ANGTOS MODE	*EDITING FORMAT*
0	*Decimal degrees*
1	*Degrees/minutes/seconds*
2	*Grads*
3	*Radian*
4	*Surveyor's units*

***Precision** is an integer number that controls the number of decimal places. Precision corresponds to the **AUPREC** system variable. The minimum value of **precision** is 0, the maximum is 8.*

eval function

Format **$(eval,string)**

The **eval** function passes the string to the DIESEL evaluator and the result obtained after evaluating the string is returned and displayed in the Status Bar.

Examples:

$(eval,welcome)	returns	welcome
$(eval,$(getvar,dimscale))	returns	value of dimscale

fix function

Format **$(fix,num)**

The **fix** function converts a real number into an integer by truncating the digits after the decimal.

Examples:

$(fix,42.573)	returns	42
$(fix,-23.50)	returns	-23

getvar function

Format **$(getvar,varname)**

The **getvar** function retrieves the value of an AutoCAD system variable.

Examples:

$(getvar,dimtad)	returns	value of dimtad
$(getvar,clayer)	returns	current layer name

rtos function

Format	**$(rtos,number)**
or	**$(rtos,number,mode,precision)**

The **rtos** function changes a given number into a real number and the format of the real number is determined by the mode and precision values.

Examples:

$(rtos,50)	returns	50
$(rtos,1.5,5,4)	returns	1 1/2

Note

The following linear unit modes are available in AutoCAD:

MODE VALUE	*STRING FORMAT*
1	*Scientific*
2	*Decimal*
3	*Engineering*
4	*Architectural*
5	*Fractional*

***Precision** is an integer number that controls the number of decimal places. Precision corresponds to the **LUPREC** system variable, and the mode corresponds to **LUNITS**. If the mode and precision values are omitted, AutoCAD uses the current values of **LUNITS** and **LUPREC** as set with the **UNITS** command.*

if function

Format **$(if,condition,then,else)**

The **if** function evaluates the first expression (then), if the value returned by the specified condition is non zero. The second expression (else) is evaluated if the specified condition returns 0.

(if condition then [else])

Where **condition** ----------------- Specified conditional statement
then -------------------- Expression evaluated if the condition returns non zero
else -------------------- Expression evaluated if the condition returns 0

Examples:

$(if,$(=,7,7),true)	returns true
$(if,$(=,5,7),true,false)	returns false
$(if,1.5,true,false)	returns true

strlen function

Format **$(strlen,string)**

The **strlen** function returns an integer number that designates the number of characters contained in the specified string.

Example:
$(strlen,Customizing AutoCAD) returns 19

upper function

Format **$(upper,string)**

The **upper** function returns the specified string in uppercase.

Example:
$(upper,Customizing) returns CUSTOMIZING

edtime function

Format **$(edtime,date,display-format)**

The **edtime** function can be used to edit the date and display it in a format specified by display-format. For example, the Julian date obtained from the **DATE** system variable is obtained in the form 2449013 . 85156759. The **edtime** function can be used to display this date in an understandable form. The following table gives the format of the phrases that can be used with the **edtime** function.

FORMAT	OUTPUT	FORMAT	OUTPUT
D	5	H	2
DD	05	HH	02
DDD	Sat	MM	23
DDDD	Saturday	SS	12
M	1	MSEC	325
MO	11	AM/PM	PM
MON	Apr	am/pm	am
MONTH	April	A/P	P
YY	03	a/p	p
YYYY	2003		

Examples:
$(edtime,$(getvar,date),DDDD","DD MONTH YY - HH:MMAM/PM)
returns Wednesday,25 May 05 - 06:52PM

In the above example, the **getvar** function retrieves the date in Julian format. The **edtime** function displays the date as specified by the display-format. The following illustration shows the corresponding fields of the display-format and the value returned by the edtime function.

DDDD","DD MONTH YY - HH:MMAM/PM

Where **DDDD** -------------------- Monday
, -------------------- ,
DD -------------------- 28
MONTH ------------------- April
YY -------------------- 03
- -------------------- -
HH -------------------- 08
MM -------------------- 52
AM/PM -------------------- PM

Self-Evaluation Test

Answer the following questions and then compare them to those given at the end of this chapter:

1. DIESEL (Direct Interpretively Evaluated String Expression Language) is a string expression language. (T/F)

2. The value assigned to the **MODEMACRO** variable is a string, however the output it generates is not a string. (T/F)

3. You cannot define a DIESEL expression in a pull-down menu. (T/F)

4. The coordinate information displayed in the Status Bar can be dynamic only. (T/F)

5. Once the value of the **MODEMACRO** variable is changed, it retains that value until you enter a new value, start a new drawing, or open an existing drawing file. (T/F)

6. The number of characters that can be displayed in the layer mode field on any system is 38. (T/F)

7. The coordinate display field cannot be changed or edited. (T/F)

8. You can write a macro expression using DIESEL to assign a value to the **MODEMACRO** system variable. (T/F)

9. In DIESEL, the drawing name can be obtained by using the macro expression **$(getvar,dwgname)**. (T/F)

10. You cannot use AutoLISP to write a DIESEL expression. (T/F)

Review Questions

Answer the following questions:

1. All macro expressions must start with a _________ sign.

2. The display-format DDD will return Monday as _________.

3. The DIESEL expression, $(upper,AutoCAD), will return _________.

4. The DIESEL expression, $(strlen,AutoCAD), will return _________.

5. The DIESEL expression, $(if,$(=,3,2),yes,no), will return _________.

6. The DIESEL expression, $(fix,-17.75), will return _________.

7. The DIESEL expression, $(eq,Customizing,customizing) will return _________.

8. The DIESEL expression, $(/,81,9,9) will return _________.

9. The MACROTRACE system variable can be used to _________ a DIESEL expression.

Exercises

Exercise 1 *General*

Using the **MODEMACRO** command, redefine the Status Bar to display the following information in it:

Your name
Name of drawing

Exercise 2 *General*

Using the **MODEMACRO** command, redefine the Status Bar to display the following information in it:

Name of the current dimension style
Dimension scale factor (dimscale)
User-elapsed time in hours

The abbreviations for dimstyle, dimscale, and user-elapsed time in hours are DIMS:, DIMFAC:, and ETH:, respectively.

Exercise 3 *General*

Using AutoLISP, redefine the Status Bar to display the following information in it:

Name of the current dimension style
Dimension scale factor (dimscale)
User-elapsed time in hours

Answers to Self-Evaluation Test

1. T, **2.** F, **3.** F, **4.** F, **5.** T, **6.** F, **7.** T, **8.** T, **9.** T, **10.** F

Chapter 13

Visual Basic for Application

Learning Objectives

After completing this chapter, you will be able to:

- *Load and run sample VBA projects.*
- *Utilize the Visual Basic Editor.*
- *Understand and use AutoCAD objects.*
- *Use object properties.*
- *Apply and use AutoCAD methods.*

ABOUT VISUAL BASIC

The original BASIC was developed in 1963. BASIC (Beginners All-purpose Symbolic Instruction Code) was intended to be an accessible, user-friendly programming language. BASIC was the language supplied with many of the early microcomputers starting in the late 1970s. It continued developing until the advent of Windows. In 1991, the first version of **Visual Basic (VB)** appeared, followed by later versions spaced about a year apart. The current version, like its predecessors, lets even a beginning programmer create a user-friendly graphical user interface (GUI) with a small fraction of the effort required previously. It used to take a team of programmers working with a language like C, which came with a stack of reference materials nearly a foot thick, to do the same thing.

Another advantage inherent to Visual Basic is that engineers may have old BASIC programs used in sizing components. This code, minus old input/output, can be used as the core of new Visual Basic modules to parametrically design and draft the component.

VISUAL BASIC FOR APPLICATION

Autodesk first licensed **Visual Basic® for Applications (VBA)** from Microsoft for use in R14. Dozens of other large software companies have similarly adopted VBA. The VBA language tools are either identical or very similar to those in the stand-alone Visual Basic. Programmers who know how to write macros for Microsoft Office only have to learn AutoCAD-specific functionality to program VBA in AutoCAD.

One of the features of VBA is its ability to communicate with other applications such as Microsoft Excel, Microsoft Word, Microsoft Access, and Visual Basic using **ActiveX Automation**. Automation lets you access and manipulate the objects and functionality of AutoCAD from Excel or another application supporting ActiveX. Alternately the objects and functionality of Excel, for example, can be used inside AutoCAD VBA programming. This cross-application macro programming capability does not exist in AutoLISP. Of course, before you can use AutoCAD's objects in some other software supporting ActiveX, you must make the application aware that AutoCAD is installed along with its **Object Library** on the computer.

There are millions of programmers using Visual Basic. A great many more people have programmed in BASIC but not VB. This chapter assumes you have some familiarity with BASIC or programming concepts such as those covered in the AutoLISP chapter of this book and are comfortable looking up commands and syntax in the help system.

In this version of AutoCAD you can now reference a macro from another project and have more than one macro loaded. You can also create libraries of common functions and macros.

USING VBA IN AUTOCAD

VB enables you to write a code that is compiled into an *.exe* file, which is an independently executable file (*.exe*). Whereas, VBA is related to the application and the document in which you write its code. VBA programs are written according to the need. For example, if your company manufactures nuts and bolts, you can use VBA to write a program that will show a dialog box when run. You can use the dialog box to specify the values and dimensions of the nuts and bolts. This enables you to reduce the designing time considerably because whenever there is a change in the dimensions of these components, you do not have to invoke the different commands to create the components.

You can save VBA programs as projects containing the information required to compile and run. VBA in AutoCAD has a separate environment in which you can do all the work related to a project. This environment is called Integrated Development Environment (IDE). This environment is discussed next.

Invoking VB Editor

The VB editor of AutoCAD allows the user to write programming codes known as modules. Collectively these modules form a project. To invoke the VB editor, choose **Manage > Applications > Visual Basic Editor** from the **Ribbon**. This window is also called **Integrated Development Environment** (IDE) window and is shown in Figure 13-1. The windows that are shown in the figure, are opened by choosing **Insert > UserForm** from the menu bar in the IDE environment. These windows are discussed next.

Note

If VB editor is not installed on your system, the ***VBA - Not Installed*** *message box will be displayed on choosing the* ***Visual Basic Editor*** *button. Click on the link given in this message box to download and install the VB editor.*

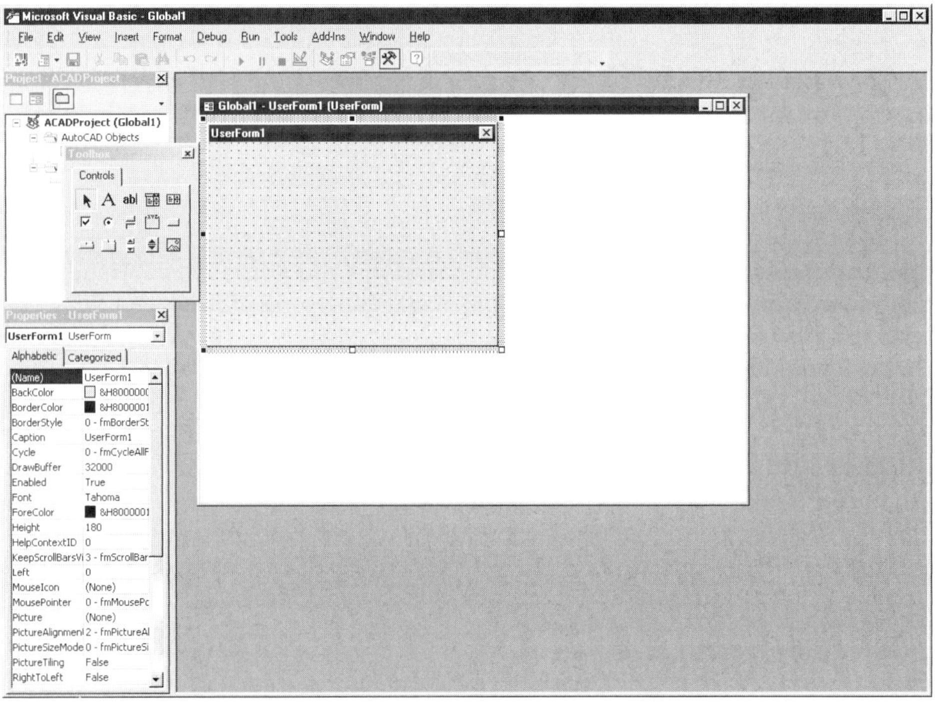

Figure 13-1 *Visual Basic Editor with Project Explorer, Properties window, Module window, User Form, and Toolbox*

Project Explorer Window

The **Project Explorer** window lists all the modules, User forms, and other objects used in a project. The items in the **Project Explorer** window are known as objects. Objects are discussed later in the chapter. This window contains three folders, **AutoCAD Objects**, **Forms**, and **Modules**, as shown in Figure 13-2. The **AutoCAD Objects** folder contains **ThisDrawing** object, which is the current AutoCAD drawing. The **Forms** folder contains Visual Basic forms used in the current project. A form is a custom dialog box you create to interact with the user. The **Modules** folder contains code modules for the current project. The code module object contains generic procedures and functions. When you add a form or code module to your project, VBA adds it to the corresponding folder in the **Project Explorer** window.

***Figure 13-2** The **Project Explorer** window*

Note

*The **Forms** and **Modules** folders are added in the **Project Explorer** window by choosing the options to add from the **Insert** menu in the menu bar.*

*If the **Project Explorer** window is not displayed, choose the **Project Explorer** button from the **Standard** toolbar.*

The **Project Explorer** window is used when the **Module** window is opened on the screen and you want to access the User form. For Example, if you want to view the code of the User form, right-click on the User form in the **Project Explorer** window to display a shortcut menu. Choose the **View Code** option from the shortcut menu. The **User Form Code** window is displayed.

Properties Window

The **Properties** window shows all the properties of a selected object. In VBA, object refers to the form and to every control that is on the form. These controls are buttons, textbox, labels, and other objects for user interface that can be added to the form. These objects have characteristics such as color, size, name, caption, background, and so on. These characteristics not only affect the appearance of the object but the way it responds to the user's action. All these characteristics of the object are called properties that are displayed in the **Properties** window.

The **Properties** window shown in Figure 13-3 is used to modify the properties of the User form, buttons, and modules. You can select an object from the drop-down list present at the top in the window. This list contains all the objects that are in the form. On the left in the **Properties** window are the names of the properties and on the right are their values.

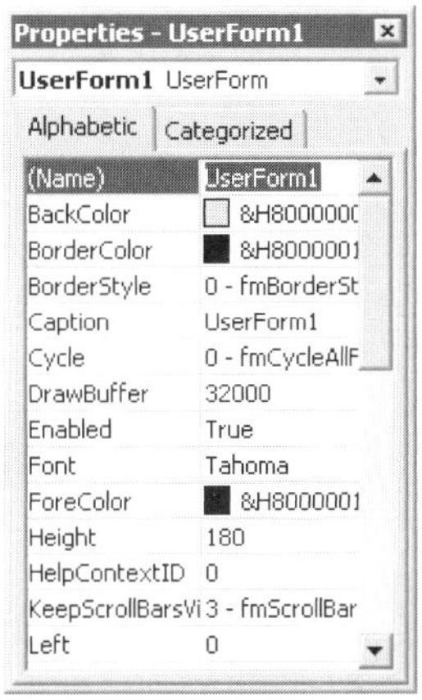

*Figure 13-3 The **Properties** window*

Module Window

The **Module** window is used to type the code. To open the **Module** window, choose **Insert > Module** from the menu bar. The **Module** window appears as shown in Figure 13-4, and you can type the code in it. You can resize the **Module** window by dragging it from one of the corners or edges. By default, the first module is called Module1.

UserForm and Toolbox

The Userform shown in Figure 13-5, is the front end of the program and provides an easy way to interact with the user. On the User form you can place various controls like buttons, check boxes, radio buttons, and so on in order to get the user input. You can insert a User form by choosing **Insert > UserForm** from the menu bar in the VBA IDE environment. When the User form is inserted, the **Toolbox** appears automatically, as shown in Figure 13-6. This box contains the various controls that are added to the form by selecting and dragging to the form.

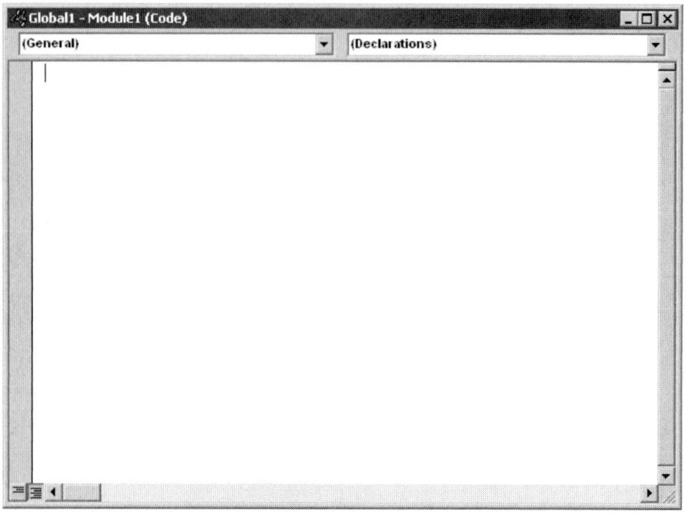

Figure 13-4 *The* ***Module*** *window*

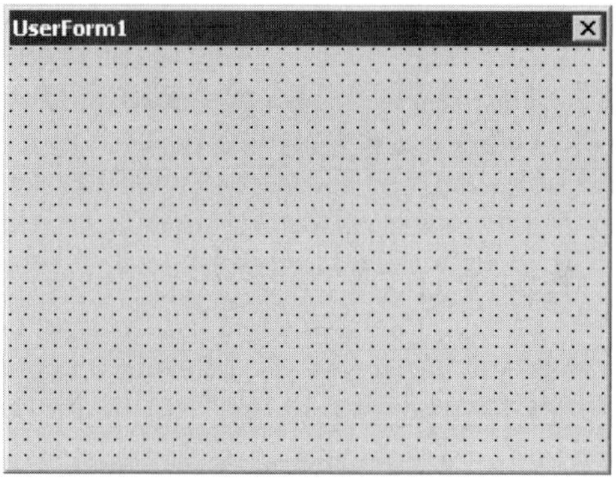

Figure 13-5 UserForm

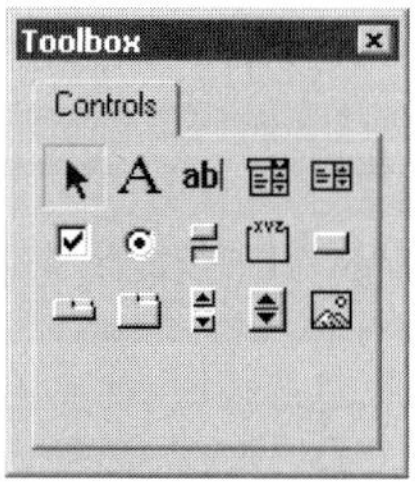

Figure 13-6 Toolbox

OBJECTS IN VBA

The primary purpose of VBA is to control the functionality of the application and automate it with VBA commands. This is made possible through **Objects**. This means that an application makes itself accessible to another application using **Objects**. The basic objects of AutoCAD are Application object and Document object.

Application Object

The application that you are working on is called an Application object. For example, AutoCAD is an Application object. There are some methods and properties associated with Application object. These methods are Quit, ZoomAll, Runmacro, and so on. The Quit method exits the application, the ZoomAll method zooms the objects and fits them in the current viewport, and Runmacro executes a macro for an application.

The properties of an Application object are height, width, caption, and so on. The height property specifies the height of the window, the width property specifies the width of the application window, and the caption is the string that appears on the application window's title bar. This property can be used to find out whether a window contains a specific document.

Document Object

The drawing that is opened in the application window is called Document object. A Document object is any document that can be opened with AutoCAD or any document or drawing that can be displayed in the AutoCAD window. All open documents are referred to by a document collection that is formed by Document objects. The documents collection consists of the following methods and properties.

1. The Count property, which returns the number of items in the collection (number of drawings opened in a session).

2. The Add method, which creates a member object (drawing) and adds it to the appropriate collection.

3. The Close method, which closes the specified drawing, or all open drawings. To access a member object, you can use the Item() method of the Documents collection, specifying the objects index as given next.

   Application.Documents.Item(1)   where 1 is the index number.

   You can also access a member object by its name as given next.

Application.Documents.Item('obj') where obj is the name of the object (filename).

You can open a drawing file as given next.

Documents.Open(filename) where filename is the path location of the object.

You can make any drawing active by calling the Activate() method of the Document Object.

ThisDrawing.Application.Documents ("piston.dwg").Activate
where piston.dwg is the drawing that will be activated.

After this statement is executed, the piston.dwg drawing will become the active document and your VB code can refer to it through This Drawing object. The hierarchy of the objects is discussed in the next section.

Hierarchy of Objects

In AutoCAD everything is an object for VBA. VBA uses hierarchy of objects in order to work.

For Example, AutoCAD is an object for VBA because it is an application. Similarly, an AutoCAD drawing is an object and so are model and paper spaces. To specify an object in a drawing, you need to:

1. Specify the application.
2. Specify the drawing.
3. Specify the object in the drawing.

VBA works in hierarchy to accomplish this. For example, you can use the following VBA code to add an arc.

Application.ActiveDocument.ModelSpace.AddArc(Start, End, Angle)

You can also use **ThisDrawing** as a shortcut for: **Application.ActiveDocument**. To achieve this, modify the above code as given below.

ThisDrawing.ModelSpace.AddArc(Start, End, Angle)

Examples of objects include drawings or documents, geometric elements such as lines or circles, and user interface controls to handle input and output to programs or macros. The "visual" part of Visual Basic refers to familiar user interface controls such as check boxes, scroll bars, and command buttons are used to open or save files in Windows applications.

These controls are screen objects that may be dragged from a Toolbox onto the background object called a form. An example of a simple form is the background window of a dialog box. In the course of dragging a control from the Toolbox onto a form, the coding that provides the functionality of the control is simultaneously added to the form object. Several aspects of true object-oriented programming (OOP) languages such as C++ are missing from Visual Basic. So VB is considered an object-based, rather than an object-oriented, programming language. The missing aspects are more than made up for by the development tools and environment available to the user.

Functions known as methods have been defined in the AutoCAD object library to perform an action on an object, for example, drawing a line in a drawing. The method AddLine adds a line object to a drawing. Properties are functions that set or return information about the state of an object. In the example of a line drawn in a drawing the Color, Layer, Linetype, and Start X would be some of the properties of the Line object.

ADD METHODS

The way to draw in paper space, model space, or in a block is to use the **Add method** such as **AddCircle**, **AddLine**, **AddArc**, and **AddText**. Instead of thinking of drawing in the model space of the current drawing, think in terms of adding a geometry. Dot notation is required to clarify on which object the method is acting. The dot notation reference starts with the most global object first narrowing to the right. Dot notation is used in a similar way with properties.

Using the AddCircle Method to Draw a Circle

The **AddCircle** method requires a predefined center point and radius. Both of these arguments are required. Other input options used in the AutoCAD circle command such as three-points, two-points, or tangent-tangent-radius require additional programming in VBA to calculate the center point and radius for the **AddCircle** method. The format of the **AddCircle** method is given next.

ThisDrawing.ModelSpace.AddCircle centerpoint, radius

or

Set Circle1=ThisDrawing.ModelSpace.AddCircle(centerpoint, radius)

where **centerpoint:** Center point, double precision vector

radius: Radius, double precision number

Examples:

1. ThisDrawing.ModelSpace.AddCircle(cnt1, 10)
   The above code allows you to draw a circle whose center point is cnt1 and radius is 10 where cnt1 is predefined in the program.

2. ThisDrawing.ModelSpace.AddCircle(0,1,3, 8)
   The above code allows you to draw a circle whose center point is (0,1,3) and radius is 8.

Note

The arguments of the first form of the ***AddCircle*** *method follow a required space. The second form of the add method, where the arguments are inside parentheses, is used where the circle object is assigned to a name for use later in the program. For the examples given below only point1 and point2 are predefined points consisting of a vector of variant/double precision or double precision coordinate values. These may be defined with assignment statements as shown in Example 1 on page 13-13.*

The pound sign (#) signifies a double precision number in Basic. Double precision, floating point number are positive numbers in the range 4.94065645841247E-324 to 1.79769313486232E308 and negative numbers in the range -1.79769313486232E308 to -4.94065645841247E-324.

Examples:
ThisDrawing.ModelSpace.AddCircle point1, 3#
Set Circle2= ThisDrawing.ModelSpace.AddCircle(point2, 4#)

Using the AddLine Method to Draw a Line

The **AddLine** method requires two predefined endpoints. The **AddLine** method has syntax similar to **AddCircle** as given next.

ThisDrawing.ModelSpace.AddLine firstpoint, secondpoint

where **firstpoint**: First point, double precision vector
secondpoint: Second point, double precision vector

1. **ThisDrawing.ModelSpace.AddLine(x1, x2)**
   The above code draws a line from point x1 to x2, where x1 and x2 are predefined in the program.

2. **ThisDrawing.ModelSpace.AddLine(0,1,1, 5,8,0)**
   The above code draws a line from coordinate (0,1,1) to (5,8,0).

Note

The second form of the Add method, where the arguments are inside parentheses, may be used for the ***AddLine*** *method or any of the other* ***Add*** *methods below where the object is assigned to a name for later use in the program. A complication for using the Add method this way is that the object defined must be declared in a Dim statement in the General declarations.*

Examples:
ThisDrawing.ModelSpace.AddLine point1, point2
Set Line1=ThisDrawing.ModelSpace.AddLine(point1, point2)

Using the AddArc Method to Draw an Arc

The **AddArc** method format is given next.

ThisDrawing.ModelSpace.AddArc ctrpt, radius, StartAng, EndAng

where	**ctrpt**:	Center point, double precision vector
	radius:	Radius, double precision number
	StartAng:	Arc starting angle in radians, double precision
	EndAng:	Arc ending angle in radians, double precision

Examples:

1. **ThisDrawing.ModelSpace.AddArc point1, 4#, 0#, 1.570796327**

2. **ThisDrawing.ModelSpace.AddArc(cnt1, rad, ang1, ang2)**
   The above code draws an arc whose center point is cnt1, radius is rad, start angle is ang1, and end angle is ang2 and all are predefined in the program.

3. **ThisDrawing.ModelSpace.AddArc(0,0,0, 8, 3.4, 1.453)**

Using the AddText Method to Write Text

The **AddText** method requires a predefined string, insertion point and text height. The **AddText** method syntax is:

ThisDrawing.ModelSpace.AddText textString$, point1, textHeight

where	**textString$:**	Actual text to be displayed
	point1:	Position point, double precision vector
	textHeight:	Text Height, positive double precision number

Example:
ThisDrawing.ModelSpace.AddText "4X .063", point1, 0.25#

Chapter 13

FINDING HELP ON METHODS AND PROPERTIES

You can find excellent help in the VBA Integrated Development Environment with good examples showing the exact syntax necessary for the **AddLine**, **AddCircle**, or any of the particular methods or properties needed to write a parametric program. To access help from the VBA IDE, choose **Help > Microsoft Visual Basic Help** from the menu bar.

You can also find general information such as on Methods, help on Visual Basic key words, and many non-AutoCAD VBA programming examples taken from Excel, Word, or PowerPoint in AutoCAD 2010. The topics under the **Help > Additional Resources** pull-down menu of AutoCAD 2010 include **Developer Help** from which you can select **ActiveX and VBA Developer's Guide**, see Figure 13-7.

Help is available through a shortcut from the AutoCAD drawing editor's help system or from the Visual Basic Editor Integrated Design Environment (VBA IDE). Look under VBA and ActiveX Automation. The **Index** and **Search** tabs of this help menu are very useful. Another way to get help in the VBA IDE, is to use the **Object Browser** that can be accessed from the IDE **View** pull-down menu. Select **AutoCAD** in the top drop-down list and then select object **AcadModelSpace** in the left window, as shown in Figure 13-8. The question mark help

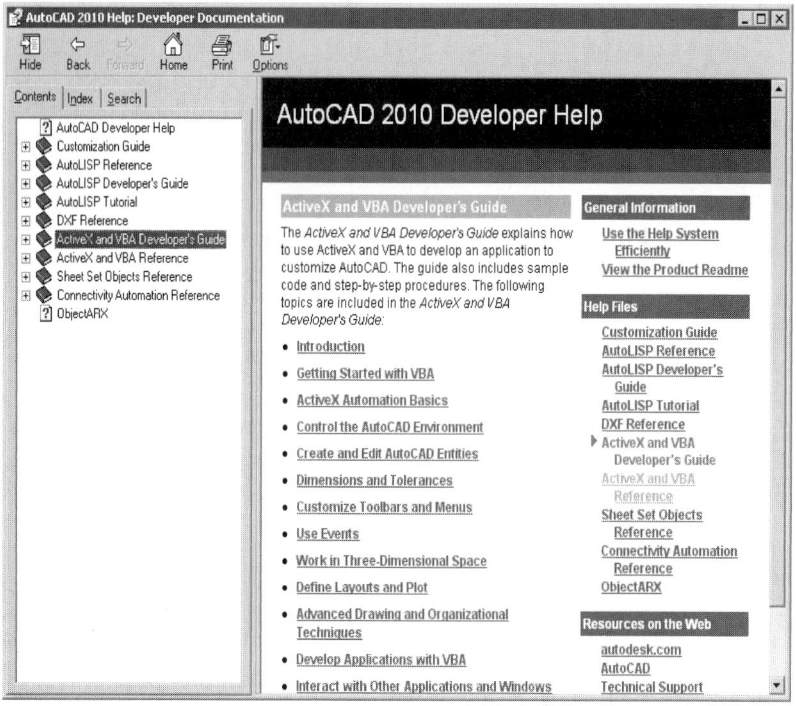

Figure 13-7 *ActiveX and VBA Developer's Guide*

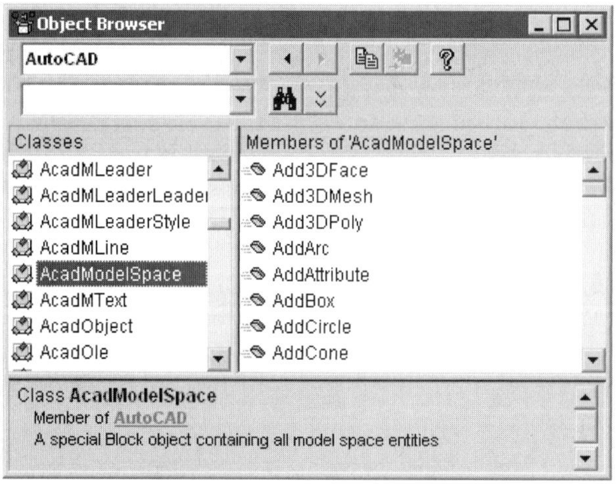

Figure 13-8 *The* ***Object Browser***

button on the toolbar or F1 takes you to a screen where you can access the ModelSpace Collection of help screens on methods, properties, and examples.

LOADING AND SAVING VBA PROJECTS

To start or load a VBA project in the AutoCAD Drawing Editor, you need to choose **Manage > Applications > Visual Basic Editor** from the **Ribbon**, see Figure 13-9. The Integrated Development Environment (IDE) will be invoked. The IDE consists of a number of useful windows that are already discussed in this chapter. These windows can be sized, shown, hidden, or otherwise customized to suit your needs.

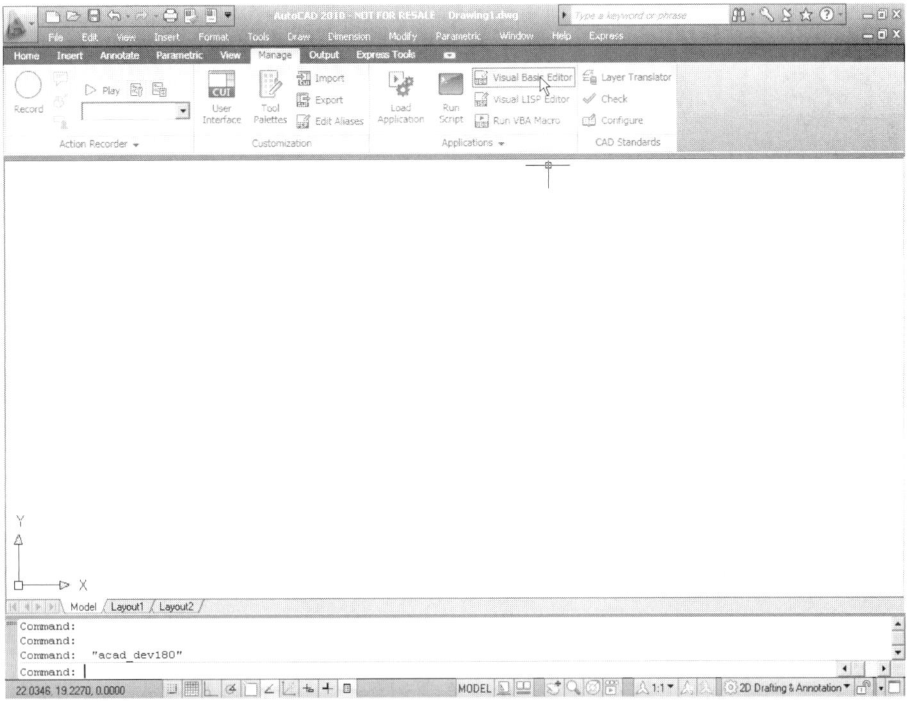

*Figure 13-9 Starting the Visual Basic Editor using the **Ribbon***

It is a good practice to save your VBA program before testing it. To save a project from the VBA IDE, choose **Save** from the **File** menu. The menu items may be more quickly accessed with the combination of the ALT key and the underlined letter in the menu. AutoCAD VBA projects have the file extension *.dvb*.

Example 1

Write a program that will draw a circle centered at 5,5,0 with radius 2, as shown in Figure 13-10. The User Interface Form is shown in Figure 13-11.

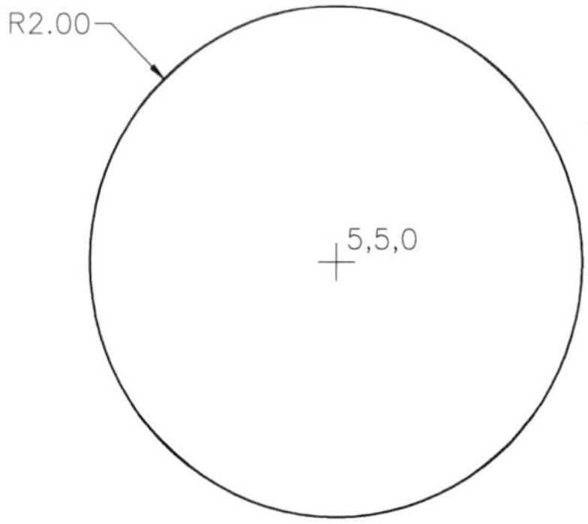

Figure 13-10 *Circle for Example 1*

Figure 13-11 *User Interface Form*

1. Start AutoCAD and open a new file using the imperial setting of the **Start from Scratch** option.

2. Choose **Manage > Applications > Visual Basic Editor** from the **Ribbon**.

3. Choose **Insert > Module** from the menu bar. If the text editor covers the entire screen, reduce the size by clicking the **Restore Window** button on the top right corner.

4. Insert a User form by choosing the **Insert > UserForm** option from the menu bar. The default names will be Module1 and UserForm1.

5. Click the form after insertion to make it the active window. To insert a control such as a command button onto the form, choose the corresponding button on the toolbox. The symbol shown as the second button from the left in the second row of the **Toolbox** in Figure 13-12 inserts the command button control.

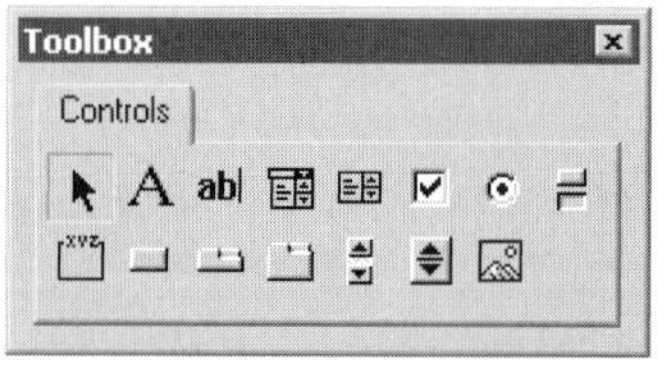

Figure 13-12 *The* ***Toolbox*** *for creating the User Interface Form*

6. Now, resize the **UserForm** and **CommandButton** to the desired size and edit the caption on the button. This can be done using the **Caption** edit box in the **Properties** window at the bottom left corner of screen. The name of the button should be **Draw Circle**. and the name of the User form should be **First Circle**.

A project is the name given to the forms, controls, modules, and programming that makes up a Visual Basic program. To finish a project, it is necessary to write the code underlying the visual control(s). One way to get into the code window for the command button control is to double-click on the button and enter the code for CommandButton1_Click(). Remember that Visual Basic is event driven. Lines 1 through 13 correspond to the code that runs when the command button is chosen.

The following is a listing of the Visual Basic program for Example 1. Enter the code given below between the lines where the cursor is blinking. The line code in line number 4 and 17 will be available in the code window by default, therefore you need not type it again. **The line numbers at the right are for reference only and are not a part of the programming.**

Command Button Code

```
1  'UserForm1 Code to Draw Circle, Radius 2
2  'Centered at 5,5,0. Trigger is mouse click on
3  'command button marked Draw Circle
4  Private Sub CommandButton1_Click()
5  Dim CenterPoint(0 To 2) As Double
6  Dim Radius As Double
7
8  'Data
9  CenterPoint(0) = 5
10 CenterPoint(1) = 5
11 CenterPoint(2) = 0
12 Radius = 2
13
14 'OLE Automation Object Call
15 ThisDrawing.ModelSpace.AddCircle CenterPoint, Radius
16 Unload Me
17 End Sub
```

Explanation

Lines 1 to 3

The first lines are comments or remarks describing the function of the program. Comments make understanding and modifying a program easier and should be used frequently. Comments start with a Rem or use an apostrophe ('). These lines are ignored when the program is run.

Line 4

Private Sub CommandButton1_Click()

This line defines where **Sub CommandButton1_Click()** starts. The subroutine is executed when **CommandButton1** is chosen. It is the code that draws the circle when the command button is chosen. There is no need to type this line because VBA generates it automatically as soon as the command button control is added to the form.

Lines 5 and 6

Dim CenterPoint(0 To 2) As Double

Dim Radius As Double

These two lines are necessary to establish the double precision variable type for the arguments needed by the **AddCircle** method.

Lines 9 through 12

CenterPoint(0) = 5

CenterPoint(1) = 5
CenterPoint(2) = 0
Radius = 2
Here the circle center and radius are assigned values. The center point is defined by *X*, *Y*, and *Z* coordinate values.

Line 15
ThisDrawing.ModelSpace.AddCircle CenterPoint, Radius
This line applies the **AddCircle** method to the **ModelSpace** object, which is part of the ThisDrawing object. Note a blank space between the keyword **AddCircle** and the first argument, CenterPoint is given.

Note
Keywords are words that have been assigned distinct meanings by the programming language. They are reserved and cannot be used as user-defined variables.

Line 16
Unload Me
This is a method that removes UserForm1 from memory and returns the focus back to AutoCAD.

Line 17
End Sub
Like line 4, this line is generated automatically.

Module1 Code

The following code is entered in the Module1 code window.

```
1 'Module 1 General Declarations
2 Sub DrawCircle()
3 UserForm1.Show
4 End Sub
```

Explanation

Line 2 through 4
Sub DrawCircle()
UserForm1.Show
End Sub
This subroutine's function is to create a Macro name, **DrawCircle**, which appears in the **Macro name** edit box of the **Macros** dialog box invoked by choosing **Macro > Macros** from the **Tools** menu in the main AutoCAD window. These four lines of code belong to the Module1 object. Another way to run a project is to choose **Run Sub/UserForm** from the **Run** menu in the VBA IDE.

7. Choose the **Run Macro** button from the **Standard** toolbar to execute the program.

GETTING USER INPUT

The successful execution of a program depends on the input from the end user. For this purpose, VBA has various Get methods that can be used in the program in order to get the input from the user.

GetPoint Method

The **GetPoint** method allows you to enter the *X*, *Y* coordinates or *X*, *Y*, *Z* coordinates of a point. The coordinates of the point can be entered using the keyboard or using pointing device. When you enter the point parameter, it draws a line from the specified point to the current position of the cursor.

The format of the **GetPoint** method is:

P = ThisDrawing.Utility.GetPoint([Point], [Prompt])
Enter a point from the keyboard or select a point in the AutoCAD graphics editor

where	**[Point]:**	Optional reference point, rubber-band origin
	[Prompt]:	Optional prompt to be displayed on screen

Examples:

1. **pnt1=ThisDrawing.Utility.GetPoint("Enter 1st Point")**
This code generates a prompt that asks you to enter the 1st Point. The value of this point is assigned to pnt1.

2. **Pt2=ThisDrawing.Utility.GetPoint(Pnt1,"Enter 2nd Point")**
This code generates a prompt that asks you to enter the 2nd point. The value of this point is assigned to Pt2.

GetDistance Method

The **GetDistance** method lets you enter a distance on the command line, a distance from a given point, or two points, and it then returns the distance as a double precision number. Use this method after the **GetPoint** method. The format of the **GetDistance** method is given next.

d = ThisDrawing.Utility.GetDistance([point], [prompt])

where	**[point]**:	Optional reference point, rubber band origin
	[prompt]:	Optional prompt to be displayed on screen

Examples:

1. **d1 = ThisDrawing.Utility.GetDistance(cen, "specify radius")**
This code generates a prompt and allows you to enter a point with respect to an existing point. The value of distance between the two points is assigned to d1.

2. **x = ThisDrawing.Utility.GetDistance(p1, "specify the height")**
   This code generates a prompt and allows you to enter a point with respect to an existing point. The value of distance between the two points is assigned to x.

GetAngle Method

The **GetAngle** method allows you to enter an angle, either using the keyboard in degrees or by selecting two points. In the case of selecting points, the positive horizontal direction is taken as one leg of the angle, the first point selected as the vertex and the second point defines the second leg. If the point argument is specified, AutoCAD uses this point as the first point or angle vertex. The **GetAngle** method returns the value of the angle in radians as a double precision value. The format of the **GetAngle** method is given next.

ang = ThisDrawing.Utility.GetAngle([point], [prompt])

where	**ang**:	Angle in radians
	[point]:	Optional vertex point
	[prompt]:	Optional screen prompt to clarify angle selection

Examples:

1. **a1 = ThisDrawing.Utility.GetAngle(, "Enter taper angle in degrees")**
   This code generates a prompt and allows you to enter a point with respect to an existing point. It assigns the value of the angle between the two points to a1.

2. **ang3 = ThisDrawing.Utility.GetAngle(p1, "Specify the center point")**
   This code generates a prompt and allows you to enter a point with respect to an existing point. It assigns the value of the angle between the two points to ang3.

Note

*The angle you enter is affected by the angle setting. The angle settings can be changed by changing the value of the **ANGBASE** and **ANGDIR** system variables. The default settings for measuring an angle are as follows:*

*The angle is measured with respect to the positive X axis (3 o'clock position). The value for 3 o'clock corresponds to the current value of the **ANGBASE** system variable, which is 0. **ANGBASE** could be set in any of four 90-degree quadrant directions*

*The angle is positive if it is measured in the counterclockwise direction and negative if it is measured in the clockwise direction. The value of this setting is saved in the **ANGDIR** system variable. The **GetOrientation** method has the same syntax as the **GetAngle** method but ignores the **ANGBASE** and **ANGDIR** system variables. The 0 angle is always at 3 o'clock, and angles are always positive counterclockwise.*

Example 2

Write a program that will draw a triangle with user-supplied vertices P1, P2, and P3, as in Figure 13-13. In this program, you will use the **GetPoint** and **AddLine** methods. The User Interface Form for this example is shown in Figure 13-14.

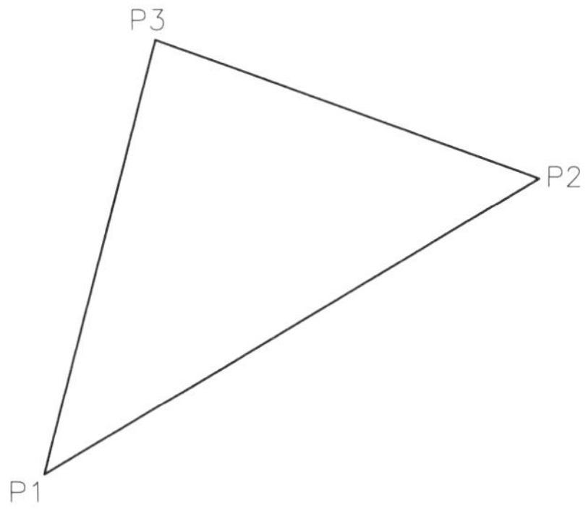

Figure 13-13 *Triangle with user-defined points*

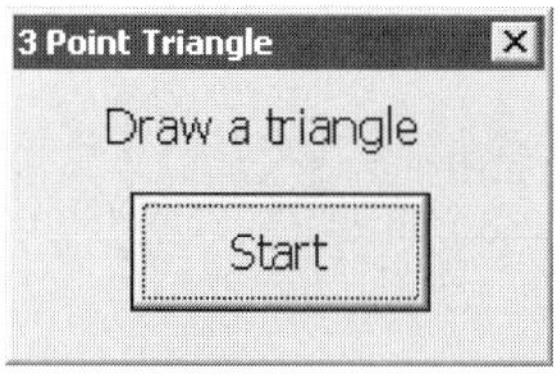

Figure 13-14 *User Interface Form for Example 2*

1. Start AutoCAD and open a new file using the imperial setting of **Start from Scratch** option.

2. Choose **Manage > Applications > Visual Basic Editor** from the **Ribbon**.

3. Choose **Insert > Module** from the menu bar to insert a module window.

4. Similarly, choose **Insert > UserForm** from the menu to insert a User form.

5. Drag and drop the **CommandButton** from the toolbar to the UserForm with the mouse to produce an interface form similar to Figure 13-14. Note that, you need to choose the **Label** button from the **Toolbox** to insert text in the User form.

6. Double-click on the command button to open the code window and type the code given below.

   The following file is a listing of the project for Example 2. **The line numbers at the right are for reference only and are not a part of the program.**

Command Button Code

```
'The function of this routine is to draw
'a triangle from 3 user specified points
'The trigger is a mouse click on the command
'button labeled Start.
'pnt1, pnt2, pnt3 are variant by default
'Returns a point in WCS
Private Sub CommandButton1_Click()
UserForm1.Hide
pnt1 = ThisDrawing.Utility.GetPoint(, "Provide the First Point: ")
'Returns a variant vector since GetPoint returns a point in WCS
'And draws a rubber-band line from the optional first point
pnt2 = ThisDrawing.Utility.GetPoint(pnt1, "Second Point? ")
```

```
13 'Draw first side of triangle
14 ThisDrawing.ModelSpace.AddLine pnt1, pnt2
15 pnt3 = ThisDrawing.Utility.GetPoint(pnt2, "3rd Point? ")
16 ThisDrawing.ModelSpace.AddLine pnt2, pnt3
17 ThisDrawing.ModelSpace.AddLine pnt3, pnt1
18 Unload Me
19 End Sub
```

Explanation

Lines 1-6, 10, 11, and 13
These lines are comments or remarks describing the function of the following line or lines. Comments make understanding and modifying a program easier and should be used liberally. Comments start with a Rem or use an apostrophe ('). These lines are ignored when the program is run.

Line 7
Private Sub CommandButton1_Click()
This line defines where Sub CommandButton1_Click() starts. The subroutine code is executed when **CommandButton1** is chosen.

Line 8
UserForm1.Hide
This statement hides the user interface form and returns focus to AutoCAD. If UserForm from the previous example was not loaded, the UserForm for this example will be **UserForm1** by default. In this event, you should change the code in line 8 to **Userform1.Hide**.

Line 9, 12 and 15
pnt1 = ThisDrawing.Utility.GetPoint(, "Provide the First Point:")
pnt2 = ThisDrawing.Utility.GetPoint(pnt1, "Second Point?")
pnt3 = ThisDrawing.Utility.GetPoint(pnt2, "3rd Point?")
The **GetPoint** method is used without a reference point in line 15. The point may be specified either with the keyboard or mouse. When it is used with a reference point we see a rubber-band line attached from the reference point to the mouse cursor.

Lines 14, 16, and 17
ThisDrawing.ModelSpace.AddLine pnt1, pnt2
ThisDrawing.ModelSpace.AddLine pnt2, pnt3
ThisDrawing.ModelSpace.AddLine pnt3, pnt1
The **AddLine** method requires the two point arguments to be double precision vectors with 3 components. Note the required space before the first point.

Line 18 and 19
Unload Me
End Sub
These lines remove UserForm1 from memory, return the focus back to AutoCAD, and end the subroutine.

Module2 Code

The following code is entered in the Module2 code window.

```
1 Option Explicit
2 Sub Triangle()
3 UserForm1.Show
4 End Sub
```

Explanation

Line 1
Option Explicit
The inclusion of this line forces explicit declaration of all variable types. This minimizes common inconsistent variable usage errors and typographic errors as they are quickly caught at run time. Otherwise, undeclared variable types would be Variant by default.

Lines 2-4
Sub Triangle()
UserForm1.Show
End Sub
This subroutine's function enables you to create a Macro name, Triangle, that can be run from AutoCAD by choosing **Manage > Applications > Run VBA Macros** from the **Ribbon**. Lines 1-4 are a part of Module1. The remaining lines of code below are the part of the UserForm2 object. Again, if UserForm1 from the previous example are not still loaded, the UserForm for this example would be named UserForm1 and similarly, Module1 would be named Module1, by default. In this event, you should change the names in lines 2-4 accordingly.

Example 3

Create a red colored solid cylinder by specifying its radius and height.

1. Start AutoCAD and open a new file using the imperial setting of **Start from Scratch** option.

2. Choose **Manage > Applications > Visual Basic Editor** from the **Ribbon**.

3. Choose **Insert > Module** from the menu bar. If the text editor covers the entire screen, reduce the size by clicking the **Restore Window** button on the top right corner.

4. Click the **Properties Window** button from the toolbar.

5. Change the default name to **CreateCylinder** in the **Name** edit box of the **Properties** window.

6. Type the following code in the CreateCylinder (Code) window:

```
Public Sub DrawCylinder()
Dim cen As Variant
Dim r As Double
Dim h As Double
Dim cyl As Acad3DSolid
cen = ThisDrawing.Utility.GetPoint(, "Specify center point: ")
r = ThisDrawing.Utility.GetDistance(cen, "Specify radius: ")
h = ThisDrawing.Utility.GetDistance(, "Specify height: ")
Set cyl = ThisDrawing.ModelSpace.AddCylinder(cen, r, h)
ThisDrawing.SendCommand ("VPOINT -1,-1,1 SHADEMODE REALISTIC ")
ThisDrawing.SendCommand ("CHPROP ")
ThisDrawing.SendCommand ("LAST  ")
ThisDrawing.SendCommand ("C RED  ")
ThisDrawing.SendCommand ("UCSICON ")
ThisDrawing.SendCommand ("NOO ")
End Sub
```

7. Choose the **Save** button on the **Standard** toolbar. The **Save As** dialog box appears. Save the code as the *cylinder.dvb* file.

8. Exit the VBA environment by choosing the **View AutoCAD** button from the toolbar.

9. Choose **Manage > Applications > Run VBA Macros** from the **Ribbon**; the **Macros** dialog box appears.

10. The **Macro name** text box of the **Macro** dialog box shows the name of the project and its location. Choose the **Run** button.

11. Follow the prompt sequences in the Command prompt and create the cylinder. You can use the keyboard to enter the values. You can also specify the values using two points on the screen.

VBA creates a shaded cylinder of red color. The viewpoint also automatically changes to the SW isometric view.

POLARPOINT AND ANGLEFROMXAXIS METHODS

PolarPoint Method

The **PolarPoint** method defines a point at a given angle and distance from a given point. It has the syntax:

P = ThisDrawing.Utility.PolarPoint(Point, Angle, Distance)

where	**Point**:	Reference point, rubber-band origin
	Angle:	Angle in radians, double precision
	Distance:	Distance from point, double precision

AngleFromXAxis Method

The **AngleFromXAxis** method calculates the angle of a line defined by two points from the horizontal axis in radians. The format of the **AngleFromXAxis method** is:

ang = ThisDrawing.Utility.AngleFromXAxis(point1, point2)

where	**point1**:	Start point of the line
	point2:	End point of the line

Exercise 1

Write a Visual Basic program that will draw a line between two points, as shown in Figure 13-15. The user may either enter coordinates or choose points with the mouse. The graphical screen should draw a rubber-band on the screen to the current position of the mouse cursor if the second point is entered with the mouse. Use a graphical user interface form similar to the one shown in Figure 13-16.

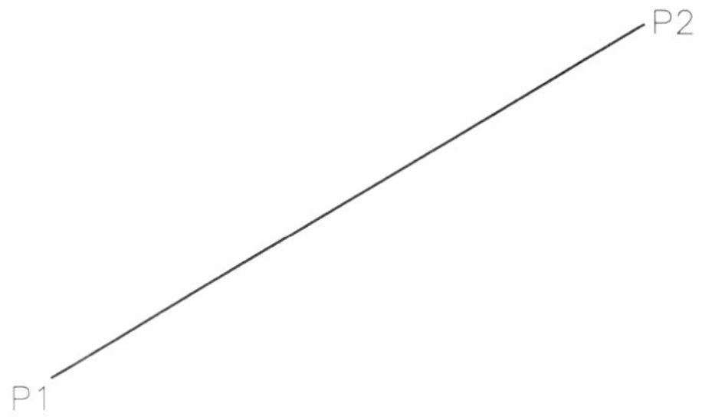

Figure 13-15 *Line with user-defined end*

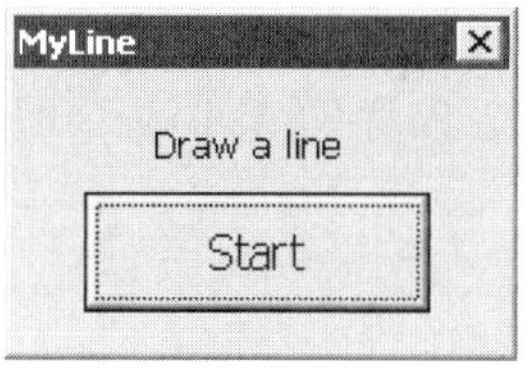

Figure 13-16 *User Interface Form for Exercise 1*

Example 4

Write a Visual Basic program that will draw a triangle based on a given line produced from two points P1 and P2, on an included angle and on a length of the second side shown in Figure 13-17. To create the **Text Boxes** on the user interface form, Figure 13-18, use the **TextBox** button. The labels and command button on the form are created by clicking or dragging the respective icons from the toolbox.

1. Start a new drawing file and enter the VBA IDE.

2. Insert the Userform and module from the **Insert** menu.

3. Create the Userform, as shown in Figure 13-18, and double-click on the **Draw SAS Triangle** button to open the code window. Type the code that is given next.

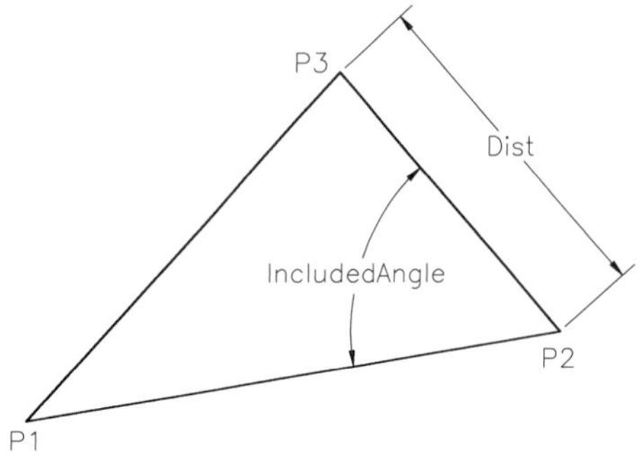

Figure 13-17 Side Angle Side Triangle

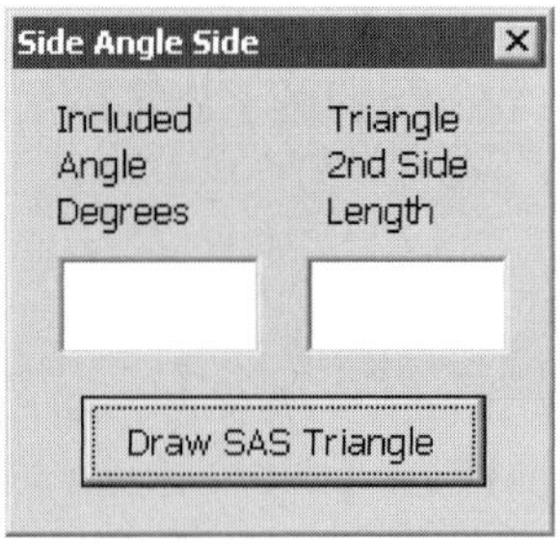

Figure 13-18 User Interface Form for Example 4

The following file is a listing of the Visual Basic program for Example 4. **The line numbers at the right are for reference only and are not a part of the program.**

Command Button Code

```
'(Declarations) (General)
Const PI = 3.141592654
Public IncludedAngle As Double 'pi-converted input angle
Public Angle As Double 'included angle
Public Dist As Double 'length of 2nd side

'This procedure draws a triangle from 2 sides and an included angle SAS
'Program trigger is the command button labeled "Draw SAS Triangle"
'Additional feature is use of text boxes for included angle, 2nd side length
'input
'Included angle is angle between base and 2nd side

Private Sub CommandButton1_Click()
UserForm1.Hide
'p1 is 1st point of base, variant type by default
'p2 is 2nd point of base, variant type by default
p1 = ThisDrawing.Utility.GetPoint(, "Enter or select 1st base point:")
p2 = ThisDrawing.Utility.GetPoint(p1, "Enter or select 2nd base point:")
ThisDrawing.ModelSpace.AddLine p1, p2 'draw base line
If TextBox1.Text = "" Then
Angle = ThisDrawing.Utility.GetAngle(p2,"Enter angle from horiz.
or select included angle:")
Else
Angle = ThisDrawing.Utility.AngleFromXAxis(p1, p2) + IncludedAngle
End If
If TextBox2.Text = "" Then
Dist = ThisDrawing.Utility.GetDistance(p2, "Enter or select dist.
```

```
25 from base point:")
26 End If
27 p3 = ThisDrawing.Utility.PolarPoint(p2, Angle, Dist)
28 ThisDrawing.ModelSpace.AddLine p2, p3
29 ThisDrawing.ModelSpace.AddLine p1, p3
30 Unload Me
31 End Sub
32
33 Private Sub textBox1_Change()
34 IncludedAngle = PI - Val(TextBox1.Text) * PI / 180
35 End Sub
36
37 Private Sub textBox2_Change()
38 Dist = Val(TextBox2.Text)
39 End Sub
```

Explanation

Line 2
Const PI = 3.141592654
The constant **pi** used in this routine is declared to demonstrate this type of statement.

Lines 3-5
Public IncludedAngle As Double
Public Angle As Double
Public Dist As Double
These variables are declared to be public, which means any procedure can access them from any module in the project (file) without passing the argument in a parameter list. Also three variables are declared to be double precision as required by the AutoCAD methods in which they will be employed.

Lines 7-12
'This module draws a triangle from 2 sides and an included angle SAS
'Program trigger is the command button labeled "Draw SAS Triangle"
'Additional feature is use of text box for included angle, 2nd side length input
'Included angle is angle between base and 2nd side
Private Sub CommandButton1_Click()
These lines give the purpose and trigger of the procedure **CommandButton1_Click**, which starts with line 8. Purpose and trigger remarks are useful for all event-driven subroutines.

Line 13
UserForm1.Hide
This statement hides the screen interface form and returns focus to AutoCAD. Otherwise the user would have to close the form with the button at the top right corner of the window to proceed with entering input in AutoCAD.

Lines 14-17

'p1 is 1st point of base, variant type by default
'p2 is 2nd point of base, variant type by default
p1 = ThisDrawing.Utility.GetPoint(, "Enter or select 1st base point:")
p2 = ThisDrawing.Utility.GetPoint(p1, "Enter or select 2nd base point:")

These lines get input from the AutoCAD graphic screen or command line for the two points defining the base side of the triangle.

Lines 19-23 and 33-35

If textBox1.Text = "" Then
Angle=ThisDrawing.Utility.GetAngle(p2,"Enter angle from Horiz. or select included angle:")
Else
Angle=ThisDrawing.Utility.AngleFromXAxis(P1,P2)+IncludedAngle
End If
Private Sub textBox1_Change()
IncludedAngle = PI - Val(textBox1.Text) * PI / 180 'in radians
End Sub

The If ... Then ... Else statement checks TextBox1 for an entry. If no angle has been entered on the UserForm, the **GetAngle** method is used. If a numeric angle was entered on the form it is converted by the subroutine textBox1_Change() to a radian-included angle. Adding the radian angle returned by the **AngleFromXAxis** method gives the AutoCAD polar angle for the vertex (third) point.

Lines 24-26

If textBox2.Text = "" Then
Dist = ThisDrawing.Utility.GetDistance(p2, "Enter or select dist. from base point:")
End If

If no numeric entry for the second side length was entered on the user form, these lines use the **GetDistance** method.

Lines 37-39

Private Sub textBox2_Change()
Dist = Val(textBox2.Text)
End Sub

If a numeric entry for the second side length was entered on the user form, these lines convert from text to numeric form.

Lines 18, 28 and 29

ThisDrawing.ModelSpace.AddLine p1, p2
ThisDrawing.ModelSpace.AddLine p2, p3
ThisDrawing.ModelSpace.AddLine p1, p3

These methods draw the base, second, and third sides of the triangle.

Module3 Code

The following code is entered in the Module3 code window.

```
Option Explicit
Sub SAS()
UserForm1.Show
End Sub
```

Exercise 2

Write a program in the AutoCAD VBA IDE that will draw a triangle based on a given line produced from two points P1 and P2, on an adjacent angle at one end, and another angle at the other end as shown in Figure 13-19. Employ a User Interface Form similar to the one shown in Figure 13-20.

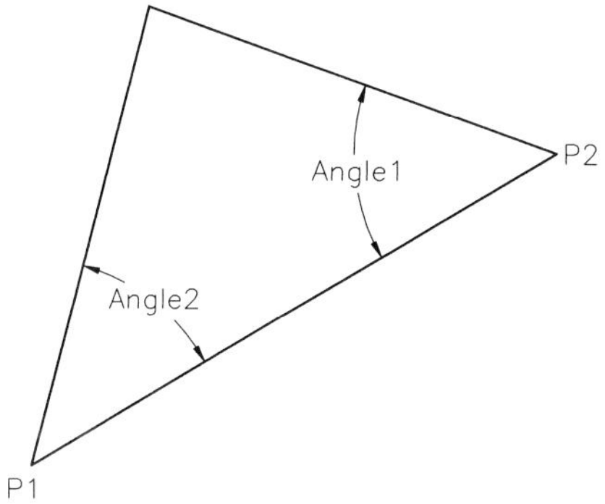

Figure 13-19 *Angle Side Angle Triangle with user-defined points defining the base*

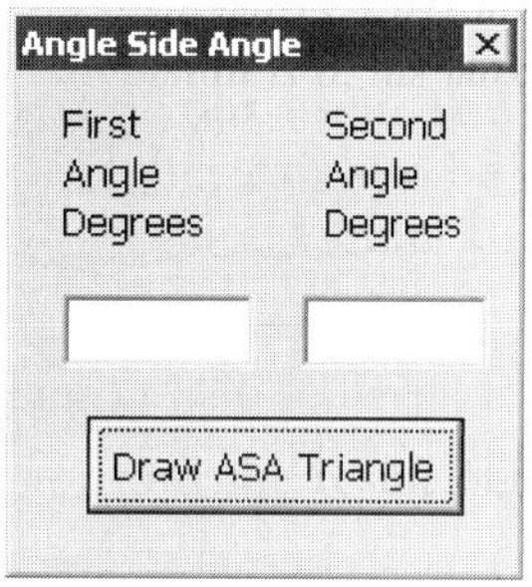

Figure 13-20 *User Interface Form for Exercise 2*

Exercise 3

Write a program in the AutoCAD VBA IDE that will draw a triangle based on a given line produced from two points P1 and P2, and two distances, which are the lengths of the sides adjoining each end as shwn in Figure 13-21. Employ a User Interface Form similar to the one shown in Figure 13-22.

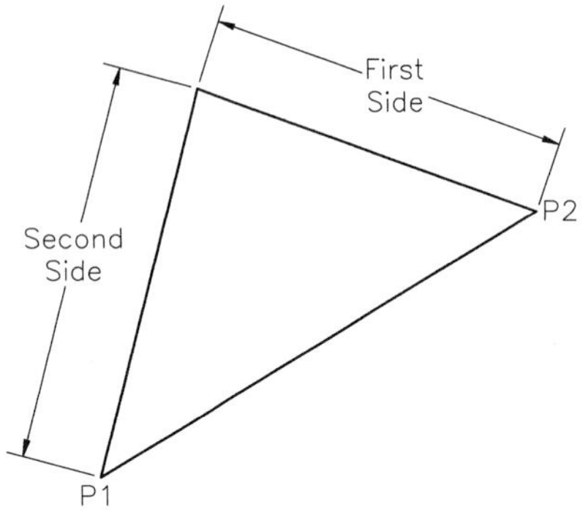

Figure 13-21 *Side Side Side with user-defined points defining the base*

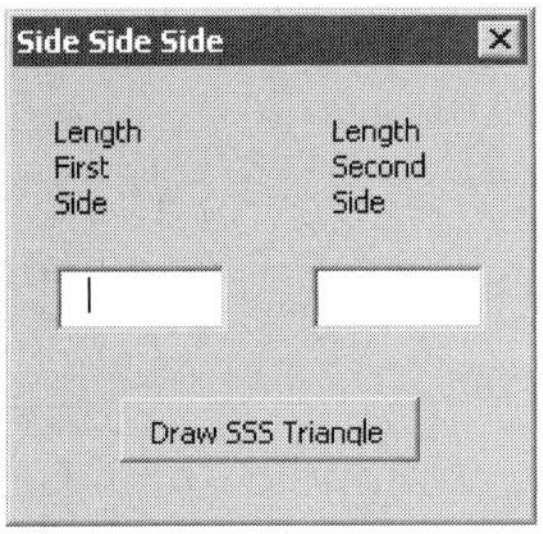

Figure 13-22 *User Interface Form for Exercise 3*

ADDITIONAL VBA EXAMPLES

More VBA examples are provided with AutoCAD 2010 beyond the large number in the help files. These are considered sample files and are located in the *Sample/VBA* subdirectory of the AutoCAD 2010 directory and the AutoCAD 2010 CD ROM.

Self-Evaluation Test

Answer the following questions and then compare them to those given at the end of this chapter:

1. VBA in AutoCAD has a separate environment in which you can do all the work related to a project. This environment is called Integrated Development Environment (IDE). (T/F)

2. To invoke the VB editor, choose **Manage > Applications > Visual Basic Editor** from the **Ribbon**. (T/F)

3. The primary purpose of VBA is to control the functionality of the application and automate it with VBA commands. (T/F)

4. The application that you are working on is called a **Document object**. (T/F)

5. Examples of objects include drawings or documents, geometric elements such as lines or circles, and user interface controls to handle input and output to programs or macros. (T/F)

6. Which one of the following methods is used to draw a line?

   (a) **AddArc** (b) **AddLine**
   (c) **AddCircle** (d) None

7. Which one of the following methods allow the user to specify the distance on the screen?

   (a) **GetPoint** (b) **GetDistance**
   (c) **GetAngle** (d) None

8. Which of the following objects come first in the hierarchy?

   (a) ActiveDocument (b) AddMethod
   (c) Application (d) None

9. ___________ notation is required to clarify on which object the method is acting.

   (a) Star (b) Dot
   (c) Blank (d) None

10. The environment that is provided in AutoCAD to work with VB is called:

    (a) Integrated Aided Environment (b) Integrated Using Environment
    (c) Integrated Design Environment (d) None

Review Questions

Answer the following questions:

1. _________ lets you access and manipulate the objects and functionality of AutoCAD from Excel or another application supporting ActiveX.

2. Before you can use AutoCAD's objects in some other software supporting ActiveX, you must make the application aware that the AutoCAD _________ Library is available on the computer.

3. There are _________ programmers using Visual Basic.

4. Visual Basic is considered an object- _______ , rather than an object-oriented, programming language.

5. Functions known as _________ have been defined in the AutoCAD object library to perform an action on an object, for example, drawing a line in a drawing.

6. _________ are functions that set or return information about the state of an object.

7. The term **IDE**, which is the Visual Basic Editor, stands for _________.

8. A _________ (extension *.dvb*) is the name given to the forms, controls, modules, and programming making up a saved AutoCAD Visual Basic file.

9. _________ forces explicit declaration of all variable types, which minimizes common inconsistent variable usage errors.

10. _________ is the variable type returned by the **GetPoint** method.

11. The _________ method always measures the angle with a positive *X* axis (3 o'clock position) and in a counterclockwise direction.

12. The _________ method allows you to enter the *X*, *Y* coordinates or *X*, *Y*, *Z* coordinates of a point.

13. The _________ method lets you retrieve the value of an AutoCAD system variable.

14. The _________ method defines a point at a given angle and distance from the given point.

15. The _________ method lets you enter a distance on the command line, a distance from a given point, or two points.

16. The _________ method allows you to enter an angle, either from the keyboard in degrees or by selecting two points.

Exercises

Exercise 4

Write a Visual Basic program that will draw an equilateral triangle inside the circle (Figure 13-23).

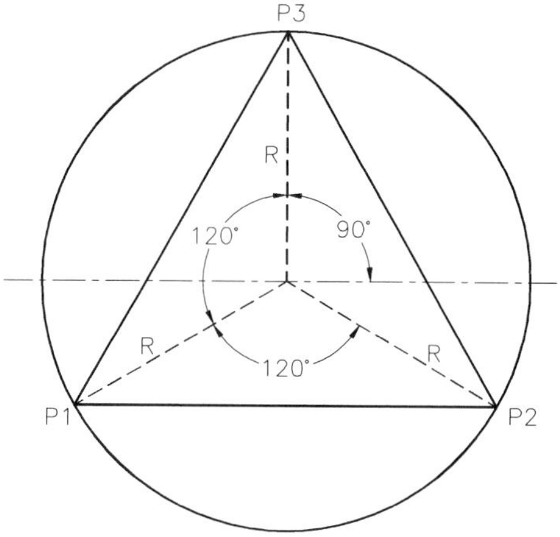

Figure 13-23 *Equilateral triangle inside a circle*

Exercise 5

Write a Visual Basic program that will draw a square of sides S and a circle tangent to the four sides of the square, as shown in Figure 13-24. The base of the square makes an angle, ANG, with the positive *X* axis. The program should allow you to enter the starting point P1, length S, and angle ANG on a user interface form or in the AutoCAD graphical interface.

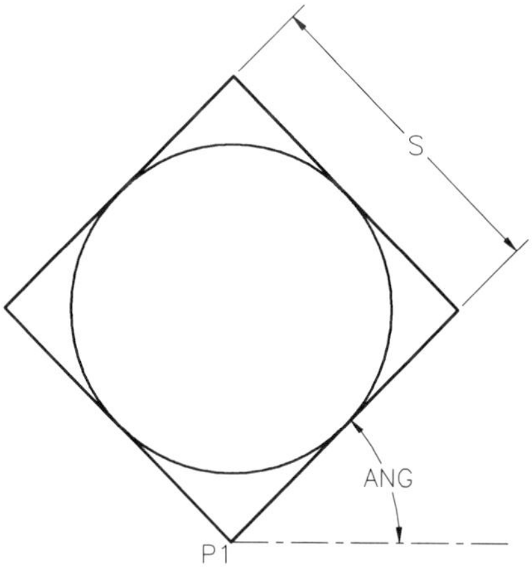

Figure 13-24 *Square of side S and at an angle ANG*

Exercise 6

Write a Visual Basic program that will draw a line at an angle A and then generate the given number of lines (N), parallel to the first line with an offset distance S (Figure 13-25) as entered on through a user form or from the keyboard.

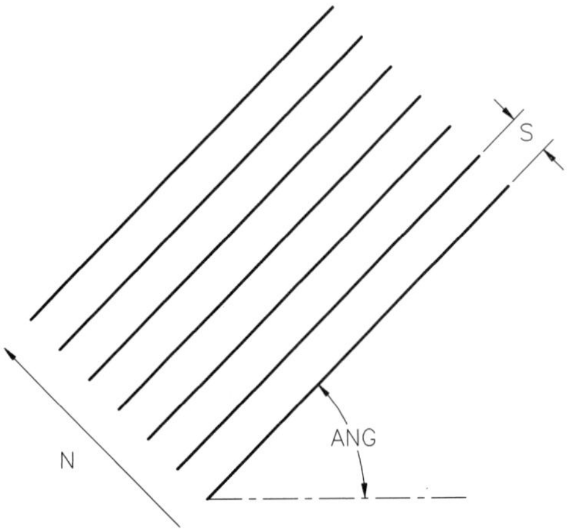

Figure 13-25 *N number of lines offset at a distance of S*

Exercise 7

Write a program that will draw a slot shown in Figure 13-26 with centerlines. The program should allow you to enter slot length, slot width, and the layer name for center lines on a user interface form or in the AutoCAD graphical interface.

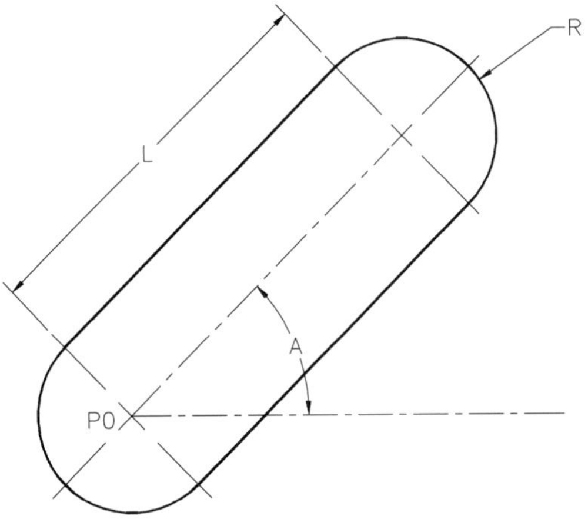

Figure 13-26 *Slot of length L and radius R*

Exercise 8

Write a Visual Basic program that will draw two lines tangent to two circles, as shown in Figure 13-27. The program should allow you to enter the circle diameters, the center distance between the circles, and the angle of the second circle with respect to X-axis on a user interface form or in the AutoCAD graphical interface.

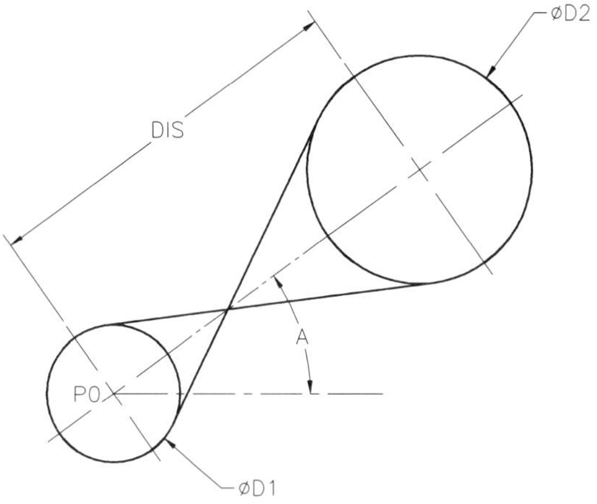

Figure 13-27 *Circles with tangent lines*

Exercise 9

Write a VBA program that will draw a hub with the key slot, as shown in Figure 13-28. The user should enter the four parameters in the note below on a user form or in the AutoCAD graphical interface. Use the program inside a circle of Diameter D1 to produce a keyed bushing with the test dimensions.

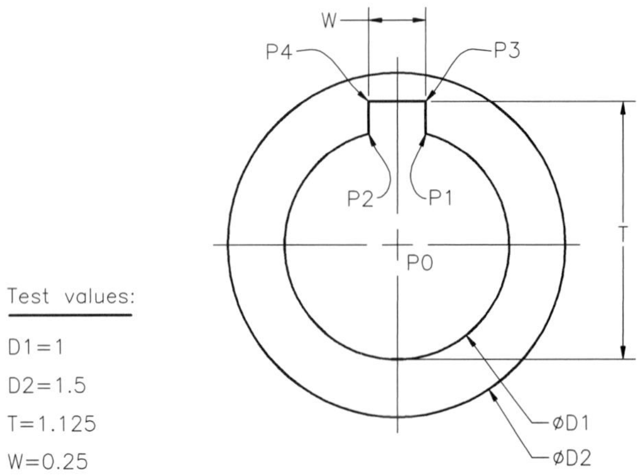

Figure 13-28 *Hub with Keyway*

Exercise 10

Write a VBA program to draw the tangent arc cam shown in Figure 13-29.

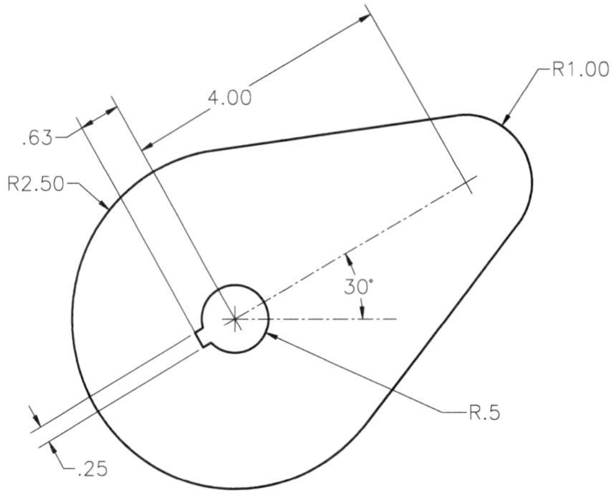

Figure 13-29 *Tangent Arc Flat Plate Cam*

Exercise 11

Write a VBA program to draw the circular arc cam shown in Figure 13-30.

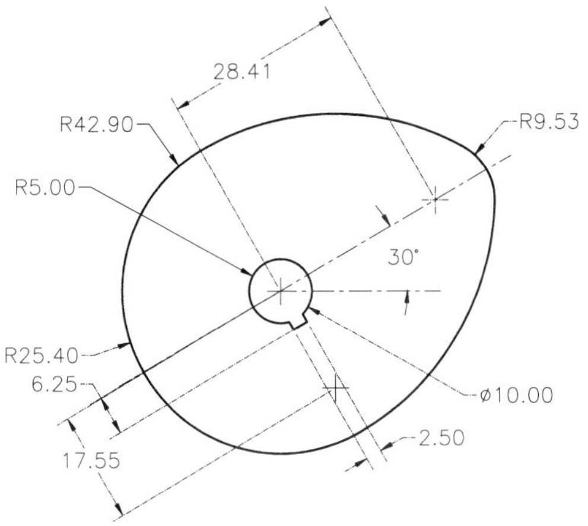

Figure 13-30 Circular Arc Cam

Project Exercise 1

Write a Visual Basic program that will draw the two views of a bushing, as shown in Figure 13-31. The program should allow you to enter the starting point P0, lengths L1, L2, and the bushing diameters ID, OD, HD on a user interface form or in the AutoCAD graphical interface. The distance between the front view and the side view of bushing is DIS (DIS = 1.25 * HD). The program should also draw the hidden lines in the HID layer and centerlines in the CEN layer. The center lines should extend 0.75 units beyond the object line.

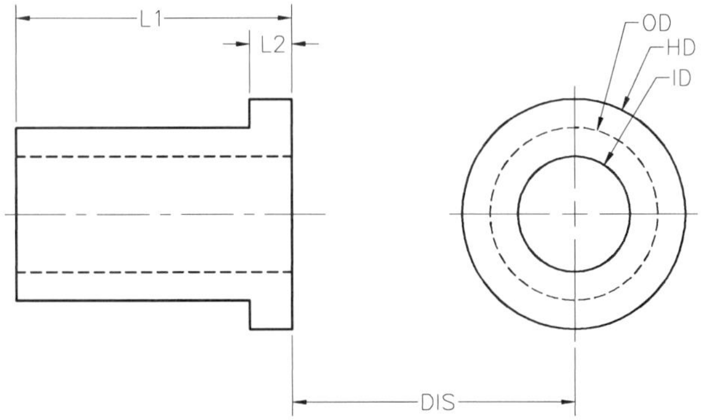

Figure 13-31 Two views of bushing

Project Exercise 2

Draw the following parametric drawing in a rectangle 320mm wide by 200mm high. The block depends on one number, Q, called the data set number. The dimensions, as shown in Figure 13-32, are given by: X = Q + 160, Z= 240 - X , Y = 120 - Z , and R = Z/4. There are 40mm between views and 20mm between a view and the surrounding rectangle at the closest points. The circle is centered in the block.

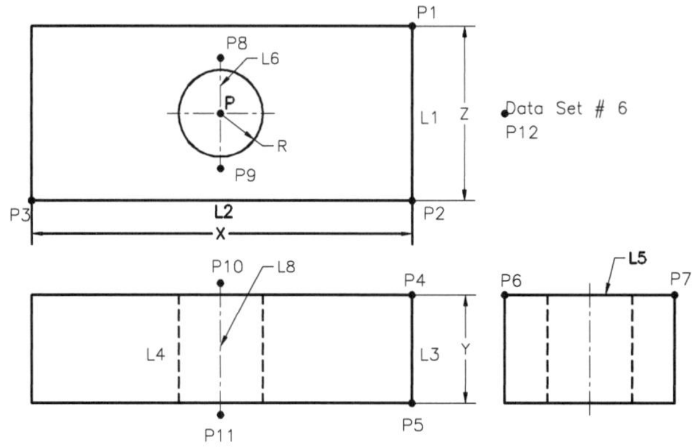

Figure 13-32 *Parametric Block with Hole*

Answers to Self-Evaluation Test

1. T, **2.** T, **3.** T, **4.** F, **5.** T, **6.** b, **7.** b, **8.** c, **9.** b, **10.** c

Chapter 14

Accessing External Database

Learning Objectives

After completing this chapter, you will be able to:
- *Understand database and the database management system (DBMS).*
- *Understand AutoCAD database connectivity feature.*
- *Configure an external database.*
- *Access and edit a database using* ***dbConnect Manager****.*
- *Create links with graphical objects.*
- *Create and display labels in a drawing.*
- *Understand* ***AutoCAD SQL Environment*** *(ASE) and create queries using* ***Query Editor****.*
- *Form selection sets using Link Select.*
- *Convert ASE links into AutoCAD 2010 format.*

UNDERSTANDING DATABASE

Database

Database is a collection of data arranged in a logical order. In a database, the columns are known as fields, individual rows are called records, and the entries in the database tables, that store data for a particular variable, are known as cells. For example, there are six computers in an office, and we want to keep a record of these computers on a sheet of paper. One of the ways of recording this information is to make a table with rows and columns, as shown in Figure 14-1. Each column can have a heading that specifies a certain feature of the computer, such as COMP_CFG, CPU, HDRIVE, or RAM. Once the columns are labeled, the computer data can be placed in the columns. By doing this, you have created a database on a sheet of paper that contains information about the computers. The same information stored on a computer, is known as a computerized database.

COMPUTER

COMP_CFG	CPU	HDRIVE	RAM	GRAPHICS	INPT_DEV
1	PENTIUM350	4300MB	64MB	SUPER VGA	DIGITIZER
2	PENTIUM233	2100MB	32MB	SVGA	MOUSE
3	MACIIC	40MB	2MB	STANDARD	MOUSE
4	386SX/16	80MB	4MB	VGA	MOUSE
5	386/33	300MB	6MB	VGA	MOUSE
6	SPARC2	600MB	16MB	STANDARD	MOUSE

Figure 14-1 *A table containing computer information*

Most of the database systems are extremely flexible and any modifications to or additions of fields or records can be done easily. Database systems also allow you to define relationships between multiple tables so that if the data of one table is altered, the corresponding values of another table with predefined relationships changes automatically.

Database Management System

The database management system (DBMS) is a program or a collection of programs (software) used to manage the database. For example, DBASE, INFORMIX, and ORACLE are database management systems.

Components of a Table

A database table is a two-dimensional data structure that consists of rows and columns, as shown in Figures 14-2 and 14-3.

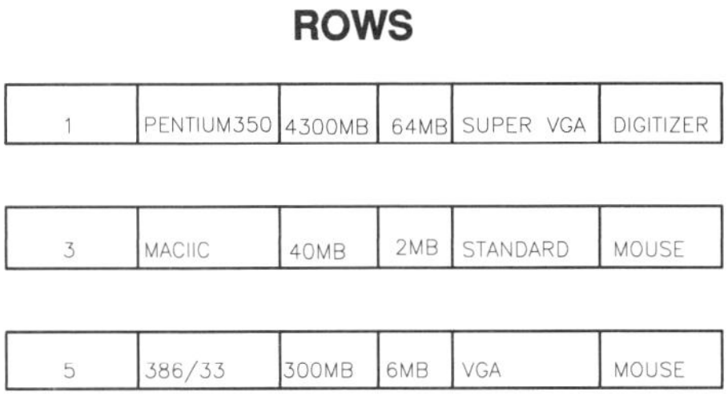

Figure 14-2 Rows in a table (horizontal group)

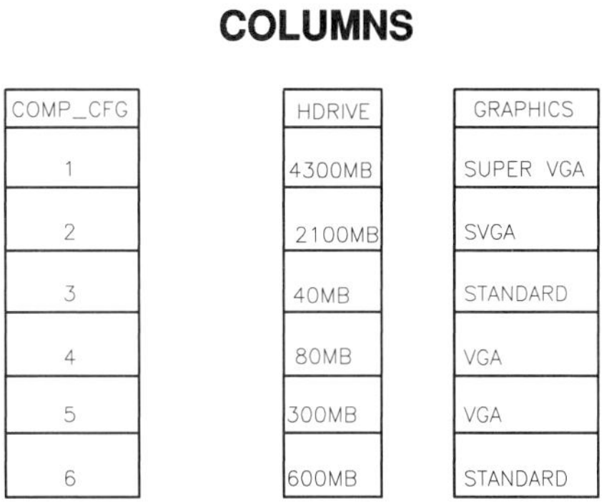

Figure 14-3 Columns in a table (vertical group)

Row

The horizontal group of data is called a row. For example, Figure 14-2 shows three rows of a table. Each value in a row defines an attribute of the item. For example, in Figure 14-2, the attributes assigned to COMP_CFG (1) include PENTIUM350, 4300MB, 64MB, and so on. These attributes are arranged in the first row of the table.

Column

A vertical group of data (attribute) is called a column. (See Figure 14-3.) **HDRIVE** is the column heading that represents a feature of a computer, and the **HDRIVE** attributes of each computer are placed vertically in this column.

AutoCAD DATABASE CONNECTIVITY

AutoCAD can be effectively used in associating data contained in an external database table with the AutoCAD graphical objects by linking. The Links are the pointers to the database tables from which the data can be referred. AutoCAD can also be used to attach Labels that will display data from the selected tables as text objects. AutoCAD database connectivity offers the following facilities:

1. A **dbConnect Manager** that can be used to associate links, labels, and queries with AutoCAD drawings.
2. An **External Configuration Utility** that enables AutoCAD to access the data from a database system.
3. A **Data View** Window that displays the records of a database table within the AutoCAD session.
4. A **Query Editor** that can be used to construct, store, and execute SQL queries. SQL is an acronym for Structured Query Language.
5. A Migration Tool that converts links and other displayable attributes of files created by earlier releases to AutoCAD 2010.
6. A Link Select Operation that creates iterative selection sets based on queries and graphical objects.

DATABASE CONFIGURATION

An external database can be accessed within AutoCAD only after configuring AutoCAD using Microsoft **ODBC** (Open Database Connectivity) and **OLE DB** (Object Linking and Embedding Database) programs. AutoCAD is capable of utilizing data from other applications, regardless of the format and the platform on which the file is stored. Configuration of a database involves creating a new **data source** that points to a collection of data and information about the required drivers to access it. A **data source** is an individual table or a collection of tables created and stored in an environment, catalog, or schema. Environments, catalogs, and schemas are the hierarchical database elements in most of the database management systems and they are analogous to Window-based directory structure in many ways. Schemas contain a collection of tables, while Catalogs contain subdirectories of schemas and Environment holds subdirectories of catalogs. The external applications supported by AutoCAD 2010 are **DBASE®**, **Oracle®**, **Microsoft Access®**, **Microsoft Visual FoxPro®**, **SQL Server**, and **Paradox®**. The configuration process varies slightly from one database system to other.

dbConnect Manager

Menu Bar: Tools > Palettes > dbConnect
Command: DBCONNECT

When you invoke this command, the **dbConnect Manager** will be displayed, as shown in Figure 14-4. Also, the **dbConnect** menu will be added to the menu bar. The **dbConnect Manager** enables you to access information from an external database more effectively. The **dbConnect Manager** is dockable as well as resizable. And, it contains a set of buttons and a tree view showing all the configured and available databases. You can use **dbConnect Manager** for associating various database objects with an AutoCAD drawing. It contains two nodes in the tree view. The Drawing nodes displays all the open drawings and the associated database objects with the drawing. The **Data Sources** node displays all the configured data on your system.

Figure 14-4** The **dbConnect Manager

AutoCAD contains several Microsoft Access sample database tables and a direct driver (*jet_dbsamples.udl*). The following example shows the procedure to configure a database with a drawing using the **dbConnect Manager**.

Example 1 — *General*

Configure a data source of Microsoft Access database with a diagram. Update and use the **jet_samples.udl** configuration file with new information.

1. Invoke the **DBCONNECT** command to display the **dbConnect Manager**. The **dbConnect** menu will be inserted between the **Parametric** and **Window** menus in the menu bar.

2. The **dbConnect Manager** will display **jet_dbsamples** under **Data Sources**. Right-click on **jet_dbsamples** to display the shortcut menu. In the shortcut menu, choose **Configure**, as shown in Figure 14-5.

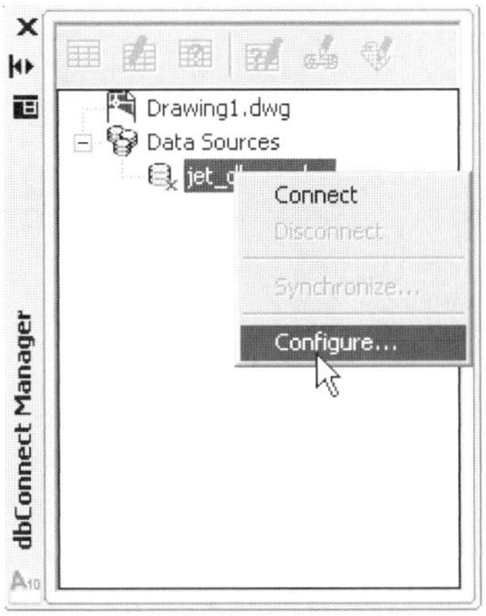

Figure 14-5 *Choosing the* ***Configure*** *option from the* ***dbConnect Manager***

3. The **Data Link Properties** dialog box will be displayed with the **Connection** tab as the current tab, see Figure 14-6.

4. In the **Data Link Properties** (**Connection** tab) dialog box, choose the [**...**] button adjacent to the **Select or enter a database name** text box; the **Select Access Database** dialog box is displayed, as shown in Figure 14-7. Select the *db_samples.mdb* file and choose the **Open** button.

Note

If you configure a driver other than ***Microsoft Jet*** *for database linking, you should consult the appropriate documentation for the configuring process.*

The other tabs in the ***Data Link Properties*** *dialog box are required for configuring various database providers supported by AutoCAD.*

5. Once the database name is selected, choose the **Test Connection** button from the **Data Link Properties** dialog box to ensure that the database source has been configured correctly. If it is configured correctly, the **Microsoft Data Link** message box is displayed, see Figure 14-8.

6. Choose the **OK** button to end the message and again, choose the **OK** button to complete the configuration process.

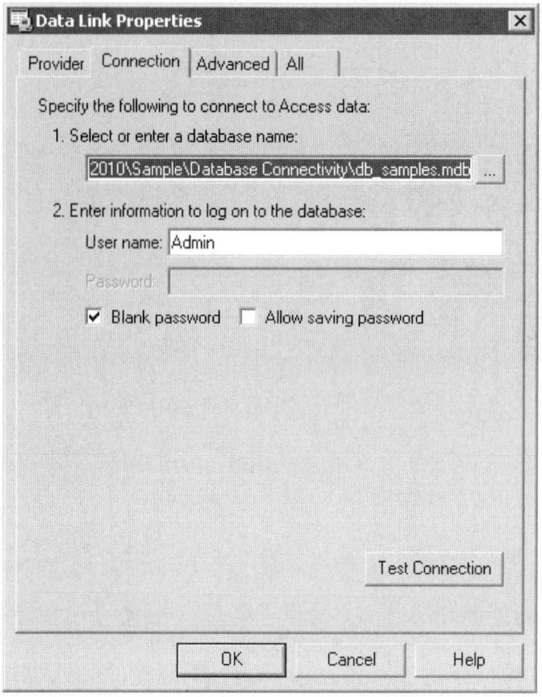

Figure 14-6 *The* ***Data Link Properties*** *dialog box*

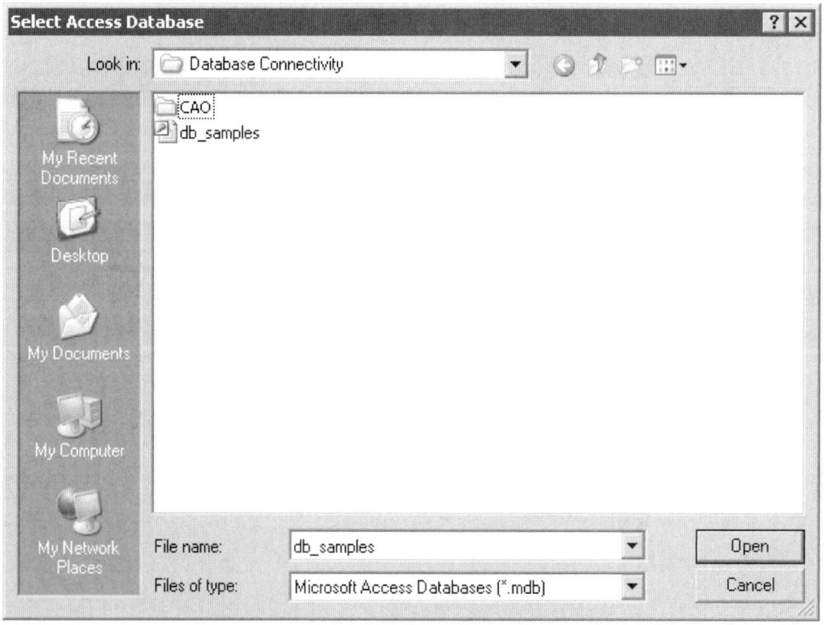

Figure 14-7 *The* ***Select Access Database*** *dialog box*

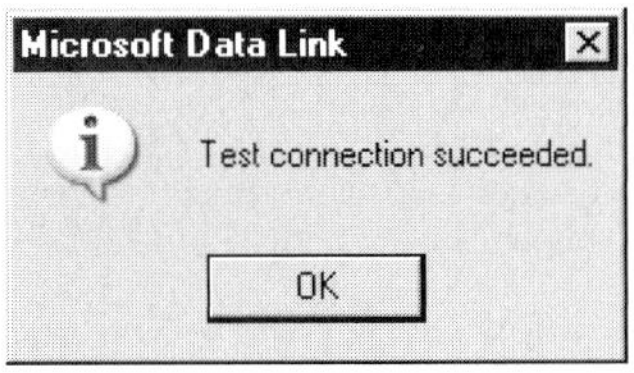

*Figure 14-8 The **Microsoft Data Link** message box*

7. After configuring the data source, double-click on **jet_dbsamples** in **dbConnect Manager**; all the sample tables will be displayed in the tree view. You can connect any table and link its records to the entities in the drawing.

VIEWING AND EDITING TABLE DATA FROM AutoCAD

After configuring a data source using the **dbConnect Manager**, you can view as well as edit its tables within the AutoCAD session using the **Data View window**. You can open tables in **Read-only** mode to view their content. But you cannot edit their records in the **Read-only** mode. Some database systems may require a valid user-name and password before connecting to AutoCAD drawing files. The records of a database table can be edited by opening in the **Edit** mode. The procedure to open a table in various modes is given next.

Read-only Mode

The tables can be opened in the **Read-only** mode by selecting the table and then choosing the **View Table** button from the **dbConnect Manager**. You can also choose **View Table** from the shortcut menu displayed upon right-clicking on the selected table. If you choose **dbConnect > View Data > View External Table** from the menu bar, the **Select Data Object** dialog box will be displayed. Select a table and choose the **OK** button. You will notice that the table that is displayed has a gray background and you cannot edit the entries in it.

Edit Mode

The tables can be opened in the **Edit** mode by choosing the **Edit Table** button from the **dbConnect Manager**. You can also choose **Edit Table** from the shortcut menu displayed upon right-clicking on the selected table. If you choose **dbConnect > View Data > Edit External Table** from the menu bar, the **Select Data Object** dialog box will be displayed. Select a table and choose the **OK** button.

Note

*The **db_samples.mdb** file is available in the directory \AutoCAD 2010\Sample\Database Connectivity.*

Example 2 *General*

In this example, you will select a **Computer** table from the **jet_dbsamples** data source and edit the rows in the table. Add a new computer entry and replace the existing entry by editing the table and save changes.

1. Select the **Computer** table from the **jet_dbsamples** in the **dbConnect Manager**, right-click to invoke the shortcut menu, and choose **Edit Table** (Figure 14-9). You can also do the same by double-clicking on the **Computer** table.

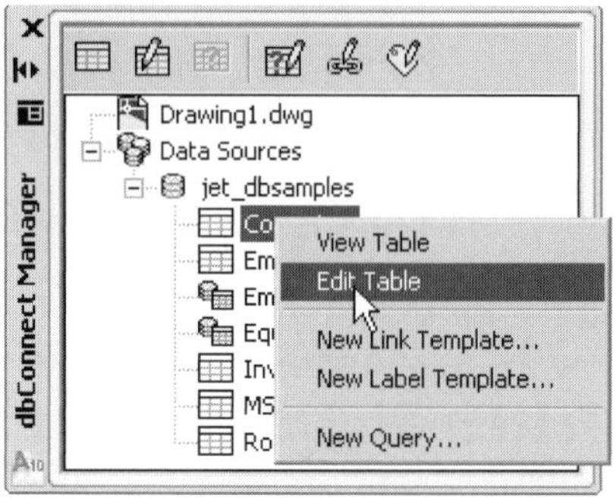

***Figure 14-9** Choosing **Edit Table** from the shortcut menu*

2. The **Data View** window is displayed with the **Computer** table in it. In the **Data View** window, you can resize, sort, hide, or freeze the columns according to your requirements.

3. To add a new item in the table, double-click on the last empty row. In the **Tag_Number** column, type **24675**. Then, type the following data into the columns:

Manufacturer:	**IBM**
Equipment_Description:	**PIV/450, 4500GL, NETX**
Item_Type:	**CPU**
Room:	**6035**

4. To edit the record for **Tag_Number 60298** and display the following in the **Data View** window (Figure 14-10), double-click on each cell and type.

Manufacturer:	**CREATIVE**
Equipment_Description:	**INFRA 6000, 40XR**
Item_Type:	**CD DRIVE**
Room:	**6996**

5. After making all changes in the database table, you have to save the changes for further use. To save the changes in the table, right-click on the **Data View grid header** (A triangle mark on the left of the **Tag_Number** column head). Choose **Commit** from the shortcut menu. The changes in the current table will be saved. The new record that you have entered does not necessarily get added at the end of the table, so if you close the table and then reopen it, you might have to search for the new record in the table. Note

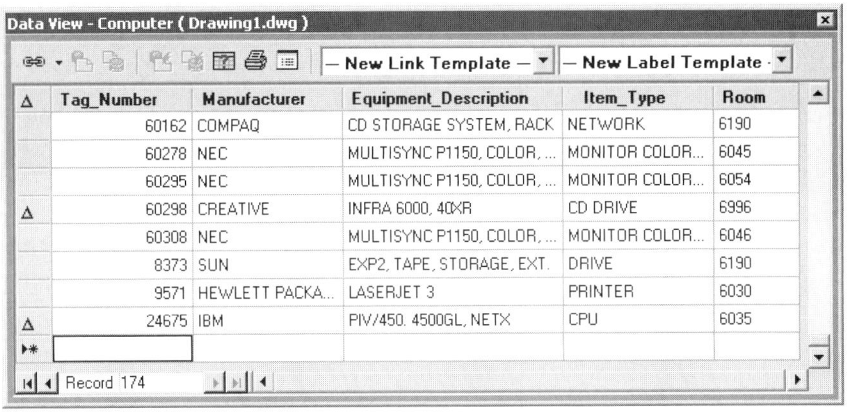

	Tag_Number	Manufacturer	Equipment_Description	Item_Type	Room
	60162	COMPAQ	CD STORAGE SYSTEM, RACK	NETWORK	6190
	60278	NEC	MULTISYNC P1150, COLOR, ...	MONITOR COLOR...	6045
	60295	NEC	MULTISYNC P1150, COLOR, ...	MONITOR COLOR...	6054
Δ	60298	CREATIVE	INFRA 6000, 40XR	CD DRIVE	6996
	60308	NEC	MULTISYNC P1150, COLOR, ...	MONITOR COLOR...	6046
	8373	SUN	EXP2, TAPE, STORAGE, EXT.	DRIVE	6190
	9571	HEWLETT PACKA...	LASERJET 3	PRINTER	6030
Δ	24675	IBM	PIV/450. 4500GL, NETX	CPU	6035

Figure 14-10 *The* ***Data View*** *window after editing*

that if you quit **Data View** window without committing, all the changes you have made during the editing session are automatically committed.

Note

After making changes, if you do not want to save the changes, choose ***Restore*** *from the above shortcut menu.*

CREATING LINKS WITH GRAPHICAL OBJECTS

The main function of the database connectivity feature of AutoCAD is to associate data from external sources with its graphical objects. You can establish the association of the database table with the drawing objects by developing a link, which will make a reference to one or more records from the table. But you cannot link nongraphical objects such as layers or linetypes with the external database. Links are very closely related with graphical objects and change in the link will change the graphical objects.

To develop links between database tables and graphical objects, you must create a link template that identifies the fields of the tables with which the links are associated to share the template. For example, you can create a link template that uses the **Tag_number** from the **COMPUTER** database table. The link template also acts as a shortcut that points to the associated database tables. You can associate multiple links to a single graphical object using different link templates. This is useful in associating multiple database tables with a single drawing object. The following example will describe the procedure of linking using the link template creation.

Example 3 *General*

Create a link template between the **Computer** database table from **jet_dbsamples** and your drawing, and then use the **Tag_Number** as the **key field** for linking.

1. Open the drawing that has to be linked with the **Computer** database table.

2. Invoke the **dbConnect Manager**. Select **Computer** from the **jet_dbsamples** data source and right-click on it to invoke the shortcut menu.

3. Choose **New Link Template** from the shortcut menu to invoke the **New Link Template** dialog box (Figure 14-11). You can also select the table in the tree view and choose the **New Link Template** button from the toolbar in the **dbConnect Manager**.

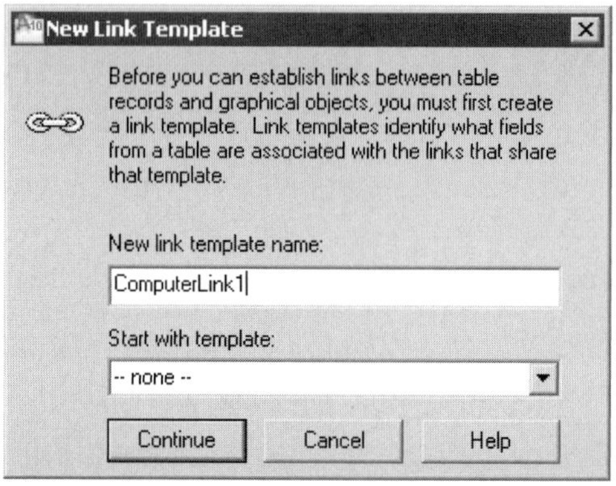

Figure 14-11 *The* ***New Link Template*** *dialog box*

4. Choose the **Continue** button to accept the default link template named **ComputerLink1**. On doing so, the **Link Template** dialog box is displayed. Select the check box adjacent to the **Tag_Number** to accept it as the **Key field** for associating the template with the block reference in the diagram. (See Figure 14-12.)

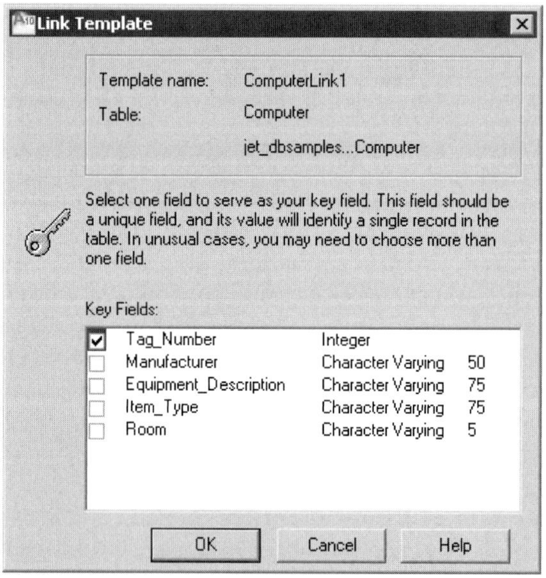

Figure 14-12 *The* ***Link Template*** *dialog box*

5. Choose the **OK** button to complete the new link template. The name of the link template will be displayed under the Drawing name node of the tree view in **dbConnect Manager**.

6. To link a record, double-click on the **Computer** table to invoke the **Data View** window (**Edit** mode). In the table, go to the **Tag_Number 24675** and highlight the record (row).

7. Right-click on the row header and choose **Link!** from the shortcut menu (Figure 14-13). Alternatively, you can link by choosing the **Create Links** option from the flyout available on choosing the down arrow on the **Link** button in the **Data View** window.

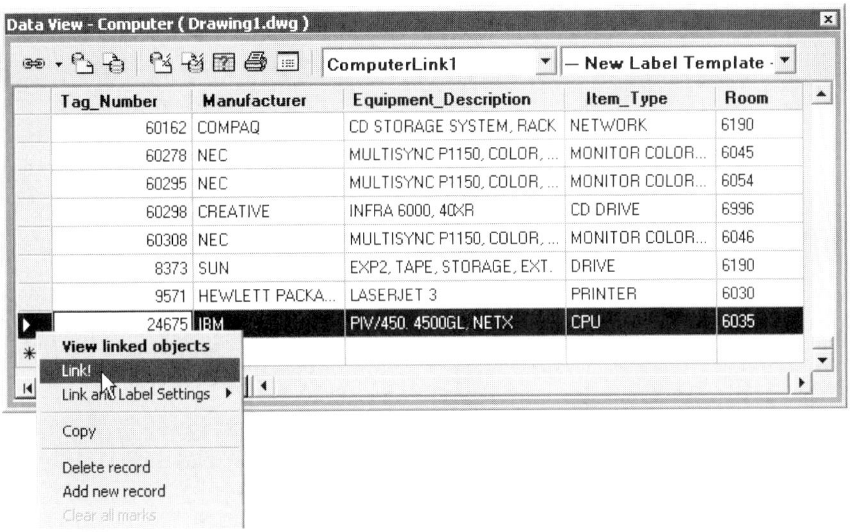

Figure 14-13 *Choosing **Link** from the shortcut menu*

8. You are prompted to select the objects. Select the required objects to link with the record and press the ENTER key. Repeat the process to link all the records to the corresponding block references in the diagram. By default, all the linked rows will get highlighted by yellow color.

9. After linking, you can view the linked objects by choosing the **View Linked Objects in Drawing** button in the **Data View** window, and the linked objects will get highlighted in the drawing area.

Additional Link View Settings

You can set a number of viewing options for linked graphical objects and linked records by using the **Data View and Query Options** dialog box (Figure 14-14). This dialog box can be invoked by choosing the **Data View and Query Options** button from the **Data View** window. You can set the **Automatically pan drawing** option so that the drawing is panned automatically to display the objects linked with the current set of selected records in the **Data View** window.

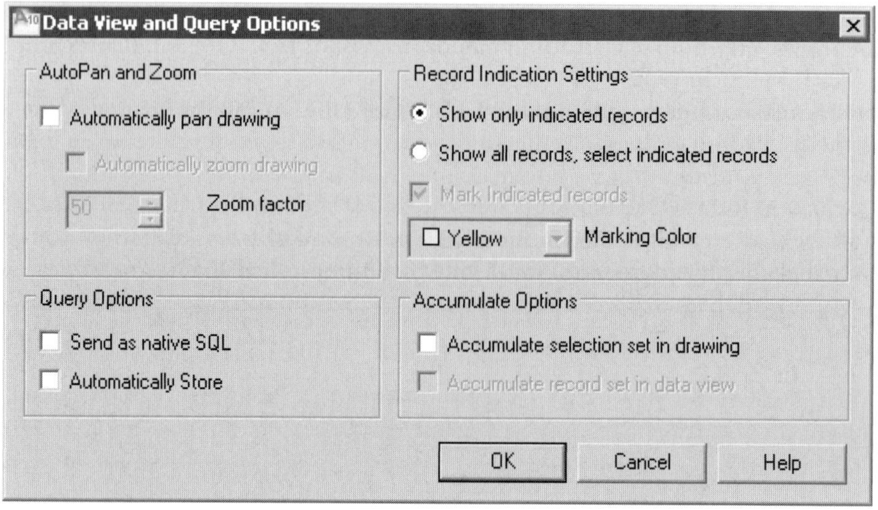

Figure 14-14 *The* ***Data View and Query Options*** *dialog box*

You can also set the options for **Automatically zoom drawing** and the **Zoom factor**. Moreover, you can change the record indication settings and the indication marking color.

Note
The ***Query Editor*** *dialog box is discussed later in this chapter.*

Editing Link Data

After linking data with the drawing objects, you may need to edit the data or update the **Key field** values. For example, you may need to reallocate the **Tag_Number** for the computer equipment or **Room** for each of the linked items. The **Link Manager** can be used for changing the key values. The next example describes the procedure of editing the linked data.

Example 4 — *General*

Use the **Link Manager** to edit the linked data from the **Computer** table and change the Key Values from **24675** to **24875**.

1. Open the diagram that was linked with the **Computer** table.

2. Select the diagram and choose **Link > Link Manager** from the shortcut menu; the **Link Manager** dialog box for the **Computer** table (Figure 14-15) is displayed.

3. Select **24675** field and choose the [**...**] button in the **Value** column to invoke the **Column Values** dialog box.

4. Select **24857** from the list (Figure 14-16) and choose the **OK** button.

5. Again, choose the **OK** button in **Link Manager** to accept the changes.

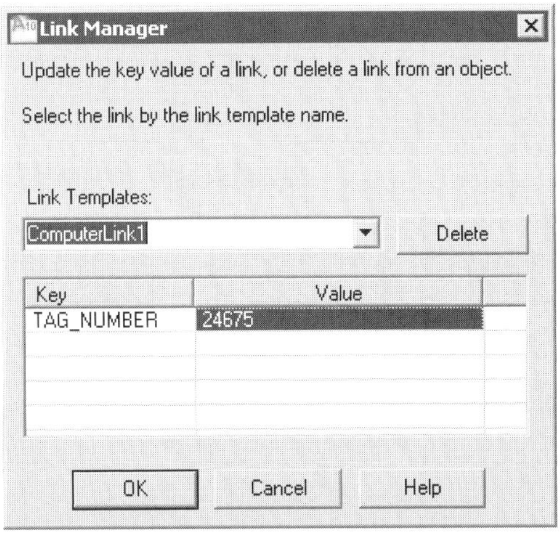

Figure 14-15 *The* ***Link Manager*** *dialog box*

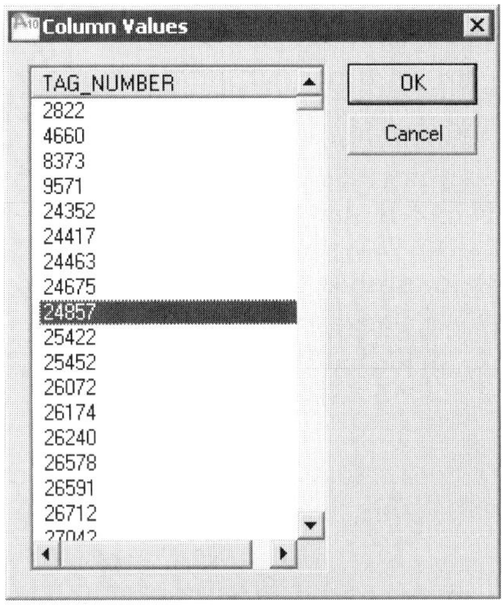

Figure 14-16 *The* ***Column Values*** *dialog box*

CREATING LABELS

You can use linking as a powerful mechanism to associate drawing objects with external database tables. You can directly access associated records in the database table by selecting linked objects in the drawing. But linking has some limitations. Consider that you want to include the associated external data with the drawing objects. Since during printing, the links are only the pointers to the external database table, they will not appear in the printed drawing. In such situations, a feature called Labels proves useful. The labels can be used for the visible representation of external data in the drawing.

Labels are the multiline text objects that display data from the selected fields in the AutoCAD drawing. The labels are of the following two types:

Freestanding labels

Freestanding labels exist in the AutoCAD drawing independent of the graphical objects. Their properties do not change with any change of graphical objects in the drawing.

Attached labels

The Attached labels are connected with the graphical objects they are associated with. If the graphical objects are moved, then the labels also move. If the objects are deleted, then the labels attached to them also get deleted.

Labels associated with the graphical objects in AutoCAD drawing are displayed with a leader. Labels are created and displayed by Label Templates and all their properties can be controlled by using the **Label Template** dialog box. The next example will demonstrate the complete procedure of creating and displaying labels in AutoCAD drawing using the label template.

Example 5 *General*

Create a new label template in the **Computer** database table and use the following specifications for the display of labels in the drawing.

a. The labels include **Tag_Number**, **Manufacturer** and **Item_Type** fields.
b. The fields in the label are **0.25** height, **Times New Roman** font, **black** (**Color 18**) color and **Middle-Left** justified.
c. The label offset starts with **Middle Center** justified and leader offset is **X=1.5** and **Y=1.5**.

1. Open the drawing where you want to attach a label with the drawing objects.

2. Choose **Tools > Palettes > dbConnect** from the menu bar to invoke the **dbConnect Manager**. Select the **Computer** table from the **jet_dbsamples** data source.

3. Right-click on **Computer** and choose **New Label Template** from the shortcut menu to invoke the **New Label Template** dialog box (Figure 14-17). You can also invoke it by choosing the **New Label Template** button from the **dbConnect Manager**.

4. In the **New Label Template** dialog box, choose the **Continue** button to accept the default label template name **Computer Label1**. The **Label Template** dialog box is displayed.

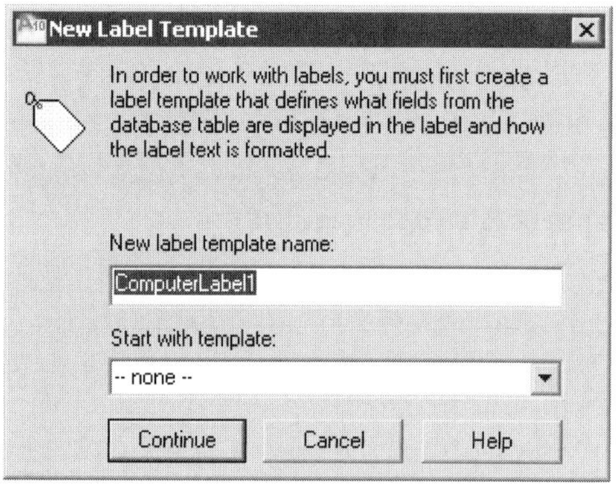

***Figure 14-17** The **New Label Template** dialog box*

5. In the **Label Template** dialog box, choose the **Label Fields** tab. From the **Field** drop-down list, one by one add **Tag_Number**, **Manufacturer**, **Item_Type** fields by using the **Add** button (Figure 14-18).

6. Choose the **Character** tab and highlight the field names by selecting them from the **fonts** drop-down list. Select the **Times New Roman** font and set the font height to **0.25**. Set the color to **Color 18**. Also, select the **Middle Left** justification in the **Properties** tab (Figure 14-19).

7. Choose the **Label Offset** tab and select **Middle Center** in the **Start** drop-down list. Set the **Leader offset** value to **X: 1.5** and **Y: 1.5** (Figure 14-20). Choose **OK**.

8. To display the label in the drawing, select the **Computer** table in the **dbConnect Manager**. Right-click on the table and choose **Edit Table** from the shortcut menu.

9. Select **ComputerLink1** (created in an earlier example) from the **Select a Link Template** drop-down list and **ComputerLabel1** from the **Select a Label Template** drop-down list in the **Data View** window.

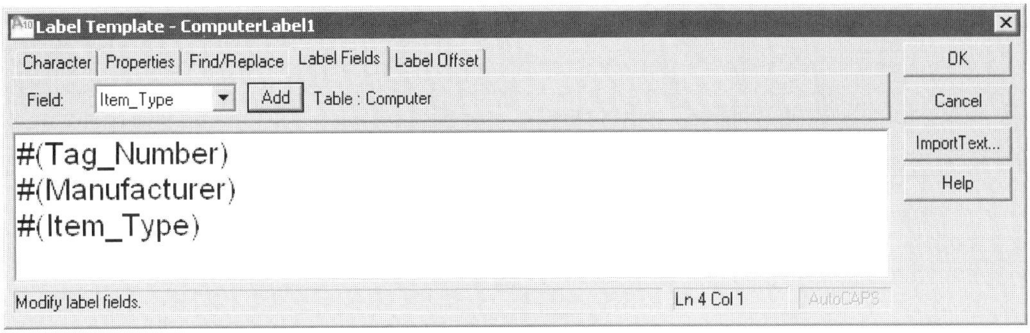

Figure 14-18 *The **Label Template** dialog box (**Label Fields** tab)*

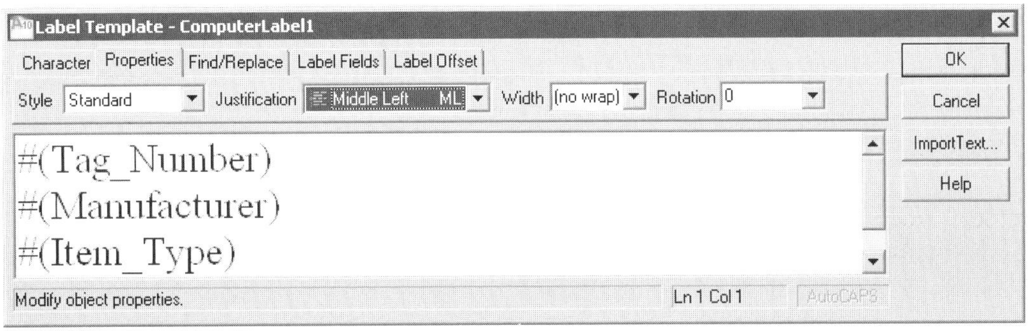

Figure 14-19 *The **Label Template** dialog box (**Properties** tab)*

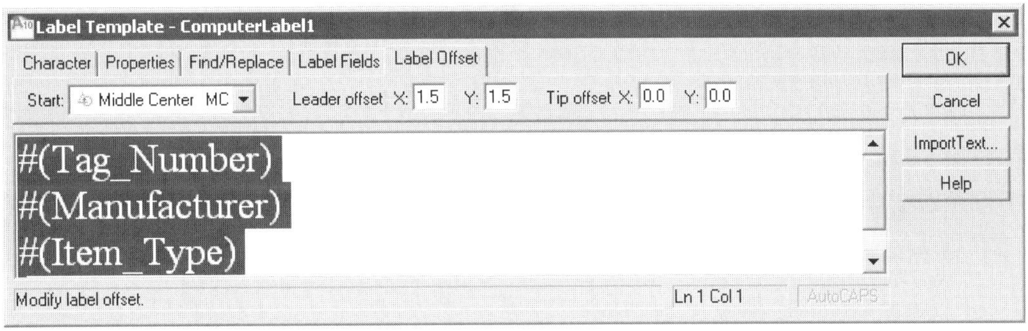

Figure 14-20 *The **Label Template** dialog box (**Label Offset** tab)*

10. Select the record (row) you want to use as a label. Then choose the down arrow button provided on the right side of the **Link** button. Choose **Create Attached Labels** from the shortcut menu (Figure 14-21). Then, choose the **Create Attached Label** button that replaces the **Link** button. Select the drawing object that you want to label. The label, with given specifications, is displayed in the drawing.

Note

*You can create **Freestanding Labels** in the same way by choosing **Create Freestanding Labels** from the **Link and Label Settings** menu in the **Data View** window.*

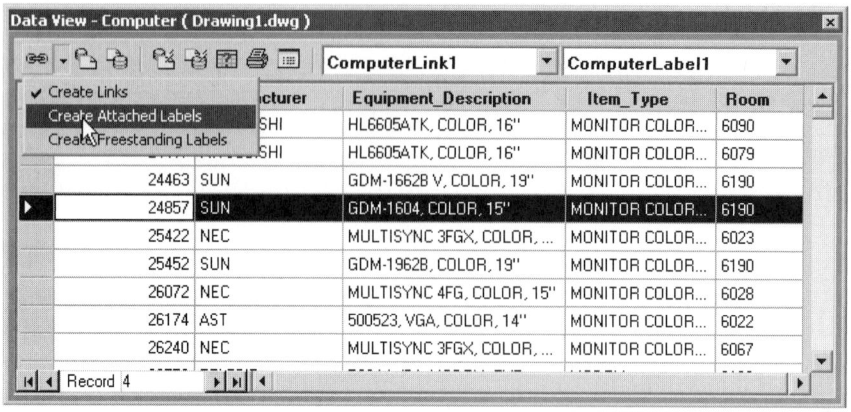

Figure 14-21 Attaching the label to the record

Updating Labels with New Database Values

You may have to change the data values in the database table after adding a label in the AutoCAD drawing. Therefore, you should update the labels in the drawing after making any alteration in the database table the drawing is linked with. The following is the procedure for updating all label values in the AutoCAD drawing:

1. After editing the database table, open the drawing that has to be updated. Select the diagram and choose **Labels > Reload** from the shortcut menu.

2. The details in the label will be modified automatically and the changes will be reflected in the label attached to the diagram.

Importing and Exporting Link and Label Templates

You may want to use the link and label templates that have been developed by some other AutoCAD users. This is very useful when developing a set of common tools to be shared by all the team members in a project. AutoCAD is capable of importing as well as exporting all the link and label templates that are associated with a drawing. The following is the procedure to export a set of templates from the current drawing:

1. Choose **dbConnect > Templates > Export Template Set** from the menu bar to invoke the **Export Template Set** dialog box. From the **Save In** drop-down list of this dialog box, select the directory to save the template set.

2. Under **File Name**, specify a name for the template set, and then choose the **Save** button to save the template in the specified directory.

The following is the procedure to import a set of templates into the current drawing:

1. Choose **dbConnect > Templates > Import Template Set** from the menu bar to invoke the **Import Template Set** dialog box, see Figure 14-22.

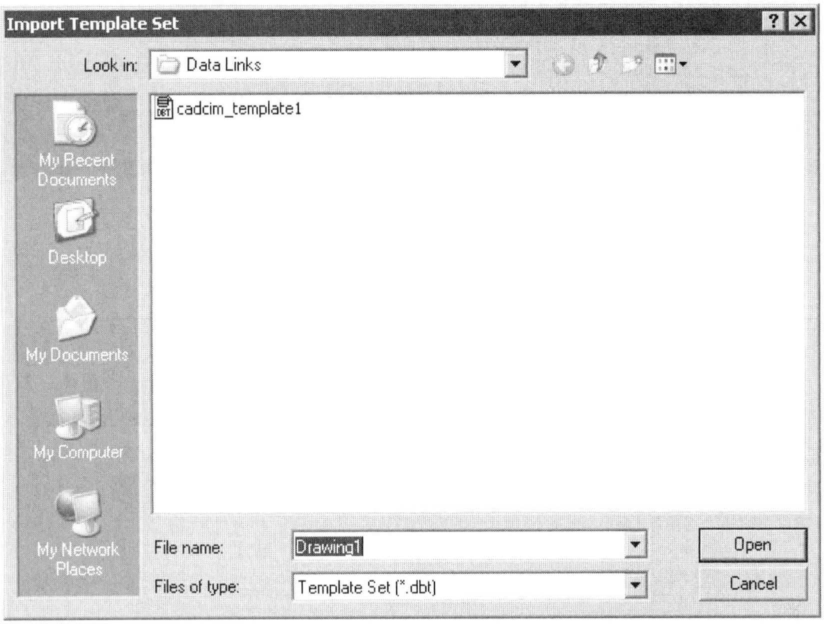

Figure 14-22 The ***Import Template Set*** *dialog box*

2. Select the template and choose **Open** to import the template set into the current drawing; AutoCAD displays an **Alert** box that can be used to provide a unique name for the template, if there is a link or label template with the same name associated with the current drawing.

AutoCAD SQL ENVIRONMENT (ASE)

SQL is an acronym for Structured Query Language. It is often referred to as Sequel. SQL is a format in computer programming that lets the user ask questions about a database according to specific rules. The AutoCAD SQL Environment (ASE) lets you access and manipulate the data that is stored in the external database table and link data from the database to objects in a drawing. Once you access the table, you can manipulate the data. The connection is made through a database management system (DBMS). The DBMS programs have their own methodology for working with the database. However, the ASE commands work the same way regardless of the database being used. This is made possible by the ASE drivers that come with AutoCAD software. In AutoCAD 2010, SQL has been incorporated. For example, you want to prepare a report that lists the computer equipment that costs more than $25. AutoCAD Query Editor can be used to easily construct a query that returns a subset of records or linked graphical objects that follow the previously mentioned criterion.

AutoCAD QUERY EDITOR

The **Query Editor** consists of four tabs that can be used to create new queries. The tabs are arranged in order of increasing complexity. For example, if you are not familiar with **SQL** (Structured Query Language), you can start with Quick Query and Range Query initially to get familiar with the query syntax.

You can start developing a query in one tab and subsequently add and refine the query conditions in the following tabs. For example, if you have created a query in the **Quick Query** tab and then decide to add an additional query using the **Query Builder** tab, when you choose the **Query Builder** tab, all the values initially selected in the previous tabs are displayed in this tab and additional conditions can be added to the query. But it is not possible to go backwards through the tabs once you have created queries with one of the advanced tabs. The reason for this is that the additional functions are not available in the simpler tabs. AutoCAD will prompt a warning indicating that the query will be reset to its default values if you attempt to move backward through the query tabs. The **Query Editor** has the following tabs for building queries:

Quick Query

This tab provides an environment where simple queries can be developed based on a single database field, single operator, and a single value. For example, you can find all records from the current table where the value of the '**Item_Type**' field equals '**CPU**'.

Range Query

This tab provides an environment where a query can be developed to return all records that fall within a given range of values. For example, you can find all records from the current table where the value of the '**Room**' field is greater than or equal to 6050 and less than or equal to 6150.

Query Builder

This tab provides an environment where more complicated queries can be developed based on a multiple search criteria. For example, you can find all records from the current table where the '**Item_Type**' equals '**CPU**' and '**Room**' number is greater than 6050.

SQL Query

This tab provides an environment where queries can be developed that confirm with the SQL 92 protocol. For example, you can select * from **Item type.Room. Tag_Number** there.

> **Item_type = 'CPU'; Room >= 6050** and **<= 6150**
> and
> **Tag_number > 26072**

The following example describes the complete procedure of creating a new query using all the tabs of the **Query Editor**:

Example 6 *General*

Create a new Query for the **Computer** database table and use all tabs of the **Query Editor** to prepare a SQL query.

1. Right-click on **Computer** in the **dbConnect Manager** to display the shortcut menu. Choose **New Query** from it to invoke the **New Query** dialog box (Figure 14-23). You can also

select **Computer** and choose the **New Query** button from the toolbar in the **dbConnect Manager** to invoke this dialog box.

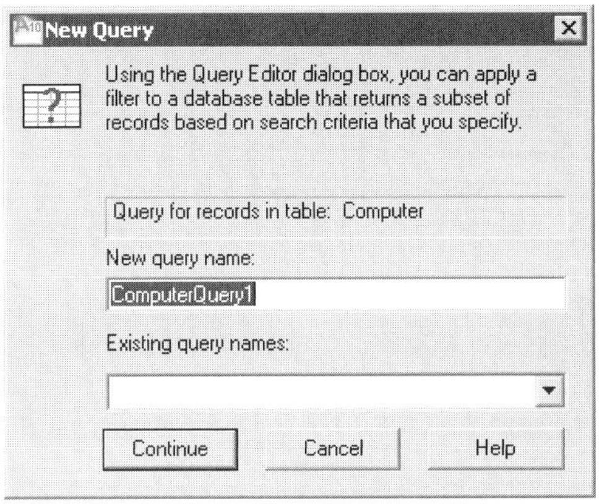

Figure 14-23 The ***New Query*** *dialog box*

2. In the **New Query** dialog box, choose the **Continue** button to accept the default query name **ComputerQuery1**; the **Query Editor** is displayed.

3. In the **Quick Query** tab of the **Query Editor**, select '**Item_Type**' from the **Field** list box, '**=Equal**' from the **Operator** drop-down list, and type '**CPU**' in the **Value** text box (Figure 14-24). You can also select '**CPU**' from the **Column Values** dialog box by choosing the **Look up values** button. Choose the **Store** button to save the query.

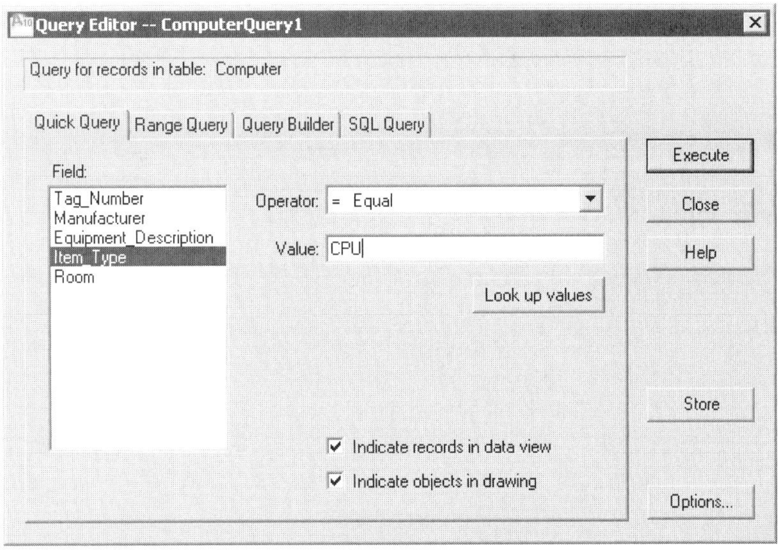

Figure 14-24 The ***Query Editor (Quick Query*** *tab)*

Chapter 14

Note

*To view the query, choose the **Execute** button from the **Query Editor**. However, you will not be able to continue with the example using the same **Query** dialog box.*

4. Choose the **Range Query** tab, and select **Room** from the **Field** list box. Enter **6050** in the **From** edit box and **6150** in the **Through** edit box (Figure 14-25). You can also select the values from the **Column Values** dialog box by choosing the **Look up values** button. Choose the **Store** button to save the query.

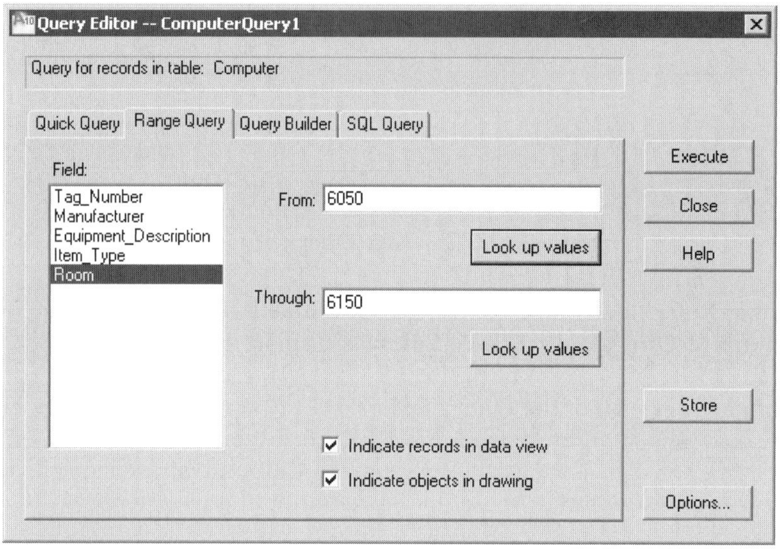

***Figure 14-25** The **Query Editor** dialog box (**Range Query** tab)*

5. Choose the **Query Builder** tab, and add **Item_Type** in **Show fields** list box by selecting **Item_Type** from the **Fields in table** list box and choosing the **Add** button. In the table area, change the entries of **Field**, **Operator**, **Value**, **Logical**, and **Parenthetical** grouping criteria, as shown in Figure 14-26, by selecting each cell and selecting the values from the respective drop-down list or options lists. Choose the **Store** button to save the query.

6. Choose the **SQL Query** tab; the query conditions specified earlier will be carried to the tab automatically. Here you can create a query using multiple tables. But select '**Computer**' from the **Table** list box, '**Tag_Number**' from the **Fields** list box, '**>= Greater than or equal to**' from the **Operator** drop-down list and enter **26072** in the **Values** edit box (Figure 14-27). You can also select from the available values by choosing the **[...]** button. Choose the **Store** button to save the query.

7. To check whether the SQL syntax is correct, choose the **Check** button; AutoCAD will display the **Information** box to determine whether the syntax is correct. Choose **OK** to exit the dialog box.

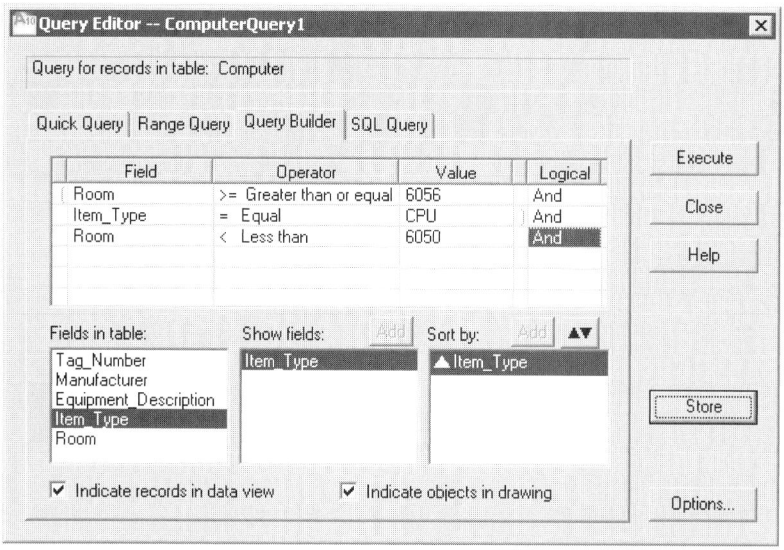

*Figure 14-26 The **Query Editor** (**Query Builder** tab)*

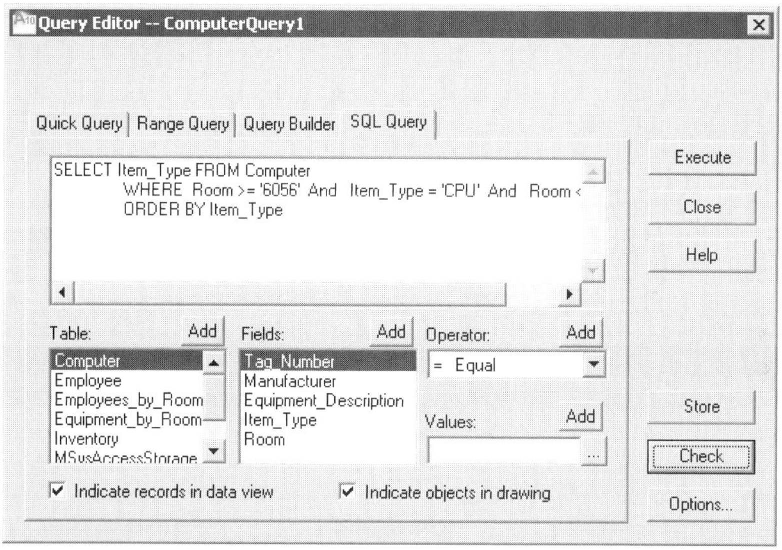

*Figure 14-27 The **Query Editor** dialog box (**SQL Query** tab)*

8. Choose the **Execute** button to display the **Data View** window showing a subset of records matching the specified query criteria.

Note

You can view the subset of records conforming to your criterion with any of the four tabs during the creation of queries.

Importing and Exporting SQL Queries

You may occasionally be required to use the queries made by some other user in your drawing or vice-versa. AutoCAD allows you to import or export stored queries. Sharing queries is very useful when developing common tools used by all team members on a project.

The following is the procedure of exporting a set of queries from the current drawing:

1. Choose **dbConnect > Queries > Export Query Set** from the menu bar to invoke the **Export Query Set** dialog box. In the dialog box, select the directory you want to save the query set from the **Save In** drop-down list.

2. Under the **File name** box, specify a name for the query set, and then choose the **Save** button.

The following is the procedure of importing a set of queries into the current drawing:

1. Choose **dbConnect > Queries > Import Query Set** from the menu bar to invoke the **Import Query Set** dialog box. In the dialog box, select the query set to be imported.

2. Choose the **Open** button to import the query set into the current drawing.

AutoCAD displays an alert box that you can use to provide a unique name for the query, if there is a query with the same name that is already associated with the current drawing.

FORMING SELECTION SETS USING THE LINK SELECT

It is possible to locate objects on the drawing on the basis of the linked nongraphic information. For example, you can locate the object that is linked to the first row of the **Computer** table or to the first and second rows of the **Computer** table. You can highlight specified objects or form a selection set of the selected objects. The **Link Select** is an advanced feature of the **Query Editor** that can be used to construct iterative selection sets of AutoCAD graphical objects and the database records. You can start constructing a query or selecting AutoCAD graphical objects for an iterative selection process. The initial selection set is referred to as set A. Now you can select an additional set of queries or graphical objects to further refine your selection set. The second selection set, is referred to as set B. To refine your final selection set you must establish a relation between set A and set B. The following is the list of available relationships or set operations (Figure 14-28).

Select

This creates an initial query or graphical objects selection set. This selection set can be further refined or modified using subsequent Link Select operations.

Union

This operation adds the outcome of a new selection set or query to the existing selection set. Union returns all the records that are members of set A or set B.

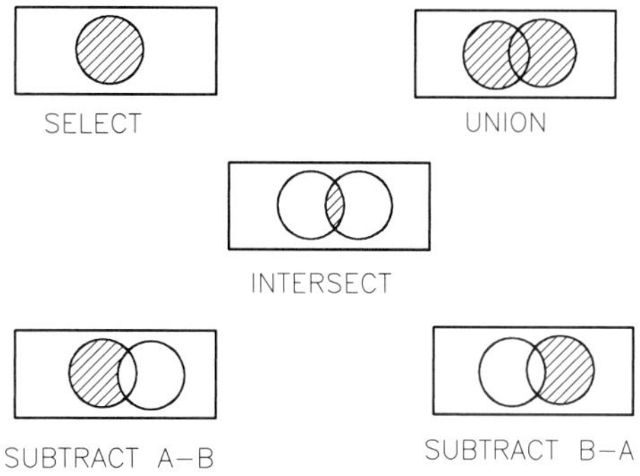

Figure 14-28 *Available relationships or Set operations*

Intersect

This operation returns the intersection of the new selection set and the existing or running selection set. Intersection returns only the records that are common to both set A and set B.

Subtract A - B

This operation subtracts the result of the new selection set or query from the existing set.

Subtract B - A

This operation subtracts the result of the existing selection set or query from the new set.

After any of the Link Selection operation is executed, the result of the operation becomes the new running selection set and is assigned as set A. You can extend refining your selection set by creating a new set B and then continuing with the iterative process.

Following is the procedure to use **Link Select** for refining a selection set:

1. Choose **dbConnect > Links > Link Select** from the menu bar to invoke the **Link Select** dialog box (Figure 14-29).

2. Select the **Select** option from the **Do** drop-down list for creating a new selection set. Also, select a link template from the **Using** drop-down list.

3. Choose either the **Use Query** or the **Select in Drawing <** radio button for creating a new query or drawing object selection set.

4. Choose the **Execute** button to execute the refining operation or choose the **Select** operation from the **Do:** drop-down list to add your query or graphical object selection set.

5. Again, choose any Link Selection operation from the **Do:** drop-down list.

6. Repeat steps 2 through 4 for creating a set B for the **Link Select** operation and then choose the **Finish** button to complete the operation.

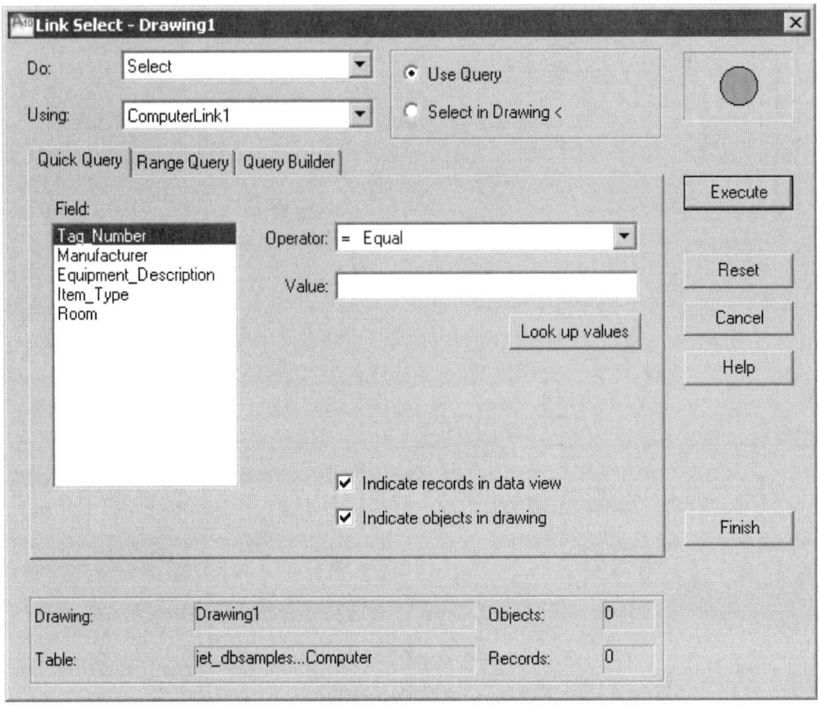

***Figure 14-29** The **Link Select** dialog box*

Note

*You can select the **Use Query** radio button to construct a new query or the **Select in Drawing <** radio button to select a graphical object from the drawing as a selection set.*

*If you select **Indicate records in data view**, then the Link Select operation result will be displayed in the **Data View** window and if you select **Indicate objects in drawing**, then AutoCAD displays a set of linked graphics objects in the drawing.*

Self-Evaluation Test

Answer the following questions and then compare them to those given at the end of this chapter:

1. The horizontal group of data is called a _________.

2. A vertical group of data (attribute) is called a _________.

3. You can connect to an external database table by using _________ **Manager**.

4. You can edit an external database table within an AutoCAD session. (T/F)

5. You can resize, dock, and hide the **Data View** window. (T/F)

6. Once you define a link, AutoCAD does not store that information with the drawing. (T/F)

7. To display the associated records with the drawing objects, AutoCAD provides__________.

8. Labels are of two types, __________ and __________.

9. It is possible to import as well as export Link and label templates. (T/F)

10. ASE stands for __________.

Review Questions

Answer the following questions:

1. The SQL statements let you search through the database and retrieve the information as specified in the SQL statements. (T/F)

2. What are the various components of a table?

3. What is a database?

4. What is a database management system (DBMS)?

5. A row is also referred to a __________.

6. A column is sometimes called a __________.

7. The __________ acts like an identification tag for locating and linking a row.

8. Only one row can be manipulated at a time. This row is called the _________.

9. The **Query Editor** has four tabs, namely __________, ___________, __________, and __________.

10. It is possible to go back to the previous query tab without resetting the values. (T/F)

11. You can execute a query after any tab in the **Query Editor**. (T/F)

12. Link select is an __________ implementation of **Query Editor**, which constructs selection sets of graphical objects or database records.

13. AutoCAD writes data source mapping information in the __________ file during the conversion process of links.

14. AutoCAD 2010 links __________ be converted into AutoCAD R12 format.

Exercise

Exercise 1

General

In this exercise, select **Employee** table from **jet_dbsamples** data source in **dbConnect Manager**, then edit the sixth row of the table (EMP_ID=1006, Keyser). Also, add a new row to the table, set the new row current, and then view it.

The row to be added has the following values:

EMP_ID	**1064**
LAST_NAME	**Joel**
FIRST_NAME	**Billy**
Gender	**M**
TITLE	**Marketing Executive**
Department	**Marketing**
ROOM	**6071**

From the **Inventory** table in **jet_dbsamples** data source, build an SQL query step-by-step using all the tabs of the **Query Editor**. The conditions to be implemented are as follows:

1. Type of Item = **Furniture**
2. Range of Cost = **200 to 650**
3. Manufacturer = **Office master**

Answers to Self-Evaluation Test

1. row, **2.** column, **3. dbConnect**, **4.** T, **5.** T, **6.** F, **7.** Leaders, **8.** Freestanding, Attached, **9.** T, **10.** AutoCAD SQL Environment

Chapter 15

Geometry Calculator

Learning Objectives

After completing this chapter, you will be able to:

- *Understand how the geometry calculator functions.*
- *Use real, integer, vector, and numeric expressions.*
- *Use snap modes in the geometry calculator.*
- *Obtain the radius of an object and locate a point on a line.*
- *Understand applications of the geometry calculator.*
- *Use AutoLISP variables and filter X, Y, and Z coordinates.*

GEOMETRY CALCULATOR

The Geometry Calculator is an ADS application that can be used as an online calculator. The calculator can be used to evaluate vector, real, and integer expressions. It can also access the existing geometry by using the first three characters of the standard AutoCAD object snap functions (MID, CEN, END). You can use the calculator to evaluate arithmetic and vector expressions. For example, you can use the calculator to evaluate an expression like 3.5^12.5*[234*log(12.5) - 3.5*cos(30)]. The results of a calculation can be returned as input to the current AutoCAD prompt.

Another application of the calculator is in assigning a value to an AutoLISP variable. For example, you can use an AutoLISP variable in the arithmetic expression, and then assign the value of the expression to an AutoLISP variable. You can invoke the **CAL** command by entering **CAL** or **'CAL** (for transparent use) at the Command prompt.

REAL, INTEGER, AND VECTOR EXPRESSIONS

Real and Integer Expressions

A real expression consists of real numbers and/or functions that are combined with numeric operators. Similarly, an integer expression consists of integers and/or functions combined with numeric operators. The following is the list of numeric operators:

Operator	Operation	Example
+	Adds numbers	2 + 3
-	Subtracts numbers	15.5 - 3.754
*	Multiplies numbers	12.34 * 4
/	Divides numbers	345.5/2.125
^	Exponentiation of numbers	25.5^2.5
( )	Used to group expressions	4.5 + (4.35^2)

Example:
Command: **CAL**
Initializing...>> Expression: **(4.5 + (4.35^2))**
23.4225

Vector Expression

A vector expression consists of points, vectors, numbers, and functions that are combined with the following operators:

Operator	Operation / Example
+	Adds vectors [a,b,c] + [x,y,z] = [a + x, b + y, c + z] [2,4,3] + [5,4,7] = [2 + 5, 4 + 4, 3 + 7] = [7.0 8.0 10.0]
-	Subtracts vectors [a,b,c] - [x,y,z] = [a - x, b - y, c - z] [2,4,3] - [5,4,7.5] = [2 - 5, 4 - 4, 3 - 7.5] = [-3.0 0.0 -4.5]
*	Multiplies a vector by a real number a * [x,y,z] = [a * x, a * y, a * z] 3 * [2,8,3.5] = [3 * 2, 3 * 8, 3 * 3.5] = [6.0 24.0 10.5]
/	Divides a vector by a real number [x,y,z] / a = [x/a, y/a, z/a] [4,8,4.5] / 2 = [4/2, 8/2, 4.5/2] = [2.0 4.0 2.25]
&	Multiplies vectors [a,b,c] & [x,y,z] = [(b * z)-(c * y), (c * x)-(a * z), (a * y)-(b * x)] [2,4,6] & [3,5,8] = [(4 * 8)-(6 * 5), (6 * 3)-(2 * 8), (2 * 5)-(4 * 3)] =
[2.0	2.0 -2.0]
( )	Used to group expressions a + (b ^ c)

Example:
Command: **CAL**
Initializing...>> Expression: **[2,4,3] - [5,4,7.5]**
-3,0,-4.5

NUMERIC FUNCTIONS

The geometry calculator (**CAL**) supports the numeric functions given next.

Function	Description
sin(angle)	Calculates the sine of an angle
cos(angle)	Calculates the cosine of an angle
tang(angle)	Calculates the tangent of an angle
asin(real)	Calculates the arcsine of a number (The number must be between -1 and 1)
acos(real)	Calculates the arccosine of a number (The number must be between -1 and 1)
atan(real)	Calculates the arctangent of a number
ln(real)	Calculates the natural log of a number
log(real)	Calculates the log, to the base 10, of a number
exp(real)	Calculates the natural exponent of a number
exp 10(real)	Calculates the exponent, to the base 10, of a number
sqr(real)	Calculates the square of a number
sqrt(real)	Calculates the square root of a number
abs(real)	Calculates the absolute value of a number
round(real)	Rounds the number to the nearest integer
trunc(real)	Returns the integer portion of a number
r2d(angle)	Converts the angle in radians to degrees
d2r(angle)	Converts the angle in degrees to radians
pi	pi has a constant value (3.14159265358 98)

Example:
Command: **CAL**
Initializing...>> Expression: **Sin (60)**
0.86602540378444

USING SNAP MODES

You can use snap modes with the **CAL** function to evaluate an expression. When you use snaps in an expression, you are prompted to select an object, and the returned value will be used in the expression. For example, if the **CAL** function is **(cen+end)/2**, the calculator will first prompt you to select an object for CENter snap mode and then select another object for ENDpoint snap mode. The corresponding coordinates of the two point values will be added and divided by **2**. The returned value is a point located midway between the center of the circle and the endpoint of the selected object. Following is the list of **CAL** snap modes and the corresponding AutoCAD snap modes:

CAL Snap Modes	AutoCAD Snap Modes
END	ENDpoint
EXT	EXTension
INS	INSert
INT	INTersection
MID	MIDpoint

CEN	CENter
NEA	NEArest
NOD	NODe
QUA	QUAdrant
PAR	PARallel
PER	PERpendicular
TAN	TANgent

Example 1

In this example, you will use the **CAL** snap modes to retrieve the point values (coordinates), and then use these values to draw a line (P3,P4). It is assumed that the circle and line (P1,P2) are already drawn (Figure 15-1).

Command: Choose the **Line** button.
Specify first point: **'CAL**
>>>> Expression: **(cen+end)/2**
>>>> Select entity for CEN snap: *Select the circle.*
>>>> Select entity for END snap: *Select one end of line (P1,P2).*
Specify next point or [Undo]: **'CAL**
>>>> Expression: **(cen+end)/2**
>>>> Select entity for CEN snap: *Select the circle.*
>>>> Select entity for END snap: *Select the other end of line (P1,P2).*

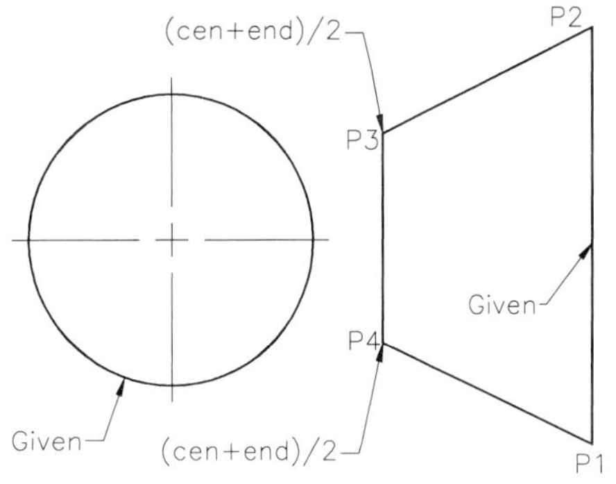

Figure 15-1 *Using the* ***CAL*** *snap modes*

Now, join point (P4) with point (P1) and point (P3) with point (P2) to complete the drawing, as shown in Figure 15-1. The '**CAL** function initializes the geometry calculator transparently. The single quote in front of the cal function (') makes the cal function transparent. The expression **(cen+end)/2** will prompt you to select the objects for **CEN snap** and **END snap**. After you select these objects, the calculator will add the corresponding coordinate of these point values and then divide the sum of each coordinate by 2. The point value that this function returns is the midpoint between the center of the circle and the first endpoint of the line.

OBTAINING THE RADIUS OF AN OBJECT

You can use the **rad** function to obtain the radius of an object. The object can be a circle, an arc, or a 2D polyline arc.

Example 2

In this example, you are given a circle of certain radius (R). You will draw a second circle whose radius is 0.75 times the radius of the given circle (0.75 * R) (Figure 15-2).

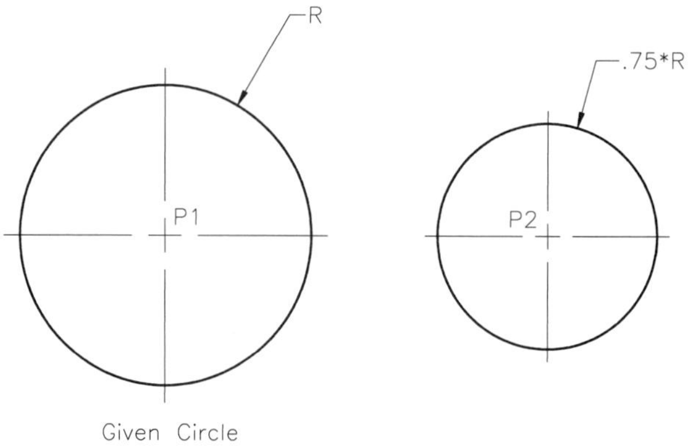

Figure 15-2 *Obtaining the radius of an object*

Command: **CIRCLE**
Specify center point for circle or [3P/2P/Ttr (tan tan radius)]: *Select a point (P2).*
Specify radius of circle or [Diameter] <current>: **'CAL**
>>>> Expression: **0.75*rad**
>>>> Select circle, arc or polyline segment for RAD function: *Select the given circle.*

Now you can enter the function name in response to the calculator prompt (**>>Expression**). In this example, the expression is **0.75*rad**. The **rad** function prompts the user to select an object, and it retrieves its radius. This radius is then multiplied by 0.75. The product of rad and 0.75 determines the radius of the new circle.

LOCATING A POINT ON A LINE

You can use the functions **pld** and **plt** to locate a point at a specified distance along a line between two points. The format of the **pld** function is **pld(p1,p2,dist)**. This function will locate a point on line (P1,P2) that is at a distance of **dist** from point (P1). For example, if the function is **pld(p1,p2,0.7)** and the length of the line is 1.5, the calculator will locate a point at a distance of 0.7 from point (P1) along line (P1,P2).

The format of the **plt** function is **plt(p1,p2,t)**. This function will locate a point on line (P1,P2) at a proportional distance as determined by the parameter, **t**. If **t = 0**, the point that this function will locate is at P1. Similarly, if **t = 1**, the point is located at (P2). However, if the value of **t** is greater than 0 and less than 1 (0> t <1), then the location of the point is determined by the value of **t**. For example, if the function is **plt(p1,p2,0.3)** and the length of the line is 1.5, the calculator will locate the point at a distance of 0.3 * 1.5 = 0.45 from point (P1).

Example 3

In this example, you will use the **pld** and **plt** functions to locate the centers of the circles. The arcs (P1,P2) and (P3,P4) are given in Figure 15-3.

Figure 15-3 illustrates the use of the **plt** and **pld** functions. The **pld** function, **pld(end,end,0.5)**, will prompt the user to select the two endpoints of the arc (P1,P2), and the function will return a point at a distance of 0.5 from point P1.

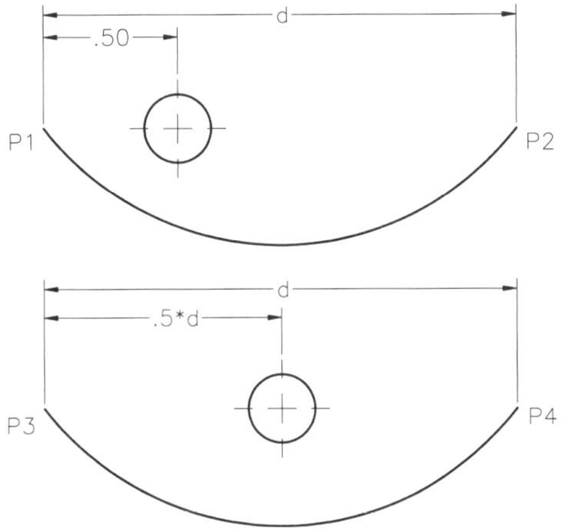

Figure 15-3 *Locating a point using calculator*

Command: Choose the **Center, Radius** button from the **Ribbon > Home > Draw**.
Specify center point for circle or [3P/2P/Ttr (tan tan radius)]: **'CAL**
>>>> Expression: **pld(end,end,0.5)**
>>>> Select entity for END snap: *Select the point P1.*
>>>> Select entity for END snap: *Select the point P2.*
Specify radius of circle or [Diameter] <current>: *Enter radius.*

Similarly, the **plt** function, **plt(end,end,0.5)**, will prompt the user to select the two endpoints of line (P3,P4); it will return a point that is located at a distance of 0.5*d units from point (P3).

Command: Choose the **Center, Radius** button from the **Ribbon > Home > Draw**.
Specify center point for circle or [3P/2P/Ttr (tan tan radius)]: **'CAL**
>>>>Expression: **plt(end,end,0.5)**
>>>>Select entity for END snap: *Select the point P3.*
>>>>Select entity for END snap: *Select the point P4.*
Specify radius of circle or [Diameter] <current>: *Enter radius.*

OBTAINING AN ANGLE

You can use the **ang** function to obtain the angle between two lines. The function can also be used to obtain the angle that a line makes with the positive *X* axis. The function has the following formats (Figure 15-4):

ang(v)
ang(p1,p2)

ang(apex,p1,p2)
ang(apex,p1,p2,p)

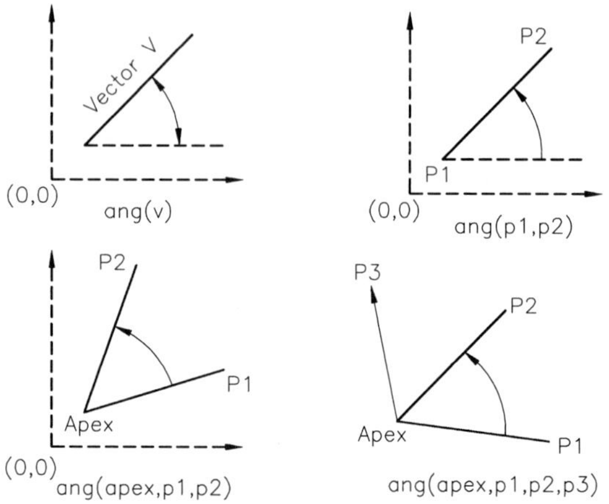

Figure 15-4 *Obtaining an angle*

ang(v)

The **ang(v)** function can be used to obtain the angle that a vector makes with the positive *X* axis. Assume a vector [2,2,0], and obtain its angle. You can use the ang(v) function to obtain the angle.

Command: **CAL**
>>Expression: **v=[2,2,0]** *(Defines a vector v.)*
Command: **CAL**
>>Expression: **ang(v)** *(v is a predefined vector.)*
45

The vector v makes a 45-degree angle with the positive *X* axis.

ang(p1,p2)

The **ang(p1,p2)** function can be used to obtain the angle that a line (P1,P2) makes with the positive *X* axis. For example, if you want to obtain the angle of a line with the positive *X* axis and having start point and endpoint coordinates at (1,1,0) and (4,4,0), respectively, follow the Command prompt given next.

Command: **CAL**
>>Expression: **p1=[1,1,0]** *(Defines a vector p1.)*
Command: **CAL**
>>Expression: **p2=[4,4,0]** *(Defines a vector p2.)*
Command: **CAL**
>>Expression: **ang(p1,p2)**
45

If the line exists, you can obtain the angle by using the following function:

Command: **CAL**
>>Expression: **ang(end,end)**
>>Select entity for END snap: *Select first endpoint of line (P1,P2).*
>>Select entity for END snap: *Select second endpoint of line (P1,P2).*
31.12480584361 *(This is the angle that the function returns.)*

ang(apex,p1,p2)

The **ang(apex,p1,p2)** function can be used to obtain the angle that a line (apex,P1) makes with (apex,P2). For example, if you want to obtain the angle between the two given lines, as shown in the third drawing of Figure 15-4, use the following commands:

Command: **CAL**
>>Expression: **ang(end,end,end)**
>>Select entity for END snap: *Select first endpoint (apex).*
>>Select entity for END snap: *Select second endpoint (P1).*
>>Select entity for END snap: *Select third endpoint (P2).*
51.414596245895 *(This is the angle that the function returns.)*

ang(apex,p1,p2,p)

The **ang(apex,p1,p2,p)** function can be used to obtain the angle that a line (apex,P1) makes with (apex,P2). The last point, **p**, is used to determine the orientation of the angle.

LOCATING THE INTERSECTION POINT

You can obtain the intersection point of two lines (P1,P2) and (P3,P4) by using the following function:

ill(p1,p2,p3,p4)

(P1,P2) are two points on the first line, and (P3,P4) are two points on the second line, as shown in Figure 15-5. It is recommended that you turn the object snap off before using this function.

Example 4

In this example, you will draw a circle whose center point is located at the intersection point of two lines (P1,P2) and (P3,P4). It is assumed that the two lines are given, as shown in Figure 15-5.

Turn the object snap off and then use the following commands to obtain the intersection point and to draw a circle with the intersection point as the center of the circle:

Command: Choose **Home > Draw > Center, Radius** button from the **Ribbon**.
Specify center point for circle or [3P/2P/Ttr (tan tan radius)]: **'CAL**

>>>>Expression: **ill(end,end,end,end)**
>>>>Select entity for END snap: *Select point P1.*
>>>>Select entity for END snap: *Select point P2.*
>>>>Select entity for END snap: *Select point P3.*
>>>>Select entity for END snap: *Select point P4.*
Specify radius of circle or [Diameter] <current>: *Enter radius.*

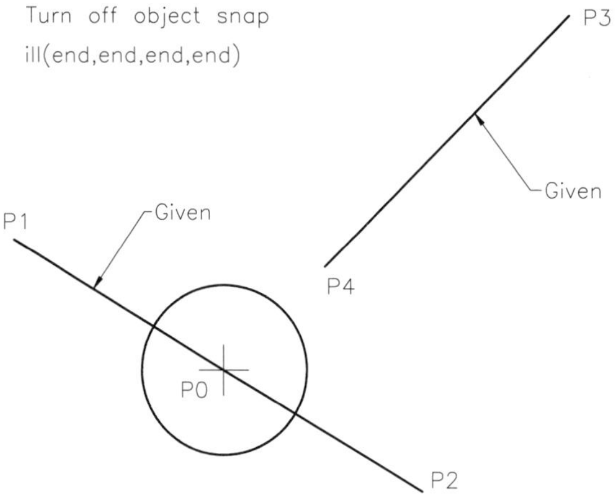

Figure 15-5 *Obtaining intersection point*

Tip
The expression ill (end,end,end,end) can be replaced by the shortcut function ***ille****. When used, it will automatically prompt for four endpoints to locate the point of intersection.*

APPLICATIONS OF THE GEOMETRY CALCULATOR

The following examples illustrate some additional applications of the geometry calculator.

Example 5

In this example, you will draw a circle whose center (P0) is located midway between endpoints (P4) and (P2), see Figure 15-6.

The center of the circle can be located by using the calculator snap modes. For example, to locate the center you can use the expression **(end+end)/2**. The other way of locating the center is by using the shortcut function **mee**, as shown here (Figure 15-6):

Command: Choose **Home > Draw > Center, Radius** button from the **Ribbon**.
Specify center point for circle or [3P/2P/Ttr (tan tan radius)]: **'CAL**
>>>>Expression: **mee**
>>>> Select one endpoint for MEE: *Select the first endpoint (P2).*
>>>> Select another endpoint for MEE: *Select the second endpoint (P4).*
Specify radius of circle or [Diameter] <current>: *Enter the radius.*

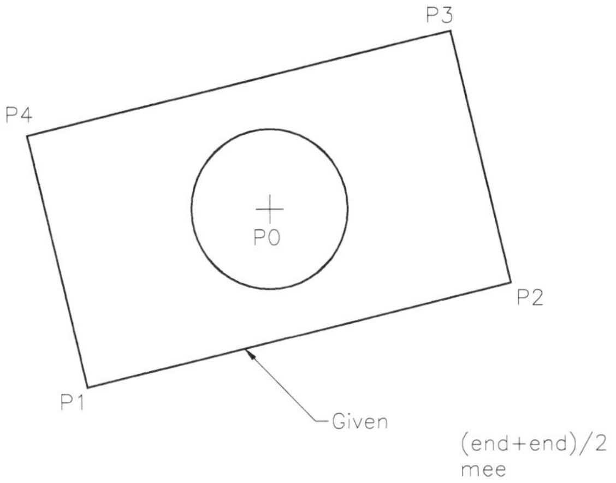

Figure 15-6 *Using the shortcut function* **mee**

Example 6

In this example, you will draw a circle that is tangent to a given line. The radius of the circle is 0.5 units, and the circle must pass through the selected point shown in Figure 15-7.

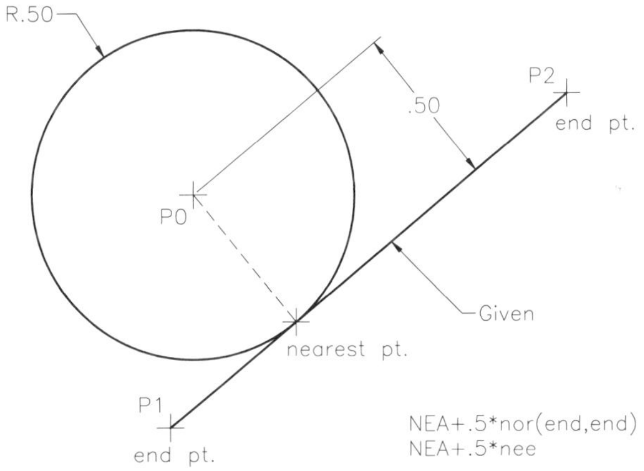

Figure 15-7 *Using the shortcut function* **nee**

To draw a circle that is tangent to a line, you must first locate the center of the circle that is at a distance of 0.5 units from the selected point. This can be accomplished by using the function **nor(p1,p2)**, which returns a unit vector normal to the line (P1,P2). You can also use the shortcut function **nee**. The function will automatically prompt you to select the two endpoints of the given line. The unit vector must be multiplied by the radius (0.5) to locate the center of the circle.

Command: Choose **Home > Draw > Center, Radius** button from the **Ribbon**.
Specify center point for circle or [3P/2P/Ttr (tan tan radius)]: **'CAL**
>>>>Expression: **NEA+0.5*nee**
>>>>Select entity for NEA snap: *Select a point on the given line.*
>>>>Select one endpoint for NEE: *Select the first endpoint on the given line.*
>>>>Select another endpoint for NEE: *Select the second endpoint on the given line.*
Specify radius of circle or [Diameter] <current>: **0.5**

Example 7

In this example, you will draw a circle with its center on a line. The radius of the circle is 0.25 times the length of the line, and it is assumed that the line is given (Figure 15-8).

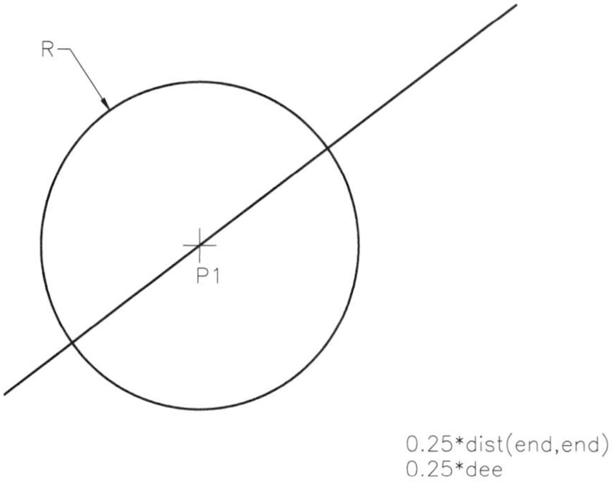

Figure 15-8 *Using the shortcut function* ***dee***

The radius of the circle can be determined by multiplying the length of the line by 0.25. The length of the line can be obtained by using the function dist(p1,p2) or by using the shortcut function **dee**. When you use the function **dee**, the calculator will automatically prompt you to select the two endpoints of the given line. It is equivalent to using the function dist(end,end).

Command: Choose the **Center, Radius** button from the **Ribbon > Home > Draw**.
Specify center point for circle or [3P/2P/Ttr (tan tan radius)]: *Select a point on the line.*
Specify radius of circle or [Diameter] <current>: **'CAL**
>>>>Expression: **0.25*dee**
>>>> Select one endpoint for DEE: *Select the first endpoint on the given line.*
>>>> Select another endpoint for DEE: *Select the second endpoint on the given line.*

USING AutoLISP VARIABLES

The geometry calculator allows you to use an AutoLISP variable in the arithmetic expression. You can also use the calculator to assign a value to an AutoLISP variable. The variables can be integer, real, or a 2D or 3D point. The next example illustrates the use of AutoLISP variables. In this example, you will draw two circles that are in the middle and are offset 0.5 units from the center. It is assumed that the other two circles are given.

Example 8

To locate the center of the top circle, first you must determine the point that is midway between the centers of the two given circles. This can be accomplished by defining a variable **midpoint** where **midpoint=(cen+cen)/2**. Similarly, you can define another variable for the offset distance: **offset=[0,0.5]**. The center of the top circle can be obtained by adding these two variables **(midpoint+offset)** (Figure 15-9).

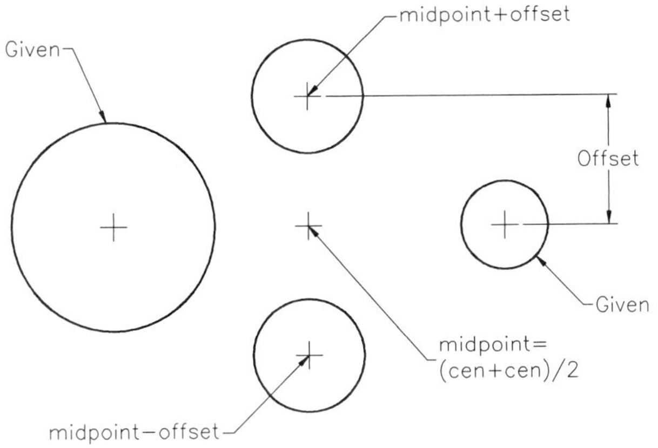

Figure 15-9 *Adding two predefined vectors*

Command: **CAL**
>>Expression: **midpoint=(cen+cen)/2**
>>Select entity for CEN snap: *Select the first given circle.*
>>Select entity for CEN snap: *Select the second given circle.*

Command: **CAL**
>>Expression: **offset=[0,0.5]**

Command: Choose **Home > Draw > Center, Radius** button from the **Ribbon**.
Specify center point for circle or [3P/2P/Ttr (tan tan radius)]: **'CAL**
>>>>Expression: **(midpoint+offset)**
Specify radius of circle or [Diameter] <current>: *Enter radius.*

To locate the center point of the bottom circle, you must subtract offset from midpoint:

Command: Choose **Home > Draw > Center, Radius** button from the **Ribbon**.
Specify center point for circle or [3P/2P/Ttr (tan tan radius)]: **'CAL**

>>>>Expression: **(midpoint-offset)**
Specify radius of circle or [Diameter] <current>: *Enter radius.*

The same results can be obtained by using AutoLISP expressions, as follows:

Command: **(Setq NEWPOINT "(CEN+CEN)/2+[0,0.5]")**
Command: Choose **Home > Draw > Center, Radius** button from the **Ribbon**.
Specify center point for circle or [3P/2P/Ttr (tan tan radius)]: **(cal NEWPOINT)**
(Recalls the expression.)
>>>>Select entity for CEN snap: *Select the first circle.*
>>>>Select entity for CEN snap: *Select the second circle.*

FILTERING X, Y, AND Z COORDINATES

The following functions are used to retrieve the coordinates of a point.

Function	Description
xyof(p)	Retrieves the *X* and *Y* coordinates of a point (p) and returns a point; the *Z* coordinate is automatically set to 0.0
xzof(p)	Retrieves the *X* and *Z* coordinates of a point (p) and returns a point; the *Y* coordinate is automatically set to 0.0
yzof(p)	Retrieves the *Y* and *Z* coordinates of a point (p) and returns a point; the *X* coordinate is automatically set to 0.0
xof(p)	Retrieves the *X* coordinate of a point (p) and returns a point; the *Y* and *Z* coordinates are automatically set to 0.0
yof(p)	Retrieves the *Y* coordinate of a point (p) and returns a point; the *X* and *Z* coordinates are automatically set to 0.0
zof(p)	Retrieves the *Z* coordinate of a point (p) and returns a point; the *X* and *Y* coordinates are automatically set to 0.0
rxof(p)	Retrieves the *X* coordinate of a point (p)
ryof(p)	Retrieves the *Y* coordinate of a point (p)
rzof(p)	Retrieves the *Z* coordinate of a point (p)

Example 9

In this example, you will draw a line by using filters to extract coordinates and points. It is assumed that the two lines are already drawn, as shown in Figure 15-10.

To draw a line, you need to determine the coordinates of the two endpoints of the line. The *X* coordinate of the first point can be obtained from point (P1), and the *Y* coordinate from point (P2). To obtain these coordinate points, you can use the filter function **rxof(end)** to extract the *X* coordinate and **ryof(end)** to extract the *Y* coordinate. To determine the coordinates of the endpoint of the line, you can filter the *XY* coordinates of point (P3) and then add the offset of 0.25 units by defining a vector [0.25,0,0].

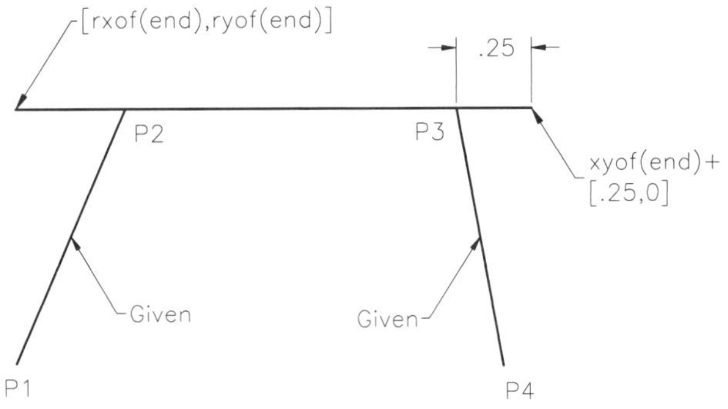

Figure 15-10 *Using filters to extract points and coordinates*

Command: Choose **Home > Draw > Center, Radius** button from the **Ribbon**.
Specify first point: '**CAL**
>>>>Expression: **[rxof(end),ryof(end)]**
>>>>Select entity for END snap: *Select point P1.*
>>>>Select entity for END snap: *Select point P2.*
Specify next point or [Undo]: **'CAL**
>>>>Expression: **xyof(end)+[0.25,0,0]**
>>>> Select entity for END snap: *Select point P3.*

CONVERTING UNITS

You can use the calculator function **cvunit** to change a given value from one system of units to another. You can also use this function to change the unit format. For example, you can change the units from feet to inches, meters to centimeters, and vice versa. The value can be a number or a point. The units available are defined in the *acad.unt* file, which is an ASCII file that you can examine. The format of the cvunit expression is:

cvunit(value, units from, units to)

Examples:
Command: **CAL**

>>Expression: **cvunit(100,cm,inch)**	**returns 39.370078740157**
>>Expression: **cvunit(100,feet,meter)**	**returns 30.48**
>>Expression: **cvunit(1,feet,inch)**	**returns 12**

ADDITIONAL FUNCTIONS

The following is the list of additional calculator functions. The description given next to the function summarizes the application of the function.

Function	Description
abs(real)	Calculates the absolute value of a number
abs(v)	Calculates the length of vector v
ang(v)	Calculates the angle between the *X* axis and vector v
ang(p1,p2)	Calculates the angle between the *X* axis and line (P1,P2)
cur	Retrieves coordinates of a point from the location of the graphics cursor
cvunit(val,from,to)	Converts the given value (val) from one unit measurement system to another
dee	Measures the distance between two endpoints; equivalent to the function dist(end,end)
dist(p1,p2)	Measures distance between the two specified points (P1,P2)
getvar(var_name)	Retrieves the value of the AutoCAD system variable
ill(p1,p2,p3,p4)	Returns the intersection point of lines (P1,P2) and (P3,P4)
ille	Returns the intersection point of lines defined by four endpoints; equivalent to the function ill(end,end,end,end)
mee	Returns the midpoint between two endpoints; equivalent to the function (end+end)/2
nee	Returns a unit vector normal to two endpoints; equivalent to the function nor(end,end)
nor	Returns a unit vector that is normal to a circle or an arc
nor(v)	Returns a unit vector in the *XY* plane that is normal to vector v
nor(p1,p2)	Returns a unit vector in the *XY* plane that is normal to line (P1,P2)
nor(p1,p2,p3)	Returns a unit vector that is normal to the specified plane defined by points (P1,P2,P3)
pld(p1,p2,dist)	Locates a point on the line (P1,P2) that is dist units from point (P1)
plt(p1,p2,t)	Locates a point on the line (P1,P2) that is t*dist units from point (P1) (Note: when t = 0, the point is (P1). Also, when t = 1, the point is (P2))
rad	Retrieves the radius of the selected object
rot(p,org,ang)	Returns a point that is rotated through angle **ang** about point org
u2w(p)	Locates a point with respect to WCS from the current UCS
vec(p1,p2)	Calculates a vector from point (P1) to point (P2)
vec1(p1,p2)	Calculates a unit vector from point (P1) to point (P2)
vee	Calculates a vector from two endpoints; equivalent to the function vec(end,end)
vee1	Calculates a unit vector from two endpoints; equivalent to the function vec1(end,end)
w2u(p)	Locates a point with respect to the current UCS from WCS

Self-Evaluation Test

Answer the following questions and then compare them to those given at the end of this chapter:

1. A real expression consists of real numbers and/or functions that are combined with numeric operators. (T/F)

2. The snap modes can be used with calculator functions to evaluate an expression. (T/F)

3. The **ill** function can be used to locate the intersection of two lines. (T/F)

4. The **dee** function is used to locate the intersection of two lines. (T/F)

5. The length of a line can be obtained by using the function **dist(p1,p2)** or by using the shortcut function __________ .

6. The function **xzof(p)** retrieves the __________ coordinates of a point (P) and returns a point. The *Y* coordinate is automatically set to (0,0).

7. The function **nor(v)** returns a ____________________ in the *XY* plane that is normal to the vector v.

8. The **ang(v)** function can be used to obtain the angle that a vector makes with the __________ axis.

9. The __________ function can be used to obtain the angle that a line (P1,P2) makes with the positive *X* axis.

10. The shortcut function **nee** is equivalent to __________ function.

Review Questions

Answer the following questions:

1. The calculator can also access the existing geometry by using standard AutoCAD object snap functions. (T/F)

2. The calculator cannot be used to assign a value to an AutoLISP variable. (T/F)

3. The **rad** function retrieves the radius of the selected object. (T/F)

4. The **ang(v)** function calculates the angle between the *X* axis and the line (P1,P2). (T/F)

5. Which function is used to calculate the absolute value of a number?

   (a) **abs(real)** (b) **abs(v)**
   (c) **abs(P1,P2)** (d) **None**

6. Which function is used to retrieve the *X* and the *Y* coordinates of a point P and return a point?

   (a) **xyof(p)** (b) **zxof(p)**
   (c) **xy(p)** (d) **None**

7. Which function is used to locate a point on line (P1,P2) that is at a distance of d from P1?

   (a) **plt(p1,p2,d)** (b) **pld(p1,p2,d)**
   (c) **plt(p2,p1,d)** (d) **None**

8. Which function is used to retrieve the *Y* coordinates of a point P and return a point?

   (a) **xyof(p)** (b) **xof(p)**
   (c) **yof(p)** (d) **None**

9. What is the short form of the function vec(end,end)?

   (a) **vee** (b) **eve**
   (c) **vem** (d) **None**

10. The __________ function returns a unit vector normal to the two endpoints.

11. The __________ can be used to obtain the radius of an object.

12. The __________ function returns a point that is rotated through angle ang about a point org.

13. The format of the **pld** function is __________.

14. The format of the **plt** function is __________.

15. The angle between two given lines can be obtained by __________.

Exercises

Exercise 1

General

Sketch the drawing shown in Figure 15-11. Use the real operators of the geometry calculator to calculate the value of L, H, and TL.

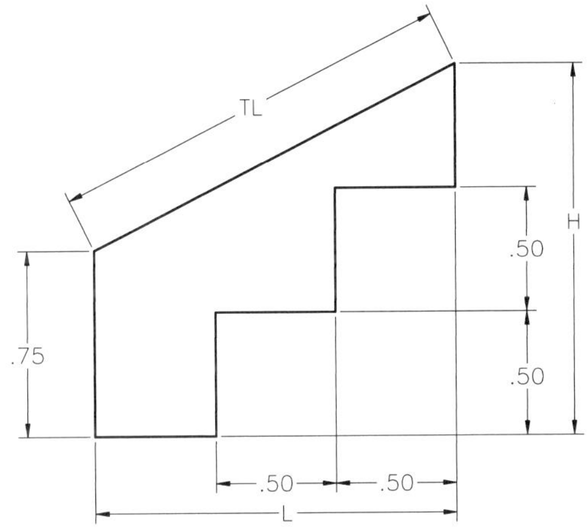

Figure 15-11 *Drawing for Exercise 1*

Exercise 2 *General*

Sketch the drawing shown in Figure 15-12; assume the dimensions. Draw a circle whose center is at point (P3). Use the calculator function to locate the center of the circle (P3) that is midway between (P1) and (P2). The points (P1) and (P2) are the midpoints on the top and bottom lines, respectively.

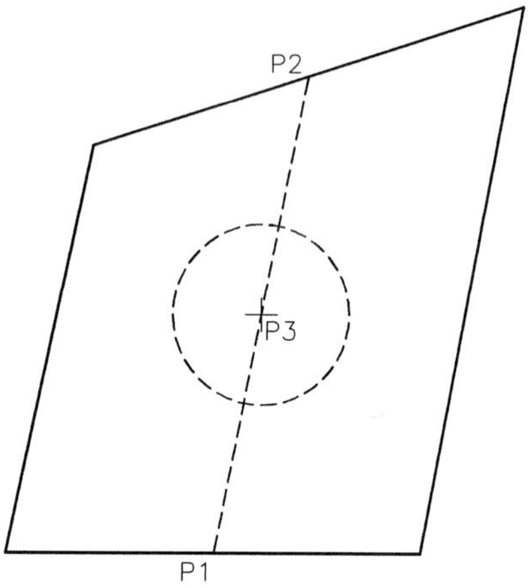

Figure 15-12 *Drawing for Exercise 2*

Exercise 3 *General*

Sketch the drawing shown in Figure 15-13; assume the dimensions.

1. Use the calculator function to locate the point (P3). Point (P3) is midway between (P1) and (P2). The points (P1) and (P2) are the midpoints on the top and the bottom lines, respectively.

2. Use the calculator function to locate the point (P4) that is normal to the line (P1,P2) at a distance of 0.25 units.

3. Draw a circle whose center is located at (P4).

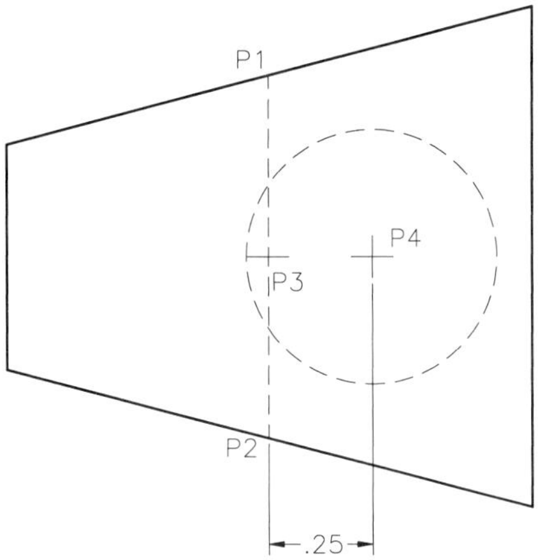

Figure 15-13 Drawing for Exercise 3

Problem-Solving Exercise 1 *General*

Sketch the drawing shown in Figure 15-14; assume the dimensions.

1. Draw a circle whose center is at point (P3). Use the calculator function to locate the center of the circle (P3) that is midway between (P1) and (P2). Points (P1) and (P2) are the midpoints on the top and bottom lines, respectively.

2. Draw a circle whose radius is 0.75 times the radius of the circle in part 1. The center of the circle (P5) is located midway between the line (P3,P4).

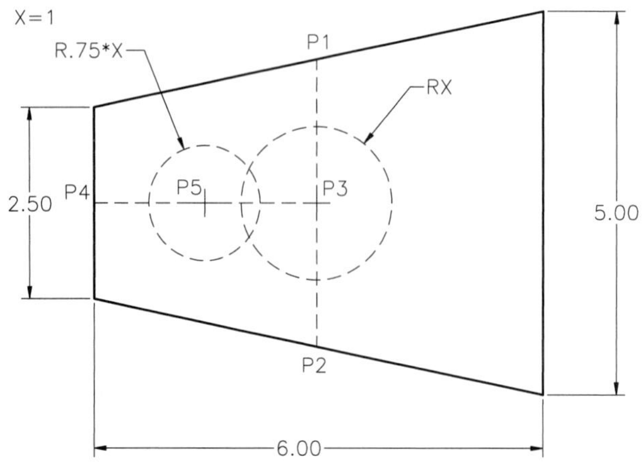

Figure 15-14 *Drawing for Problem-Solving Exercise 1*

Answers to Self-Evaluation Test

1. T, **2.** T, **3.** T, **4.** F, **5. dee**, **6.** *X* and *Z*, **7.** unit vector, **8.** positive *X*, **9.** ang(p1,p2), **10.** nor(p1,p2)

Chapter 16

Image Tile Menus

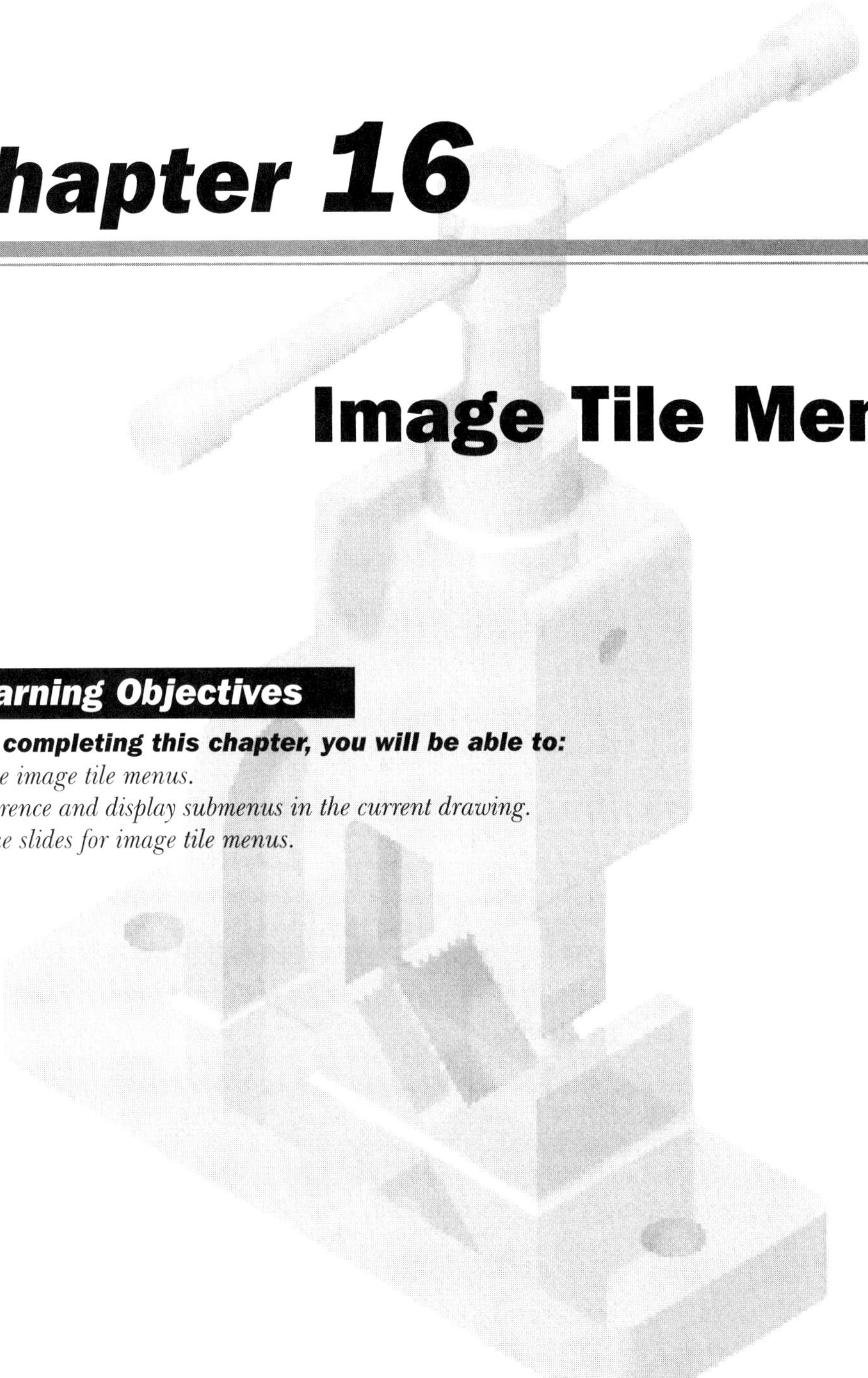

Learning Objectives

After completing this chapter, you will be able to:

- *Write image tile menus.*
- *Reference and display submenus in the current drawing.*
- *Make slides for image tile menus.*

IMAGE TILE MENUS

The image tile menus, also known as icon menus, are extremely useful for inserting a block, selecting a text font, or drawing a 3D object. You can also use the image tile menus to load an AutoLISP routine or a predefined macro. Thus, the image tile menu is a powerful tool for customizing AutoCAD.

The image tile menus can be accessed from the pull-down, tablet, button, or screen menu. However, the image tile menus cannot be loaded by entering a command. When you select an image tile, a dialog box that contains 20 image tiles is displayed on the screen (Figure 16-1). The names of the slide files associated with image tiles appear on the left of the dialog box with a scroll for the file names. The title of the image tile menu is displayed at the top of the dialog box (Figure 16-1). When you activate the image tile menu, the cursor is replaced by the arrow cursor and is used to select any image tile that appears in the image tile menu. You can select the required image tile by selecting the slide file name from the dialog box and then choosing the **OK** button or double-clicking on the slide file name.

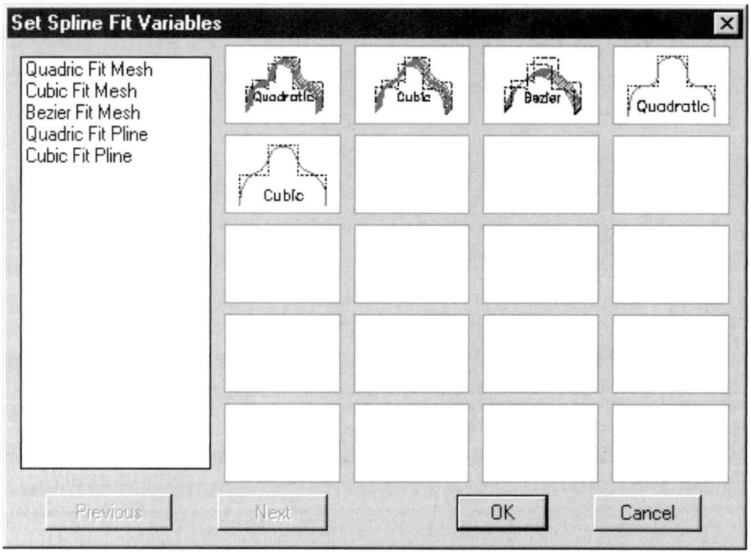

Figure 16-1 *Sample image tile menu display*

When you select the slide file, the corresponding image tile is highlighted in the image tile menu. You can also select an image tile by moving the arrow to the desired image tile, and clicking on it. The corresponding slide file name will be automatically highlighted. If you choose the **OK** button or double-click on the image tile, the command associated with that menu item will be executed. Press the ESC key, or choose the **Cancel** button to exit the image tile menu.

You can define an unlimited number of menu items in the image tile menu, but only 20 image tiles will be displayed at a time. If the number of items exceeds 20, you can use the **Next** and **Previous** buttons of the dialog box to page through the various pages of image tiles.

SLIDES FOR IMAGE TILE MENUS

The idea behind creating slides for the image tile menus is to display graphical symbols in the image tiles. This symbol makes it easier for you to identify the operation that the image tile will perform. Any slide can be used for the image tile. However, the following guidelines should be kept in mind when creating slides for the image tile menu:

1. When you make a slide for an image tile menu, draw the object so that it fits the entire screen. The **MSLIDE** command makes a slide of the existing screen display. If the object is small, the picture in the image tile menu will be small. Use **ZOOM Extents** or **ZOOM Window** command to display the object before making a slide.

2. When you use the image tile menu, it takes some time to load the slides for display in the image tiles. The more complex the slides, the more time it will take to load them. Therefore, the slides should be kept as simple as possible and at the same time give enough information about the object.

3. Do not fill the object, because it takes a long time to load and display a solid object. If there is a solid area in the slide, it is not displayed in the image tile.

4. If the objects are too long or too wide, it is better to center the image with the AutoCAD **PAN** command before making a slide.

5. The space available on the screen for the image tile display is limited. Make the best use of this small area by giving only the relevant information in the form of a slide.

6. The image tiles that are displayed in the dialog box have the length-to-width ratio (aspect ratio) of 1.5:1. For example, if the length of the image tile is 1.5 units, the width is 1 unit. If the drawing area of your screen has an aspect ratio of 1.5 and the slide drawing is centered in the drawing area, the slide in the image tile will also be centered.

CREATING IMAGE TILE MENUS

To understand the process of creating an image tile menu, consider the following example.

Example 1

Create an image tile menu for the **ARC** command with the following options:

ARC, 3 POINTS	ARC, SEA
ARC, SCE	ARC, SED
ARC, SCA	ARC, SER
ARC, SCL	ARC, CONTINUE

The images that you need to use for these commands are given in Figure 16-2.

Step 1: Creating Slides

1. Create drawings similar to those shown in the images in Figure 16-2.

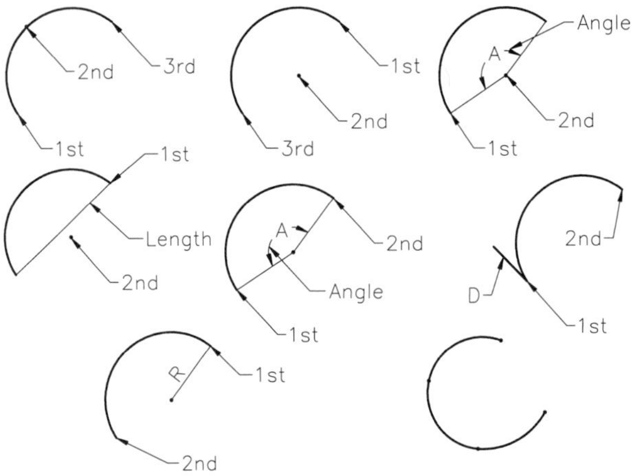

Figure 16-2 *Images for the image tile menu for Example 1*

2. Using the **MSLIDE** command, create eight slides of the eight images shown in Figure 16-2. Name the slides based on those given in the example description. Save the slides in the */customizing_2010/c16/example-1* folder.

3. Add the path of this folder to the AutoCAD support file search path using the **Options** dialog box.

Step 2: Creating the Menu

As mentioned earlier, the menus are created using the **Customize User Interface** dialog box. The following steps explain the procedure of creating this menu.

1. Invoke the **Customize User Interface** dialog box by choosing **Manage > Customization > User Interface** from the **Ribbon**. Alternatively, choose **Tools > Customize > Interface** from the menu bar to invoke this dialog box.

2. Scroll up in the **Customizations in All Files** rollout and expand the **Legacy** node. Then expand the **Image Tile Menus** sub-node under the **Legacy** node. This will display all image menus.

3. To add a new menu, right-click on the **Image Tile Menus** element and choose **New Image Tile Menu** from the shortcut menu; a new menu with the name **ImageTileMenu1** is added to the list and its properties are displayed in the right pane.

4. Rename the menu to **Tiled Arc Menu** by selecting the **Name** field in the **Properties** rollout.

5. Click on the field on the right of **Aliases** in the right pane; a swatch [...] is displayed in the field.

6. Click on the swatch [...] to display the **Aliases** dialog box. Type **IMAGE_ARC** and choose **OK** to close the **Aliases** dialog box.

7. Choose **Apply** button of **Customize User Interface** dialog box to save the changes.

   Next, you need to add the image tiles to this menu.

8. Select **Draw** from the **Filter the command list by category** drop-down list in the **Command List** rollout; all the **Draw** commands are displayed in the list box.

9. Scroll to the **Arc, 3 Points** item in the list box available in the **Command List** rollout. Drag this item and drop it on **Tiled Arc Menu**; the **Arc, 3 Points** tool is added to the menu.

Note

If you face any problem in dragging and dropping the command directly on the ***Tiled Arc Menu*** *element, expand the* ***Tiled Viewport Layout*** *node under the* ***Image Tile Menus*** *node and drop the command under this node. Next, drag the command from this node and drop it on the* ***Tiled Arc Menu*** *element. After adding the command to the* ***Tiled Arc Menu*** *node, delete the command dropped under the* ***Tiled Viewport Layout*** *node.*

10. Now, click on the **Arc, 3 Points** item you have added to **Tiled Arc Menu** in **Customizations in All Files** rollout of the **Customize User Interface** dialog box to display properties of this tool in the right pane.

    You might have noticed that **Slide library** and **Slide label** fields are empty. Here you should enter the name of slide library and slide name you want to be displayed as an image tile.

11. Type **3-points** against the **Slide label** field.

Note

To display slides in the image tile menu, slide library must be saved in any support file search path.

The name of slide and its corresponding slide label should be same. Otherwise, the image tile will not link with the slide.

12. Similarly, drag and drop the remaining commands from **Command List** and add the name of corresponding slide to the properties of the image tiles.

13. Choose **Apply** and then choose the **OK** button from the **Customize User Interface** dialog box. The image tile menu will be saved.

Displaying the Image Tile Menu

Image tile menu created in Example 1 can be invoked only if it is linked with the pull-down menu, shortcut menu, toolbar, tablet, screen menu, or mouse button. The method of linking the image tile menus with these items is discussed next.

Linking the Image Tile Menus

To display an image tile menu, you have to add a nonstandard command to the **Command List** rollout of the **Customize User Interface** dialog box. The following steps describe how to do this:

1. Invoke the **Customize User Interface** dialog box.

2. Choose the **Create a new command** button on the right of the **Filter the command list by category** drop-down list in the **Command List** rollout; a new command with the name **Command1** is added to the list of commands. The **Properties** rollout in the right pane displays the properties of the new command.

3. Modify the name of the new command in the **Properties** rollout to **CALLITM**.

4. In the description field, enter **This command will display Tiled Arc Image tile menu**.

5. In the **Macro** filed, enter **$I=IMAGE_ARC $I=***.

   This is called the submenu reference. It consists of a $ sign followed by a letter that specifies the menu section. The letter that specifies an image tile menu section is **I**. The menu section is followed by an equal sign (=) and the name of the image menu you want to invoke. The following is the format of a submenu reference.

   $Section=Submenu

   Where
   **$** ——————— "$" sign
   Section ——— Menu section specifier
   = ——————— "=" sign
   Submenu ——— Name of image submenu

   $I=IMAGE_ARC

   Where
   I ——————— I specifies image tile menu section
   IMAGE_ARC — Name of image submenu

   Remember that you entered **IMAGE_ARC** as the alias when you defined the **Tiled Arc Menu**.

6. Click on any standard image for the new command from the **Button Image** rollout of the **Customize User Interface** dialog box or choose the **Edit** button to create or select a new image.

7. Choose **Ribbon > Panels > Home 2D - Draw** node in the **Customizations in All Files** rollout. Next, drag and drop the **CALLITM** command from the **Command List** rollout in **Row4** under **Home 2D - Draw** node. On doing so, you will be able to invoke the image tile menu from the **Home > Draw** panel of the **Ribbon** in **2D Drafting & Annotation** workspace. Similarly, you can drag and drop this command on any other menu or toolbar also.

8. Choose **Apply** and then choose **OK** from the **Customize User Interface** dialog box; the image tile menu will be saved.

Invoking the Image Tile Menu

1. Choose **Home > Draw > CALLITM** from the **Ribbon**; the **Tiled Arc Menu** dialog box will be displayed with the image tiles menu for Example 1, as shown in Figure 16-3.

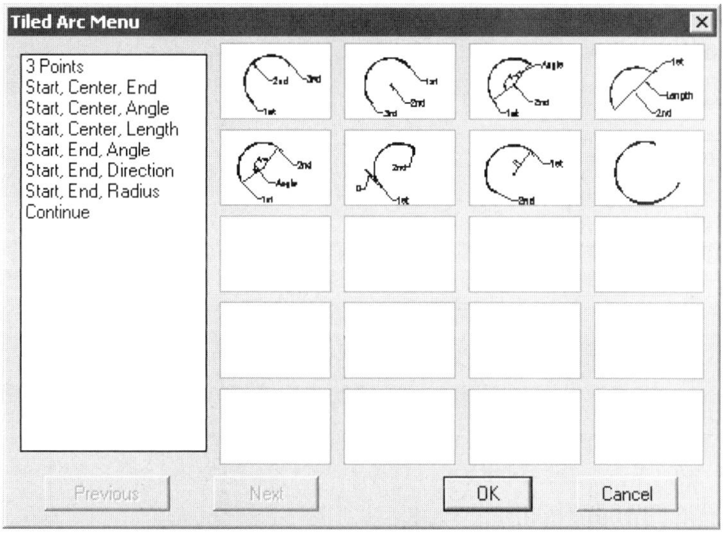

***Figure 16-3** The **Tiled Arc Menu** dialog box*

2. Double-click on any image tile to invoke that command. Alternatively, you can select the image tile and choose the **OK** button to invoke that command.

WRITING IMAGE TILE MENUS

AutoCAD also allows you to write an image tile menu. Similar to the other files, the image tile menu files are also saved in *.mnu* and *.mns* formats. The image tile menu consists of the section label ***IMAGE followed by the image tiles or image tile submenus. The menu file can contain only one image tile menu section (***IMAGE); therefore, all image tiles must be defined in this section.

*****IMAGE**

Where ******* ---------------- Three asterisks designate a section label
IMAGE ---------- Section label for an image tile

You can define any number of submenus in the image tile menu. All submenus have two asterisks followed by the name of the submenu (**PARTS or **IMAGE1):

****IMAGE1**

Where ****** ----------------- Two asterisks designate a submenu
IMAGE1 -------- Name of submenu

The first item in the image tile menu is the title of the image tile menu, which is also displayed at the top of the dialog box. The image tile dialog box title has to be enclosed in brackets ([PLC-SYMBOLS]) and should not contain any command definition. If it contains a command definition, the same is ignored. The remaining items in the image tile menu file contain slide names in the brackets and the command definition outside the brackets.

***IMAGE	Image tile menu section
**BOLTS	Image tile submenu (BOLTS)
[HEX-HEAD BOLTS]	Image tile title
[BOLT1]^C^CINSERT;B1	BOLT1 is slide file name; B1 is block name

The following example explains the procedure of writing the image tile menu.

Example 2

Write an image tile menu that will enable you to insert the block shapes shown in Figure 16-4 in a drawing by selecting the corresponding image tile from the dialog box. Use the menu to load the image tile menu.

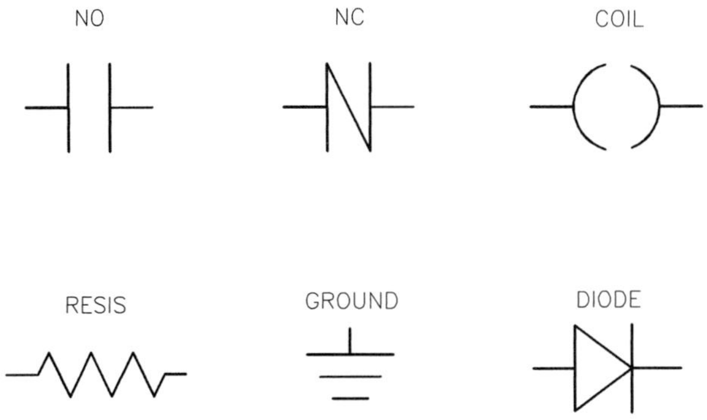

***Figure 16-4** Block shapes for the image tile menu*

<u>PLC SYMBOLS</u>
NO (NORMALLY OPEN)
NC (NORMALLY CLOSED)
COIL

<u>ELECTRIC SYMBOLS</u>
RESIS (RESISTANCE)
DIODE
GROUND

Step 1: Converting the given shapes in wblocks and slides

Draw the shapes shown in the example and then use the **WBLOCK** command to create wblocks. Use the **MSLIDE** command to create the slides that will be used to show the preview of images in the dialog box that has the image tiles. Make sure you add the folder, in which you saved the slides, to the AutoCAD support file search path. Next, use the **INSERT** command to once insert the blocks into the drawing to make sure these blocks are saved in the memory of the current drawing.

Step 2: Designing the image tile menu

Design the menu so that the commands are arranged in a desired configuration. Figure 16-5 shows one possible design for the menu and the image tile menu for Example 2.

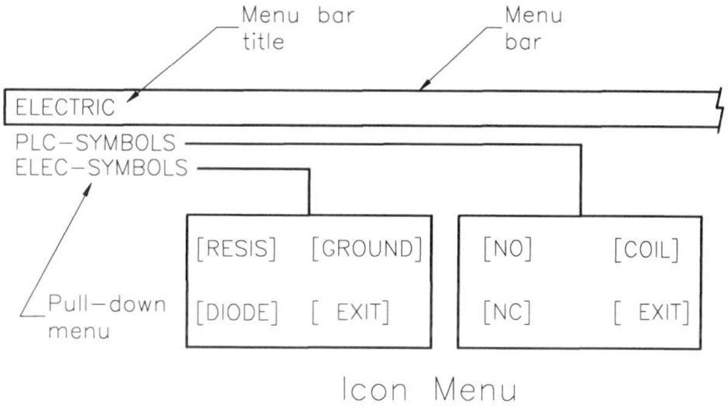

***Figure 16-5** Design of the menu and image tile menu for Example 2*

Step 3: Writing the image tile menu

You can use any text editor like Notepad to write the file. **The line numbers in the following file are only for reference and are not a part of the menu file.**

```
***POP1
[ELECTRIC]
[PLC-SYMBOLS]$I=IMAGE1 $I=*
[ELEC-SYMBOLS]$I=IMAGE2 $I=*
***IMAGE
**IMAGE1
[PLC-SYMBOLS]
[NO]^C^C-INSERT;NO;\1.0;1.0;0
[NC]^C^C-INSERT;NC;\1.0;1.0;0;
[COIL]^C^C-INSERT;COIL
[ No-Image]
```

```
12 **IMAGE2
13 [ELEC-SYMBOLS]
14 [RESIS]^C^C-INSERT;RESIS;\\\\
15 [DIODE]^C^C-INSERT;DIODE;\1.0;1.0;\
16 [GROUND]^C^C-INSERT;GROUND;\1.0;1.0;0;;
```

Explanation

Line 1
*****POP1**
In this menu item, ***POP1 is the section label and defines the first section of the menu.

Line 2
[ELECTRIC]
In this menu item, [ELECTRIC] is the menu bar label for the POP1 menu. It will be displayed in the menu bar.

Line 3
[PLC-SYMBOLS]$I=IMAGE1 $I=*
In this menu item, $I=IMAGE1 loads the submenu IMAGE1; $I=* displays the current image tile menu on the screen.

[PLC-SYMBOLS]$I=IMAGE1 $I=*
Where **Image1** ---------- Loads submenu IMAGE1
$ ------------------- Forces display of current menu

Line 5
*****IMAGE**
In this menu item, ***IMAGE is the section label of the image tile menu. All the image tile menus have to be defined within this section; otherwise, AutoCAD cannot locate them.

Line 6
****IMAGE1**
In this menu item, **IMAGE1 is the name of the image tile submenu.

Line 7
[PLC-SYMBOLS]
When you select line 3([PLC-SYMBOLS] $I=IMAGE1$I=*), AutoCAD loads the submenu IMAGE1 and displays the title of the image tile at the top of the dialog box (Figure 16-6). This title is defined in Line 7. If this line is missing, the next line will be displayed at the top of the dialog box. Image tile titles can be of any length, as long as they fit in the length of the dialog box.

Line 8
[NO]^C^C-INSERT;NO;\1.0;1.0;0
In this menu item, the first NO is the name of the slide and has to be enclosed within brackets. The name should not have any trailing or leading blank spaces. If the slides are not present,

no graphical symbols are displayed in the image tiles. However, the menu items will be loaded, and if you select this item, the command associated with the image tile will be executed. The second NO is the name of the block that is to be inserted. The backslash (\) pauses for user input; in this case, it is the block insertion point. The first 1.0 defines the X scale factor. The second 1.0 defines the Y scale factor, and the following 0 defines the rotation.

[NO]^C^C-INSERT;NO;\1.0;1.0;0

Where **NO** --------------- Block name
\\-------------------- Pause for block insertion point
1.0 ---------------- X scale factor
1.0 ---------------- Y scale factor
0 ------------------- Rotation angle

When you select this item, it will automatically enter all the prompts of the **INSERT** command and insert the NO block at the given location. The only input you need to enter is the insertion point of the block.

Line 10
[COIL]^C^C-INSERT;COIL
In this menu item the block name is given, but you need to define other parameters when inserting this block.

Line 11
[No-Image]
Notice the blank space before No-Image. If there is a space following the open bracket, AutoCAD does not look for a slide. Instead the text is displayed enclosed within the brackets in the slide file list box of the dialog box.

Line 14
[RESIS]^C^C-INSERT;RESIS;
This menu item inserts the block RESIS, see Figure 16-7. The first backslash (\) is for the block insertion point. The second and third backslashes are for the X scale and Y scale factors. The fourth backslash is for the rotation angle. This menu item can also be written as:

[RESIS]^C^C-INSERT;RESIS;
or
[RESIS]^C^C-INSERT;RESIS

If a macro runs out before the command completes, then the command reverts to asking for user input.

Line 15
[DIODE]^C^C-INSERT;DIODE;\1.0;1.0;
If you select this menu item, you are prompted to enter the block insertion point and the rotation angle. The first backslash is for the block insertion point; the second backslash is for the rotation angle.

[DIODE]^C^C-INSERT;DIODE;\1.0;1.0;

Where \-------------------- Pause for insertion point
\-------------------- Pause for rotation angle

Line 16
[GROUND]^C^C-INSERT;GRD;\1.0;1.0;0;;
This menu item has two semicolons (;) at the end. The first semicolon after 0 is for RETURN and completes the block insertion process. The second semicolon enters a RETURN and repeats the **INSERT** command. However, when the command is repeated, you will have to respond to all of the prompts. It does not accept the values defined in the menu item.

Note
*The slide files must be in the support file search path. If they are not in the support directory, specify the directory location in the **Support File Search Path** located in the **Files** tab of the **Options** dialog box.*

The menu item repetition feature can be used with the image tile menus. For example, if the command definition starts with an asterisk ([GROUND]^C^CINSERT;GRD;\1.0;1.0;0;;), the command is automatically repeated, as is the case with a pull-down menu.*

A blank line in an image tile menu terminates the menu and clears the image tiles.

The menu command $I= that displays the current menu cannot be entered using the keyboard.*

If you want to cancel or exit an image tile menu, press the ESC (Escape) key. AutoCAD ignores all other entries from the keyboard.

You can define any number of image tile menus and submenus in the image tile menu section of the menu file.

Step 4: Saving the image tile menu file
Save the file as *example-2.mnu*.

Step 5: Writing the *.mns* file
The following is the content of the *.mns* file for Example 2.

```
//
//     AutoCAD menu file - C:\customizing_2010\c16\example-2.mnu
//

***MENUGROUP=C:\customizing_2010\c16\example-2.mnu

***POP1
        [ELECTRIC]
        [PLC-SYMBOLS]$I=Image1 $I=*
        [ELEC-SYMBOLS]$I=Image2 $I=*
```

```
***IMAGE
**IMAGE1
[PLC-SYMBOLS]
[NO]^C^C-INSERT;NO;\1.0;1.0;0
[NC]^C^C-INSERT;NC;\1.0;1.0;0;
[COIL]^C^C-INSERT;COIL
[ No image]

**IMAGE2
[ELEC-SYMBOLS]
[RESIS]^C^C-INSERT;RESIS;\\\\
[DIODE]^C^C-INSERT;DIODE;\1.0;1.0;\
[GROUND]*^C^C-INSERT;GROUND;\1.0;1.0;0;;


//
//      End of AutoCAD menu file - C:\customizing_2010\c16\example-2.mnu
//
```

Save the file with the name *example-2.mns*.

Step 6: Loading the menu

The method of loading the image tile menu using an *mnu* file is different. In this case, you need to use the **MENU** command. When you invoke this command, the **Select Customization File** dialog box will be displayed. Select **Legacy menu file (*.mns, *.mnu)** from the **Files of type** drop-down list and then select the *example-2.mns* file; the default AutoCAD menu will be removed and this menu will be loaded. The menu bar now shows only the **ELECTRIC** menu. You can select the **PLC-SYMBOLS** or **ELEC-SYMBOLS** from the **ELECTRIC** menu to display its respective dialog box, as shown in Figures 16-6 and 16-7.

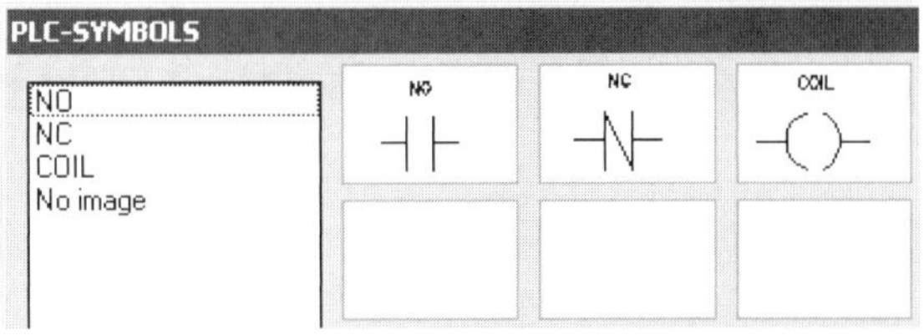

Figure 16-6** The image tile box for **PLC-SYMBOLS

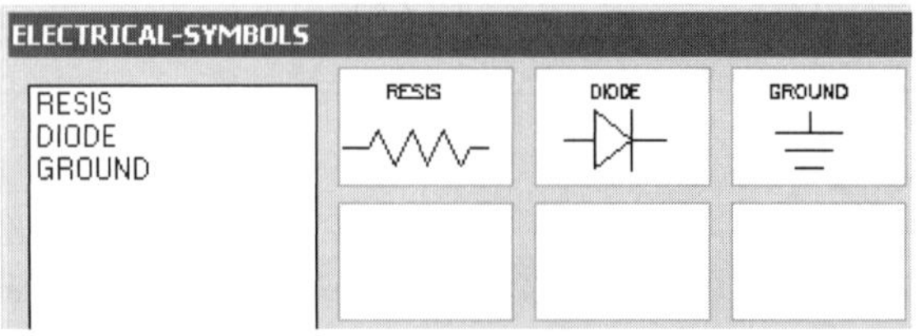

Figure 16-7** The image tile box for **ELECTRICAL-SYMBOLS

To load the original *acad.cui* file, set the value of the **FILEDIA** system variable to **0** and then invoke the **MENU** command. Enter **ACAD** at the command prompt; the default menu will be loaded again. Remember to restore the value of the **FILEDIA** system variable to **1**.

RESTRICTIONS

The pull-down and image tile menus are very easy to use and provide quick access to some of the frequently used AutoCAD commands. However, the menu bar, the pull-down menus, and the image tile menus are made unavailable while using the following commands:

TEXT Command
After you assign the text height and the rotation angle to a **TEXT** command, the menu is automatically made unavailable.

SKETCH Command
The menus are made unavailable after you set the record increment in the **SKETCH** command.

Exercise 1 *General*

Write an image tile menu for inserting the blocks shown in Figure 16-8. Arrange the blocks in two groups so that you have two submenus in the image tile menu.

PIPE FITTINGS	ELECTRIC SYMBOLS
GLOBE-P	BATTERY
GLOBE	CAPACITOR
REDUCER	COUPLER
CHECK	BREAKER

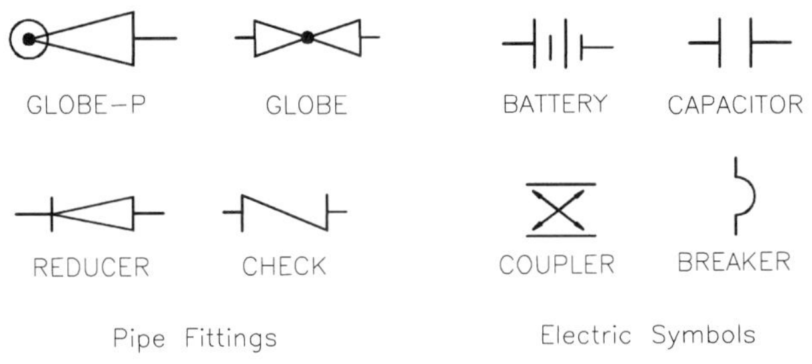

Figure 16-8 *Block shapes for Exercise 1*

IMAGE TILE MENU ITEM LABELS

As with screen and menus, you can use menu item labels in the image menus. However, the menu item labels in the image tile menus use different formats and each format performs a particular function in the image tile menu. The menu item labels appear in the slide list box of the dialog box. The maximum number of characters that can be displayed in this box is 23. The characters in excess of 23 are not displayed in the list box. However, this does not affect the command that is defined with the menu item.

Menu Item Label Formats

[slidename]. In this menu item label format, **slidename** is the name of the slide displayed in the image tile. This name (slidename) is also displayed in the list box of the corresponding dialog box.

[slidename,label]. In this menu item label format, **slidename** is the name of the slide displayed in the image tile. However, unlike the previous format, the **slidename** is not displayed in the list box. Instead the **label** text is displayed. For example, if the menu item label is **[BOLT1,1/2-24UNC-3LG], BOLT1** is the name of the slide and **1/2-24UNC-3LG** is the label that will be displayed in the list box.

[slidelib(slidename)]. In this menu item label format, **slidename** is the name of the slide in the slide library file **slidelib**. The slide (slidename) is displayed in the image tile, and the slide file name (slidename) is also displayed in the list box of the corresponding dialog box.

[slidelib(slidename,label)]. In this menu item label format, **slidename** is the name of the slide in the slide library file **slidelib**. The slide (slidename) is displayed in the image tile, and the label text is displayed in the list box of the corresponding dialog box.

[label]. If the **label** text is preceded by a space, AutoCAD does not look for a slide. The label text is displayed in the list box only. For example, if the menu item label is [EXIT]^C, the label text (EXIT) will be displayed in the list box. If you select this item, the **CANCEL** command (^C) defined with the item will be executed. The **label** text is not displayed in the image tile of the dialog box.

Example 3

Write the pull-down and image tile menus for inserting the following commands. B1 to B15 are the block names.

BLOCK
WBLOCK
ATTDEF
LIST
INSERT

BL1	BL6	BL11
BL2	BL7	BL12
BL3	BL8	BL13
BL4	BL9	BL14
BL5	BL10	BL15

Step 1: Making wblocks and slides

Draw suitable figures for different blocks names shown in the example. Convert all figures into wblocks and make slides with the given name.

Step 2: Designing the image tile menu

The second step in writing a menu is to design it. Figure 16-9 shows the design of the menu. If you choose **Insert** from the menu, the image tiles and block names will be displayed in the dialog box.

Step 3: Writing the image tile menu

Use any text editor to write the menu. The following file is a listing of the menu file for Example 3. The file contains the pull-down and image tile menu sections. **The line numbers are not a part of the menu file; they are given for reference only.**

```
***POP1
[INSERT]
[BLOCK]^C^CBLOCK
[WBLOCK]^C^CWBLOCK
[ATTRIBUTE DEFINITION]^C^CATTDEF
[LIST BLOCK NAMES]^C^C-INSERT;?
[INSERT]^C^C$I=IMAGE1 $I=*
[—]
[ATTDIA-ON]^C^CSETVAR;ATTDIA;1
```

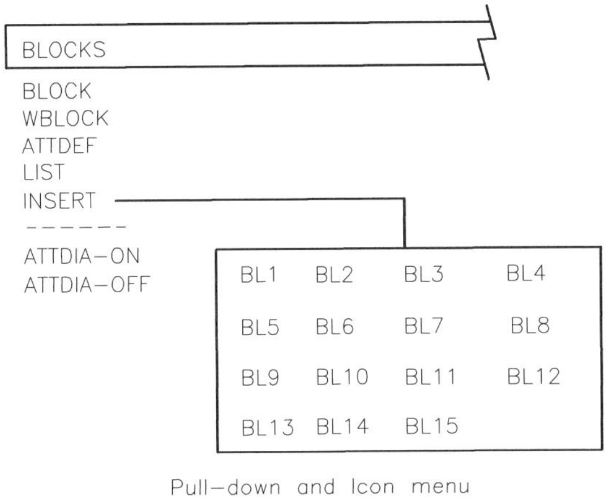

Figure 16-9 *Design of screen and menus for Example 3*

```
10 [ATTDIA-OFF]^C^CSETVAR;ATTDIA;0
11
12 ***IMAGE
13 **IMAGE1
14 [BLOCK INSERTION FOR EXAMPLE-3]
15 [BL1]^C^C-INSERT;BL1;\1.0;1.0;\
16 [BL2]^C^C-INSERT;BL2;\1.0;1.0;0
17 [BL3]^C^C-INSERT;BL3;\;;\
18 [BL4]^C^C-INSERT;BL4;\;;;
19 [BL5]^C^C-INSERT;*BL5;\1.75
20 [BL6]^C^C-INSERT;BL6;\XYZ
21 [BL7]^C^C-INSERT;BL7;\XYZ;;;\0
22 [BL8]^C^C-INSERT;BL8;\XYZ;;;;;
23 [BL9]^C^C-INSERT;BL9;\XYZ;;;;\
24 [BL10]^C^C-INSERT;*BL10;\XYZ;\
25 [BL11]^C^C-INSERT;BL11;\XYZ;1;1.5;2;45
26 [BL12]^C^C-INSERT;BL12;\XYZ;\\;;
27 [BL13]^C^C-INSERT;*BL13;\\45
28 [BL14]^C^C-INSERT;BL14;\C;@1.0,1.0;0
29 [BL15]^C^C-INSERT;BL15;\C;@1.0,2.0;\
```

Explanation

Line 1

*****POP1**

This is the section label of the first menu. The menu items defined on lines 2 through 10 are defined in this section.

Line 12
*****IMAGE**
This is the section label of the image tile menu.

Line 13
****IMAGE1**
IMAGE1 is the name of the submenu. The items on lines 14 through 29 are defined in this submenu.

Line 15
[BL1]^C^C-INSERT;BL1;\1.0;1.0;
In this menu item, BL1 is the name of the slide and **-INSERT** is AutoCAD's **-INSERT** command.

```
[BL1]^C^C-INSERT;BL1;\1.0;1.0;\
          Where  -INSERT-------- AutoCAD -INSERT command
                 BL1 ----------- Slide file name
                 1.0,1.0 ------- X and Y scale factors
```

Step 4: Saving the menu file
Save the menu with the name *example-3.mnu.*

Step 5: Writing the *.mns* file
The following is the *.mns* file for this example:

```
//
//    AutoCAD menu file - C:\customizing_2010\c16\example-3.mnu
//

***MENUGROUP=C:\customizing_2010\c16\example-3.mnu

***POP1
        [INSERT]
        [BLOCK]^C^CBLOCK
        [WBLOCK]^C^CWBLOCK
        [ATTRIBUTE DEFINITION]^C^CATTDEF
        [LIST BLOCK NAMES]^C^C-INSERT;?
        [INSERT]^C^C$I=IMAGE1 $I=*
        [--]
        [ATTDIA-ON]^C^CSETVAR;ATTDIA;1
        [ATTDIA-OFF]^C^CSETVAR;ATTDIA;0

***IMAGE
**IMAGE1
[BLOCK INSERTION FOR EXAMPLE-3]
```

```
[BL1]^C^C$S=INSERT1 INSERT;BL1;\1.0;1.0;\
[BL2]^C^C$S=INSERT1 INSERT;BL2;\1.0;1.0;0
[BL3]^C^C$S=INSERT1 INSERT;BL3;\;;\
[BL4]^C^C$S=INSERT1 INSERT;BL4;\;;;
[BL5]^C^C$S=INSERT1 INSERT;*BL5;\1.75
[BL6]^C^C$S=INSERT1 INSERT;BL6;\XYZ
[BL7]^C^C$S=INSERT1 INSERT;BL7;\XYZ;;;\0
[BL8]^C^C$S=INSERT1 INSERT;BL8;\XYZ;;;;;
[BL9]^C^C$S=INSERT1 INSERT;BL9;\XYZ;;;;\
[BL10]^C^C$S=INSERT1 CINSERT;*BL10;\XYZ;\
[BL11]^C^C$S=INSERT2 INSERT;BL11;\XYZ;1.0;1.5;2.0;45
[BL12]^C^C$S=INSERT2 INSERT;BL12;\XYZ;\\;;
[BL13]^C^C$S=INSERT2 INSERT;*BL13;\\45
[BL14]^C^C$S=INSERT2 INSERT;BL14;\C;@1.0,1.0;0
[BL15]^C^C$S=INSERT2 INSERT;BL15;\C;@1.0,2.0;\

//
//     End of AutoCAD menu file - C:\customizing_2010\c16\example-3.mnu
//
```

Step 6: Loading the menu file

Load the menu with the **MENU** command. Figure 16-10 shows the menu for this example and Figure 16-11 shows the image tile menu for this example.

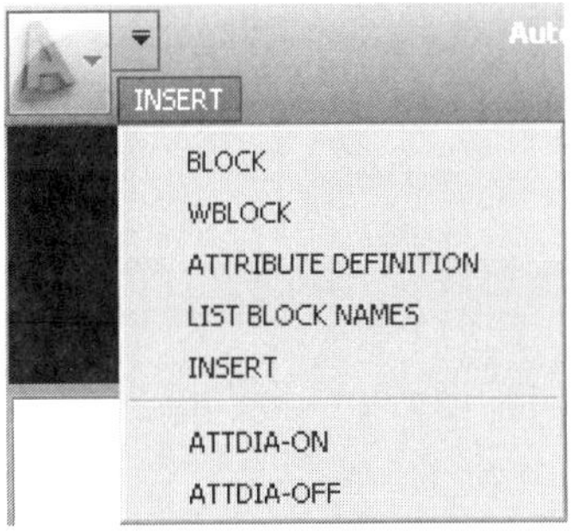

Figure 16-10 *Menu for Example 3*

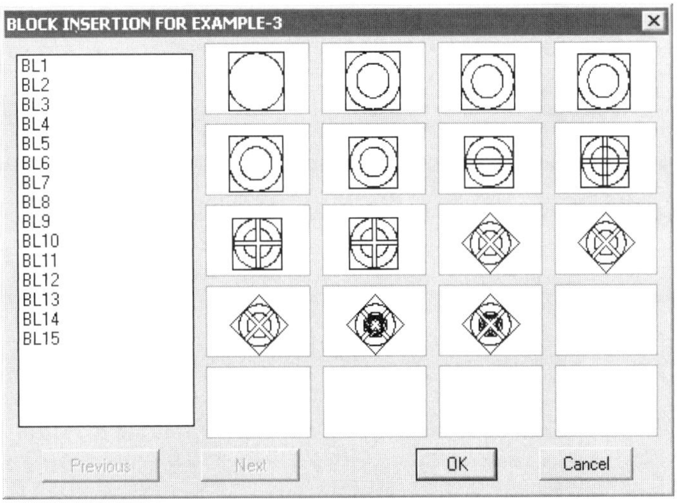

Figure 16-11 *The image tile menu for Example 3*

Self-Evaluation Test

Answer the following questions and then compare them to those given at the end of this chapter:

1. The image tile menu is also known as _________.
2. The image tile menu _________ be linked with the commands.
3. The submenu in an image tile menu starts with _________.
4. In order to load an image tile menu written in a Notepad file, AutoCAD uses a special command _________.
5. The image tile menu consists of the section label _________.

Review Questions

Answer the following questions:

1. The maximum number of characters displayed in the menu item label is _________.
2. The image tiles are displayed in the _______________ box.
3. An image tile menu can be canceled by entering ___________ using the keyboard.
4. The image title dialog box can contain a maximum of _________ image tiles.
5. A blank line in an image tile menu __________ the image tile menu.
6. The drawing for a slide should be __________ on the entire screen before making a slide.
7. You _________________ fill a solid area in a slide for an image tile menu.
8. An image tile menu _________ be accessed from a tablet menu.

Exercises

Exercise 2 *General*

Write an image tile menu for inserting the following blocks.

B1	B4	B7
B2	B5	B8
B3	B6	B9

Exercise 3 *General*

Create an image tile menu for the following commands. Make the slides that will graphically illustrate the function of the command. The image tiles to be used are shown in Figure 16-12.

LINE	**CIRCLE C,R**
PLINE	**CIRCLE C,D**
	CIRCLE 2P

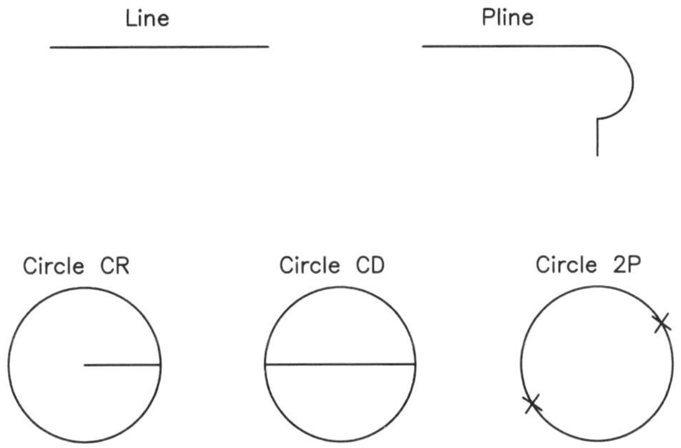

Figure 16-12 *Image tiles for Exercise 3*

Exercise 4 *General*

Write an image tile menu for inserting the following blocks. The B blocks and C blocks should be in separate image dialog boxes.

B1	B2	B3	C1	C2	C3
B4	B5	B6	C4	C5	C6
B7	B8	B9	C7	C8	C9

Answers to Self-Evaluation Test

1. Icon menu, **2.** cannot, **3.** **, **4. MENU**, **5. ***IMAGE**

Chapter 17

Button Menus

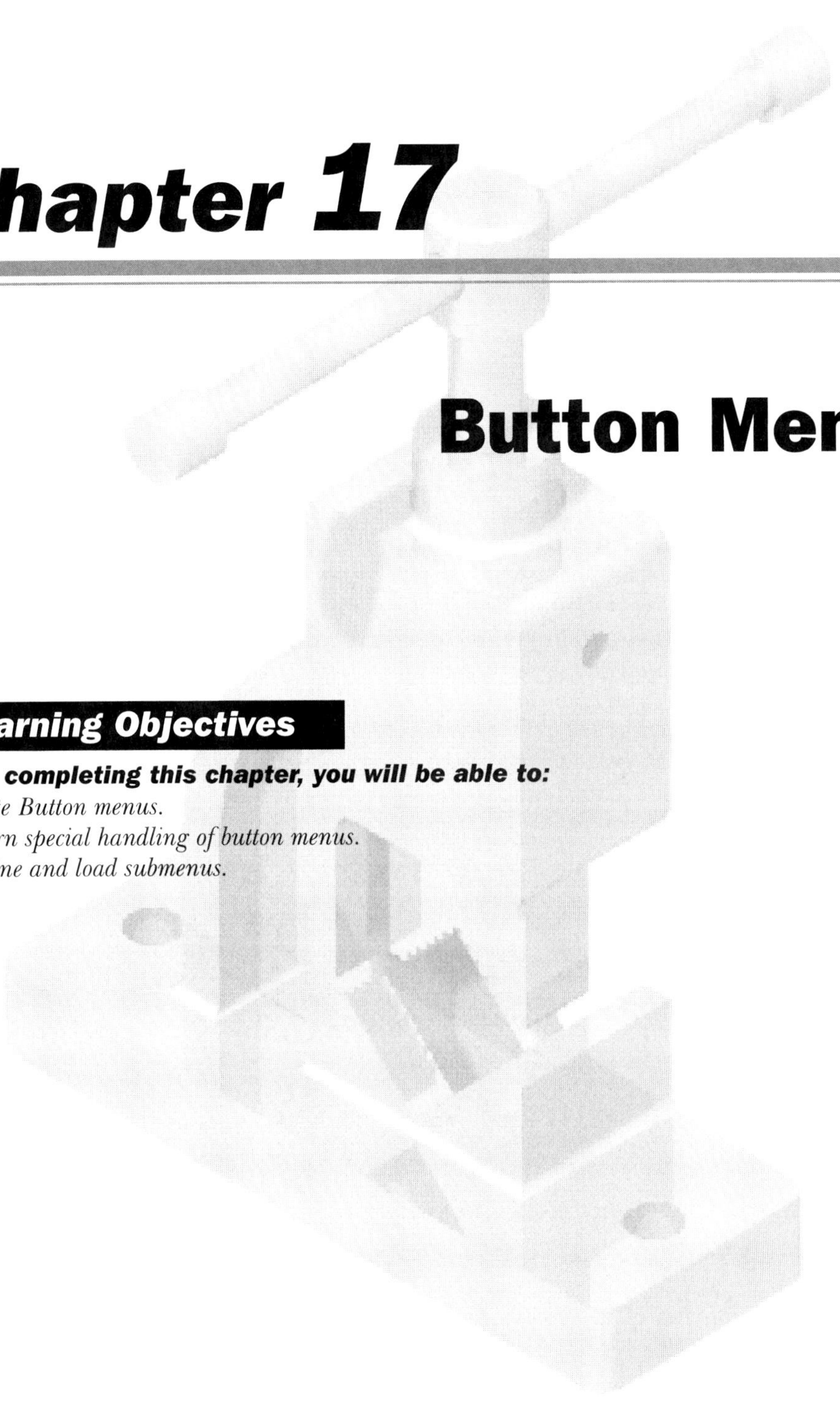

Learning Objectives

After completing this chapter, you will be able to:

- *Write Button menus.*
- *Learn special handling of button menus.*
- *Define and load submenus.*

BUTTON MENUS

The vast majority of AutoCAD installations use the mouse as the pointing device. However, you can also use a multibutton pointing device to specify points, select objects, and execute commands. These pointing devices come with different numbers of buttons, but four button and twelve button pointing devices are very common. In addition to selecting points and objects, the multibutton pointing devices can be used to provide access to frequently used AutoCAD commands and macros. The commands are selected by pressing the desired button; AutoCAD automatically executes the command or the macro that is assigned to that button. Figure 17-1 shows one such pointing device with 12 buttons.

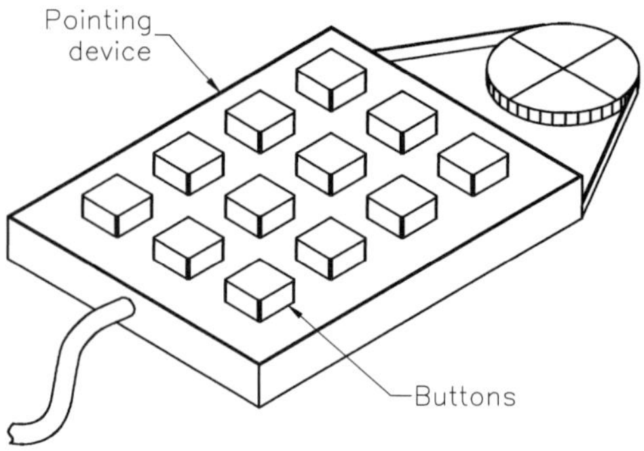

Figure 17-1 *Pointing device with 12 buttons*

AutoCAD has an in-built standard button menu that is part of the *acad.cui* file. The standard menu is automatically loaded when you start AutoCAD and enter the drawing editor. As a result, some commands and functionalities are automatically assigned to the buttons. To view these commands and functionalities, invoke the **Customize User Interface** dialog box and expand the **Mouse Buttons** tree view in the **Customization in All Files** rollout of the **Customize** tab. When you expand the **Click** tree view, you will notice the default commands and functionalities assigned to various buttons, see Figure 17-2.

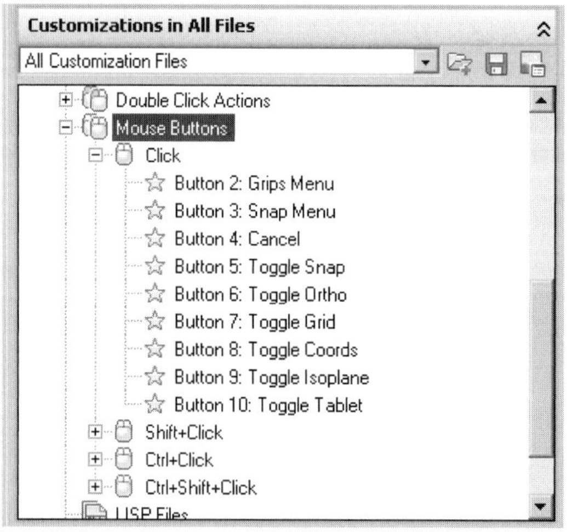

Figure 17-2 *Default commands and functionalities assigned to the buttons in the* ***Customize User Interface*** *dialog box*

You can also assign different commands and functionalities to these buttons or add more buttons, based on the requirement of your pointing device. The following example explains the process of doing this.

Example 1

Assign the following commands and functionalities to the pointing device with 12 buttons (Figure 17-3). The first button is used as a pick button.

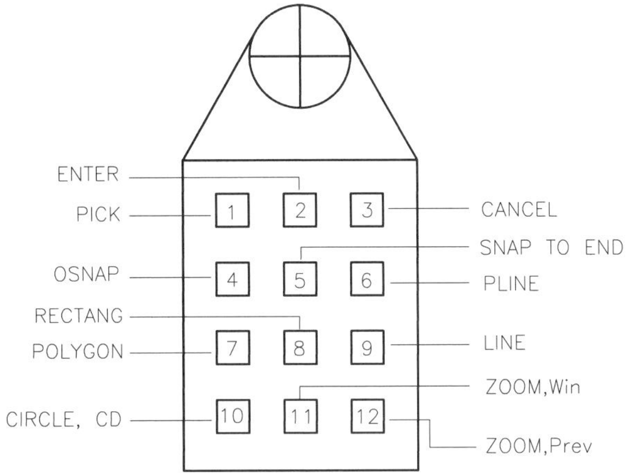

Figure 17-3 *Pointing device*

Button	Function	Button	Function
2	**ENTER**	3	**CANCEL**
4	**SNAP, OSNAP SETTINGS**	5	**SNAP, ENDPOINT**
6	**PLINE**	7	**POLYGON**
8	**RECTANG**	9	**LINE**
10	**CIRCLE, CD**	11	**ZOOM Win**
12	**ZOOM Prev**		

Assigning commands and functionalities to buttons

1. Invoke the **Customize User Interface** dialog box and expand the **Mouse Buttons** tree view.

2. Next, expand the **Click** tree view; you will notice that there are nine buttons by default, and some commands and functionalities are assigned to them. You need to clear these commands and functionalities and then assign those given in the example description.

3. One by one, right-click on all the buttons in the **Click** tree view and choose **Clear Assignment** from the shortcut menu.

4. Right-click on any of the buttons and choose **New Button** from the shortcut menu to add button 11. Similarly, add button 12.

5. Next, drag the commands and functionalities one by one from the **Commands List** rollout and drop them on the mouse buttons. The partial view, after adding the commands, is shown in Figure 17-4.

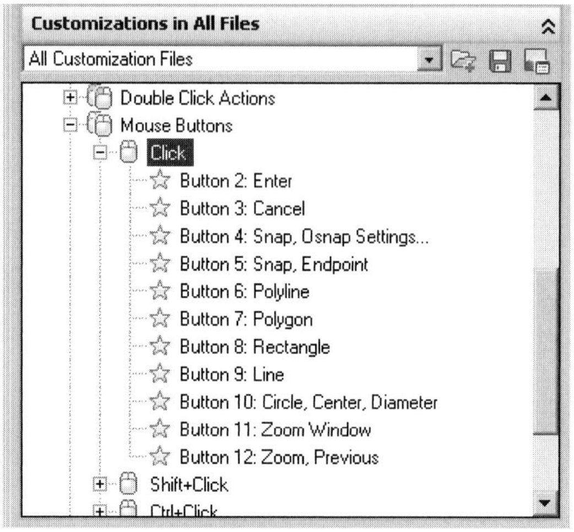

Figure 17-4 *Commands and functionalities assigned to the buttons*

6. Choose **Apply** and then exit the dialog box; the buttons in the pointing device will now invoke the assigned commands and functionalities.

WRITING BUTTON MENUS

You can write your own button menu and assign the desired commands or macros to various buttons of the pointing device. In a menu file, you can have up to four button menus (BUTTONS1 through BUTTONS4) and four auxiliary menus (AUX1 through AUX4). The buttons and the auxiliary menus behave identically. However, they are OS dependent. If your system has a pointing device (digitizer puck), AutoCAD automatically assigns the commands defined in the BUTTONS sections of the customization file to the buttons of the pointing device. When you load the customization file, the commands defined in the BUTTONS1 section of the customization file are assigned to the pointing device (digitizer puck) and if your computer has a system mouse, the mouse will use the auxiliary menus. You can also access other button menus (BUTTONS2 through BUTTONS4) by using the following keyboard and button (buttons of the pointing device-digitizer puck) combinations.

Aux Menu	Buttons Menu	Keyboard + Button Sequence
AUX1	BUTTONS1	Press the button of the pointing device.

AUX2	**BUTTONS2**	Hold down the SHIFT key and press the button of the pointing device.
AUX3	**BUTTONS3**	Hold down the CTRL key and press the button of the pointing device.
AUX4	**BUTTONS4**	Hold down the SHIFT and CTRL keys and press the button of the pointing device.

One of the buttons, generally the first, is used as a pick button to specify the coordinates of the screen crosshairs and send that information to AutoCAD. This button can also be used to select commands from various other menus, such as the tablet menu, screen menu, and image tile menu. This button cannot be used to enter a command. However, you can assign the commands to the other buttons of the pointing device. Before writing a button menu, you should decide the commands and options you want to assign to different buttons, and know the prompts associated with those commands. The following example illustrates the working of the button menu and the procedure for assigning commands to different buttons.

Note

*The first line after the menu section label ***AUX1 or ***BUTTONS1 is used only when the SHORTCUTMENU system variable is set to 0. If SHORTCUTMENU is set to a value other than 0, the in-built menu is used. Similarly, the second line after the ***AUX1 or ***BUTTONS1 label is used only when the MBUTTONPAN system variable is set to 0.*

Example 2

Write a button menu for the following AutoCAD commands. The pointing device has 12 buttons. The first button should be used as the pick button.

Button	Function	Button	Function
2	**ENTER**	3	**CANCEL**
4	**CURSOR MENU**	5	**SNAP TO ENDPOINT**
6	**ORTHO**	7	**AUTO**
8	**INT,END**	9	**LINE**
10	**CIRCLE**	11	**ZOOM Win**
12	**ZOOM Prev**		

Step 1: Writing the menu file

You can use the Notepad or any other text editor to write the menu file. The following file is a listing of the button menu for Example 2. **The line numbers are for reference only and are not a part of the menu file.**

```
***BUTTONS
;
^C^C
$P0=*
_ENDP
^L
```

```
7  AUTO
8  INT,ENDP
9  ^C^CLINE
10 ^C^CCIRCLE
11 'ZOOM;Win
12 'ZOOM;Prev
```

Explanation

Line 1
*****BUTTONS**
***BUTTONS is the section label for the first button menu. When the menu is loaded, the menu file is compiled and the commands are assigned to the buttons of the pointing device.

Line 2
;
This menu item assigns a semicolon (;) to button number 2. When you specify the second button on the pointing device, it enters a Return. It is like entering Return using the keyboard or the digitizer.

Line 3
^C^C
This menu item cancels the existing command twice (^C^C). This command is assigned to button number 3 of the pointing device. When you pick the third button on the pointing device, it cancels the existing command twice.

Line 4
$P0=*
This menu item loads and displays the cursor menu POP0, which contains various object snap modes. It is assumed that the POP0 menu has been defined in the menu file. This command is assigned to button number 4 of the pointing device. If you press this button, it will load and display the shortcut menu on the screen near the crosshairs location.

Line 5
_ENDP
This menu item overrides running Object Snaps to Endpoint Object Snap; it is assigned to button number 5 of the pointing device. When you pick the fifth button on the pointing device, it temporarily turns off all running Object Snaps, except the Endpoint Object Snap.

Line 6
^L
This menu item changes the ORTHO mode; it is assigned to button number 6. When you pick the sixth button on the pointing device, it turns the ORTHO mode on or off.

Line 7
AUTO

This menu item selects the **AUTO** option for creating a selection set; this command is assigned to button number 7 on the pointing device.

Line 8
INT,ENDP
In this menu item, INT is for the Intersection Osnap, and ENDP is for the Endpoint Osnap. This command is assigned to button number 8 on the pointing device. When you pick this button, AutoCAD looks for the intersection point. If it cannot find an intersection point, it then starts looking for the endpoint of the object that is within the pick box.

INT,ENDP
Where **INT** ------------- Intersection object snap
ENDP ----------- Endpoint object snap

Line 9
^C^CLINE
This menu item defines the **LINE** command; it is assigned to button number 9. When you select this button, the existing command is cancelled and the **LINE** command is selected.

Line 10
^C^CCIRCLE
This menu item defines the **CIRCLE** command; it is assigned to button number 10. When you pick this button, the **CIRCLE** command is selected and you are prompted for the user input.

Line 11
'ZOOM;Win
This menu item defines a transparent **ZOOM** command with the **Window** option; it is assigned to button number 11 of the pointing device.
'ZOOM;Win
Where **'** ------------------- Single quote makes **ZOOM** command transparent
ZOOM ----------- AutoCAD **ZOOM** command
; ------------------ Semicolon for RETURN
Win -------------- Window option of **ZOOM** command

Line 12
'ZOOM;Prev
This menu item defines a transparent **ZOOM** command with the **Previous** option; it is assigned to button number 12 of the pointing device.

Save this file with the name *example-2.mnu* in the *c17* folder.

Step 2: Writing the *.mns* file
The following is the *.mns* file for this menu.

```
//
//    AutoCAD menu file - C:\customizing_2010\c17\example-2.mnu
//

***MENUGROUP=C:\customizing_2010\c17\example-2.mnu

***BUTTONS
;
^C^C
$P0=*
_ENDP
^L
AUTO
INT,ENDP
^C^CLINE
^C^CCIRCLE
'ZOOM;Win
'ZOOM;Prev

//
//    End of AutoCAD menu file - C:\customizing_2010\c17\example-2.mnu
//
```

Save this file as *example-2.mns* in the *c17* folder.

Step 3: Loading the menu file

Load the *c17/example-2.mns* file using the **MENU** command.

Note

If the button menu has more menu items than the number of buttons on the pointing device, the menu items in excess of the number of buttons are ignored. This does not include the pick button. For example, if a pointing device has three buttons, in addition to the pick button, the first three menu items will be assigned to the three buttons (buttons 2, 3, and 4). The remaining lines of the button menu are ignored.

The commands are assigned to the buttons in the same order in which they appear in the file. For example, the menu item that is defined on line 3 will automatically be assigned to the fourth button of the pointing device. Similarly, the menu item that is on line 4 will be assigned to the fifth button of the pointing device. The same is true of other menu items and buttons.

SPECIAL HANDLING OF BUTTON MENUS

When you press any button on the multibutton pointing device, AutoCAD receives the following information:

1. The button number
2. The coordinates of the screen crosshairs

You can write a button menu that uses one or both pieces of information. The following example uses only the button number and ignores the coordinates of the screen crosshairs:

Example:
^C^CLINE

If this command is assigned to the second button of the pointing device and you select this button, AutoCAD will receive the button number and the coordinates of the screen crosshairs. AutoCAD will execute the command that is assigned to the second button, but it will ignore the coordinates of the crosshairs. The following example uses both the button number and the coordinates of the screen crosshairs:

Example:
^C^CLINE;\

In this menu item, the **LINE** command is followed by a semicolon (;) and a backslash (\). The semicolon inputs an ENTER and the backslash normally causes a pause for user input. However, in the buttons menu, AutoCAD will not pause for the user input. The backslash (\) in this menu item will use the coordinates of the screen crosshairs supplied by the pointing device as the starting point (From point) of the line. AutoCAD will then prompt for the second point of the line (To point).

Example 3

Write a button menu for the following AutoCAD commands for a pointing device with seven buttons, as shown in Figure 17-5. The menu items should use the information about the coordinate points of the screen crosshairs, where applicable.

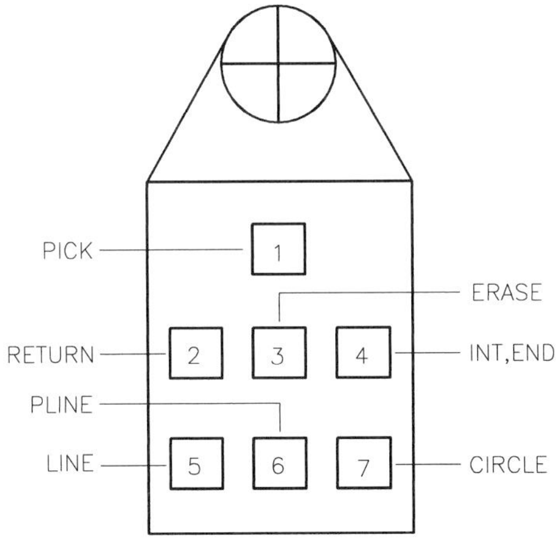

Figure 17-5 *Pointing device with seven buttons*

Button	Function
1	PICK
2	ENTER (RETURN)
3	ERASE (with SI and NEAR options)
4	INT,END
5	LINE
6	PLINE
7	CIRCLE

Step 1: Writing the menu file

The following file is a listing of the button menu for Example 3. **The line numbers are not a part of the file; they are for reference only.**

```
1 ***BUTTONS1
2 ;
3 ^C^CERASE;SI;NEAR;\
4 INT,ENDP;\
5 LINE;\
6 PLINE;\
7 CIRCLE;\
```

Explanation

Line 3

^C^CERASE;SI;NEAR;

This menu item defines an **ERASE** command with a single object selection option (SI) and near object snap (NEAR). The backslash is used to accept the coordinates supplied by the screen crosshairs. This command is assigned to button number 3 of the pointing device. If you point to an object and press the third button on the pointing device, the object will be erased and AutoCAD will automatically return to the command prompt.

```
^C^CERASE;SI;NEAR;\
        Where  SI ----------------- SIngle object selection mode
               NEAR------------ NEAR object snap
               \------------------- Accepts coordinates of screen crosshairs
```

Line 4

INT,ENDP;

In this menu item INT is for the intersection snap and ENDP is for the endpoint snap. This command is assigned to button number 4 on the pointing device. When you pick this button, AutoCAD looks for the intersection point. If it cannot find an intersection point, it starts looking for the endpoint of the object that is within the pick box. The backslash (\) is used here to accept the coordinates of the screen crosshairs. If the crosshairs are near an object within the pick box AutoCAD will snap to the intersection's point, or the endpoint of the object, when you press button number 4 on the pointing device.

INT,ENDP;

Where **INT** -------------- Intersection object snap
EBDP ------------ Endpoint object snap
\\-------------------- Accepts the coordinates of the screen crosshairs

Line 7
CIRCLE;
This menu item generates a circle. The backslash (\) is used here to accept the coordinates of the screen crosshairs as the center of the circle. If you press button number 7 of the pointing device to draw a circle, you do not need to enter the center of the circle because the current position of the screen crosshairs automatically becomes the center of the circle. You only have to define the radius to generate the circle.

Save the file with the name *c17/example-3.mnu.*

Step 2: Writing the *.mns* file

```
//
//    AutoCAD menu file - C:\customizing_2010\c17\example-3.mnu
//

***MENUGROUP=C:\customizing_2010\c17\example-3.mnu

    ***BUTTONS1
    ;
    ^C^CERASE;SI;NEAR;\
    INT,ENDP;\
    LINE;\
    PLINE;\
    CIRCLE;\

//
//    End of AutoCAD menu file - C:\customizing_2010\c17\example-3.mnu
//
```

Save this file as *example-3.mns in c17 folder.*

Step 3: Loading the menu file
Load the *example-3.mns* file using the **MENU** command.

Note

The coordinate information associated with the button can be used with the first backslash only. Remember that if a menu item in a button menu has more than one backslash (\), the remaining backslashes are ignored. For example, in the following menu item the first backslash uses the coordinates of screen crosshairs as the insertion point and the remaining backslashes are ignored and do not cause a pause as in other menus.

Example:

INSERT;B1\\\0

Where \ -------------------- *Insertion point*
\ -------------------- *X scale factor*
\ -------------------- *Y scale factor*
0 ------------------- *Rotation*

SUBMENUS

The facility to define submenus is not limited to screen, pull-down, and image menus only. You can also define submenus in the button menu.

Submenu Definition

A submenu label consists of two asterisk signs (**) followed by the name of the submenu. The buttons menu can have any number of submenus and every submenu should have a unique name. The items that follow a submenu, up to the next section label or submenu label, belong to that submenu. The format of a submenu label is:

****Name**

Where **** ----------------- Two asterisk signs (**) designate a submenu
Name ----------- Name of the submenu

Note

The submenu name can be up to 31 characters long and can consist of letters, digits, and special characters such as $ (dollar), - (hyphen), and _ (underscore). However, the submenu names should be unique in a menu file. Also, the submenu name should not have any embedded blanks.

Submenu Reference

The submenu reference is used to reference or load a submenu. It consists of a "$" sign followed by a letter that specifies the menu section. The letter that specifies a button menu section is B. The menu section is followed by "=" sign and the name of the submenu that the user wants to activate. The submenu name should be without "**". The following is the format of a submenu reference:

$Section=Submenu

Where **$** ------------------ "$" sign
Section ---------- Menu section specifier
= ------------------ "=" sign
Submenu ------- Name of submenu

Example:
$B=BUTTON1

Where **B** ------------------ B specifies buttons menu section
BUTTON1 ----- Name of submenu

LOADING MENUS

From the buttons menu, you can load any menu defined in the screen, pull-down, or image menu sections by using the appropriate load commands. It may not be needed in most of the applications, but if you want, you can load the menus that are defined in other menu sections.

Loading Screen Menus

You can load any menu that is defined in the screen menu section from the button menu by using the following load commands:

$S=X $S=05_INSERT

Where **S** ------------------- S specifies screen menu
X ------------------ Submenu name defined in screen menu section
05_INSERT ---- Submenu name defined in screen menu section

The first load command ($S=X) loads the submenu X that has been defined in the screen menu section of the customization file. The X submenu can contain 21 blank lines, so that when it is loaded, it clears the screen menu. The second load command (**$S=05_INSERT**) loads the submenu **05_INSERT** that has been defined in the screen menu section of the customization file.

Loading a Menu

You can load a menu from the button menu by using the following command:

$P1=P1A $P1=*

Where **$P1=P1A** ------ Loads the submenu P1A
$P1=* ----------- Forces the display of new menu items

The first load command **$P1=P1A** loads the submenu P1A that has been defined in the POP1 section of the customization file. The second load command **$P1=*** forces the new menu items to be displayed on the screen.

Loading an Image Menu

You can also load an image menu from the button menu by using the following load command:

$I=IMAGE1 $I=*

Where **$I=IMAGE1** --- Load the submenu IMAGE1
$I=* ------------- To display the dialog box

This menu item consists of two load commands. The first load command $I=IMAGE1 loads the image submenu IMAGE1 that has been defined in the image menu section of the file. The second load command $I=* displays the new dialog box on the screen.

Example 4

Write a button menu for a pointing device that has six buttons. The functions assigned to different buttons are shown in the following table:

SUBMENU B1	SUBMENU B2
1. PICK	1. PICK
2. ENTER	2. ENTER
3. LOAD OSNAPS SUBMENU	3. LOAD IMAGE1 SUBMENU
4. LOAD ZOOM > WINDOW	4. EXPLODE
5. LOAD BUTTON SUBMENU B1	5. LOAD BUTTON SUBMENU B1
6. LOAD BUTTON SUBMENU B2	6. LOAD BUTTON SUBMENU B2

Note

OSNAPS submenu is defined in the POP1 section of the menu and the ZOOM1 submenu is defined in the screen section of the customization file.

IMAGE1 submenu is defined in the IMAGE section of the customization file. The IMAGE1 submenu contains four images for inserting the blocks.

In this example, there are two submenus that are loaded by picking button 5 for submenu B1 and button 6 for submenu B2. When the submenu B1 is loaded, AutoCAD assigns the commands that are defined under the submenu B1 to the buttons of the pointing device. Similarly, when the submenu B2 is loaded, AutoCAD assigns the commands that are defined under the submenu B2 to the buttons of the pointing device. Figure 17-6 shows the commands that are assigned to the buttons after loading submenu B1 and submenu B2.

Step 1: Writing the menu file

The following file is the listing of the buttons menu for Example 4. **The line numbers are not a part of the file. They are shown here for reference only.**

```
***BUTTONS
**B1
;
$P1=*
$'ZOOM;Win
$B=B1
$B=B2
**B2
;
^C^C$I=IMAGE1 $I=*
```

```
11 EXPLODE;\
12 $B=B1
13 $B=B2
```

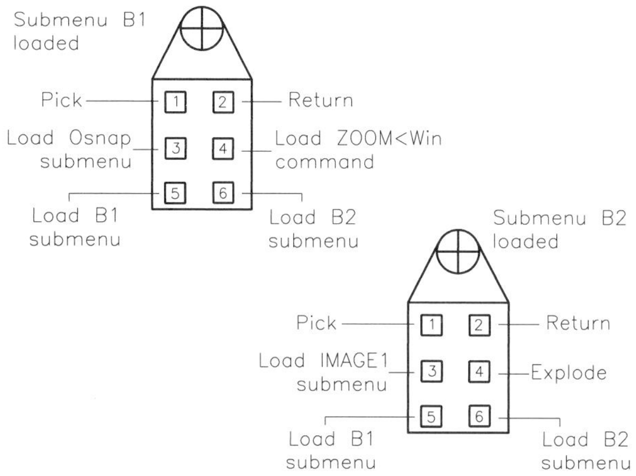

Figure 17-6 *Commands assigned to different buttons of the pointing device*

Explanation

Line 2

****B1**

This menu item defines a submenu. The name of the submenu is B1.

Line 4

$P1=*

This menu item loads and displays the pull-down menu that has been defined in the POP1 section of the menu.

Line 5

'ZOOM;Win

This menu item **'ZOOM;Win** is a transparent **ZOOM** command with the **Window** option.

Line 8

****B2**

This menu item defines the submenu B2.

Line 10

^C^C$I=IMAGE1 $I=*

This menu item cancels the existing command twice and then loads the submenu IMAGE1 that has been defined in the IMAGE menu. $I=* displays the current dialog box on the screen.

^C^C$I=IMAGE1 $I=*

Where **^C^C** ---------- Cancels the existing command twice
$I=IMAGE1 --- Loads IMAGE1 submenu
$I=* ------------- Displays the dialog box with images

Line 11
EXPLODE;
This menu item will explode an object. It utilizes the special feature of the pointing device buttons that supply the coordinates of the screen crosshairs. When you select this button, it will explode the object where the screen crosshairs is located.

EXPLODE;

Where **Explode** --------- AutoCAD's **EXPLODE** command
; ------------------- Semicolon (;) for ENTER
\\-------------------- Utilizes the coordinates of screen
-------------------- crosshairs as input

Line 12
$B=B1
This menu item loads the submenu B1 and assigns the functions defined under this submenu to different buttons of the pointing device.

Line 13
$B=B2
This menu item loads the submenu B2 and assigns the functions defined under this submenu to different buttons of the pointing device.

Save this file with the name *c17/example-4.mnu*.

Step 2: Writing the *.mns* file

```
//
//      AutoCAD menu file - C:\customizing_2010\c17\example-4.mnu
//

***MENUGROUP=C:\customizing_2010\c17\example-4.mnu

    ***BUTTONS
    **B1
    ;
    $P1=*
    $'ZOOM;Win
    $B=B1
    $B=B2
    **B2
    ;
```

```
^C^C$I=IMAGE1 $I=*
EXPLODE;\
$B=B1
$B=B2

//
//    End of AutoCAD menu file - C:\customizing_2010\c17\example-4.mnu
//
```

Save this file with the name *c17/example-4.mns*.

Step 3: Loading the menu file
Load the *example-4.mns* file using the **MENU** command.

Self-Evaluation Test

Answer the following questions and then compare them to those given at the end of this chapter:

1. __________ button and __________ button pointing devices are very common.

2. In a menu file, you can have up to __________ auxiliary menu sections.

3. Submenus can also be __________ in the button menu.

4. The submenu name can be up to __________ characters long.

5. A multibutton pointing device can be used to specify or select ______________, or enter AutoCAD ______________.

Review Questions

Answer the following questions:

1. AutoCAD receives the button _______________ and _______________ of screen crosshairs when a button is activated on the pointing device.

2. If the number of menu items in the button menu is more than the number of buttons on the pointing device, the excess lines are _______________.

3. Commands are assigned to the buttons of the pointing device in the _______________ order in which they appear in the buttons menu.

4. The format of displaying a loaded submenu that has been defined in the image menu is _______________.

5. The format of the **LOAD** command for loading a submenu that has been defined in the image menu is _______________.

Exercises

Exercise 1 *General*

Write a button menu for the following AutoCAD commands. The pointing device has 10 buttons (Figure 17-7), and button number 1 is used for specifying the points. The blocks are to be inserted with a scale factor of 1.00 and a rotation of 0-degree.

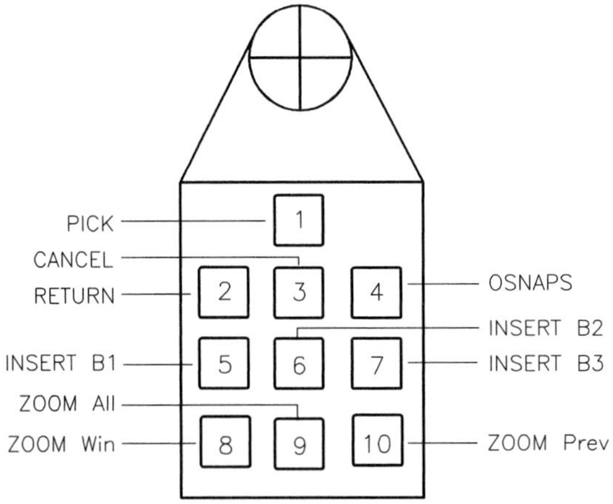

Figure 17-7 *Pointing device with 10 buttons*

1. PICK BUTTON	2. RETURN	3. CANCEL
4. OSNAPS	5. INSERT B1	6. INSERT B2
7. INSERT B3	8. ZOOM Window	9. ZOOM All
10. ZOOM Previous		

1. B1, B2, B3 are the names of the blocks or Wblocks that have already been created.
2. Assume that the Osnap submenu has already been defined.
3. Use the transparent **ZOOM** command for **ZOOM Previous** and **ZOOM Window.**

Exercise 2 *General*

Write a button menu for a pointing device that has 10 buttons. The functions assigned to different buttons are shown in the following table and also in Figure 17-8:

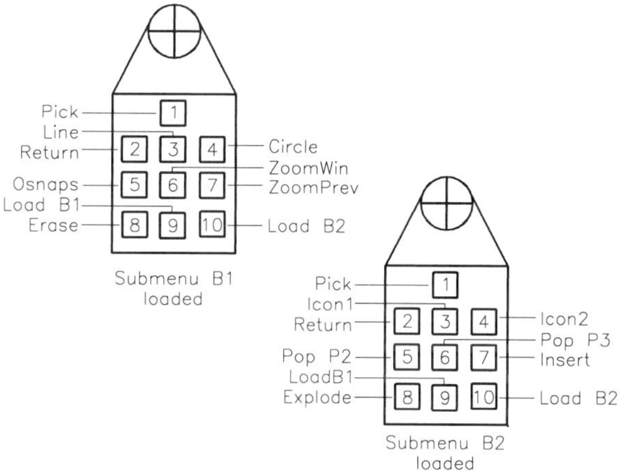

Figure 17-8 *Commands assigned to different buttons of the pointing device*

SUBMENU B1	SUBMENU B2
1. PICK	1. PICK
2. ENTER	2. ENTER
3. LINE	3. LOAD IMAGE1
4. CIRCLE	4. LOAD IMAGE2
5. LOAD OSNAPS	5. LOAD P2 (Pull-down)
6. ZOOM Win	6. LOAD P3 (Pull-down)
7. ZOOM Prev	7. INSERT
8. ERASE	8. EXPLODE
9. LOAD BUTTON MENU B1	9. LOAD BUTTON MENU B1
10. LOAD BUTTON MENU B2	10. LOAD BUTTON MENU B2

Assume that:

1. OSNAPS submenu is defined in the POP1 or POP0 (Cursor menu) section of the menu.
2. Menus P2 and P3 are defined in the POP2 and POP3 sections of the menu.
3. IMAGE1 and IMAGE2 submenus are defined in the IMAGE section of the menu file. The IMAGE1 and IMAGE2 submenus contain four images each for inserting the blocks.

Answers to Self-Evaluation Test

1. Four, Twelve, **2.** Four, **3.** Defined, **4.** Thirtyone, **5.** Points, Command

Chapter 18

Tablet Menus

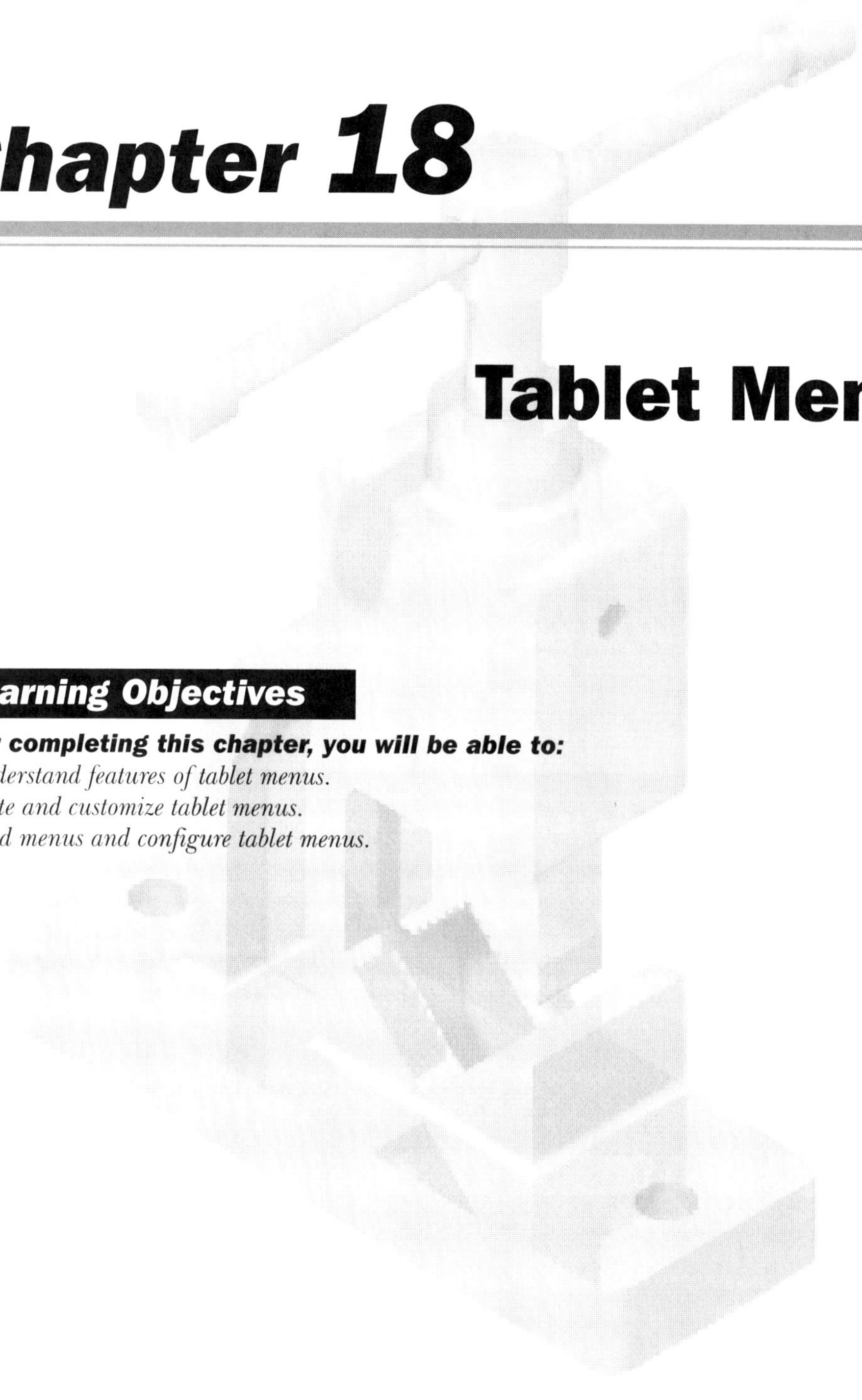

Learning Objectives

After completing this chapter, you will be able to:

- *Understand features of tablet menus.*
- *Write and customize tablet menus.*
- *Load menus and configure tablet menus.*

STANDARD TABLET MENU

The tablet menu provides another alternative for entering commands. In the tablet menu, the commands are selected from the template that is secured on the surface of a digitizing tablet. To use the tablet menu, you need a digitizing tablet and a pointing device. You also need a tablet template (Figure 18-1) that contains AutoCAD commands arranged in various groups for easy identification.

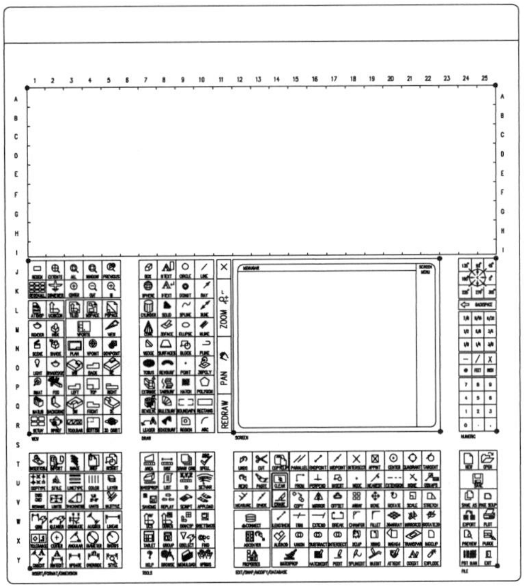

Figure 18-1 *Sample tablet template*

The standard AutoCAD menu file has four tablet menu sections: TABLET1, TABLET2, TABLET3, and TABLET4. When you start AutoCAD and get into the drawing editor, the tablet menu sections TABLET1, TABLET2, TABLET3, and TABLET4 are automatically loaded. The commands defined in these four sections are then assigned to different blocks of the template.

TABLET1 has 9 rows and 25 columns, TABLET2 has 9 rows and 11 columns, TABLET3 has 13 rows and 9 columns, and TABLET4 has 7 rows and 25 columns.

FEATURES OF A TABLET MENU

A tablet menu has the following features:

1. In the tablet menu, the commands can be arranged so that the most frequently used commands can be accessed directly. This can save considerable time in entering the commands. In the screen menu or the pull-down menu, some of the commands cannot be accessed directly. For example, to generate a horizontal dimension you have to go

through several steps. You first select Dimension from the root menu and then Linear. In the tablet menu, you can select the linear dimensioning command directly from the digitizer. This saves the time that is required to page through different screens.

2. You can have the graphical symbols of the AutoCAD commands drawn on the tablet template. This makes it much easier to recognize and select commands. For example, if you are not an expert in AutoCAD dimensioning, you may find the baseline and continue dimensioning confusing. But if the command is supported by the graphical symbol illustrating what a command does, the chances of selecting a wrong command are minimized.

3. You can assign any number of commands to the tablet overlay. The number of commands you can assign to a tablet is limited only by the size of the digitizer and the size of the rectangular blocks.

CUSTOMIZING A TABLET MENU

As with any other menu, you can customize the AutoCAD tablet menu also. This is a powerful customizing tool to make AutoCAD more efficient.

The tablet menu can contain a maximum of four sections: TABLET1, TABLET2, TABLET3, and TABLET4. Each section represents an area on the digitizing tablet. These rectangular areas can be further divided into any number of rectangular blocks. The size of each block depends on the number of commands that are assigned to the tablet area. Also, the rectangular tablet areas can be located anywhere on the digitizer and can be arranged in any order. The AutoCAD **TABLET** command configures the tablet.

Before writing a tablet menu file, it is very important to design the layout of the design of the tablet template. A well-thought-out design can save a lot of time in the long run. The following points should be considered when designing a tablet template:

1. Understand the AutoCAD commands that you use in your profession.

2. Group the commands based on their function, use, or relationship with other commands.

3. Draw a rectangle representing a template so that it is easy for you to move the pointing device around. The size of this area should be appropriate to your application. It should not be too large or too small. Also, the size of the template depends on the active area of the digitizer.

4. Divide the remaining area into four different rectangular tablet areas for TABLET1, TABLET2, TABLET3, and TABLET4. It is not necessary to use all four areas; you can have fewer tablet areas, but four is the maximum.

5. Determine the number of commands you need to assign to a particular tablet area; then

determine the number of rows and columns you need to generate in each area. The size of the blocks does not need to be the same in every tablet area.

6. Use the **TEXT** command to print the commands on the tablet overlay, and draw the symbols of the command, if possible.

7. Plot the tablet overlay on good-quality paper or a sheet of Mylar. If you want the plotted side of the template to face the digitizer board, you can create a mirror image of the tablet overlay and then plot the mirror image.

MODIFYING AN EXISTING TABLET MENU

As mentioned earlier, when you start AutoCAD, the default tablet menus are loaded. You can modify these menus by assigning different commands and functionalities to them. To understand this process, consider the following example.

Example 1

Modify the default tablet menu and assign commands to Row A and Row B of the tablet. The commands to be added are given below:

Row A	Row B
LINE	TRIM
PLINE	EXTEND
RECTANGLE	FILLET
POLYGON	CHAMFER
CIRCLE, CR	HATCH

Step 1: Assigning commands to tablet menus

1. Invoke the **Customize User Interface** dialog box and expand the **Legacy** tree view.

2. Next, expand the **Tablet Menus** tree view; the four default tablet menus are displayed.

3. Expand **Tablet Menu 1**; 9 rows are displayed. Expand **Row A**; 25 columns are displayed in a tree view.

4. Assign the commands shown in the example description to **Columns 1** through **5** of **Row A**.

5. Similarly, assign the commands shown in the example description to **Columns 1** through **5** of **Row B**, see Figure 18-2.

6. Choose **Apply** and then choose **OK** to exit the dialog box.

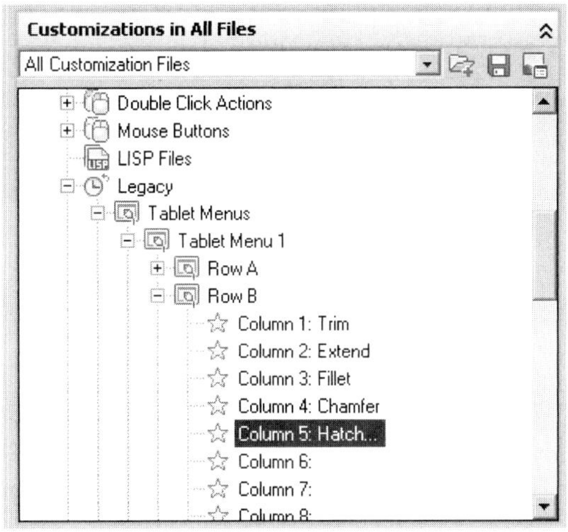

Figure 18-2 *The Tablet Menu after assigning the commands*

WRITING A TABLET MENU

Similar to other menus, you can also write tablet menus. When writing a tablet menu, you must understand the AutoCAD commands and the prompt entries required for each command. Equally important is the design of the tablet template and the placement of various commands on it. Give considerable thought to the design and layout of the template, and if possible, invite suggestions from AutoCAD users in your trade.

As mentioned earlier, by default, there are different number of rows and columns in each tablet menu. Therefore, if you want to add some commands in Row A & some in Row B, then you need to leave spaces accordingly by "\" with Row A. The following listing shows an example of this for TABLE1.

```
***TABLET1
^C^CLINE
^C^CCIRCLE
^C^CARC
^C^CDONUT
^C^CPOINT
\
\
\
\
\
\
\
\
```

```
\
\
\
\
\
\
\
\
\
\
\
\
^C^CZOOM;E;
^C^CPAN
^C^CVPOINT
```

In the above example, **ZOOM;E**, **PAN**, and **VPOINT** commands will be assigned to Row B of tablet menu 1.

To understand the process involved in developing and writing a tablet menu, consider Example 2.

Example 2

Write a tablet menu for the following commands. The commands are to be arranged, as shown in Figure 18-3. Make a tablet menu template for configuration and command selection (filename *example-2.mnu*).

LINE	**CIRCLE C,R**	**PLINE**
CIRCLE C,D	**ERASE**	**CIRCLE 2P**

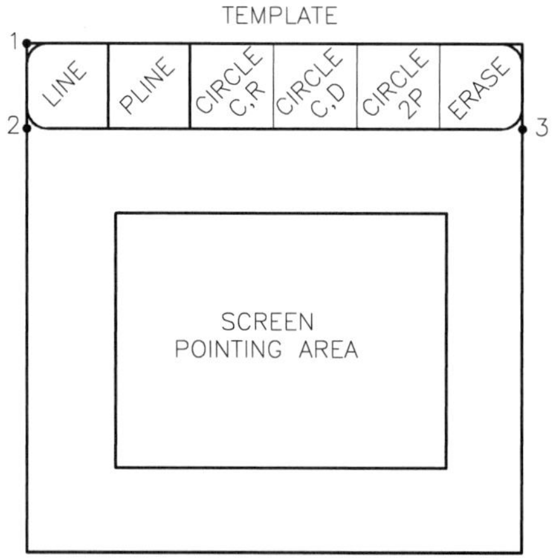

Figure 18-3 *Design of the tablet template*

Step 1: Writing the tablet menu

Figure 18-3 represents one of the possible template designs, where the commands are in one row at the top of the template, and the screen pointing area is in the center. There is only one area in this template; therefore, you can place all these commands under the section label TABLET1. To write a menu file, you can use any text editor like Notepad.

The name of the file is example-2, and the extension of the file is *.MNU*. **The line numbers are not part of the file; they are shown here for reference only.**

```
1 ***TABLET1
2 ^C^CLINE
3 ^C^CPLINE
4 ^C^CCIRCLE
5 ^C^CCIRCLE;\D
6 ^C^CCIRCLE;2P
7 ^C^CERASE
```

Explanation for writing the tablet menu

Line 1

*****TABLET1**

TABLET1 is the section label of the first tablet area. All the section labels are preceded by three asterisks (***).

*****TABLET1**

Where *** ---------------- Three asterisks designate the section label.
TABLET1 ------ Section label for TABLET1

Line 2

^C^CLINE

^C^C cancels the existing command twice; **LINE** is a command. There is no space between the second ^C and **LINE**.

^C^C LINE

Where **^C^C** ---------- Cancels the existing command twice
Line -------------- AutoCAD **LINE** command

Line 3

^C^CPLINE

^C^C cancels the existing command twice; **PLINE** is a command.

Line 4

^C^CCIRCLE

^C^C cancels the existing command twice; **CIRCLE** is a command. The default input for the **CIRCLE** command is the center and the radius of the circle; therefore, no additional input is required for this line.

Line 5
^C^CCIRCLE;\D
^C^C cancels the existing command twice; **CIRCLE** is a command like the previous line. However, this command definition requires the diameter option of the circle command. This is accomplished by using \D in the command definition. There should be no space between the backslash (\) and the D, but there should always be a space or semicolon **before** the backslash (\). The backslash (\) lets the user enter a point, and in this case it is the center point of the circle. After you enter the center point, the diameter option is selected by the letter D, which follows the backslash (\).

^C^CCIRCLE;\D
Where **CIRCLE** -------- **CIRCLE** command
; ------------------- Semicolon or Space for RETURN
\\-------------------- Pause for input
D ------------------ Diameter option

Line 6
^C^CCIRCLE;2P
^C^C cancels the existing command twice; **CIRCLE** is a command. Semicolon is for ENTER or RETURN. The 2P selects the two-point option of the **CIRCLE** command.

Line 7
^C^CERASE
^C^C cancels the existing command twice; **ERASE** is a command that erases the selected objects.

Step 2: Writing the *.mns* file
The following is the *.mns* file for this menu.

```
//
//    AutoCAD menu file - C:\customizing_2010\c18\example-2.mnu
//

***MENUGROUP=C:\customizing_2010\c18\example-2.mnu

    ***TABLET1
    ^C^CLINE
    ^C^CPLINE
    ^C^CCIRCLE
    ^C^CCIRCLE;\D
    ^C^CCIRCLE;2P
    ^C^CERASE

    //
    //    End of AutoCAD menu file - C:\customizing_2010\c18\example-2.mnu
    //
```

Save this file with the name *example-2.mns*.

Step 3: Loading the menu file

Load the *example-2.mns* file using the **MENU** command.

Note

In the tablet menu, the part of the menu item that is enclosed in the brackets is used for screen display only. For example, in the following menu item, T1-6 will be ignored and will have no effect on the command definition.

[T1-6]^C^CCIRCLE;2P

Where **[T1-6]** ------------ For reference only and has no effect on the command definition

T1 ----------------- Tablet area 1

6 ------------------- Item number 6

The reference information can be used to designate the tablet area and the line number.

Before you can use the commands from the new tablet menu, you need to configure the tablet and load the tablet menu.

TABLET CONFIGURATION

To use the new template that will select the commands, you need to install Wintab driver and configure the tablet. This allows AutoCAD to know the location of the tablet template and the position of the commands assigned to each block. This is accomplished by using the **TABLET** command. Secure the tablet template (Figure 18-3) on the digitizer with the edges of the overlay approximately parallel to the edges of the digitizer. Enter the **TABLET** command, select the **Configure** option, and respond to the following prompts:

Command: **TABLET**
Enter an option [ON/OFF/CAL/CFG]:**CFG**
Enter number of tablet menus desired (0–4) <current>: **1**
Do you want to realign tablet menus? [Yes/No] <N>: **Y**
Digitize upper left corner of menu area 1; **P1**
Digitize lower left corner of menu area 1; **P2**
Digitize lower right corner of menu area 1; **P3**
Enter the number of columns for menu area 1: **6**
Enter the number of rows for menu area 1: **1**
Do you want to respecify the screen pointing area? [Yes/No]<N>: **Y**
Digitize lower left corner of pointing area: **P4**
Digitize upper right corner of pointing area: **P5**
Do you want to specify the Floating Screen Pointing area? [Yes/No]<N>: **N**

Note

The three points P1, P2, and P3 should form a 90-degree angle. If the selected points do not form a 90-degree angle, AutoCAD will prompt you to enter the points again until they do.

The tablet areas should not overlap the screen pointing area.

The screen pointing area can be of any size and located anywhere on the tablet as long as it is within the active area of the digitizer. The screen pointing area should not overlap other tablet areas. The screen pointing area you select will correspond to the monitor screen area. Therefore, the length-to-width ratio of the screen pointing area should be the same as that of the monitor, unless you are using the screen pointing area to digitize a drawing.

Exercise 1

Write a tablet menu for the following AutoCAD commands. Make a tablet menu template for configuration and command selection (filename *tme1.mnu*).

LINE	**TEXT-Center**
CIRCLE C,R	**TEXT-Left**
ARC C.S.E	**TEXT-Right**
ELLIPSE	**TEXT-Aligned**
DONUT	

Use the template in Figure 18-4 to arrange the commands. The draw and text commands should be placed in two separate tablet areas.

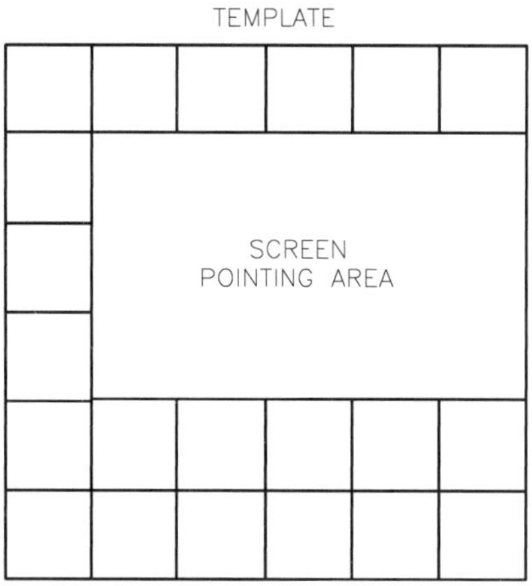

Figure 18-4 *Template for Exercise 1*

Self-Evaluation Test

Answer the following questions and then compare them to those given at the end of this chapter:

1. To use the tablet menu, you need a _________ and a _________.
2. TABLET1 has _________ rows and _________ columns.
3. In the tablet menu, the commands can be arranged so that the _________ used commands can be accessed directly..
4. The tablet area should not _________ the screen pointing area.
5. The tablet menus can be loaded using the _________ command.

Review Questions

Answer the following questions:

1. The maximum number of tablet menu sections is _________.
2. A tablet menu area is _________ in shape.
3. The blocks in any tablet menu area are _________ in shape.
4. A tablet menu area can have _________ number of rectangular blocks.
5. The _________ command is used to configure the tablet menu template.

Exercises

Exercise 2 *General*

Design a template and write a tablet menu to insert the following user-defined blocks:

BX1	BX5	BX9
BX2	BX6	BX10
BX3	BX7	BX11
BX4	BX8	BX12

Exercise 3 *General*

Design the tablet menu for the following AutoCAD commands:

LINE	**ZOOM-Win**	**DIM-Horz**
PLINE	**ZOOM-Dyn**	**DIM-Vert**
ARC	**ZOOM-All**	**DIM-Alig**
CIRCLE	**ZOOM-Pre**	**DIM-Ang**
ELLIPSE	**ZOOM-Ext**	**DIM-Rad**
POLYGON	**ZOOM-Scl**	**DIM-Cen**

Exercise 4 *General*

Write a tablet menu for the commands shown in the tablet menu template of Figure 18-5. Make a tablet menu template for configuration and command selection.

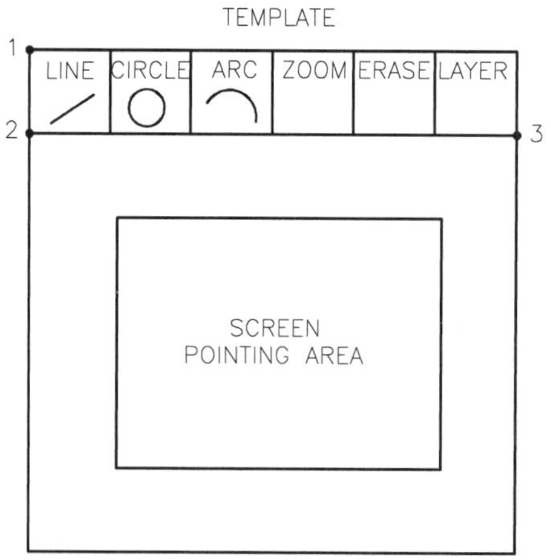

Figure 18-5 *Tablet menu template for Exercise 4*

Exercise 5 *General*

Write a tablet menu for the commands given in the template of the Tablet menu shown in Figure 18-6. Configure the tablet and then load the new menu. Make a tablet menu template for configuration and command selection.

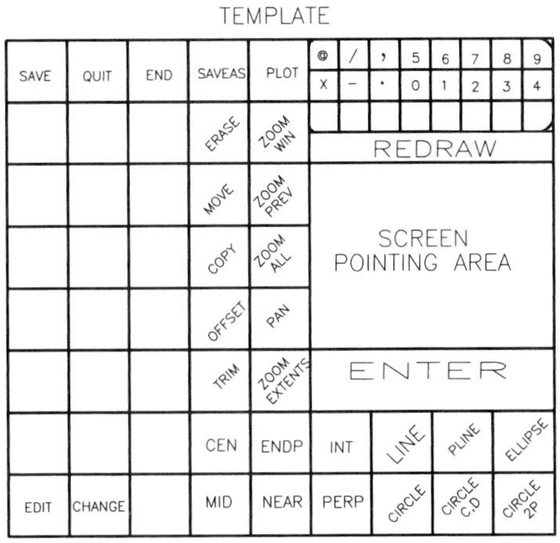

***Figure 18-6** Tablet menu template for Exercise 5*

Exercise 6 *General*

Write a tablet menu file for the commands shown in the template of Figure 18-7. Make a tablet menu template for configuration and command selection.

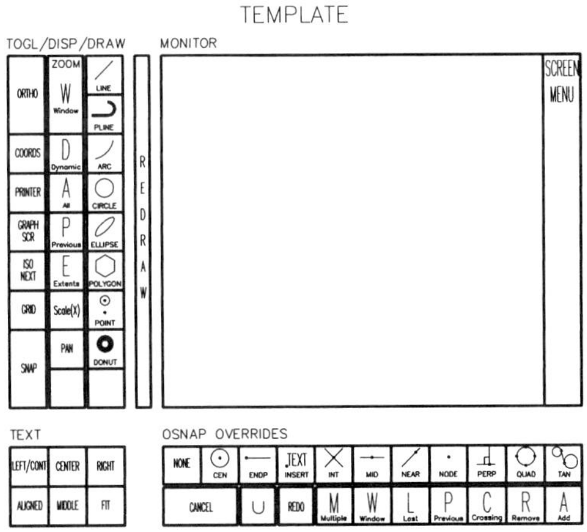

***Figure 18-7** Tablet menu template for Exercise 6*

Exercise 7 *General*

Write a combined pull-down and tablet menu file for the commands shown in the tablet menu template of Figure 18-8. Make a tablet menu template for configuration and command selection.

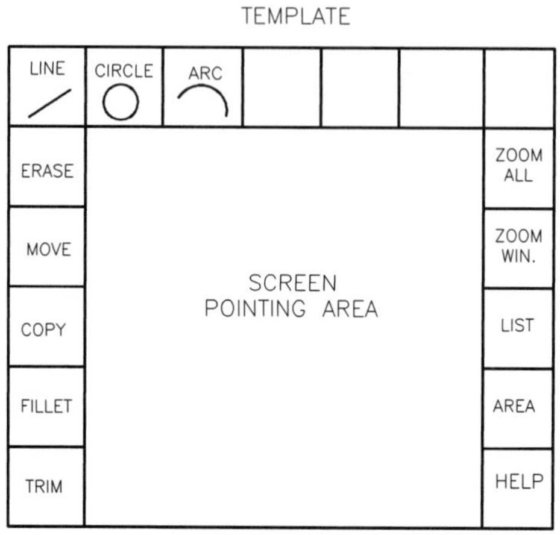

Figure 18-8 *Tablet menu template for Exercise 7*

TEMPLATES FOR TABLET CONFIGURATIONS

Given below are the templates for Tablet menu configuration for various examples and exercises.

Template for Example 2

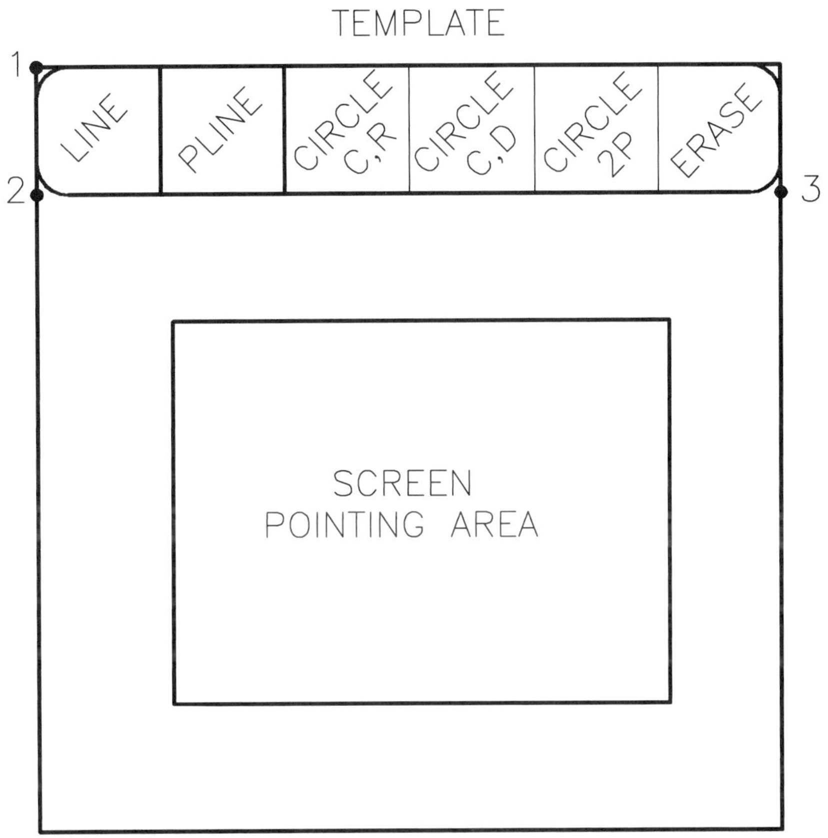

Note

This template is for tablet configuration. You may make a copy of this page and then secure it on the digitizer surface for configuration.

Template for Exercise 1

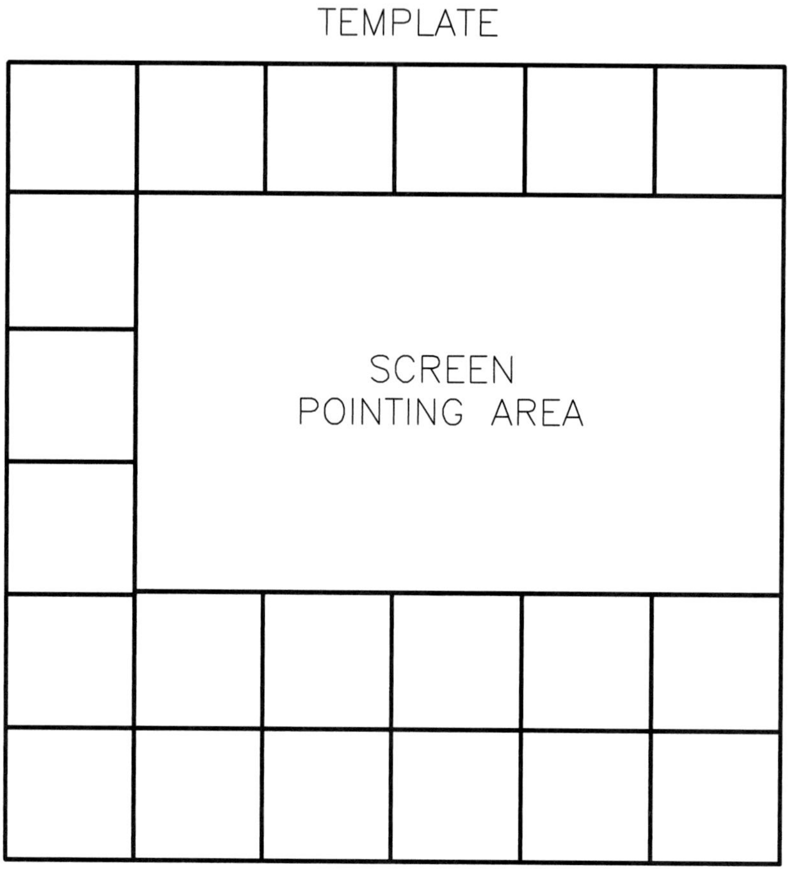

Note
This template is for tablet configuration. You may make a copy of this page and then secure it on the digitizer surface for configuration.

Template for Exercise 4

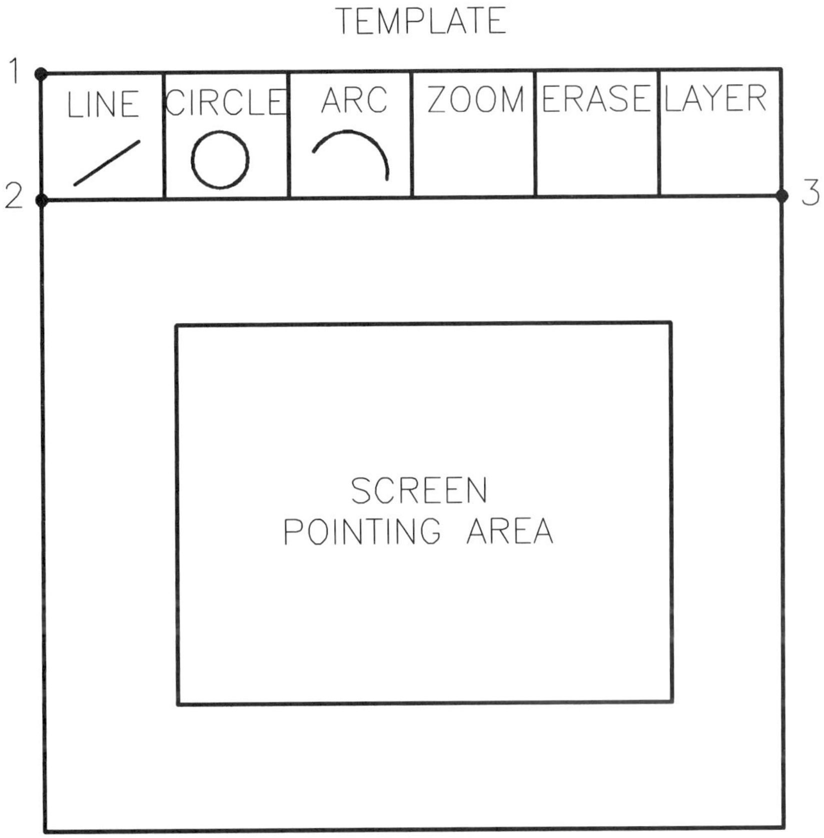

Note

This template is for tablet configuration. You may make a copy of this page and then secure it on the digitizer surface for configuration.

Template for Exercise 5

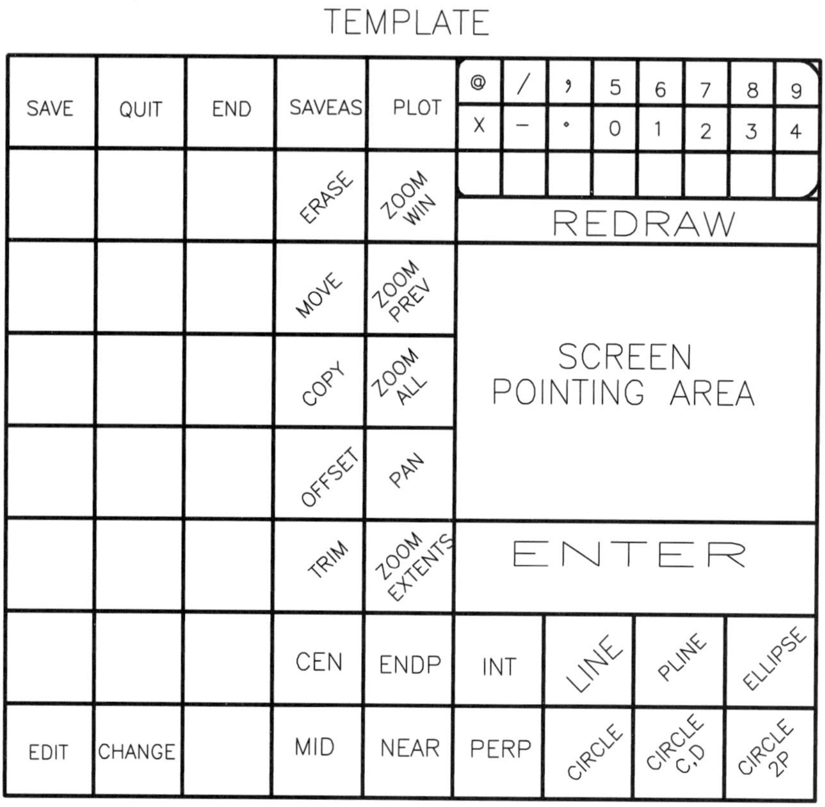

Note

This template is for tablet configuration. You may make a copy of this page and then secure it on the digitizer surface for configuration.

Template for Exercise 6

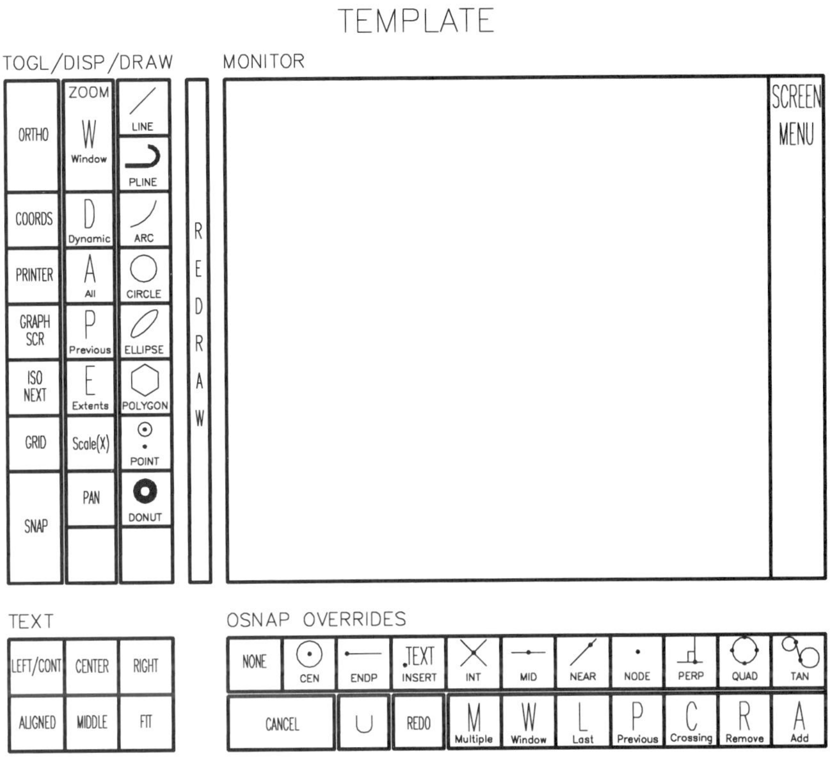

Note

This template is for tablet configuration. You may make a copy of this page and then secure it on the digitizer surface for configuration.

Template for Exercise 7

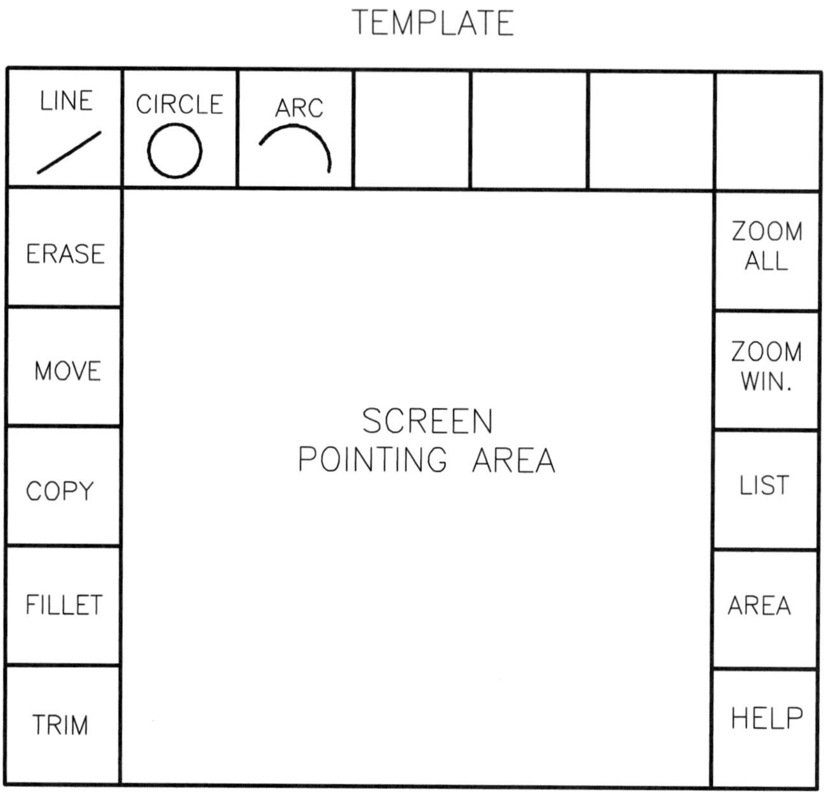

Note

This template is for tablet configuration. You may make a copy of this page and then secure it on the digitizer surface for configuration.

Answers to Self-Evaluation Test

1. digitizing tablet, pointing device, **2.** 9, 25, **3.** frequently, **4.** overlap, **5.** **MENU**

Chapter 19

Customizing Screen Menus

Learning Objectives

After completing this chapter, you will be able to:

- *Create a customized screen menu.*
- *Write screen menus.*
- *Load screen menus.*
- *Write submenus and referencing submenus.*
- *Write menu files with multiple submenus.*
- *Use menu command repetition in menus.*
- *Write menus for foreign languages.*
- *Use control and special characters in menu items.*
- *Use command definition without return or space.*
- *Use menu items with single object selection mode.*
- *Use AutoLISP expressions in menus.*

SCREEN MENU

When you are in the AutoCAD drawing editor, by default, the screen menu is not displayed on the screen. To display the screen menu, choose the **Options** button from the **Application Menu**; the **Options** dialog box will be displayed. Choose the **Display** tab and select the **Display screen menu** check box. Next, choose the **OK** button to exit the **Options** dialog box. The AutoCAD screen menu is displayed on the right of the screen with **AutoCAD** at the top, followed by asterisk signs (* * * *) and a list of commands (Figure 19-1).

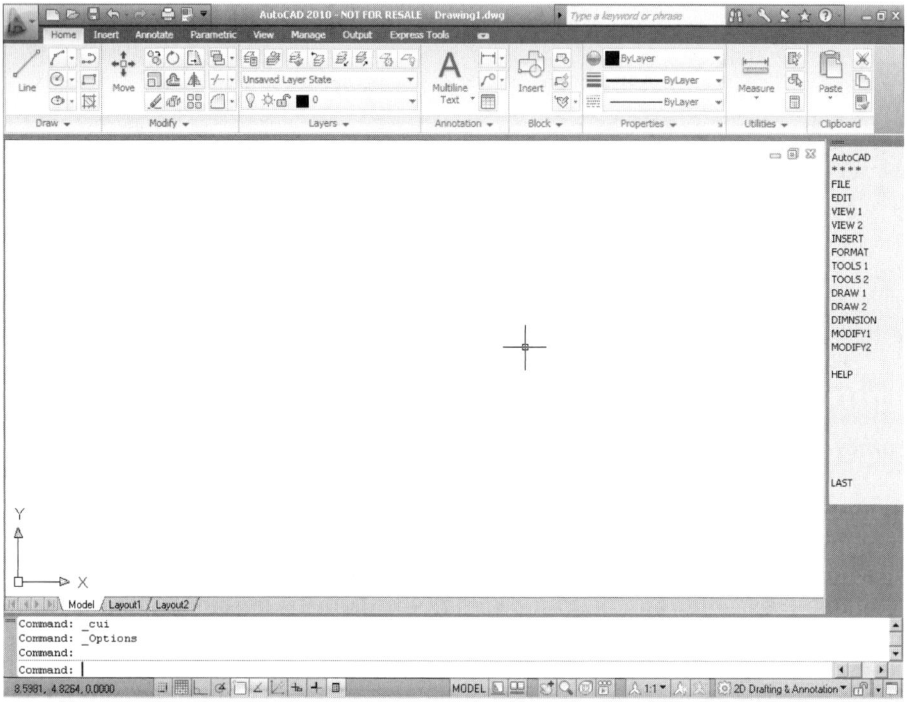

Figure 19-1 *AutoCAD window display with screen menu*

CREATING A CUSTOMIZED SCREEN MENU

Before creating a screen menu, you need to design it and arrange the commands the way you want them to appear on the screen. To design a screen menu, you should select and arrange the commands in a way that provides easy access to the most frequently used commands. A careful design saves a lot of time in the long run.

To understand the process of developing a screen menu, consider the example given next.

Example 1

Create a screen menu for the AutoCAD commands given next. Save the screen menu as a separate *.cuix* file and load it in AutoCAD.

LINE
POLYLINE
MOVE
COPY
REDRAW
ZOOM EXTENTS
SAVE
EXIT

The layout of these commands is shown in Figure 19-2. Figure 19-3 shows the screen menu that is displayed on the AutoCAD screen.

Figure 19-2 *Layout of the screen menu*

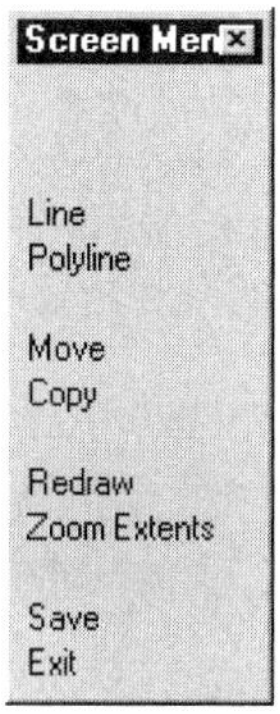

Figure 19-3 *Screen menu after loading the file menu*

Step 1: Creating the screen menu

The customized screen menus are created using the **Customize User Interface** dialog box. The following steps explain the procedure for creating this menu.

1. Invoke the **Customize User Interface** dialog box by choosing **Manage > Customization > User Interface** from the **Ribbon**. Alternatively, choose **Tools > Customize > Interface** from the menu bar.

2. Scroll down the **Customization in All Files** rollout and expand the **Legacy** node in the tree view. Then, expand the **Screen Menus** node in the tree view. This will display all the screen menus present in AutoCAD.

3. To add a new screen menu, right-click on the **Screen Menus** element and choose **New Screen Menu** from the shortcut menu; a new screen menu with the name **ScreenMenu1** is added to the list and its properties are displayed in the right pane.

4. Click on the plus sign (+) located on the left of the **ScreenMenu1** to expand the tree view. This will display the lines from 1 to 22. These lines define the screen submenus.

Next, you need to add the commands to the screen submenus. This is done by dragging the commands from the **Command List** rollout.

5. Select **All Commands** from the **Filter the command list by category** drop-down list in the **Command List** rollout; all commands are displayed in the list box.

6. Scroll to the **Line** item in the list box available in the **Command List** rollout. Drag this item and drop it in front of **Line 3**, as shown in Figure 19-4. The unassigned lines are displayed as an empty space in the screen menu.

7. Similarly, drag the other commands to be added to the screen menu from the **Command List** and drop them in front of the lines at the desired locations.

8. Choose the **Apply** button to add to the new screen menu.

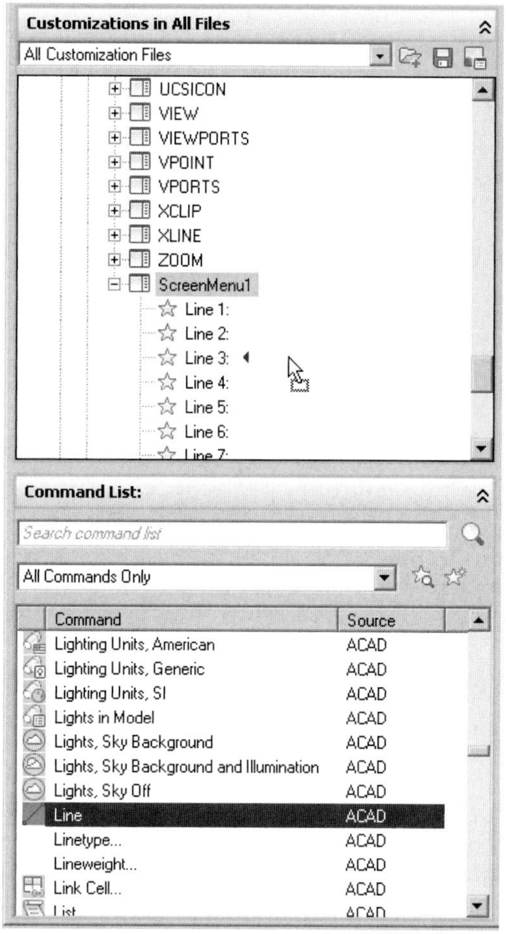

***Figure 19-4** Dragging and adding a command to the screen menu*

Step 2: Exporting the screen menu

1. Choose the **Transfer** tab. This tab also has two panes. The left pane shows the user interface elements of the default cuix file and the right pane shows the headings under a **New File**.

2. By default, **New File** is displayed in the drop-down list in the right pane. If not, choose the **Create a new customization file** button on the right of the drop-down list in the right pane to make sure a blank customization group is displayed in the right pane.

3. Expand the **Legacy** and then the **Screen Menus** tree views in the main customization file in the left pane. Similarly, expand the **Legacy** tree view in the **Customizations in New File** rollout.

4. Select the **ScreenMenu1** screen menu and drag it to the **Screen Menus** heading in the right pane; the selected screen menu is displayed in the right pane also.

5. Right-click on the selected screen menu on the left pane and delete it to make sure that it is removed from the default customization file.

6. Choose the **Save the current customization file** button, which is the third button on the right of the drop-down list in the right pane; the **Save As** dialog box is displayed.

7. Save the customization file with the name *example1.cuix* in your folder.

8. Choose **Apply** and then choose **OK** button to accept and exit the **Customize User Interface** dialog box.

This saves the customized menus in a separate customization file.

Step 3: Loading the CUIx file
To load the *example1.cuix* file, use the **MENU** command. When you invoke this command, the **Select Customization File** dialog box is displayed. Select **Customization files (*.cuix)** from the **Files of type** drop-down list and then open the *example1.cuix* file from your folder. The default AutoCAD menu will be removed and this screen menu will be loaded.

If the screen menu is not displayed on the screen, invoke the **Options** dialog box. Choose the **Display** tab and select the **Display screen menu** check box. Choose the **OK** button to exit the dialog box.

To load the original *acad.cuix* file again, set the value of the **FILEDIA** system variable to **0** and then invoke the **MENU** command. Enter **ACAD** at the command prompt; the default menu will be loaded again. Remember to restore the value of the **FILEDIA** system variable to **1**.

WRITING SCREEN MENUS

Depending on the scope of the menu, the size of the menu file can vary from a few lines to several hundred. A menu file consists of section labels, submenus, and menu items. A menu item consists of an item label and a command definition. The menu item label is enclosed in brackets, and the command definition (menu macro) is outside the brackets.

The menu item label that is enclosed in the brackets is displayed in the screen menu area of the monitor and is not a part of the command definition. The command definition, the part of the menu item outside the bracket, is the executable part of the menu item. To understand the process of developing and writing a screen menu, consider the following example.

[LINE] $S=X $S=LINE ^C^CLINE

Where **[LINE] $S=X $S=LINE ^C^CLINE** ------- Menu item
[LINE] ------- Item label or menu command name
$S=X $S=LINE ^C^CLINE Command definition

Example 2

Write a screen menu for the following AutoCAD commands (file name *sm1.mnu*).

LINE
CIRCLE C,R
CIRCLE C,D
CIRCLE 2P
ERASE
MOVE

The layout of these commands is shown in Figure 19-5 and Figure 19-6 shows the screen menu to be displayed on the AutoCAD screen. This menu is named MENU-1 and it should be displayed at the top of the screen menu. It lets you know the menu you are using. You can use any text editor to write the file. The file extension must be *.mnu*. For Example 1, *sm1.mnu*. *sm1* is the name of the screen menu file, and *.mnu* is the extension of this file. All menu files have the extension *.mnu*.

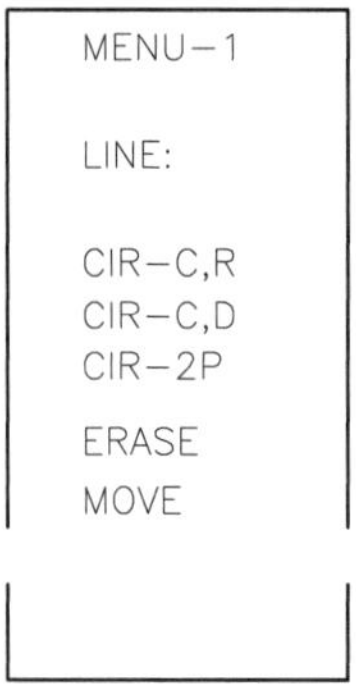

Figure 19-5 *Layout of the screen menu*

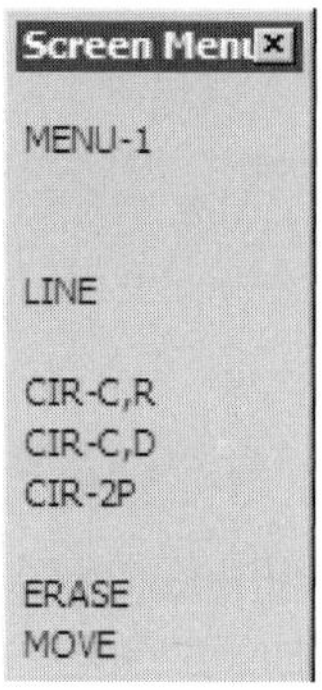

Figure 19-6 *Screen menu after loading the menu file*

The following file is a listing of the screen menu for Example 2. **The line numbers on the right are not a part of the file. They are shown here for reference only.**

```
***SCREEN                                1
[ MENU-1       ]                         2
[              ]                         3
[              ]                         4
[LINE          ]^C^CLINE                 5
[              ]                         6
[CIR-C,R       ]^C^CCIRCLE               7
[CIR-C,D       ]^C^CCIRCLE;\D            8
[CIR- 2P       ]^C^CCIRCLE;2P            9
```

```
10 [            ]
11 [ERASE       ]^C^CERASE
12 [MOVE        ]^C^CMOVE
```

Explanation

Line 1
*****SCREEN**
***SCREEN is the section label for the screen menu. The lines that follow the screen menu are treated as a part of this menu. The screen menu definition will be terminated by another section label, such as ***TABLET1 or ***POP1.

Line 2
[MENU-1]
This menu item displays MENU-1 on the screen. Anything that is inside the brackets is for display only and has no effect on the command. The item names can be any length. However, it is recommended that the names be short and meaningful to save the valuable screen space.

Lines 3-4
[   ]
These menu items print a blank line on the screen menu. There are eight blank spaces inside the bracket. You could also use the brackets with no spaces between the brackets ([]). This line does not contain anything outside the bracket; therefore, no command is executed. To provide the space in the menu you can also leave a blank space in the menu file or have two brackets ([ ]). The next line, line 4, also prints a blank line.

Line 5
[LINE]^C^CLINE
This menu item displays LINE on the screen. The first ^C (caret C) cancels the existing command, and the second ^C cancels the command again. The two **CANCEL** (^C^C) commands are required to make sure that the existing commands are canceled before a new command is executed. Most AutoCAD commands can be canceled with just one **CANCEL** command. However, some commands for dimensioning and polyline edit need to be canceled twice to get out of the command. **LINE** will prompt the user to enter the points to draw a line. Since there is nothing after the LINE, it automatically enters a Return.

[LINE]^C^CLINE
Where **[LINE]---------** For screen display only
^C --------------- ^C is the first **CANCEL** command
^C --------------- ^C is the second **CANCEL** command
LINE------------- AutoCAD's **LINE** command

Line 7
[CIR-C,R]^C^CCIRCLE
The part of the menu item that is enclosed within the brackets is for screen display only. The part of the menu item that is outside the brackets is executed when this line is selected. ^C^C (caret C) cancels the existing command twice. **CIRCLE** is a command that generates

a circle. A space after **CIRCLE** automatically causes a RETURN. The space acts like pressing the SPACEBAR on the keyboard.

[CIR-C,R]^C^CCIRCLE

Where **[CIR-C,R]** ----- For screen display only
^C -------------- ^C for first **CANCEL** command
^C -------------- ^C for second **CANCEL** command
CIRCLE -------- AutoCAD's **CIRCLE** command

Line 8
[CIR-C,D]^C^CCIRCLE;\D
Part of the menu item that is enclosed in the brackets is for screen display only. The part that is outside the brackets is the executable part. ^C^C will cancel the existing command twice.

[CIR-C,D]^C^CCIRCLE;\D

Where **^C^C** ---------- ^C^C cancels the existing command
CIRCLE -------- **CIRCLE** command
; ------------------ Semicolon (;) for RETURN
\\------------------- AutoCAD pauses for user input
D ----------------- Diameter option

The **CIRCLE** command is followed by a semicolon, a backslash (\\), and a D for diameter option. The semicolon (;) after the **CIRCLE** command causes a RETURN, which has the same effect as entering RETURN at the keyboard. The backslash (\\) pauses for user input. In this case it is the center point of the circle. D is for the diameter option and it is automatically followed by RETURN. The semicolon in this example can also be replaced by a blank space, as shown in the following line. However, the semicolon is easier to spot.

[CIR-C,D]^C^CCIRCLE \D

Where Space between **CIRCLE** and \\ -- Blank space for RETURN

Line 9
[CIR- 2P]^C^CCIRCLE;2P
In this menu item ^C^C cancels the existing command twice. The semicolon after **CIRCLE** enters a RETURN. 2P is for the two-point option, followed by the blank space that causes a RETURN. You will notice that the sequence of the command and input is the same as discussed earlier. Therefore, it is essential to know the exact sequence of the command and input; otherwise, the screen menu is not going to work.

[CIR- 2P]^C^CCIRCLE;2P

Where **^C^C** ---------- ^C^C Cancels existing command twice
CIRCLE -------- **CIRCLE** command
; ------------------ Semicolon (;) for RETURN
2P ---------------- 2-Point option for **CIRCLE**

The semicolon after the **CIRCLE** command can be replaced by a blank space, as shown in the following line. The blank space causes a RETURN, like a semicolon (;).

[CIR- 2P]^C^CCIRCLE 2P
Where Space between **CIRCLE** and 2P - Blank space or semicolon for RETURN

Line 11
[ERASE]^C^CERASE
In this menu item ^C^C cancels the existing command twice, and **ERASE** is an AutoCAD command that erases the selected objects.

[ERASE]^C^CERASE
Where **ERASE** ------------ AutoCAD **ERASE** command

Line 12
[MOVE]^C^CMOVE
In this menu item ^C^C cancels the existing command twice, and the **MOVE** command will move the selected objects.

[MOVE]^C^CMOVE
Where **Move** ------------- AutoCAD **MOVE** command

LOADING SCREEN MENUS

AutoCAD automatically loads the *acad.CUIx* file when you get into the AutoCAD drawing editor. However, you can also load a different menu file by using the **MENU** command.

Command: **MENU**

When you enter the **MENU** command, AutoCAD displays the **Select Customization File** dialog box (Figure 19-7) on the screen. Select the **Legacy menu files (*.mns, *.mnu)** option from the **Files of type** drop-down list. Then, select the menu file that you want to load and choose the **Open** button.

You can also load the menu file from the command line.

Command: **FILEDIA**
Enter new value for FILEDIA <1>: **0**

Command: **MENU**
Enter customization file name or [. (for none)] <current>: **PDM1.MNU**

Where **<current>** ----- current menu file
PDM1.MNU --- Name of menu file

After you enter the **MENU** command, AutoCAD will prompt for the file name. Enter the name of the menu file with the file extension (*.mnu*).

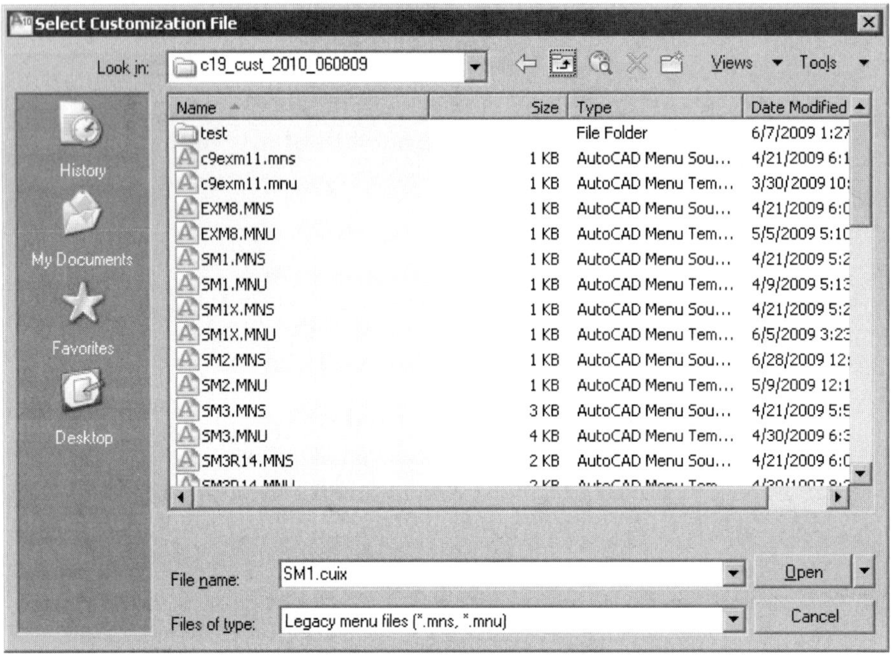

Figure 19-7 *The* ***Select Customization File*** *dialog box*

Note

1. After you load the new menu, you cannot use the screen menu, buttons menu, or digitizer because the original menu, acad.CUIx, is not present and the new menu does not contain these menu areas.

2. To activate the original menu again, load the menu file by using the command line:

Command: ***MENU***
Enter customization file name or [. (for none)] <current>: ***ACAD.MNU***

3. If you need to use input from a keyboard or a pointing device in a menu macro, use the backslash (\). The system will pause for you to enter data.

4. There should be no space after the backslash (\).

5. The menu items, menu labels, and command definition can be uppercase, lowercase, or mixed.

6. You can introduce spaces between the menu items to improve the readability of the menu file.

7. The number of lines displayed on the screen depends on your system and are limited by the setting of the ***SCREENBOXES*** *system variable. If there are more items in the menu than the number of spaces available, the excess items are not displayed on the screen. For example, if the display device limits the number of items to 28, items in excess of 28 will not be displayed on the screen and are therefore inaccessible.*

Exercise 1 *General*

Design and write a screen menu for the following AutoCAD commands. Save the file with the name *sme1.mnu*.

PLINE
ELLIPSE (Center)
ELLIPSE (Axis endpoint)
ROTATE
OFFSET
SCALE

SCREEN SUBMENUS

The screen menu file can be so large that all items cannot be accommodated on one screen. The number of lines displayed on the screen depends on your system and are limited by the setting of the **SCREENBOXES** system variable. For example, the maximum number of items that can be displayed on your screen is 28. If the screen menu has more than 28 items, the menu items in excess of 28 are not displayed on the screen and therefore, cannot be accessed. You can overcome this problem by using submenus that enable the user to define smaller groups of items within a menu section. When a submenu is selected, it loads the submenu items and displays them on screen. However, depending on your system, the number of items you can display can be higher and you may not need submenus.

Submenu Definition

A submenu definition consists of two asterisks (**) followed by the name of the submenu. A menu can have any number of submenus, and every submenu should have a unique name. The items that follow a submenu, up to the next section label or submenu label, belong to that submenu. The format of a submenu definition is:

****Name**

Where ** ----------------- Two asterisks (**) designate a submenu
Name ------------ Name of the submenu

****DRAW1**

Where **DRAW1** --------- Name of the submenu
** ----------------- ** designates that DRAW1 is a submenu

Note

1. The submenu name can be up to 33 characters long.

2. The submenu name can consist of letters, digits, and such special characters as $ (dollar sign), - (hyphen), and _ (underscore).

3. The submenu name should not have any embedded blanks (spaces).

4. The submenu names should be unique in a menu file.

Submenu Reference

The submenu reference is used in a menu item to reference or load a submenu. It consists of a "$" sign followed by a letter that specifies the menu section. The letter that specifies a screen menu section is S. The section is followed by an "=" sign and the name of the submenu the user wants to activate. The submenu name should be without the **. The following is the format of a submenu reference:

$Section=Submenu

Where **$** ------------------- "$" sign
Section ---------- Menu section specifier
= ------------------ "=" sign
Submenu ------- Name of submenu

$S=EDIT

Where **S** ------------------- S specifies screen menu section
Edit --------------- Name of submenu

Note

$ — *A special character code used to load a submenu in a menu file.*
$M= — *This is used to load a DIESEL macro from a menu item.*

The following are the section specifiers:

S	*Specifies the* ***SCREEN*** *menu.*
P0 - P16	*Specifies the POP menus, POP0 through POP16.*
I	*Specifies the Image Tile menu.*
B1 - B4	*Specifies the BUTTONS menu, B1 through B4.*
T1 - T4	*Specifies the TABLET menus, T1 through T4.*
A1 - A4	*Specifies the AUX menus A1 through A4.*

Nested Submenus

When a submenu is activated, the current menu is copied to a stack. If you select another submenu, the submenu that was current will be copied or pushed to the top of the stack. The maximum number of menus that can be stacked is eight. If the stack size increases to more than eight, the menu at the bottom of the stack is removed and forgotten. You can call the previous submenu by using the nested submenu call. The format of this call is as follows:

$S=

Where **$** ------------------- "$" sign
S ------------------- Screen menu specifier
= ------------------ "=" sign

The maximum number of nested submenu calls is eight. Each time you call a submenu (issue $S=), this pops the last item from the stack and reactivates it.

Example 3

Design a menu layout and write a screen menu for the following commands.

LINE	**ERASE**
PLINE	**MOVE**
ELLIPSE-C	**ROTATE**
ELLIPSE-E	**OFFSET**
CIR-C,R	**COPY**
CIR-C,D	**SCALE**
CIR- 2P	

As mentioned earlier, the first and the most important part of writing a menu is the design of the menu and knowing the commands and the prompts associated with those commands. You should conceptualize the way you want the menu to look and the way you want to arrange the commands for maximum efficiency. Write out the menu on a piece of paper, and check it thoroughly to make sure you have arranged the commands the way you want them. Use submenus to group the commands based on their use, function, and relationship with other submenus. Make provisions to access other frequently used commands without going through the root menu.

Figure 19-8 shows one of the possible arrangements of the commands and the design of the screen menu. It has one main menu and two submenus. One of the submenus is meant for draw tools, the other is for edit tools. The colon (:) at the end of the commands is not required. It is used here to distinguish the commands from those items that are not used as commands. For example, **DRAW** in the root menu is not a command; therefore, it has no colon at the end. On the other hand, if you select **ERASE** from the **EDIT** menu, it executes the **ERASE** command, so it has a colon (:) at the end of the command.

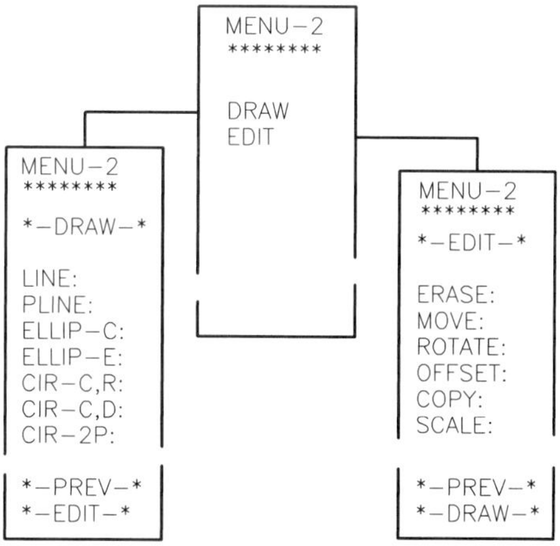

Figure 19-8 *Screen menu design*

The following file is a listing of the menu file of Example 3. **The line numbers on the right are not a part of the file; they are given here for reference only.**

```
***SCREEN
[ MENU-2      ]
[********     ]
[             ]
[             ]
[             ]
[             ]
[DRAW         ]^C^C$S=DRAW
[EDIT         ]^C^C$S=EDIT

**DRAW
[ MENU-2      ]^C^C$S=SCREEN
[********     ]
[
[*-DRAW-*     ]
[             ]
[LINE:        ]^C^CLINE
[PLINE:       ]^C^CPLINE;\W;0.1;0.1
[ELLIP-C:     ]^C^CELLIPSE;C
[ELLIP-E:     ]^C^CELLIPSE
[CIR-C,R:     ]^C^CCIRCLE
[CIR-C,D:     ]^C^CCIRCLE;\D
[CIR-2P:      ]^C^CCIRCLE;2P
[             ]
[             ]
[             ]
[             ]
[             ]
[             ]
[*-PREV-*     ]^C^C$S=
[*-EDIT-*     ]^C^C$S=EDIT

**EDIT
[ MENU-2      ]^C^C$S=SCREEN
[********     ]
[             ]
[*-EDIT-*     ]
[             ]
[ERASE:       ]^C^CERASE
[MOVE:        ]^C^CMOVE
[ROTATE:      ]^C^CROTATE
[OFFSET:      ]^C^COFFSET
[COPY:        ]^C^CCOPY
```

```
[SCALE:      ]^C^CSCALE                                    44
[            ]                                             45
[            ]                                             46
[            ]                                             47
[            ]                                             48
[            ]                                             49
[            ]                                             50
[            ]                                             51
[*-PREV-*    ]^C^C$S=                                      52
[*-DRAW-*    ]^C^C$S=DRAW                                  53
```

Explanation

Line 1

*****SCREEN**

***SCREEN is the section label for the screen menu.

Line 2

[MENU-2]

This menu item displays MENU-2 at the top of the screen menu.

Line 3

[********]

This menu item prints eight asterisk signs (********) on the screen menu.

Lines 4-7

[   ]

These menu items print four blank lines on the screen menu. The brackets are not required. They could be just four blank lines without brackets.

Line 8

[DRAW   ]^C^C$S=DRAW

[DRAW   ] displays DRAW on the screen, letting you know that by selecting this function you can access the draw commands. ^C^C cancels the existing command, and $S=DRAW loads the DRAW submenu on the screen.

```
[DRAW    ]^C^C$S=DRAW
        Where  ^C^C ---------- Cancels the existing command twice
               = ------------------ Loads submenu DRAW
               DRAW ----------- Name of submenu
```

Line 9

[EDIT   ]^C^C$S=EDIT

[EDIT   ] displays EDIT on the screen. ^C^C cancels the current command, and $S=EDIT loads the submenu **EDIT**.

Line 10
The blank lines between the submenus or menu items are not required. It just makes it easier to read the file.

Line 11
****DRAW**
**DRAW is the name of the submenu; lines 12 through 31 are defined under this submenu.

Line 15
[*-DRAW-*]
This prints *-DRAW-* as a heading on the screen menu to let the user know that the commands listed on the menu are draw commands.

Line 18
[PLINE:]^C^CPLINE;\W;0.1;0.1
[PLINE:] displays PLINE: on the screen. ^C^C cancels the command, and **PLINE** is a command. The semicolons are for RETURN. The semicolons can be replaced by a blank space. The backslash (\) is for user input. In this case it is the start point of the polyline. W selects the width option of the polyline. The first 0.1 is the starting width of the polyline, and the second 0.1 is the ending width. This command will draw a polyline of 0.1 width.

[PLINE:]^C^CPLINE;\W;0.1;0.1
Where **PLINE** ------------ **PLINE** (Polyline) command
; ------------------- RETURN
\\------------------- Pause for input (start pt.)
W ----------------- Width
; ------------------- RETURN
0.1 ---------------- Starting width
; ------------------- RETURN
0.1 ---------------- Ending width

Line 19
[ELLIP-C:]^C^CELLIPSE;C
[ELLIP-C:] displays ELLIP-C: on the screen menu. ^C^C cancels the existing command, and **ELLIPSE** is an AutoCAD command to generate an ellipse. The semicolon is for RETURN, and C selects the center option of the **ELLIPSE** command.

[ELLIP-C:]^C^CELLIPSE;C
Where **CELLIPSE** ------ **ELLIPSE** command
; ------------------- RETURN
C ------------------ Center option for ellipse

Line 20
[ELLIP-E:]^C^CELLIPSE
[ELLIP-E:] displays ELLIP-E: on the screen menu. ^C^C cancels the existing command,

and **ELLIPSE** is an AutoCAD command. Here, the **ELLIPSE** command uses the default option, axis endpoint, instead of center.

Line 30
[*-PREV-*]^C^C$S=
[*-PREV-*] displays *-PREV-* on the screen menu, and ^C^C cancels the existing command. $S= restores the previous menu that was displayed on the screen before loading the current menu.

[*-PREV-*]^C^C$S=
Where **$S** ---------------- Restores the previous screen menu

Line 31
[*-EDIT-*]^C^C$S=EDIT
[*-EDIT-*] displays *-EDIT-* on the screen menu, and ^C^C cancels the existing command. $S=EDIT loads the submenu EDIT on the screen. This lets you access the **EDIT** commands without going back to the root menu and selecting **EDIT** from there.

[*-EDIT-*]^C^C$S=EDIT
Where **$S** ---------------- Loads the submenu EDIT
EDIT ------------ Name of the submenu

Line 33
****EDIT**
**EDIT is the name of the submenu, and lines 34 to 53 are defined under this submenu.

Line 34
[MENU-2]^C^C$S=SCREEN
[MENU-2] displays MENU-2 on the screen menu, and ^C^C cancels the existing command. $S=SCREEN loads the root menu SCREEN on the screen menu.

Line 39
[ERASE:]^C^CERASE
[ERASE:] displays ERASE: on the screen menu, and ^C^C cancels the existing command. **ERASE** is an AutoCAD command for erasing the selected objects.

Line 53
[*-DRAW-*]^C^C$S=DRAW
$S=DRAW loads the **DRAW** submenu on the screen. It lets you load the **DRAW** menu without going through the root menu. The root menu is the first menu that appears on the screen when you load a menu or start AutoCAD.

When you select DRAW from the root menu, the submenu DRAW is loaded on the screen. The menu items in the draw submenu completely replace the menu items of the root menu. If you select **MENU-2** from the screen menu now, the root menu will be loaded on the screen, but some of the items are not cleared from the screen menu (Figure 19-9). This is because the

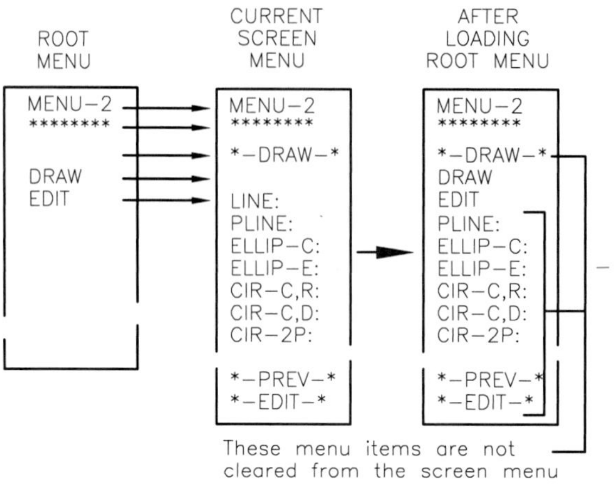

Figure 19-9 *Screen menu display after loading the root menu*

root menu does not have enough menu items to completely replace the menu items of the DRAW submenu.

One of the ways to clear the screen is to define a submenu that has 21 blank lines. When this submenu is loaded, it will clear the screen. If you then load another submenu, there will be no overlapping, because the screen menu has already been cleared and there are no menu items left on the screen (Figure 19-10). Example 3 illustrates the use of such a submenu.

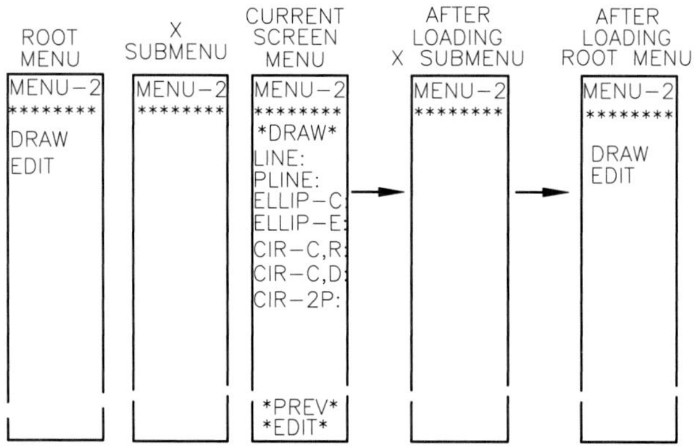

Figure 19-10 *Screen menu display after loading the root menu*

Another way to avoid overlapping the menu items is to define every submenu so that they all have the same number of menu items. The disadvantage with this approach is that the menu file will be long because every submenu will have 21 lines.

Exercise 2 *General*

Write a screen menu file for the following AutoCAD commands (file name *sme2.mnu*).

ARC	**MIRROR**
-3P	**BREAK-F**
-SCE	**BREAK-@**
-SCA	**EXTEND**
-SCL	**STRETCH**
-SEA	**FILLET-0**
POLYGON-C	**FILLET**
POLYGON-E	**CHAMFER**

MULTIPLE SCREEN SUBMENUS

A menu file can have any number of submenus. All the submenu names have two asterisks (**) in front of them, even if there is a submenu within a submenu. Also, several submenus can be loaded in a single statement. If there is more than one submenu in a menu item, they should be separated from one another by a space. Example 4 illustrates the use of multiple submenus.

Example 4 *General*

Design the layout of the menu, and then write the screen menu for the following AutoCAD commands.

Draw	**ARC**	**Edit**	**Display**
LINE	**3Point**	**EXTEND**	**ZOOM**
Continue	**SCE**	**STRETCH**	**REGEN**
Close	**SCA**	**FILLET**	**SCALE**
Undo	**CSE**		**PAN**
.X	**CSA**		
.Y	**CSL**		
.Z			
.XY			
.XZ			
.YZ			

There are several different ways of arranging the commands, depending on your requirements. The following layout is one of the possible screen menu designs for the given AutoCAD commands (Figure 19-11).

The name of the following menu file is *sm3.mnu*. You can use a text editor to write the file. The following file is the listing of the menu file *sm3.mnu*. **The line numbers on the right are not a part of the file. They are shown here for reference only.**

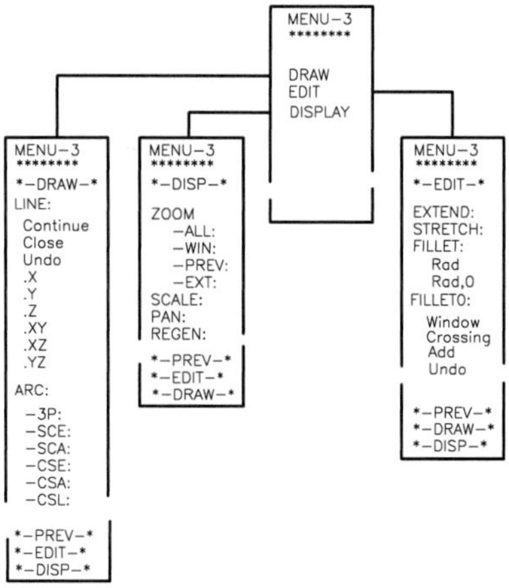

Figure 19-11 *Screen menu design with submenus*

```
***SCREEN
**S
[ MENU-3            ]^C^C$S=X $S=S
[********           ]$S=OSNAP
[                   ]
[DRAW               ]^C^C$S=X $S=DRAW
[EDIT               ]^C^C$S=X $S=EDIT
[DISPLAY            ]^C^C$S=X $S=DISP
**DRAW 3
[*-DRAW-*           ]
[                   ]
[LINE:              ]^C^CLINE
[ Continue          ]^C^CLINE;;
[ Close             ]CLOSE
[ Undo              ]UNDO
[ .X                ].X
[ .Y                ].Y
[ .Z                ].Z
[ .XY               ].XY
[ .XZ               ].XZ
[ .YZ               ].YZ
[                   ]
[ARC:               ]^C^CArc
[   -3P:            ]^C^CARC \\DRAG
[   -SCE:           ]^C^CARC \C \DRAG
[   -SCA:           ]^C^CARC \C \A DRAG
[   -CSE:           ]^C^CARC C \\DRAG
[   -CSA:           ]^C^CARC C \\A DRAG
[   -CSL:           ]^C^CARC C \\L DRAG
```

```
[                   ]
[*-PREV-*           ]$S= $S=
[*-EDIT-*           ]^C^C$S=X $S=EDIT
[*-DISP-*           ]$S=X $S=DISP
**EDIT 3
[*-EDIT-*           ]
[                   ]
[EXTEND:            ]^C^CEXTEND
[STRETCH:           ]^C^CSTRETCH;C
[FILLET:            ]^C^CFILLET
[ Rad               ]R;\Fillet
[ Rad 0             ]R;0;Fillet
[FILLET0:           ]^C^CFillet;R 0;;
 Window
 Crossing
 Add
 Undo
[                   ]
[*-PREV-*           ]$S= $S=
[*-DRAW-*           ]^C^C$S=X $S=DRAW
[*-DISP-*           ]$S=X $S=DISP
**DISP 3
[*-DISP-*           ]
[                   ]
[ZOOM:              ]'ZOOM
[  -ALL             ]A
[  -WIN             ]W
[  -PREV            ]P
[  -EXT             ]E
[                   ]
[SCALE:             ]'ZOOM
[PAN:               ]'PAN
[REGEN:             ]^C^CREGEN
[                   ]
[*-PREV-*           ]$S= $S=
[*-EDIT-*           ]^C^C$S=X $S=EDIT
[*-DRAW-*           ]$S=X $S=DRAW
**X 3
[                   ]
[                   ]
[                   ]
[                   ]
[                   ]
[                   ]
[                   ]
[                   ]
```

```
76  [                    ]
77  [                    ]
78  [                    ]
79  [                    ]
80  [                    ]
81  [                    ]
82  [                    ]
83  [                    ]
84  [                    ]
85  [                    ]
86  [                    ]
87  **OSNAP 2
88  [-OSNAPS-            ]
89  [                    ]
90  [Center              ]CEN $S=
91  [Endpoint            ]END $S=
92  [Insert              ]INS $S=
93  [Intersec            ]INT $S=
94  [Midpoint            ]MID $S=
95  [Nearest             ]NEA $S=
96  [Node                ]NOD $S=
97  [Perpend             ]PER $S=
98  [Quadrant            ]QUA $S=
99  [Tangent             ]TAN $S=
100 [None                ]NONE $S=
101 [                    ]
102 [                    ]
103 [                    ]
104 [                    ]
105 [                    ]
106 [*-PREV-*            ]$S=
```

Explanation

Line 3

[MENU-3]^C^C$S=X $S=S

In this menu item, [MENU-3] displays MENU-3 at the top of the screen menu. ^C^C cancels the command, and $S=X loads submenu X. Similarly, $S=S loads submenu S. Submenu X, defined on line 171, consists of 18 blank lines. Therefore, when this submenu is loaded it prints blank lines on the screen and clears the screen menu area. After loading submenu X, the system loads submenu S, and the items that are defined in this submenu are displayed on the screen menu.

[MENU-3]^C^C$S=X $S=S

Where **^C^C** ---------- Cancels the existing command
$S=X ------------ Loads submenu X (clears screen)
$S=S ------------- Loads submenu S

Line 4
[********]$S=OSNAP
This menu item prints eight asterisks on the screen menu, and $S=OSNAP loads submenu OSNAP. The OSNAP submenu, defined on line 191, consists of object snap modes.

Line 6
[DRAW]^C^C$S=X $S=DRAW
This menu item displays DRAW on the screen menu area, cancels the existing command, and then loads submenu X and submenu DRAW. Submenu X clears the screen menu area, and submenu DRAW, as defined on line 11, loads the items defined under the DRAW submenu.

Line 9
****DRAW 3**
This line is the submenu label for DRAW; 3 indicates that the first line of the DRAW submenu will be printed on line 3. Nothing will be printed on line 1 or line 2. The first line of the DRAW submenu will be on line 3, followed by the rest of the menu. All the submenus except submenus S and OSNAP have a 3 at the end of the line. Therefore, the first two lines (MENU-3) and (********) will never be cleared and will be displayed on the screen all the time. If you select MENU-3 in any menu, it will load submenu S. If you select ********, it will load submenu OSNAP.

****DRAW 3**
Where **DRAW** ----------- Submenu name
3 ------------------- Leaves two blank lines and prints from third line

Line 12
[LINE:]^C^CLINE
^C^C cancels the existing command, and **LINE** is the AutoCAD **LINE** command.

Line 13
[Continue]^C^CLINE;;
In this menu item, the **LINE** command is followed by two semicolons that continue the **LINE** command. To understand it, look at the **LINE** command to see how the **Continue** option is used in this command:

Command: **LINE**
Specify first point: **RETURN (Continue)**
Specify next point or [Undo]:

Following is the command and prompt entry sequence for the **LINE** command, if you want to start a line from the last point:

LINE
RETURN
RETURN
SELECT A POINT

Therefore, in the screen menu file the **LINE** command has to be followed by two Returns to continue from the previous point.

Line 16
[.X].X
In this menu item, .X extracts the *X* coordinate from a point. The bracket, .X, and spaces inside the brackets ([.X]) are not needed. The line could consist of .X only. The same is true for lines 39 through 43.

Example

[.X	].X	can also be written (without brackets) as	.X
[.Y	].Y		.Y
[.Z	].Z		.Z

Line 31
[*-PREV-*]$S= $S=
This menu item recalls the previous menu twice. AutoCAD keeps track of the submenus that were loaded on the screen. The first **$S=** will load the previously loaded menu, and the second **$S=** will load the submenu that was loaded before that. For example, in line 16 ([ARC:]$S=X $S=ARC), two submenus have been loaded: first X and then ARC. Submenu X is stacked before loading submenu ARC, which is current and is not stacked yet. The first $S= will recall the previous menu, in this case submenu X. The second $S= will load the menu that was on the screen before selecting the item in line 16.

[*-PREV-*]$S= $S=
Where **$S=** --------------- Loads last menu
$S= --------------- Loads next to last menu

Line 33
[*-DISP-*]$S=X $S=DISP
In this menu item, $S=X loads submenu X, and $S=DISP loads submenu DISP. Notice that there is no CANCEL (^C^C) command in the line. There are some menus you may like to load without canceling the existing command. For example, if you are drawing a line, you may want to zoom without canceling the existing command. You can select [*-DISP-*], select the appropriate zoom option, and then continue with the **LINE** command. However, if the line has a CANCEL command ([*-DISP-*]^C^C$S=X $S=DISP), you cannot continue with the **LINE** command, because the existing command will be canceled when you select *-DISP-* from the screen. In line 28 ([*-EDIT-*]^C^C$S=X $S=EDIT), ^C^C cancels the existing command, because you cannot use any of the editing commands unless you have canceled the existing command.

Line 38
[STRETCH:]^C^CSTRETCH;C
^C^C cancels the existing command, and **STRETCH** is an AutoCAD command. The C is for the crossing option that prompts you to enter two points to select the objects.

[STRETCH:]^C^CSTRETCH;C

Where **STRETCH** ----- **STRETCH** command
; ------------------- RETURN
C ------------------ Center O'SNAP

Line 40
[RAD]R;\FILLET
This menu item selects the radius option of the **FILLET** command, and then waits for you to enter the radius. After entering the radius, execute the **FILLET** command again to generate the desired fillet between the two selected objects.

[RAD]R;\FILLET

Where **R** ------------------ Radius option of **FILLET** command
; ------------------- Space or semicolon for RETURN
\------------------- Pause for user input
FILLET --------- **FILLET** command

Line 41
[RAD 0]R;0;FILLET
This menu item selects the radius option of the **FILLET** command and assigns it a 0 value. Then it executes the **FILLET** command to generate a 0 radius fillet between the two selected objects.

[RAD 0]R;0;FILLET

Where **R** ------------------ Radius option of **FILLET** command
; ------------------- Space or semicolon for RETURN
0 ------------------- 0 radius
; ------------------- Space or semicolon for RETURN
FILLET --------- **FILLET** command

Line 42
[FILLET0:]^C^CFILLET;R 0;;
This menu item defines a **FILLET** command with 0 radius and then generates 0 fillet between the two selected objects.

[FILLET0:]^C^CFILLET;R 0;;

Where **FILLET** --------- **FILLET** command
; ------------------- Semicolon or space for RETURN
R ------------------ R for radius option
Space ------------- Semicolon or space for RETURN
0 ------------------- 0 radius
; ------------------- Semicolon for RETURN
; ------------------- Repeats the **FILLET** command

Chapter 19

Line 54
[ZOOM:]'ZOOM
This menu item defines a transparent **ZOOM** command.

[ZOOM:]'ZOOM
Where **'** -------------------- Single quote (') makes **ZOOM** transparent
ZOOM ----------- **ZOOM** command

Line 90
[Center]CEN $S=
In this menu item, CEN is used for center object snap and $S= automatically recalls the previous screen menu after selecting the object.

[Center]CEN $S=
Where **CEN** -------------- Center object snap
$S= --------------- Loads previous menu

Note

If any menu item on a screen menu includes more than one load command, the commands must be separated by a space.

Example:
[LINE:]$S=X $S=LINE
Note that there is a space between X and $S.

Similarly, if a menu item includes both a load command and an AutoCAD command, they should be separated by a space.

Example:
[LINE:]$S=LINE ^C^CLINE

Exercise 3 *General*

Write a screen menu for the AutoCAD commands shown in Figure 19-12 (file name *sme3.mnu*). The ******** are to access the object snap submenu.

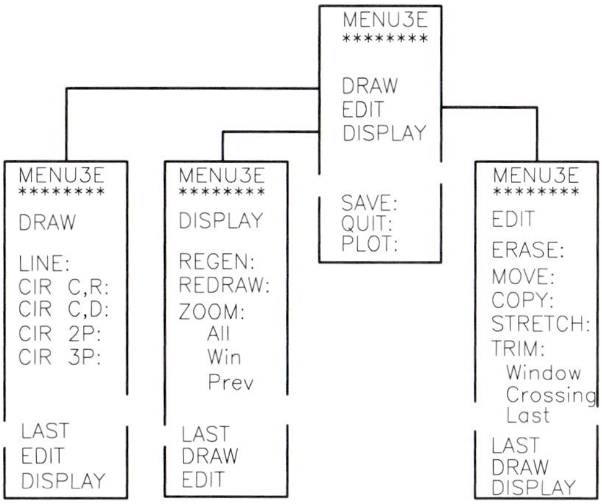

Figure 19-12 *Screen menu design with submenus*

Example 5 *General*

Write a combined screen and tablet menu for the commands shown in the tablet menu template shown in Figure 19-13(a) and the screen menu shown in Figure 19-13(b). When the user selects a command from the digitizer template, it should automatically load the corresponding screen menu on the screen (filename *tm3.mnu*).

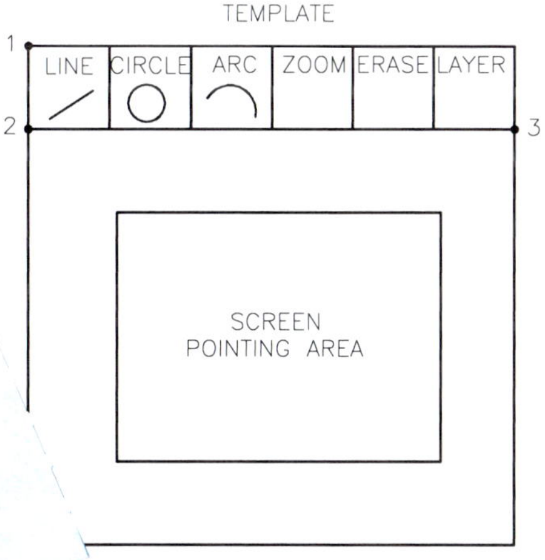

19-13(a) *Tablet menu template for Example 5*

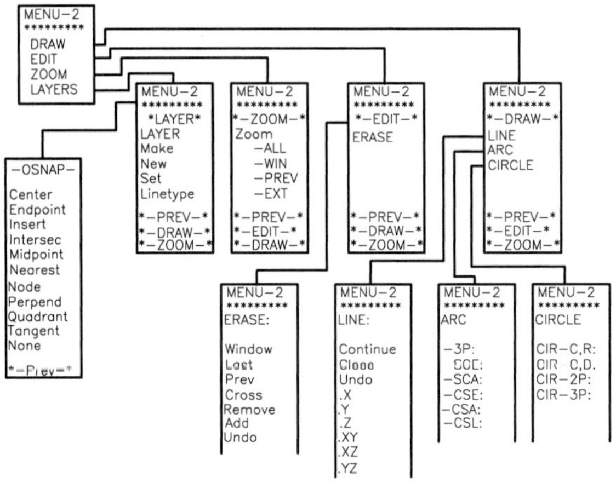

Figure 19-13(b) *Screen menu display*

The following file is a listing of the combined menu for Example 5. **The line numbers on the right are not a part of the file; they are shown here for reference only.**

```
1  ***SCREEN
2  **S
3  [ MENU-2  ]^C^C$S=X $S=S
4  [********  ]$S=OSNAP
5  [          ]
6  [          ]
7  [DRAW      ]^C^C$S=X $S=DRAW
8  [EDIT      ]^C^C$S=X $S=EDIT
9  [ZOOM      ]^C^C$S=X $S=ZOOM
10 [LAYER     ]^C^C$S=X $S=LAYER
11 
12 **DRAW 2
13 [*-DRAW-* ]
14 [          ]
15 [          ]
16 [LINE:     ]$S=X $S=LINE ^C^CLINE
17 [ARC:      ]$S=X $S=ARC
18 [CIRCLE:   ]$S=X $S=CIRCLE ^C^CCIRCLE
19 [          ]
20 [          ]
21 [          ]
22 [          ]
23 [          ]
24 [          ]
25 [          ]
```

```
[            ]
[            ]
[*-PREV-*   ]$S= $S=
[*-EDIT-*   ]^C^C$S=X $S=EDIT
[*-ZOOM-* ]$S=X $S=ZOOM

**LINE 2
[LINE:       ]^C^CLINE
[            ]
[            ]
[Continue   ]^C^CLINE;;
[Close       ]CLOSE
[Undo        ]UNDO
[.X          ].X
[.Y          ].Y
[.Z          ].Z
[.XY         ].XY
[.XZ         ].XZ
[.YZ         ].YZ
[            ]
[            ]
[            ]
[*-PREV-*   ]$S= $S=
[*-EDIT-*   ]^C^C$S=X $S=EDIT
[*-ZOOM-* ]$S=X $S=ZOOM

**ARC 2
[ARC         ]
[            ]
[   -3P:     ]^C^CARC \\DRAG
[   -SCE:    ]^C^CARC \C \DRAG
[   -SCA:    ]^C^CARC \C \A DRAG
[   -CSE:    ]^C^CARC C \\DRAG
[   -CSA:    ]^C^CARC C \\A DRAG
[   -CSL:    ]^C^CARC C \\L DRAG
[            ]
[            ]
[            ]
[            ]
[            ]
[            ]
[            ]
[*-PREV-*   ]$S= $S=
[*-EDIT-*   ]^C^C$S=X $S=EDIT
[*-ZOOM-* ]$S=X $S=ZOOM
```

```
**CIRCLE 2
[CIRCLE:   ]
[          ]
[  CIR-C,R:]^C^CCIRCLE
[  CIR-C,D:]^C^CCIRCLE \D
[  CIR-2P: ]^C^CCIRCLE 2P
[  CIR-3P: ]^C^CCIRCLE 3P
[          ]
[          ]
[          ]
[          ]
[          ]
[          ]
[          ]
[          ]
[          ]
[*-PREV-*  ]$S= $S=
[*-EDIT-*  ]^C^C$S=X $S=EDIT
[*-ZOOM-* ]$S=X $S=ZOOM


**EDIT 2
[*-EDIT-*  ]
[          ]
[          ]
[ERASE:    ]$S=X $S=ERASE ^C^CERASE
[          ]
[          ]
[          ]
[          ]
[          ]
[          ]
[          ]
[          ]
[          ]
[          ]
[          ]
[*-PREV-*  ]$S= $S=
[*-DRAW-* ]^C^C$S=X $S=DRAW
[*-ZOOM-* ]$S=X $S=ZOOM

**ERASE 2
[ERASE:    ]^C^CERASE
[          ]
Window
Last
```

```
Prev
Cross
Remove
Add
Undo
[         ]
[         ]
[         ]
[         ]
[         ]
[         ]
[*-PREV-*  ]$S= $S=
[*-DRAW-* ]^C^C$S=X $S=DRAW
[*-ZOOM-* ]$S=ZOOM

**ZOOM 2
[*-ZOOM-* ]
[         ]
[         ]
[ZOOM:    ]'ZOOM
[ -ALL    ]A
[ -WIN    ]W
[ -PREV   ]P
[ -EXT    ]E
[         ]
[         ]
[         ]
[         ]
[         ]
[         ]
[         ]
[*-PREV-*  ]$S= $S=
[*-EDIT-*  ]^C^C$S=X $S=EDIT
[*-DRAW-* ]$S=X $S=DRAW

***LAYER 2
[*-LAYER* ]
[         ]
[         ]
[LAYER:   ]^C^CLAYER
Make
New
Set
Linetype
```

```
[          ]
[          ]
[          ]
[          ]
[          ]
[*-PREV-*  ]$S= $S=
[*-DRAW-*  ]$S=X $S=DRAW
[*-ZOOM-*  ]$S=ZOOM

**X 3
[          ]
[          ]
[          ]
[          ]
[          ]
[          ]
[          ]
[          ]
[          ]
[          ]
[          ]
[          ]
[          ]
[          ]
[          ]
[          ]
[          ]
[          ]

**OSNAP 2
[-OSNAPS-  ]
[          ]
[Center    ]CEN $S=
[Endpoint  ]END $S=
[Insert    ]INS $S=
[Intersec  ]INT $S=
[Midpoint  ]MID $S=
[Nearest   ]NEA $S=
[Node      ]NOD $S=
[Perpend   ]PER $S=
[Quadrant  ]QUA $S=
[Tangent   ]TAN $S=
[None      ]NONE $S=
[          ]
[          ]
[          ]
```

```
208 [            ]
209 [            ]
210 [*-PREV-*    ]$S=
211
212 ***TABLET1
213 $S=X $S=LINE ^C^CLINE
214 $S=X $S=CIRCLE ^C^CCIRCLE
215 $S=X $S=ARC ^C^CARC
216 $S=X $S=ZOOM ^C^CZOOM
217 $S=X $S=ERASE ^C^CERASE
218 $S=X $S=LAYER ^C^CLAYER
```

Line 1
*****SCREEN**
This is the section label for the screen menu. The line numbers 1 through 213 are defined in this section of the menu file.

Line 3
[MENU-3]^C^C$S=X $S=S
In this menu item, $S=X loads the submenu X and $S=S loads the submenu S.

Line 7
[DRAW   ]^C^C$S=X $S=DRAW
In this menu item, $S=DRAW loads the submenu DRAW that is defined in the menu file.

Line 12
****DRAW 3**
In this menu item DRAW is the name of the submenu. The 3 forces the submenu to be displayed on line 3. Since nothing is printed on the first two lines, [MENU-3] and [********] will be displayed on the screen all the time. If you select [MENU-3] from any menu, it will load the submenu S and display it on the screen. Similarly, if you pick [********] from any menu, it will load and display the OSNAP submenu.

Line 28
[*-PREV-*]$S= $S=
In this menu item, $S= $S= loads the previous two submenus. One of them is the submenu X and the second submenu is the one that was displayed on the screen before the current menu.

Line 212
*****TABLET1**
TABLET1 is the section label for the tablet area 1. The line numbers 212 through 218 are defined in this section.

Line 213
$S=X $S=LINE ^C^CLINE

$S=X, defined in the screen menu section, loads the submenu X on the screen menu area. The purpose of loading the submenu X is to clear the screen so that there is no overlapping of the screen menu items. $S=LINE, defined in the screen menu section, loads the **LINE** submenu on the screen menu. ^C^CLINE cancels the existing command twice and then executes the **LINE** command.

When you select the LINE block from the digitizer, AutoCAD automatically clears the screen menu, loads the LINE submenu, and enters the **LINE** command. This makes it easier for the user to select the command options from the screen menu, since they are not on the digitizer template.

$S=X $S=LINE ^C^CLINE

Where **$S=X** ----------- Loads submenu X
LINE------------- Loads submenu LINE
^C^C ---------- Cancels the existing command
LINE------------- AutoCAD's **LINE** command

You can also use the automatic menu swapping feature of AutoCAD to load the corresponding screen menus. If you want to utilize this feature, the TABLET1 section must be written as follows:

```
***TABLET1
$S=X ^C^CLINE
$S=X ^C^CCIRCLE
$S=X ^C^CARC
$S=X ^C^CZOOM
$S=X ^C^CERASE
$S=X ^C^CLAYER
```

LONG MENU DEFINITIONS

You can put any number of commands in one screen menu line. There is no limit to the number of commands and the order in which they appear in the line, as long as they satisfy all the requirements of the command and the sequence of prompt entries. Again, you need to know the AutoCAD commands, the options in the command, the command prompts, and the entries for various prompts. If the statement cannot fit on one line, you can put a plus (+) sign at the end of the first line and continue with a second line. The command definition that involves several commands put together in one line is also called a macro. The next example illustrates a long menu item or a macro.

Example 6 *General*

Write a screen menu command definition that performs the following functions (filename *sm4.mnu*).

Draw a border using polyline

Width =	0.01
Point-1	0,0
Point-2	12,0
Point-3	12,9
Point-4	0,9
Point-5	0,0

Initial drawing setup

Snap	0.25
Grid	0.5
Limits	12,9
Zoom	All

Before writing a menu, you should know the commands, options, prompts, and the prompt entries required for the commands. Therefore, you need to first study the commands involved in the previously mentioned drawing setup.

POLYLINE
Command: **PLINE**
Specify start point: **0,0**
Specify next point or [Arc/Close/Halfwidth/Length/Undo/Width]: **W**
Specify starting width <0.0000>: **0.01**
Specify ending width <0.01>: *Press ENTER*
Specify next point or [Arc/Close/Halfwidth/Length/Undo/Width]: **12,0**
Specify next point or [Arc/Close/Halfwidth/Length/Undo/Width]: **12,9**
Specify next point or [Arc/Close/Halfwidth/Length/Undo/Width]: **0,9**
Specify next point or [Arc/Close/Halfwidth/Length/Undo/Width]: **C**

Screen menu command definition for **PLINE**
PLINE;0,0;W;0.01;;12,0;12,9;0,9;C

SNAP
Command: **SNAP**
Specify snap spacing or [ON/OFF/Aspect/Rotate/Style/Type] <default>: **0.25**

Screen menu command definition for SNAP
SNAP;0.25

GRID
Command: **GRID**

Specify grid spacing(X) or [ON/OFF/Snap/Aspect] <default>: **0.5**

Screen menu command definition
GRID;0.5

LIMITS
Command: **LIMITS**
Specify lower left corner or [ON/OFF] <current coordinates>: **0,0**
Specify upper right corner <current coordinates>: **12,9**

Screen menu command definition
LIMITS;0,0;12,9

ZOOM
Command: **ZOOM**
Specify corner of window, enter a scale factor (nX or nXP), or
[All/Center/Dynamic/Extents/Previous/Scale/Window] <real time>: **A**

Screen menu command definition
ZOOM;A

Now you can combine these individual command definitions to form a single screen menu command definition that will perform all the functions when you select NSETUP from the new screen menu.

Combined screen menu line

```
[-NSETUP-]PLINE;0,0;W;0.01;;12,0;12,9;0,9;C;+
SNAP;0.25;GRID;0.5;LIMITS;+
0,0;12,9;ZOOM;A
```

Exercise 4 *General*

Write a screen menu item that will set up the following parameters for the **UNITS** command (filename *sme4.mnu*).

System of units: Scientific
Number of digits to right of decimal point: 2
System of angle measure: Decimal
Number of fractional places for display of angles: 2
Direction for angle: 0
Angles measured counterclockwise

MENU COMMAND REPETITION

AutoCAD has a provision for repeating a menu item until it is canceled by the user by pressing the ESC key from the keyboard or selecting another menu item. It is particularly useful when you are editing a drawing and using the same command several times. A command can be repeated if the menu command definition starts with an asterisk sign (*).

[ERASE,W:]*^C^CERASE W

Where ***** ------------------- Asterisk for command repetition
^C^C ---------- Cancels existing command
Erase ------------ AutoCAD's **ERASE** command
W ---------------- Window option

If you select this menu item, AutoCAD will prompt you to enter two points to select the objects because the window option for the **ERASE** command requires two points. When you press ENTER, the objects you selected will be erased and the command will be repeated.

Example 7 *General*

Write a screen menu for the following AutoCAD commands. Provide for automatically repeating the commands (filename *sm6.mnu*).

LINE	**LIST**
ERASE	**INSERT**
TRIM	**DIST**

The following file is a listing of the screen menu for Example 7 that will repeat the selected command until canceled by pressing the CTRL and C keys.

```
[-REPEAT-      ]
[              ]
[              ]
[LINE:         ]*^C^CLINE
[ERASE:        ]*^C^CERASE
[TRIM:         ]*^C^CTRIM
[LIST:         ]*^C^CLIST
[INSERT:       ]*^C^CINSERT
[DIST:         ]*^C^CDIST
```

Note

If the command definition is incorrect and you happen to select that item, the screen prompt display will be repeated indefinitely. To get out of this infinite loop, press the ESC key. On some systems, pressing the ESC key or selecting another command may not stop the display of the prompt. You may need to reboot the system or close the program. In Windows, press CTRL+ALT+DEL to close the program.

One of the major drawbacks of the menu command repetition is that it does not let you select a different command option. In the following menu item, the ***ERASE*** *command uses the* ***C*** *(****Crossing****) option to select the objects. When the command is repeated, it does not allow you to use a different option for object selection.*

[ERASE,C:]*^C^CERASE C

AUTOMATIC MENU SWAPPING

Screen menus can be automatically swapped by using the system variable **MENUCTL**. When the system variable **MENUCTL** is set to 1, AutoCAD automatically issues the **$S=CMDNAME** command, where **CMDNAME** is the command that loads the submenu. For example, if you select the **LINE** command from the digitizer or pull-down menu, or enter the **LINE** command at the keyboard, the **CMDNAME** command will load the **LINE** submenu and display it in the screen menu area. To use this feature, the command name and the submenu name must be the same. For example, if the name of the arc submenu is ARC and you select the **ARC** command, the **Arc** submenu will be automatically loaded on the screen. However, if the submenu name is different (for example, MYARC), AutoCAD will not load the **Arc** submenu on the screen. The default value of the **MENUCTL** system variable is 1. If you set the **MENUCTL** variable to 0, AutoCAD will not use the **$S=CMDNAME** command feature to load the submenus.

MENUECHO SYSTEM VARIABLE

If the system variable **MENUECHO** is set to 0, the commands that you select from the digitizer, screen menu, pull-down menu, or buttons menu will be displayed in the Command prompt area. For example, if you select the **CIRCLE** command from the menu, AutoCAD will display **_circle Specify center point for circle or [3P/2P/Ttr (tan tan radius)]:**. If you set the value of the **MENUECHO** variable to 1, AutoCAD will suppress the echo of the menu item and display **Specify center point for circle or [3P/2P/Ttr (tan tan radius)]:** only. You will notice that the **_circle** is not displayed when **MENUECHO** is set to 1. You can use **^P** in the menu item to turn the echo on or off. The **MENUECHO** system variable can also be assigned a value of 2, 4, or 8, which controls the suppression of system prompts, disabling the ^P toggle, and debugging aid for DIESEL macros, respectively.

MENUS FOR FOREIGN LANGUAGES

Besides English, AutoCAD has several foreign language versions. If you want to write a menu that is compatible with other foreign language versions of AutoCAD, you must precede each command and keyword in the menu file with the underscore (_) character.

Examples:

```
[New    ]^C^C_New
[Open   ]^C^C_Open
[Line   ]^C^C_Line
[Arc-SCA]^C^C_Arc;\_C;\_A
```

Any command or keyword that starts with the underscore character will be automatically translated. If you check the *acad.CUIx* file, you will find that AutoCAD has made extensive use of this feature.

Example 8 *General*

Rewrite the screen menu of Example 2 so that the menu file is compatible with other foreign language versions of AutoCAD.

The following file is the listing of the menu file for Example 8:

```
***SCREEN
[ MENU-1      ]
[             ]
[             ]
[LINE         ]^C^C_LINE
[             ]
[CIR-C,R      ]^C^C_CIRCLE
[CIR-C,D      ]^C^C_CIRCLE;\_D
[CIR- 2P      ]^C^C_CIRCLE;_2P
[             ]
[ERASE        ]^C^C_ERASE
[MOVE         ]^C^C_MOVE
```

USE OF CONTROL CHARACTERS IN MENU ITEMS

You can use ASCII control characters in the command definition by using the caret sign (^) followed by the control character. For example, if you want to write a menu item that will toggle the SNAP off or on, you can use a caret sign followed by the control character B as shown in the following example.

[SNAP-TOG]^B

Where **SNAP-TOG** ----- Command label for SNAP toggle
^ ----------------- Caret sign (^)
B ------------------ Control character for SNAP

Examples:

^C	(Cancel)
^G	(Grid on/off)
^H	(Backspace)

^O	(Ortho on/off)
^T	(Tablet on/off)
^E	(Isoplane top/left/right)

^B is equivalent to pressing the CTRL and B keys from the keyboard that toggle the SNAP mode. **SNAP-TOG** is the menu item label that will be displayed on the screen menu. You can use any ASCII control character in the command definition. Following are some of the ASCII control characters:

^@	(ASCII code 0)
^[	(ASCII code 27)
^\	(ASCII code 28)
^]	(ASCII code 29)
^^	(ASCII code 30)
^-	(ASCII code 31)

SPECIAL CHARACTERS

The following is a list of the special characters that you can use in the AutoCAD menus:

Character	Description
***	Three asterisks indicate a section title
**	Two asterisks indicate a submenu
[ ]	Used to enclose a menu item label
;	Causes an ENTER
Space	Space is equivalent to pressing the spacebar
\	AutoCAD pauses for user input
_	(Underscore character) translates the AutoCAD commands and keyword that follow it into English.
+	Used to continue the menu item definition to next line
=*	Forces the display of pull-down, cursor or image tile menus
*	Menu item repetition
$M=	Special character to load a DIESEL macro expression
$S=CMDNAME	Special command that loads a screen submenu
^B	Toggles snap (On/Off)
^C	Cancels the existing command
^D	Toggles coordinate dial (On/Off)
^E	Changes the isometric plane (Left/Right/Top)
^G	Toggles grid (On/Off)
^H	Causes a backspace
^O	Toggles ortho mode (On/Off)
^P	Toggles MENUECHO on or off
^Q	Echoes all prompts to the printer
^T	Toggles tablet mode (On/Off)
^V	Changes current viewport

^Z	Suppresses the automatic addition of spacebar at the end of a menu item

Example 9 *General*

Write a screen menu for the following toggle functions (filename *sm8.mnu*).

ORTHO	**SNAP**
GRID	**COORDINATE DIAL**
TABLET	**ISOPLANE**
CURRENT VIEWPORT	

Before writing a screen menu for these toggle functions, it is important to know the control characters that are used to turn these functions on or off. The following is a list of the control characters for the functions in this example:

ORTHO	CTRL O
SNAP	CTRL B
GRID	CTRL G
COORDINATE DIAL	CTRL D
TABLET	CTRL T
ISOPLANE	CTRL E
CURRENT VIEWPORT	CTRL V

The following file is a listing of the screen menu that toggles the given functions. You can use **EDIT** command to write this file.

```
[-TOGGLE-    ]
[            ]
[            ]
[            ]
[ORTHO       ]^O          ———————    turns ORTHO on/off
[SNAP        ]^B          ———————    turns SNAP on/off
[GRID        ]^G          ———————    turns GRID on/off
[CO-ORDS     ]^D          ———————    turns COORDINATE dial on/off
[TABLET      ]^T          ———————    turns TABLET on/off
[            ]
[ISOPLANE    ]^E          ———————    turns ISOPLANE on/off
[CURRENT VIEWPORT]^V      ———————    Changes the current viewport
```

Chapter 19

COMMAND DEFINITION WITHOUT ENTER OR SPACE

All command definitions that have been discussed so far use a semicolon (;) or a space for Enter. However, sometimes a user might want to define a menu item that is not followed by an Enter or a space. This is accomplished by using the backspace ASCII control character (^H). It is especially useful when you want to write a screen menu for a numeric keypad. ^H control character can be used with any character or a group of characters. The following is the format of menu item definition without an ENTER or space:

[9]9X^H

Where **9** ------------------- Menu item label
9 ------------------- Character returned by selecting this item
X ------------------- X is erased because of ^H (backspace)
(X could be substituted by any character)
^H --------------- ^H for backspace

^H is the ASCII control character for backspace. When you select this item, ^H erases the previous character X and therefore returns 9 only. It is not necessary to have X preceding ^H. X is a dummy character; it could be any character. You can continue selecting more characters. When you are done, use the keyboard or the digitizer to press ENTER.

Example 10 — *General*

Write a screen menu for the following characters (filename *sm9.mnu*).

0	5	.
1	6	,
2	7	X
3	8	Y
4	9	Z

The following file is a listing of the screen menu of Example 10 that will enable the user to pick the characters without an ENTER.

```
[-KEYPAD-    ]
[            ]
[            ]
[            ]
[0]0Y^H              ————    returns 0
[1]1Y^H              ————    returns 1
[2]2Y^H              ————    returns 2
[3]3Y^H
[4]4Y^H
[5]5Y^H
[6]6Y^H
[7]7Y^H
```

```
[8]8Y^H
[9]9Y^H
[.].Y^H          ————————    returns period (.)
[,],Y^H          ————————    returns comma (,)
[]               ————————    for space in the screen menu
[X]XX^H          ————————    returns X
[Y]YX^H          ————————    returns Y
[Z]ZZ^H          ————————    returns Z
```

Note

You can also use a backslash after the character to continue selecting more characters. For example, [2]2\ will return 2 and then pause for the user input, which could be another character. The following file uses a backslash (\) to let the user enter another character.

```
[-KEYPAD- ]
[             ]
[             ]
[             ]
[0]0\
[1]1\
[2]2\
[3]3\
[4]4\
[5]5\
[6]6\
[7]7\
[8]8\
[9]9\
[.].\
[,],\
[]
[X]X\
[Y]Y\
[Z]Z\
```

MENU ITEMS WITH SINGLE OBJECT SELECTION MODE

The Single options for object selection, combined with the menu item repetition, can provide a powerful editing tool.

[ERASE]*^C^CERASE Single

Where * ------------------- Menu item repetition

Single ------------ Single object selection option

In this menu item, the asterisk that follows the menu item label repeats the command. ^C^C cancels the existing command. **ERASE** is an AutoCAD command that erases the selected objects. The Single option enables the user to select the objects and then AutoCAD automatically terminates the object selection process. The **ERASE** command then erases the selected objects.

If the point you pick is in a blank area (if there is no object), AutoCAD automatically defaults to crossing or window option. If you pull the window to the left, it is the crossing option and if you pull the box to the right, it is the window option for selecting the objects. The asterisk in front of the menu item repeats the command.

USE OF AutoLISP IN MENUS

You can combine AutoLISP variables and expressions with the menu items as a command definition. When you select that item from the menu, it will evaluate all the expressions and generate the necessary output. The following examples illustrate the use of AutoLISP variables and expressions. (For more information on AutoLISP, see Chapters 12 and 13.)

Example 11 *General*

Write an AutoLISP program that draws a square and then write a screen menu that utilizes the AutoLISP variables and expressions to draw a square.

The following file is the listing of the AutoLISP program that generates a square. The program prompts the user to enter the starting point and the length of the side.

```
(DEFUN C:SQR()
(SETVAR "CMDECHO" 0)
(SETQ P1 (GETPOINT "\n ENTER STARTING POINT: "))
(SETQ S (GETDIST "\n ENTER LENGTH OF SIDE: "))
(SETQ P2 (LIST (+ (CAR P1) S) (CADR P1)))
(SETQ P3 (POLAR P2 (/ PI 2) S))
(SETQ P4 (POLAR P1 (/ PI 2) S))
(COMMAND "PLINE" P1 P2 P3 P4 "C")
(SETVAR "CMDECHO" 1)
(PRINC)
)
```

The following file is the listing of the screen menu file that utilizes the AutoLISP variables and expressions to draw a square:

```
[-SQUARE- ]
[         ]
[         ]
[SQUARE:  ](SETQ P1(GETPOINT "ENTER STARTING POINT:+
"));\+
(SETQ S (GETDIST "ENTER LENGTH OF SIDE: "));\+
(SETQ P2 (LIST (+ (CAR P1) S) (CADR P1)))+
```

```
(SETQ P3 (POLAR P2 (/ pi 2) S))+
(SETQ P4 (POLAR P1 (/ pi 2) S));+
PLINE !P1 !P2 !P3 !P4 C
```

If you compare the AutoLISP program with the screen menu, you will notice that the statements that prompt the user to enter the information and the statements that do actual calculations are identical. Therefore, you can utilize AutoLISP variables and expressions in the menu file to customize AutoCAD.

Self-Evaluation Test

Answer the following questions and then compare them to those given at the end of this chapter:

1. The command definition that involves several commands put together in one line is also called a __________.
2. A command can be repeated if the menu command definition starts with an __________.
3. The screen menus can be automatically swapped by using the system variable __________.
4. __________ command will load the submenu.
5. __________ system variable is used to display the new status line.
6. You can use __________ in the menu item to turn the echo on or off.
7. If the statement canot fit on one line, you can put a __________ at the end of the first line and then continue with a second line.
8. The __________ is used in a menu item to refer to or load a submenu.
9. Command definition without enter or space can be accomplished by using __________.

Review Questions

Answer the following questions:

1. The AutoCAD menu file can have up to __________ main sections.
2. A submenu is designated by _______________.
3. The part of the menu item that is inside the brackets is for _______________ only.
4. Only the first _______________ characters can be displayed on the screen menu.
5. If the number of characters inside the bracket exceeds eight, the screen menu __________ work.
6. In a menu file you can use _______________ to cancel the existing command.
7. _______________ is used to pause for input in a screen menu definition.
8. The menu item label _______________ can be a combination of uppercase and lowercase characters.
9. Submenu names can be __________ characters long.
10. The maximum number of accessible menu items in a screen menu depends on the __________.

Exercises

Exercise 5 *General*

Write a screen menu for the following AutoCAD commands (file name *sme5.mnu*).

POLYGON (Center)
POLYGON (Edge)
ELLIPSE (Center)
ELLIPSE (Axis End point)
CHAMFER
EXPLODE
COPY

Exercise 6 *General*

Write a screen menu file for the following AutoCAD commands. Use submenus if required (file name *sme6.mnu*).

ARC	**ROTATE**
-3P	**ARRAY**
-SCE	**DIVIDE**
-CSE	**MEASURE**
BLOCK	
INSERT	**LAYER**
WBLOCK	**SET**
MINSERT	**LIST**

Exercise 7 *General*

Write a screen menu item that will set up the following layers, linetypes, and colors (file name *sme7.mnu*).

Layer Name	Color	Linetype
0	WHITE	CONTINUOUS
OBJECT	RED	CONTINUOUS
HIDDEN	YELLOW	HIDDEN
CENTER	BLUE	CENTER
DIM	GREEN	CONTINUOUS

Exercise 8 *General*

Write a screen menu for the commands in Figure 19-14. Use the menu item ******** to load Osnaps, and use the menu item MENU-8 to load the root menu (file name *sme8.mnu*).

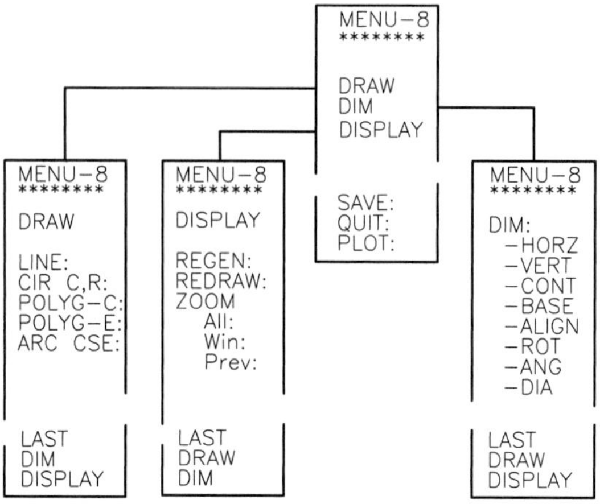

Figure 19-14 *Screen menu displays*

Exercise 9 *General*

Write a coordinated pull-down and screen menu for the following AutoCAD commands. Use a cascading menu for the **LINE** command options in the menu. When the user selects an item from the menu, the corresponding screen menu should be automatically loaded on the screen.

LINE	**ZOOM All**	**TIME**
Continue	**ZOOM Win**	**LIST**
Close	**ZOOM Pre**	**DISTANCE**
Undo	**PAN**	**AREA**
.X		**DBLIST**
.Y		**STATUS**
.Z		

The layouts of the screen menu and the pull-down menu are shown in Figures 19-15(a) and 19-15(b).

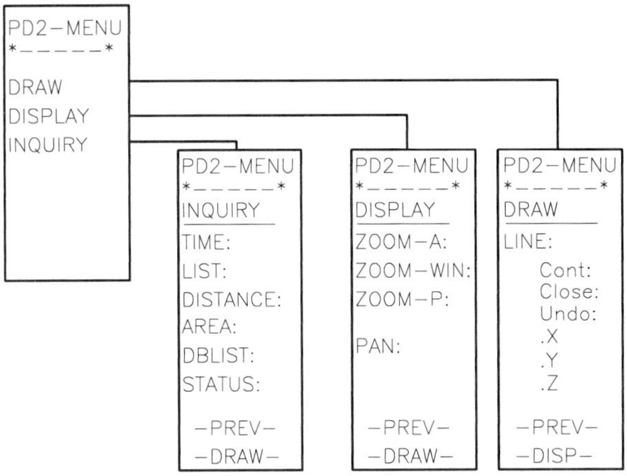

Figure 19-15(a) *Design of screen menu for Exercise 9*

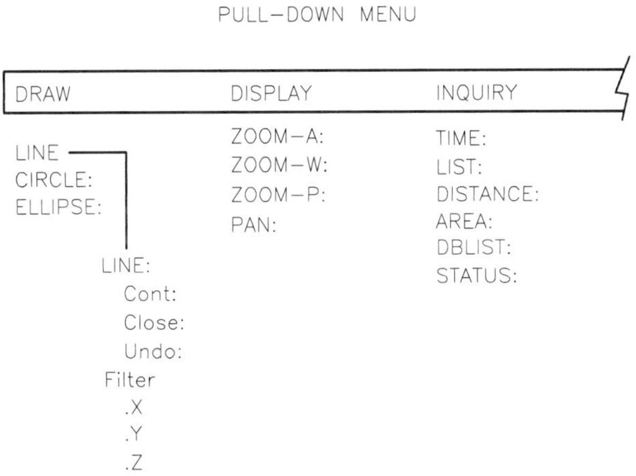

Figure 19-15(b) *Design of screen menu for Exercise 9*

Exercise 10 *General*

Write a menu, a screen menu, and a tablet menu for the AutoCAD commands given next. When you select a command from the template or the menu, the corresponding screen menu should be automatically loaded on the screen. The layout of the screen is shown in Figure 19-16.

Chapter 19

LINE	**BLOCK**
PLINE	**WBLOCK**
CIRCLE C,R	**INSERT**
CIRCLE C,D	**BLOCK LIST**
ELLIPSE AXIS ENDPOINT	**ATTDEF**
ELLIPSE CENTER	**ATTEDIT**

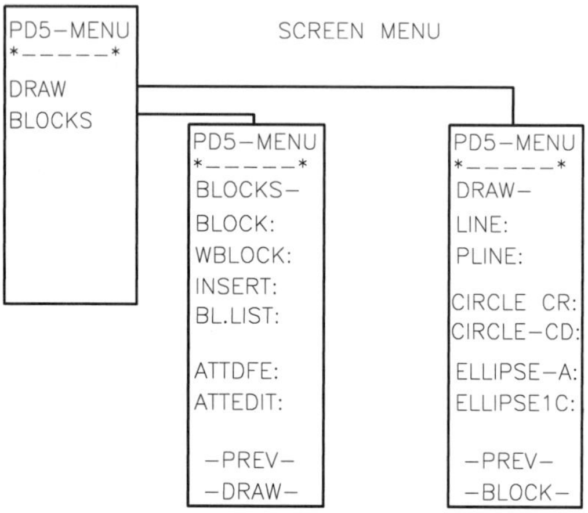

Figure 19-16 *Design of screen menu for Exercise 10*

Exercise 11 *General*

Write a screen, pull-down, and image tile menu for the commands shown in Figure 19-17. When the user selects a command from the menu, it should automatically load the corresponding menu on the screen.

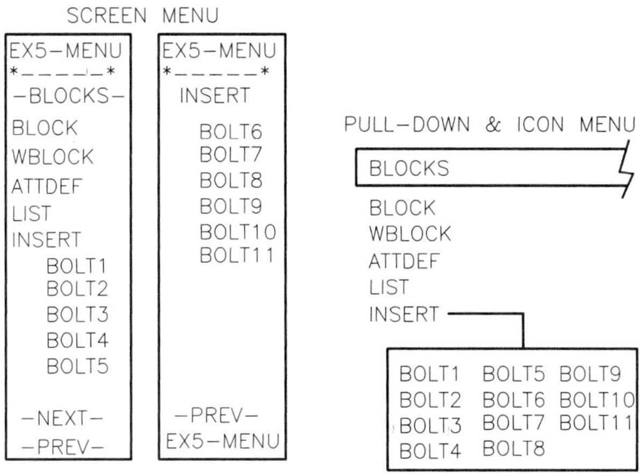

Figure 19-17 *Design of screen, pull-down, and image tile menus for Exercise 11*

Exercise 12 *General*

Write a screen, tablet, pull-down, and image tile menu for inserting the following commands. B1 to B15 are the block names.

BLOCK
WBLOCK
ATTDEF
LIST
INSERT

BL1	**BL6**	**BL11**
BL2	**BL7**	**BL12**
BL3	**BL8**	**BL13**
BL4	**BL9**	**BL14**
BL5	**BL10**	**BL15**

Exercise 13 *General*

Write a pull-down and screen menu for the following AutoCAD commands. Use submenus for the **ARC** and **CIRCLE** commands. When the user selects an item from the menu, the corresponding screen menu should be automatically loaded on the screen.

LINE	**BLOCK**	**QUIT**
PLINE	**INSERT**	**SAVE**
ARC	**WBLOCK**	——
ARC 3P		**PLOT**
ARC SCE		

ARC SCA
ARC CSE
ARC CSA
ARC CSL
CIRCLE
CIRCLE C,R
CIRCLE C,D
CIRCLE 2P

The layout shown in Figure 19-18(a) is one of the possible designs for this menu. The **ARC** and **CIRCLE** commands are in separate groups that will be defined as submenus in the menu file. The layout of the screen menu is shown in Figure 19-18(b).

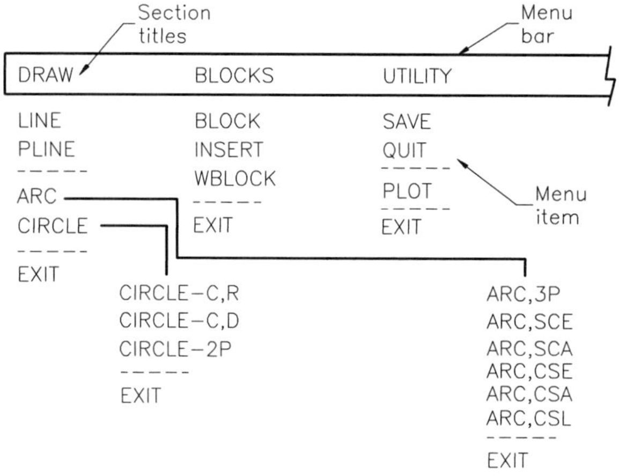

Figure 19-18(a) Design of pull-down menu for Exercise 13

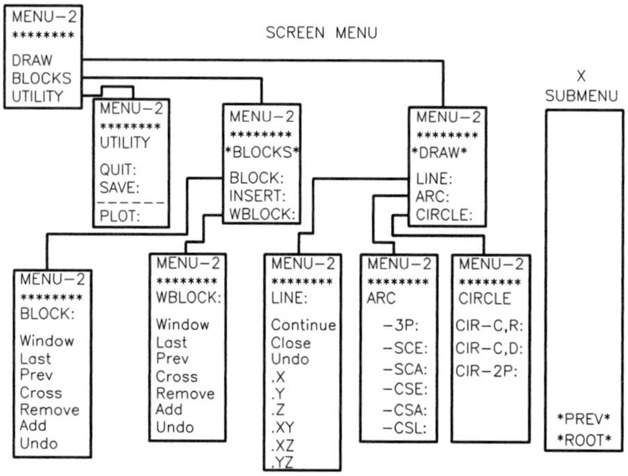

Figure 19-18(b) Design of screen menu

Answers to Self-Evaluation Test

1. MACRO, **2.** asterisk(*), **3. MENUCTL**, **4. CMDNAME**, **5. MODEMACRO**, **6.** ^p, **7.** plus (+), **8.** submenu reference, **9.** ^H

Index

D

E

F

G

H

I

K

L

M

N

O

P

R

S